Automated Pattern Analysis in Petroleum Exploration

Ibrahim Palaz   Sailes K. Sengupta
Editors

Automated Pattern Analysis in Petroleum Exploration

With 213 Illustrations, 23 in Full Color

Springer-Verlag
New York Berlin Heidelberg London Paris
Tokyo Hong Kong Barcelona Budapest

Ibrahim Palaz
Geophysicist
Amoco Production Company
501 West Lake Park Boulevard
Houston, TX 77253, USA

Sailes K. Sengupta
Lawrence Livermore National Laboratory
Livermore, CA 94511, USA;
formerly professor at:
South Dakota School of Mines and Technology
Rapid City, SD 57701, USA

Cover illustration: Random distributions of the views from a thin section, Figure 13.2b, page 252.

Library of Congress Cataloging-in-Publication Data
Automated pattern analysis in petroleum exploration / Ibrahim Palaz,
   Sailes Sengupta, editors.
      p. cm.
   Includes bibliographical references and index.
   ISBN 0-387-97468-7 (New York). — ISBN 3-540-97468-7 (Berlin)
   1. Petroleum—Prospecting—Data processing. 2. Expert systems
   (Computer science) 3. Pattern recognition systems. I. Palaz,
   Ibrahim. II. Sengupta, Sailes, 1935–
   TN271.P4A86 1991
   622′.1828—dc20   91-2814

Printed on acid-free paper.

Production coordinated by Chernow Editorial Services, Inc. and managed
by Linda H. Hwang.
Typeset by Publishers Service of Montana Inc., Bozeman, MT.
Printed and bound by Edwards Brothers Inc., Ann Arbor, MI.
Printed in the United States of America.

9 8 7 6 5 4 3 2 1

ISBN 0-387-97468-7 Springer-Verlag New York Berlin Heidelberg
ISBN 3-540-97468-7 Springer-Verlag Berlin Heidelberg New York

Preface

Computers contributed greatly to the evolution of petroleum exploration. Today the complexity or size of an exploration task is no longer a limiting factor for most computers. From field geology to enhanced oil recovery, every aspect of finding hydrocarbons involves the use of computers at varying levels.

The impact of computers on technologies such as pattern recognition (PR), image analysis (IA), and artificial intelligence (AI) has been even greater than on petroleum exploration. These technologies did not have meaningful applications until the arrival of faster and more sophisticated computers. Since the 1960s there has been an increasing number of applications of PR, IA, and AI in scientific and engineering disciplines as they were proved to be very powerful tools. In the early 1980s there were few applications of these technologies in petroleum exploration and they were mostly in research laboratories. In the late 1980s there were special sessions dedicated to the application of these technologies at international petroleum meetings. This was a clear reflection of the growing interest among explorationists to utilize one or more of these technologies to solve old problems.

This book is a collection of carefully selected papers. In each chapter PR, IA, or AI is applied to some petroleum exploration task. This book is not intended to be a discussion of the pros and cons of these technologies. Readers who are interested in the theory of these techniques can refer to publications listed in the reference section of each chapter.

The fields in which PR, IA, and AI are applied in this book are not limited to geology, geophysics, and petroleum engineering. Chapters cover topics from sand grain shape analysis to well test analysis. We had two objectives in collecting such a wide range of applications. The first was to illustrate that every aspect of exploration can potentially use these technologies. The second was that petroleum exploration is an integrated effort of the geologist, geophysicist, and engineer, and topics in this book reflect this. Most of our problems are common, but tools are different, and there is the distinct possibility of integrating a number of tools to tackle common problems. PR, IA, and AI can help us not only to solve problems but also to make the integration of exploration efforts a reality.

The first chapter, by Allain and Horne, is titled "The Use of Artificial Intelligence for Model Identification in Well Test Interpretation." The authors describe the well test interpretation as an inverse problem. Their aim is to determine a system in which the input and the system response are known. This chapter is an excellent example of full automation. By using artificial intelligence, automation is accomplished in model identification. Such a system has clear advantages not only in the interpretation of well test data but also in actually monitoring the test. The authors illustrate an example using real data.

The second chapter is titled "Artificial Intelligence in Formation Evaluation." Kuo et al. share their wealth of experience in the use of artificial intelligence for formation evaluation. The authors describe basic concepts in AI and in formation evaluation. They review expert systems developed in their field and present their approach to the problem. Similarly they cover the topics of edge detection and pattern recognition and present their approaches to both.

In the third chapter, I. Williamson addresses some basic issues in developing an intelligent knowledge-based system, in his chapter "Intelligent Knowledge Based Systems and Seismic Interpretation." Issues such as languages and structuring knowledge bases are discussed and Williamson specifically talks about possible uses and benefits of utilizing a knowledge-based system in seismic sequence analysis. He presents a simple example of a knowledge-based system.

The fourth chapter by Davis is an excellent illustration of the use of an expert system to solve an important problem. The title of the chapter is "An Expert System for the Design of Array Parameters for Onshore Seismic Surveys." This expert system can be used in the office as well as in the field. It is designed to determine parameters for proper acquisition of seismic data. The author briefly introduces the array theory and then describes the system in detail. His system is particularly interesting because it was developed in Quickbasic rather than in LISP or PROLOG. He demonstrates on an example run how his expert system works.

Crisi, in his chapter "An Expert System to Assist in Processing Vertical Seismic Profiles," illustrates how he acquired seismic information and coded it into a knowledge base. Starting with field tapes, his system can advise what would be the most appropriate processing flow along with the processing parameters which would help produce the best final section. His system hints that it is possible to develop a much needed expert system for seismic data processing.

Huang, who has long been involved with the application of AI and PR to petroleum exploration, illustrates two expert systems for seismic exploration in his chapter, "Expert Systems for Seismic Interpretations and Validation of Simulation Stacking Velocity Functions." The first is for velocity analysis of seismic data processing. His VELXPERT utilizes an inference engine with forward chaining to select the rules and uses of amended transition trees to implement parsing of the questions and the formation of answers in natural language. In his seismic data interpretation expert system, SIES, Huang illustrates how pattern matching, backward chaining, and augmented transition trees can be effectively used. SEIS is a prototype system that needs to be tested thoroughly; the author illustrates test runs of both systems.

"Pattern Recognition to Seismic Exploration" by Huang is an odyssey through a gamut of both classical and relatively modern pattern recognition methods based on the author's long experience in real and synthetic seismograph analysis. The study is geared to the detection and recognition of structural seismic patterns, including the detection of physical anomalies leading to the possible discovery of hydrocarbon deposits. The theoretical techniques include linear and quadratic discriminant analysis, tree classification, and different variants or syntactic pattern recognition. The application of these techniques to seismic data analysis is well documented in this chapter.

In her chapter "Pattern Recognition for Marine Seismic Exploration," F. El-Hawary presents a scholarly exposition of how an expert system approach can be employed for marine seismic identification of hydrocarbon formation. The task involves image acquisition, processing, pattern recognition, and, above all, a great deal of expert knowledge and judgment. The author carefully examines each step in the process and its incorporation in the overall framework of a proposed expert system.

Projection pursuit is a technique that has been in use in the exploratory analysis of multivariate data since the pioneering work in 1974. Several variants of this technique have been proposed since then, and A.T. Walden in his chapter "Clustering of Attributes by Projection Pursuit for Reservoir Characterization" outlines a version suitable for use as an aid to reservoir characterization. In projection pursuit, clustering of data is facilitated by projecting the multidimensional data along a direction that, at least locally, maximizes a certain "entropy index," which in turn is a measure of the multimodal characteristics or "non-Gaussianness" of the projected computation of the entropy index and its derivative, which, in turn, is done by extensive use of fast Fourier transform, making it computationally efficient. The author brings this powerful tool from multivariate analysis to within the reach of quantitative geoscientists.

In their chapter "Exploring the Fractal Mountains," Klinkenberg and Clarke explore in a leisurely fashion a timely topic, bringing it within the access of geoscientists. The importance of fractal geometry in the study of scientific phenomena has been well documented in the scientific literature. The authors present the topic from a geomorphological perspective. They also point out the importance of a comparative study of the methods for determining the fractal dimensions based on "truly fractal" data sets, indicating some practical difficulties encountered in topographic studies. Several caveats as well as some useful recommendations coupled with a good body of references for fractal application in geosciences make this work particularly useful.

Particle size and shape have been under study in various disciplines associated with the mineral industries. The chapter "Image Analysis of Particle Shape" by Starkey and Rutherford is the culmination of several earlier studies on image analysis by these authors. Digitized images of thin sections under the petrographic microscope are first subjected to standard gray level thresholding to delineate the particle boundaries. Then a best fit ellipse is used to approximate its shape and size in an automated manner to provide reliable estimates for the corresponding statistics of the aggregate. The emphasis is on automation with accuracy.

In the chapter "Interactive Image Analysis of Borehole Televiewer Data," Barton, Tesler, and Zoback provide a valuable tool with which the practical geophysicist can analyze both large and fine scale features in a televiewer image by permitting access to a graphics window. This is implemented by a popular product, MacApp, written in an object-oriented language Object Pascal, supported by C subroutines for the image analysis. The software has been used very effectively in the analysis of data from the Cajon Pass well in California and is, currently, being used for the KTB (Germany) well site data. The software is flexible enough to allow easy extension of the analytical tools to a wide variety of other types of geophysical image data.

Standard image processing techniques coupled with some basic models in spatial statistics can aid in handling problems in petrophysical analysis of difficult pore complexes. This is demonstrated by Gerand et al. in their chapter "Petrographic Image Analysis: An Alternate Method for Determining Petrophysical Properties" with a case study of the successful classification of hydrocarbon reservoirs by a quantitative characterization of its pore complexes. This, in turn, permits a ranking of such reservoirs as an aid to decision making in exploration. The authors achieve their objective in two steps. They first segment the cross-sectional image by rendering it as a binary image representing pores and rock materials only. Next, they use a "sizing" technique to derive the three-dimensional petrophysical properties from two-dimensional fractal/geometric properties derived from the resulting image. Other potential uses of their technique are indicated.

Some standard image processing algorithms such as smoothing, edge enhancement, and histogram equalization can be employed in a variety of instances for greater ease in scientific data interpretation. For the past two decades they have been used in remote sensing, biomedical, and robotics applications. The use of these techniques for magnetic data processing is fairly recent and is illustrated by Wu Chaojun in his chapter "Image Processing of Magnetic Data and Application of Integrated Interpretation for Mineral Resources Detection in Yieshan Area, East China." He presents his techniques of potential transforms in an integrated fashion. The author has found these techniques useful in the detection of mineral resources. He also indicates their potential usefulness in the interpretation of gravity data. His methods can clearly be applied to any gravity and magnetic data for petroleum exploration.

In their chapter "Interactive Three-Dimensional Seismic Display by Volumetric Rendering," Wolfe and Liu provide us with an extremely useful display technique for seismic data. For years geophysicists have had to tackle the dual problem of displaying simultaneously the spatial and waveform attributes of such data. The standard way had been to display the data in three dimensions, the waveform in one dimension and two of the three coordinates of the wave position in the other two dimensions. Then animation helped provide a mental picture of the third coordinate. The authors' approach is different. They choose to display the waveform attribute by a thresholded color coding, while considering all three dimensions of the wave position simultaneously in the display. By avoiding having to go through a multitude of two-dimensional sections, the method enables one to gain three-dimensional views of an underground structure in an effective manner, even with modest computing resources. In addition, the flexibility made available in the preprocessing stage makes it a valuable tool in the hands of exploration geophysicists.

Ibrahim Palaz
Sailes K. Sengupta

Contents

Preface v
Contributors xiii

Chapter 1
The Use of Artificial Intelligence for Model Identification
in Well Test Interpretation 1
Olivier Allain and Roland N. Horne

Chapter 2
Artificial Intelligence in Formation Evaluation 33
Tsai-Bao Kuo, Steven A. Wong, and Richard A. Startzman

Chapter 3
Intelligent Knowledge Based Systems and Seismic Interpretation .. 61
I. Williamson

Chapter 4
An Expert System for the Design of Array Parameters
for Onshore Seismic Surveys 71
Barrie K. Davis

Chapter 5
An Expert System to Assist in Processing
Vertical Seismic Profiles 81
Peter Crisi

Chapter 6
Expert Systems for Seismic Interpretations and Validation
of Simulated Stacking Velocity Functions 99
Kou-Yuan Huang

Chapter 7
Pattern Recognition to Seismic Exploration 121
Kou-Yuan Huang

Chapter 8
Pattern Recognition for Marine Seismic Exploration 155
Ferial El-Hawary

Chapter 9
Clustering of Attributes by Projection Pursuit
for Reservoir Characterization 173
A.T. Walden

Chapter 10
Exploring the Fractal Mountains 201
Brian Klinkenberg and Keith C. Clarke

Chapter 11
Image Analysis of Particle Shape 213
John Starkey and Sandra Rutherford

Chapter 12
Interactive Image Analysis of Borehole Televiewer Data 223
Colleen A. Barton, Lawrence G. Tesler, and Mark D. Zoback

Chapter 13
Petrographic Image Analysis: An Alternate Method
for Determining Petrophysical Properties 249
*R.E. Gerard, C.A. Philipson, F.M. Manni,
and D.M. Marschall*

Chapter 14
Image Processing of Magnetic Data and Application
of Integrated Interpretation for Mineral Resources Detection
in Yiesan Area, East China 265
Wu Chaojun

Chapter 15
Interactive Three-Dimensional Seismic Display
by Volumetric Rendering 285
Robert H. Wolfe, Jr. and C.N. Liu

Index 293

Contributors

Olivier Allain, Petroleum Engineering Department, Stanford University, Stanford, CA 94305, USA

Colleen A. Barton, Geophysics Department, Stanford University, Stanford, CA 94305, USA

Wu Chaojun, Department of Applied Geophysics, China University of Geosciences, Wuhan, China

Keith C. Clarke, Department of Geology and Geography, Hunter College, City University of New York, New York, NY 10021, USA

Peter Crisi, Geophysicist, Mobil E&P Service Inc., 3000 Pegaus, Dallas, TX 75247, USA

Barrie K. Davis, 17 Lynmouth Road, Fortos Green, London N29 NR, United Kingdom

Ferial El-Hawary, Signal Analysis Laboratory, Technical University of Nova Scotia, Halifax, Nova Scotia B3J 2X4, Canada

R.E. Gerard, Core Laboratories, 10201 Westheimer, Houston, TX 77042, USA

Roland N. Horne, Petroleum Engineering Department, Stanford University, Stanford, CA 94305, USA

Kou-Yuan Huang, Institute and Department of Information Science, National Chiao Tung University, Hsinchu, Taiwan 30050, Republic of China

Brian Klinkenberg, Department of Geography, University of British Columbia, Vancouver, British Columbia V6T 1W5, Canada

Tsai-Bao Kuo, ARCO Oil and Gas Company, Plano, TX 75075, USA

C.N. Liu, Computer Science Department, T.J. Watson Research Center, IBM, P.O. Box 704, Yorktown Heights, NY 10598, USA

F.M. Manni, Core Laboratories, 10201 Westheimer, Houston, TX 77042, USA

D.M. Marschall, Core Laboratories, 10201 Westheimer, Houston, TX 77042, USA

Ibrahim Palaz, Geophysicist, Amoco Production Company, Houston, TX 77253, USA

C.A. Philipson, Core Laboratories, 10201 Westheimer, Houston, TX 77042, USA

Sandra Rutherford, Department of Geology, University of Western Ontario, London, Ontario N6A 5B7, Canada

Sailes K. Sengupta, Lawrence Livermore National Laboratory, Livermore, CA 94550, USA; formerly professor at South Dakota School of Mines and Technology, Rapid City, SD 57701, USA

John Starkey, Department of Geology, University of Western Ontario, London, Ontario N6A 5B7, Canada

Richard A. Startzman, Petroleum Engineering Department, Texas A&M University, College Station, TX 77843-3116, USA

Lawrence G. Tesler, Geophysics Department, Stanford University, Stanford, CA 94305, USA

A.T. Walden, Department of Mathematics, Imperial College of Science, Technology, and Medicine, Huxley Building, 180 Queen's Gate, London SW7 2BZ, United Kingdom

I. Williamson, Department of Geology, Imperial College of Science, Technology, and Medicine, London SW7 2BP, United Kingdom

Robert H. Wolfe, Jr., Computer Science Department, T.J. Watson Research Center, IBM, P.O. Box 704, Yorktown Heights, NY 10598, USA

Steven A. Wong, ARCO Oil and Gas Company, Plano, TX 75075, USA

Mark D. Zoback, Geophysics Department, Stanford University, Stanford, CA 94305, USA

Automated Pattern Analysis in Petroleum Exploration

1
The Use of Artificial Intelligence for Model Identification in Well Test Interpretation

Olivier Allain and Roland N. Horne

Introduction

Pressure transient testing is used to determine characteristic properties of an oil or gas reservoir by interpreting its dynamic behavior. This dynamic behavior is represented at a given well by two different quantities: pressure and flow rate. During a well test, a perturbation is imposed on the rate, and the resulting pressure variation is measured. Reservoir properties are then obtained from the interpretation of this variation.

In recent years, there have been a number of advances in the field of transient pressure testing, which have resulted in major improvements of the interpretation methods (Ramey, 1976, 1982). Theoretical developments have been greatly stimulated by the advent of high-precision pressure gauges and increased computing power. Moreover, the availability of inexpensive microcomputers has induced a rapid and widespread use of interpretation programs, that not only speed up the analysis, but also improve its reliability.

Well Test Interpretation

A systematic approach to well test interpretation was defined by Gringarten (1982) as follows. Well test interpretation can be considered as an instance of what is known in mathematics as an "inverse problem," where an unknown system S must be identified from its response O to a given input I. In well testing, the system comprises the well and the reservoir, the input is the flow rate history, and the response is the pressure variation recorded during the test. If we are given a set of mathematical models that represent the possible descriptions for the system S, then the inverse problem becomes a recognition problem. The adequate description for S is found as the theoretical system which response to I is the closest to O. A model can be described by an analytical relation between I and O that involves specific parameters. Solving the inverse problem requires finding not only the most appropriate model, but also the most appropriate value for its parameters. Those values are then considered to be representative of the real system.

The mathematical models used in well testing are usually called "interpretation models." They are constituted by three different components that describe the basic behavior of the reservoir, the well and its surroundings, and the outer boundaries of the reservoir. Graphical methods are used to identify an interpretation model for given data. Often, this identification is performed on a log–log scale presentation of the pressure variation versus elapsed time (called "log–log plot"). The shape of the real response on a log–log plot is compared to "type curves," which describe typical responses of the various models. Because a log–log plot of the pressure does not emphasize all the flow regimes that successively dominate the response, it is necessary to use other diagnostic tools. Traditionally, these tools are graphical presentations of the data on different scales, where a specific regime exhibits a straight line. Bourdet et al. (1983, 1984) proposed an alternative approach that unifies all the different tools into a single plot. In this approach, the derivative of the pressure is computed with respect to an appropriate time function, and plotted versus elapsed time on a log–log scale, along with the pressure. With this "augmented log–log plot," the interpretation of a set of data is broken down into two stages: (1) model identification, and (2) type curve matching.

In step 1, the different shapes recognized on the log–log plot direct the choice for a particular interpretation model. After a model has been chosen, a match is attempted with the data in step 2. A successful match confirms the choice of model and provides the parameter values for the reservoir, such as permeability and skin. Until the recent application of nonlinear regression methods to the automation of this match (Earlougher, 1977; Padmanabhan and Woo, 1976; Tsang et al., 1977; McEdwards, 1981; Rosa and Horne, 1983; Barua et al., 1988), the match was obtained by trial and error and probably constituted the most exacting part of the interpretation (see Clark and Van Golf-Racht, 1984; Proano and Lilly, 1986; and Houzé et al., 1988, for examples). Beyond a speed-up of the analysis, automated type curve matching improves the confidence in the results by quantifying the goodness of the match obtained, which a human expert could not do. Also, the match obtained is fully objective and is guaranteed to be the best solution achievable for the model considered.

Toward a Fully Automated Interpretation

Because type curve matching is now automated, the logical orientation of the interpretation procedure is toward a full automation. A completely automatic procedure would present several important advantages. First, interpretations could be performed at any location, for instance at the well site during the monitoring of the test. The procedure could be used to decide at any moment of the acquisition whether enough data have been obtained to determine a specific property of the reservoir. Therefore, such problems as shutting-in a well for too long, or repeating a test that was originally too short could be avoided. With an automatic interpretation, it would also be possible to ensure that all possible solutions to an interpretation problem are objectively proposed, thus improving the consistency of the analysis.

Model Identification and Artificial Intelligence

Identifying the adequate interpretation model(s) for given data requires the identification of correspondences or similarities between those data and the models. A model was defined earlier as a function involving some parameters. Equivalently, a model can be defined as a set of functions. A set of data is a set of numbers and cannot be compared as such with a model.

The first step involved in model identification is to represent the data and the models in a similar form, thereby allowing comparison. In a numerical procedure, the model is first represented by all its responses to the considered flow rate history, and then by only its closest response to the data. Because this representation can be compared directly to the data, the data are not changed. This representation is justified because a numerical procedure needs to execute operations on numbers.

Now, consider model identification based on the pressure derivative, as performed by a human expert. What representation of the models and the data does the expert use? What in the procedure allows the expert to construct those representations?

Representations of models and data are constructed unconsciously, as merely a result of seeing a graphical presentation of the responses. Seeing involves two steps. The first step is "sensing," i.e., transforming light energy reflected from images into an electrical signal. The second step is "perception," i.e., understanding the sensing. Perception involves the transformation of the signal into "symbols," which describe features of what is seen. Those symbolic descriptions then serve as a basis for understanding what is seen.

Therefore, the graphical methods used in well test interpretation allow us to represent models and data symbolically. All that is involved in model identification is the ability to manipulate symbols.

The manipulation of symbols (as opposed to operation on numbers) has been emphasized by Artificial Intelligence (AI), in which the central goal is to develop programs that exhibit intelligent behavior (Genesereth and Nilsson, 1987). An intelligent system can be characterized by its strength in information handling in addition to some other capabilities. The most important capability is the ability to capture information in a representation that is appropriate for decision making. The idea that symbols can be used to represent information is at the heart of AI. The ability to manipulate symbols has even been explicitly stated as a sufficient reason to explain intelligence, in the "Physical Symbol System Assumption" expressed by Newell and Simon, (1976). The need to manipulate symbols in AI led to creation of special computer languages such as LISP or Prolog.

In the AI framework, we develop methods for model identification that allow a computer to reproduce the visual diagnosis performed by a human expert on a log–log plot of the data derivative. Those methods are characterized by the use of symbolic representations for models and data, and inference procedures using those representations. The methodology described here has been implemented both in LISP (Allain, 1987) and in Prolog (Allain, 1988).

For a given set of data, the methods presented here can be used to propose, along with the set of possible interpretation models, a first estimate of their parameters. This first estimate can then be used as a starting point for an automated type curve matching analysis.

With the tools described here, there is only one step left in the automation of well test interpretation: the choice of the most appropriate model when several possible solutions are proposed. Since the models proposed are all adequate, the confidence intervals obtained from automated type curve matching could be used to make the decision. This decision should also take into account the geological and petrophysical information about the reservoir.

Components of Model Identification

A log–log plot of a derivative curve provides us with visual information. When we perceive this information, we unconsciously construct an internal representation of the curve. This representation can be expressed in terms of symbols, to describe the features that we see on the curve. The representation is not unique, for an expert and a novice may not be sensitive to the same features.

The choice of a particular representation is directed by the task for which the perception of the curve is intended. For model identification, the representation used by an expert must allow differentiation between the various models. Since perception is not open to introspection, however, we do not know the exact nature of this representation.

To have a computer perform model identification in the same way as an expert, we need to choose a particular representation for the curves that we think is similar to the one used by the expert. This representation is chosen by defining a specific language, in which we will describe the shapes occurring on a derivative curve. Defining a particular language determines how much information about the curves will be taken into account and, therefore, how much information will be available to discriminate between the models.

Now imagine that we have chosen a particular language for describing curves. We assume that the description of a given curve in this language is similar to the representation of this curve constructed through perception. With this assumption, we identify the components that must be achieved to allow a computer to perform model identification.

Knowledge of the Interpretation Models

The knowledge of the models is available to the expert as a representation of the information pertaining to the model. We need to provide the computer with a similar representation.

We said earlier that we consider model identification using the pressure derivative. There are two sources of information for the derivative of a given model: (1) type curves, and (2) regime properties.

Type curves describe typical responses of the model. This information is provided to the expert as visual information. The representation of this information used by the expert is the description in terms of symbols, of a typical derivative produced by the model. In our case, this description will be done in the language we have chosen.

Regime properties are derived from the study of the analytical expression associated with the model. They give quantitative information about sections of the response, such as fixed slope value for instance. We will discuss later how this information might be represented.

Observation

The language we have chosen for describing curves defines the result of perception. For the models, we use this language to describe a typical derivative. In other words, it is *we* who perceive *models* and provide the computer with the resulting representation. On the other hand, for a given set of *data*,

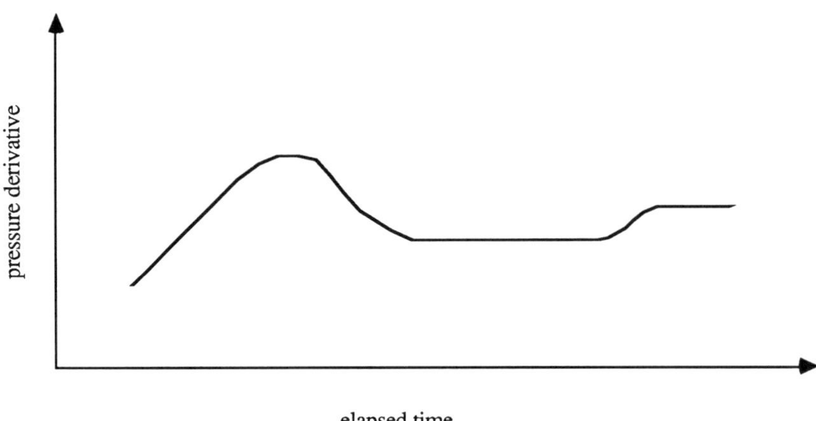

FIGURE 1.1. Example derivative.

the computer needs to perceive the derivative curve on its own.

When real data are perceived by the expert, the human brain does not take into account all the information present on the curve. The representation of the curve used by the expert is a description of the true reservoir response only, as opposed to noise. To allow a computer to perceive the curve in the same way, methods must be developed to distinguish the true reservoir response from the noise in the data. Once this response is obtained, perception amounts to the description of it in the language chosen.

Matching

With the observation step, the data are represented in a form that is equivalent to the one used for models. We can then look for correspondences between data and models to find the appropriate interpretation models. We said earlier that the representation used for models should also take into account quantitative properties of certain flow regimes. Those properties need to be checked on the data when a model is chosen.

Language for Curves

The language we use for describing curves is given in this section. This language is constituted by a vocabulary and a syntax.

The vocabulary is defined by the following symbols:

$$up,\ down,\ maximum,\ minimum,\ plateau,\ valley \tag{1}$$

Where a *plateau* is a flat section preceded by an *up*. A *valley* is a flat section preceded by a *down*. A flat section is a section whose slope is less than 0.1 in absolute value. This style of syntax follows that used for describing well log traces by Startzman and Kuo (1986).

The syntax is defined by the rules:

1. An *up* can be followed by a *maximum* or a *plateau*.
2. A *down* can be followed by a *minimum* or a *valley*.
3. A *maximum* can be followed by a *down*.
4. A *minimum* can be followed by an *up*.
5. A *plateau* can be followed by an *up* or a *down*.
6. A *valley* can be followed by an *up* or a *down*.

For example, with this language, we can describe the curve shown in Figure 1.1 as

up, maximum, down, valley, up, plateau

One consequence of the rules given for the syntax is that a curve description will never contain consecutive identical symbols (e.g., *up, up*). This constraint implies that in most cases, changes in slope cannot be represented. This limitation emphasizes the earlier statement about the importance of choosing the language to use. The language chosen determines how much information about the responses can be taken into account.

The next three sections describe the methods developed for achieving respectively, *observation*, *knowledge of the interpretation models*, and *matching*. The methods presented use a representation of curves based on the language just defined. Allain (1988) considered the use of more information about the derivatives and explained the problems encountered. Solutions were also proposed to cope with those problems.

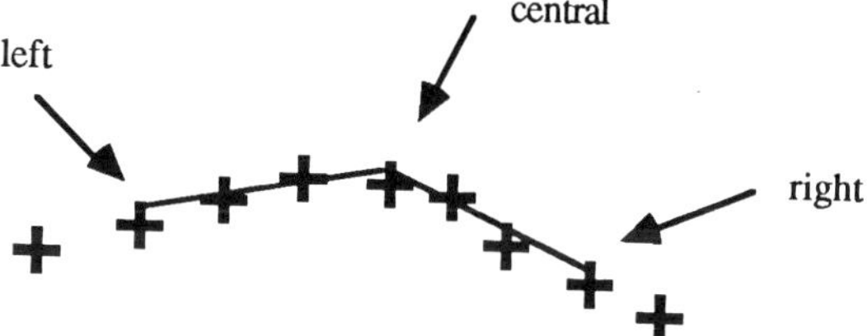

FIGURE 1.2. Central point method for derivative algorithm.

Observation

The purpose of observation is to describe the shapes that are present on the derivative of the data using the language defined in the previous section. The main obstacle to producing this description is the distinction between the true reservoir response and the noise present on the curve. This distinction can be done almost unconsciously by a human expert but it needs to be made explicit in a computerized procedure. Before presenting the methods we developed to identify the true reservoir response, we emphasize how to compute the pressure derivative, since this can greatly affect how noisy the resulting curve is.

Computation of the Derivative

The derivative is obtained with the central scheme described in Bourdet et al. (1984). This scheme uses a point somewhere before (left) and one somewhere after (right) the point of interest (central), computes the two corresponding slopes, and finally takes their weighted mean (Figure 1.2).

The three points, left, central, and right, do not need to be contiguous. We can actually impose a minimum value for the distance of the central point to the two other points, and this minimum value is usually called "differentiation interval." If the three points are very close to each other, a small error in the pressure values may induce a large error in the derivative. This situation often occurs at late time since the logarithmic scale will tend to cluster the last data points. To avoid introducing large scatter due to the proximity of the points, it is necessary to choose a strictly positive differentiation interval. Based on our experience with examples of real well test data, we used a value of 0.2 log cycles.

Perceptual Organization

There are two possible approaches to distinguishing the true reservoir response from the noise on the derivative. The first is to remove the noise from the curve by applying a smoothing algorithm. The features that the smoothed curve exhibits can then be considered as representative of the true response. The second approach is to extract significant features directly on the original derivative, without prior smoothing. The second approach is more appealing for it seems closer to what human perception can actually achieve.

The human visual system has a highly developed capability to detect the relevant groupings and structures among a pattern of dots or a list of points, even without knowledge of their nature. This capability, called "perceptual organization," is considered of prime importance to the understanding of the visual system, and has been studied both in psychology and in computer vision. The ultimate goal of this research is to find a common theme to the different functions of perceptual organization, and to come up with a single principle that would unify the various experimental observations.

The first attempt to formulate such a principle was by the Gestalt school of psychology in the 1920s. The Gestalt psychologists thought that this principle was the ability of the visual system to perceive a pattern as a whole and not as constituted of individuals (the word "Gestalt" itself means "whole"). The Gestalt movement tried to formulate "laws" for the formation of these wholes but failed because their results were not sufficiently concrete to provide any quantitative theory. Later attempts in the 1950s led to the "minimum principle," which stated that people perceive the simplest possible interpretation for given data (Hochberg, 1957). This idea was then exploited in the frame of information theory, defining specific languages for patterns such that the simplest pattern is represented with the least information (Leeuwenberg and Buffart, 1983).

In 1983, Witkin and Tennenbaum proposed as the underlying principle for perceptual organization the "nonaccidentalness argument," which states that the adequacy of a structural description for given data is based on the degree to which the structure is unlikely to have arisen by accident. They also presented an application of this principle to the task of describing a one-dimensional signal, noting that the description obtained seemed to capture perceptually significant features much better than the conventional linear filters. The only requirement for using this principle is to have a way of measuring the significance of a given type of curve description for a list of points.

Perception of Linear Structures

Lowe (1983) defined the following measure for the significance of a straight line fit to a list of points. If l is the length of the line, and d the maximum deviation of a point to the line, then the significance is given by

$$\frac{l}{d} \tag{2}$$

Based on this measure, Lowe presented a method for determining the relevant linear sections of a one-dimensional signal. Since it is not possible to tell a priori what the length of those sections is going to be, groupings of several sizes are attempted on the curve. The different sizes are referred to as possible scales for the description of the curve. At each scale, groupings are realized at several locations on the curve, and their significance is computed with the measure defined by Eq. (2). The most significant structures are chosen as the ones with a locally maximum significance value as the range of scales is traversed. Since it is quite possible for a curve to exhibit significant linearity at more than one scale, several segments may be found at a given location.

To emphasize this property and study the way the significance varies as the size of the groupings changes at a given location, we consider the following example. For a curve with n data points, we search the most significant structures at the first point. These structures could be supported by any of the following subsets (referred to by the indices of the points in them):

$$\{1,2,3\}$$
$$\cdots$$
$$\{1,2,\ldots,n-2,n-1\}$$
$$\{1,2,\ldots,n-1,n\}$$

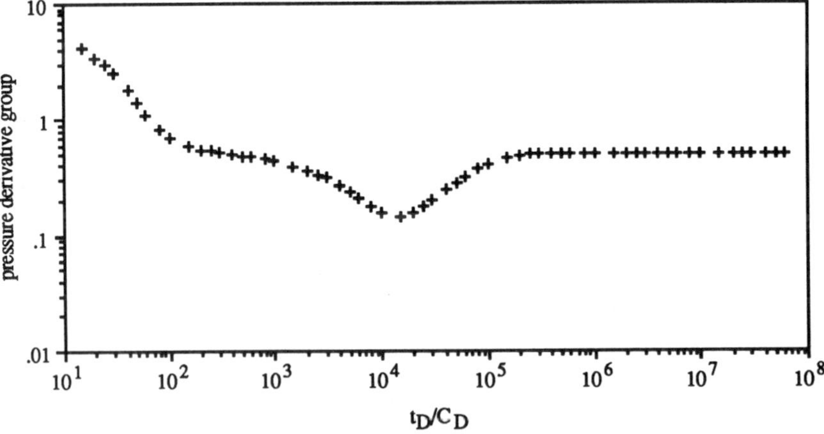

FIGURE 1.3. Example derivative 1.

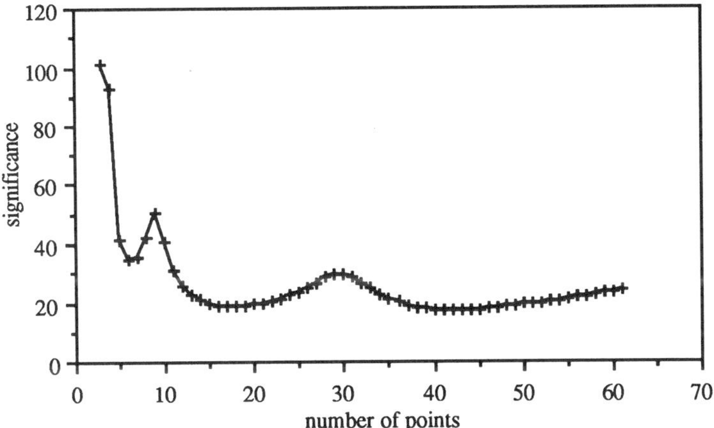

FIGURE 1.4. Significance of linear structures. Examples derivative 1.

To evaluate the significance of a straight line fit to a given subset, we use the following measure, slightly different from the one suggested by Lowe. If l is the length of the line and σ the standard deviation of a point to the line, the significance is computed as:

$$\frac{l}{\sigma} \qquad (3)$$

To keep the significance from becoming infinite, it is necessary to set a minimum nonzero value for standard deviation. Also, since nothing can be said on the significance of a straight line fit obtained with only two points, $\{1,2\}$ is not considered among the subsets. This remark implies that the method cannot be successfully applied on a curve that would exhibit a trend supported by only two points.

We study the behavior of the significance first on analytical data, and then on real ones. The method we outlined is applied to the pressure derivative vs. time in log–log coordinates. Based on the observations made with those examples we describe the basis for an application of the method to the description of derivative curves.

Analytical Example

We consider the derivative shown in Figure 1.3, generated for a well with wellbore storage and skin in a reservoir with pseudosteady-state double porosity, with the following parameter values:

$$C_D e^{2S} = 1.12 \times 10^7$$
$$\omega = 0.0873$$
$$\lambda e^{-2S} = 1.35 \times 10^{-2}$$

The significance values obtained for the different subsets described earlier are plotted in Figure 1.4, as a function of the number of points grouped.

The curve on Figure 1.4 exhibits two local maxima, for 9 and 29 points grouped. As pointed out earlier, the corresponding groupings are therefore among the most significant structures.

For small groupings (less than 7 points), the significance increases as the number of points grouped decreases. In that case should the first grouping be considered as locally maximum? We can derive an answer to that question from a qualitative explanation of this behavior of the significance. The trend of the significance curve is justified since in the absence of noise, and with a sufficiently high number of points per log cycles (here 9), a straight line becomes a better approximation to a section of the curve as the length of this line decreases. As we will see in the next example, this is not true for real data. This behavior can be seen as a capability of the measure to detect the absence of noise and to indicate in that case that small segments are more significant. For this reason, we will treat the grouping corresponding to the first point on the curve shown on Figure 1.4 as a relevant structure.

For large groupings (more than 40 points), the significance increases as the number of points increases and we need to decide whether to consider the last grouping as significant or not. As before, we base this decision on the qualitative explanation of the behavior observed. Because most of the vertical variations on the response occur at early time, the standard deviations for

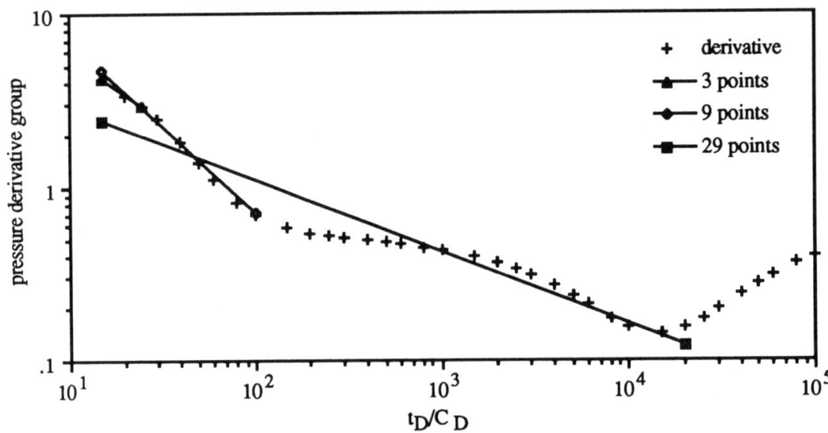

FIGURE 1.5. Most significant structures. Example derivative 1.

all the large groupings are not much different. But the length of the lines increases. Therefore, the significance value increases. Since this behavior occurs because the standard deviation has reached a maximum value, the last groupings cannot be considered as relevant structures.

The three significant segments (first point and two local maxima) are plotted in Figure 1.5 along with the supporting section of the data. They clearly correspond to the description at three different scales of the beginning of the data, as expected.

Real Example

We consider the derivative shown in Figure 1.6. As in the previous example, the significance values are plotted as a function of the number of points grouped (Figure 1.7).

The curve in Figure 1.7 exhibits two local maxima, for 5 and 29 points respectively. The behavior for small size groupings is different from the one observed with analytical data. The noise present on the curve keeps the significance from increasing as the number of points grouped decreases.

The explanation given in the previous example for the late time trend still holds here and the last grouping will therefore not be considered as significant. The only two relevant scales detected are plotted in Figure 1.8 with the supporting section of the curve.

For the interpretation purpose, we are usually interested in recognizing on the derivative curve only one shape at a given location. For this reason we should select only one among the most significant segments at this location. On the basis of the

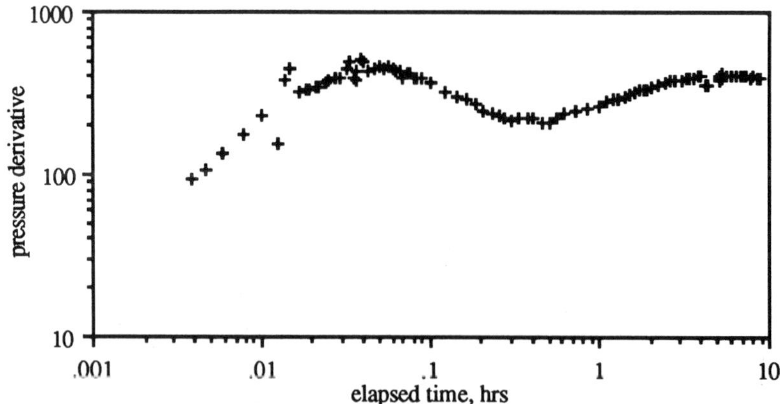

FIGURE 1.6. Example derivative 2.

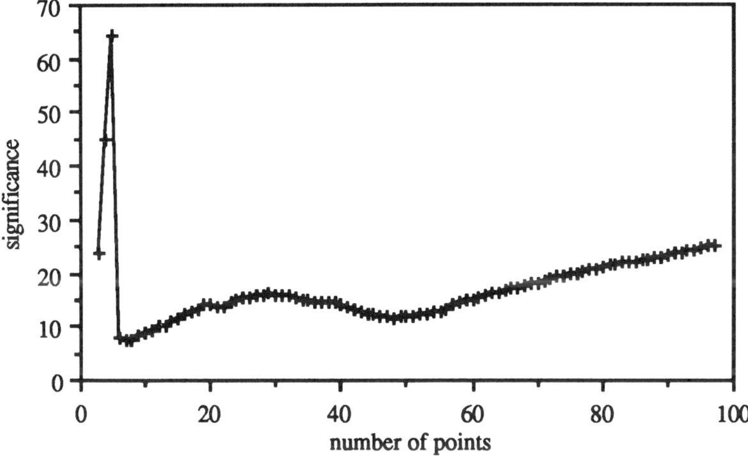

FIGURE 1.7. Significance of linear structures. Example derivative 2.

previous examples, it is clear that this segment should be the smallest one as it is the most faithful to the actual trend of the curve, while ignoring the noise. Thus, we define *the* most significant segment as the smallest one with a locally maximum significance value. With this modification and a simple extension to the method used in the examples above, we develop the algorithm that will be used to describe the derivative curves.

Sketch of the Derivative

In the previous section, we presented a method for finding the first significant linear structure on a derivative curve, that is the first group of points that can be approximated by a straight line. On the remaining data points, we can apply the same method to find the second significant structure. By recursively applying this procedure we can eventually substitute the original derivative with a list of straight line segments that represent the main trends of the curve. The consecutive segments can then be intersected to produce a simplified version of the derivative, which we call the "sketch."

The algorithm we developed contains slight variations to just a recursive application of the method presented in the previous section. As for the previous method, this algorithm is applied to the derivative in log–log coordinates, and the significance is measured with Eq. (4).

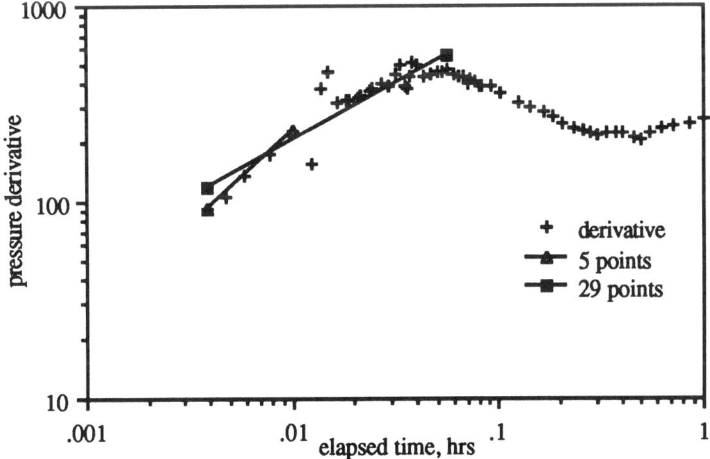

FIGURE 1.8. Most significant structures. Example derivative 2.

To find the most significant structure at a given point of index i, we do the following:

1. Define an initial grouping G_0 with the points of index i through k, such that it contains at least 3 points and is the smallest one with length greater than 0.2 log cycles. Compute the significance of S_0 of G_0.
2. Constitute a new grouping G_1 by including in G_0 the first point on its left. Compute the significance of S_1 of G_1.
3. Constitute a new grouping G_r by including in G_0 the first point on its right. Compute the significance, S_r of G_r.
4. Take the maximum among S_0, S_1, S_r.

If this maximum is S_0, then the most significant structure is defined by G_0. Otherwise, take G_0 to be the grouping with the maximum significance value, take S_0 to be this maximum value. This lengthens the original grouping in one direction. Finally, go to step 2 to seek a still longer grouping.

If we applied recursively the method described in the previous section, we would try to extend the groupings at a given point, to the right only (forward). Instead, we consider extensions in both directions, forward and backward. This was found to produce better results in preserving the trends of the derivatives. Of course, extensions in a given direction are attempted only when they are possible. For instance at the first data point, groupings will be extended only forward.

In the previous section, we noted that in case of noise-free data, the significance value increased when the groupings became small. We also remarked that this was not the case for real data. Actually, if the points are very close to each other, as can happen at late time, very small groupings will always have very high significance values, even on noisy data. Because this behavior is due only to the proximity of the points, we impose a minimum length of 0.2 log cycles for any grouping. This is the reason why we define the initial grouping at any location as the smallest one with at least 3 points, that is at least 0.2 log cycle long.

The overall description of the algorithm producing the sketch is given as

1. $i = 1$.
2. Find the most significant grouping at point of index i. If there is no more point on the right of this grouping, then go to 3. Otherwise, take i to be the first point following the grouping. Go to 2.
3. Consecutive segments are intersected.

Several additional features were included in the algorithm but before describing them, we show the sketch obtained for the example derivative 1 and 2, in Figures 1.9 and 1.10, respectively.

Intersections

The sketch is produced by intersecting the consecutive straight line segments found on the derivative. In an ideal case, if those segments are indexed from 1 to n, then the sketch is constituted by the points $I(1,2), \ldots, I(n - 1, n)$, where $I(j, j + 1)$

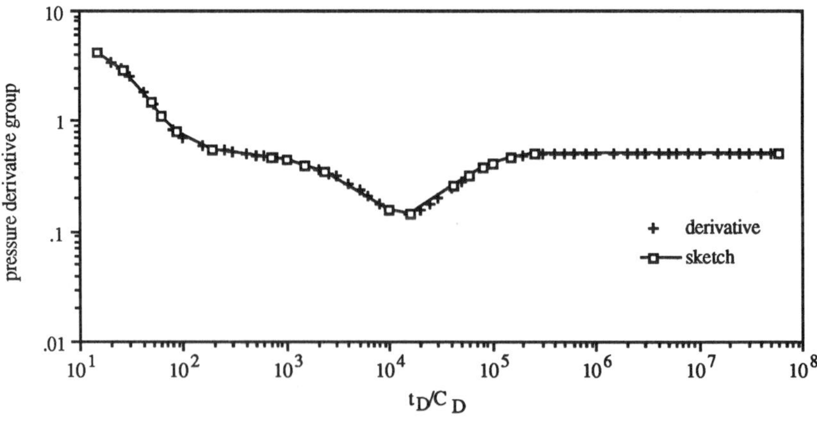

FIGURE 1.9. Sketch for example derivative 1.

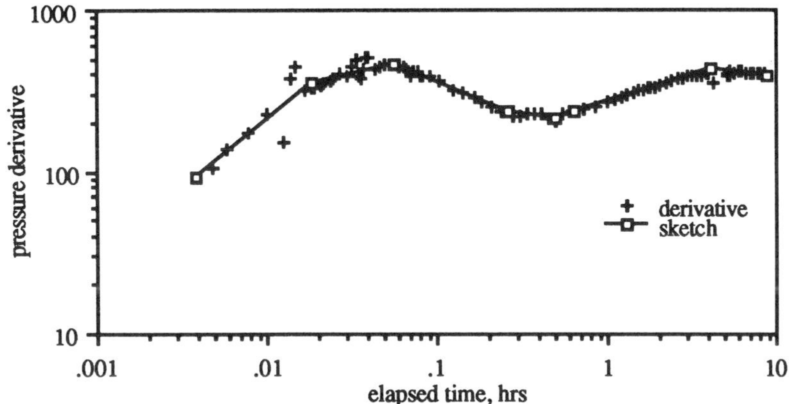

FIGURE 1.10. Sketch for example derivative 2.

denotes the intersection of the segment of index j with the segment of index $j + 1$. For this definition to work, the intersection points must be such that, for all j in $[0, n - 2]$:

$$abscissa\ [(I(j, j + 1))] \\ < abscissa\ [(I(j + 1, j + 2))] \qquad (4)$$

In some cases this relation does not hold. An example of anomalous intersection is given in Figure 1.11, where $I(j, j + 1)$ (point B) has a higher abscissa than $I(j + 1, j + 2)$ (point C).

This situation can occur when a trend on the derivative is interrupted by a region with a different trend. In this case, we do not retain either point B or C and we represent the curve between A and D with only one straight line segment, as shown in Figure 1.11.

Peaks

Earlier, we defined the most significant grouping at a given point, as the first grouping with a locally maximum significance value. This definition is justified because the decrease of the significance value is symptomatic of a change in trend. However, if there is a peak on the curve it will also be perceived by the algorithm as a change in trend. We consider the example derivative shown in Figure 1.12. The sketch obtained for this derivative is plotted in Figure 1.13, and can be seen to reproduce the peaks that were present on the data. This persistence of the peaks is not acceptable.

To cope with this problem, it is first necessary to define a peak in a way that differentiates it from an actual change in trend. Such a definition

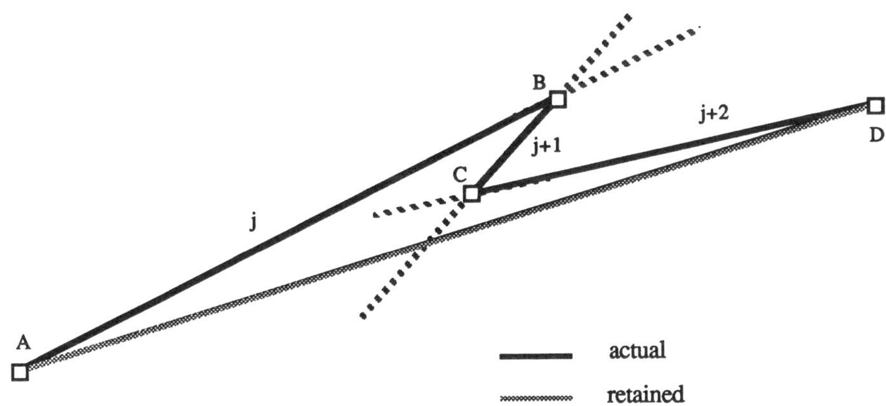

FIGURE 1.11. Example of anomalous intersections.

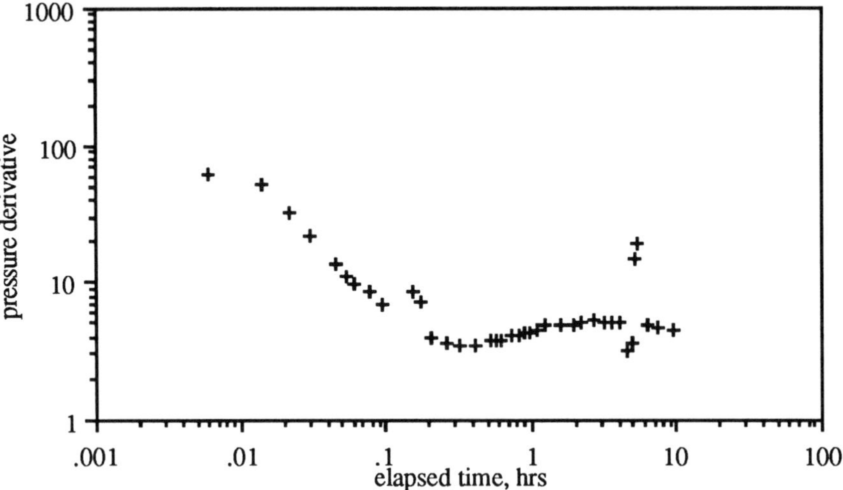

FIGURE 1.12. Example derivative 3.

must exist since *we* as humans are able to make the distinction.

Our visual system is very sensitive to continuity. If short sections of a curve interrupt an otherwise well-defined trend, we can instantaneously detect the continuity of the different pieces and perceive a single trend, rather that a succession of separate ones. For instance, when we look at a dashed line, we detect instantaneously the underlying line, rather than the different segments (the name "dashed line" is besides quite significant). When the interruptions contain points, those points are ruled out in the reconstruction of the original trend, and this leads us to the following definition for a peak. A section of a curve is a peak if it causes a short interruption in an otherwise well-defined trend.

We now consider the case of a linear trend interrupted by a peak, as shown in Figure 1.14. We refer to the points before the peak as group 1. The first point in the peak is called A, the first point after, B. The most significant segment at the first data point (according to our definition so far) is a straight line

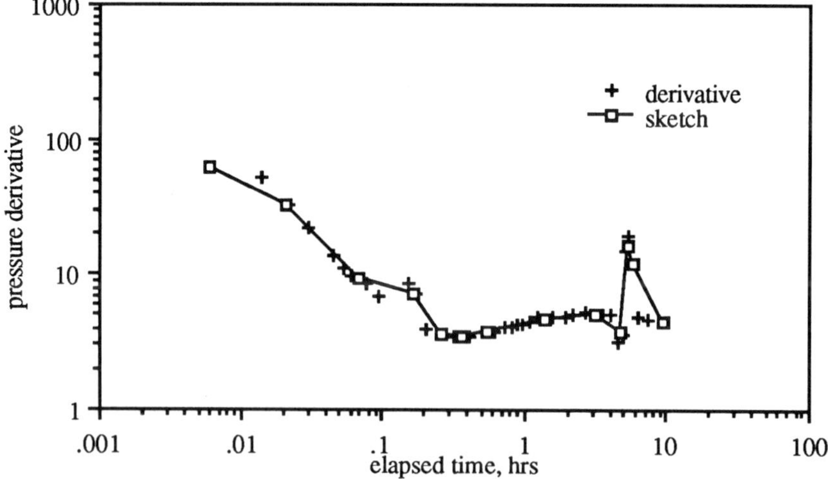

FIGURE 1.13. Sketch for example derivative 3.

fit on group 1. As we emphasized earlier, this is justified by the decrease of the significance when we try to include one more point (A in this case). However, a straight line fit on the points in {group 1 and B} should have a higher significance value than the straight line fit on group 1 only. In other words, the significance can still increase when we extend after group 1, *provided that we skip the points in the peak*. This property can be used to modify the definition of most significant structure. Rather than stopping the extension of a grouping as soon as the significance decreases, we check whether this significance would still increase if some points were skipped. It is necessary to limit the number of points that can be skipped, and since a peak is assumed to be short we set this limit arbitrarily to 4.

In Figure 1.15, we present the sketch obtained with this modification on the data that were shown in Figure 1.12. This time, the peaks have been successfully ignored.

The modification we have made to the algorithm allows the removal of peaks when they occur in a linear trend. If a peak occurs in a section with high curvature, the algorithm is still going to reproduce it. However, segments that are constructed in a peak are likely to have a large slope (in absolute value). Because derivative curves never exhibit very sharp trends, it is salient to set a range for the slope value of the segments to be considered in the sketch. Any segment whose slope falls outside this

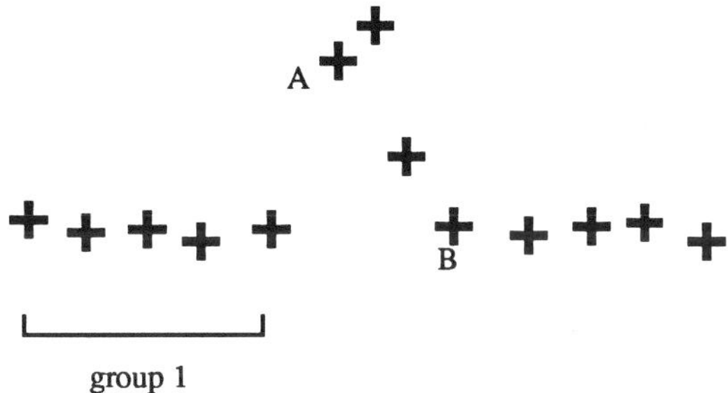

FIGURE 1.14. Peak in a linear trend.

range can be discarded, and the effect of the remaining peaks can therefore in most cases be eradicated. In our work we fixed the range of acceptable slopes from -4 to 4.

Description of the Derivative

The algorithm presented in the previous section allows us to make explicit the true reservoir response (as opposed to noise). From the sketch, it is now possible to produce the description of this response with the language defined in the section entitled "Components of Model Identification."

The description is realized in three steps:

1. Represent each segment in the sketch by the appropriate symbol according to its slope (and the slope of the previous segment if necessary).
2. Insert extrema.

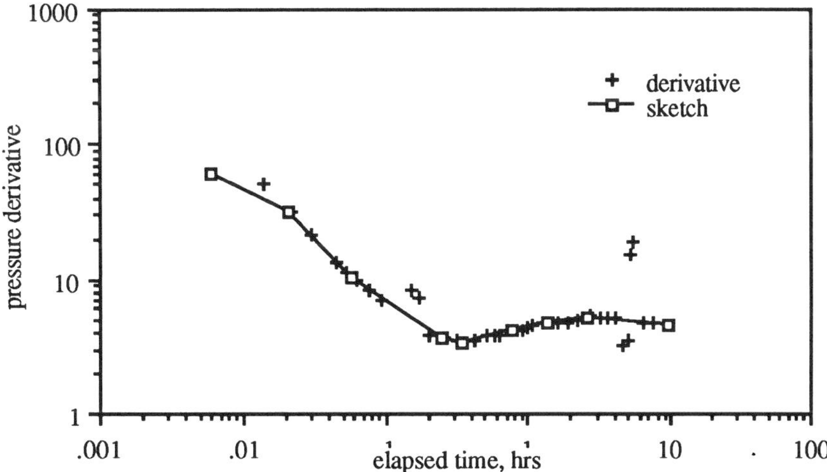

FIGURE 1.15. Final sketch for example derivative 3.

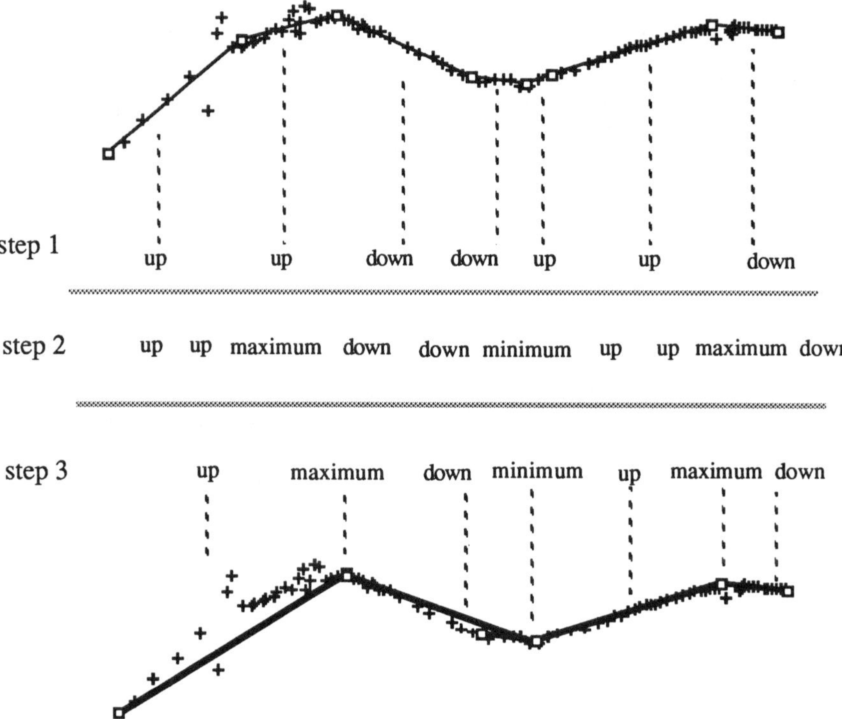

FIGURE 1.16. Description for example derivative 2.

3. Regroup consecutive identical symbols (e.g., *up, up* is described by *up*).

The last two steps are necessary to ensure that the description obtained is syntactically correct (the syntax rules were given in a previous section).

In Figure 1.16, we illustrate the application of this procedure to the sketch obtained for the example derivative 2 (Figure 1.10). At the last step, regrouping consecutive identical symbols is equivalent to merging the corresponding segments of the sketch. The last plot shown in Figure 1.16 gives the curve obtained after those merges have been performed (thick line). It is clear from this figure that a given symbol in the description may correspond to several segments of the sketch.

To emphasize what information we extract about the data and how we can access this information, we explain in the next section the implementation of the description procedure.

Implementation

In an implementation of these procedures described by Allain (1988), the sketch is produced by a FORTRAN algorithm and constitutes the input to the model identification program written in Prolog. The sketch is input to this program as a list of segments, where a segment is defined as

$$[slope, length, log(t_1), log(p_1)] \qquad (5)$$

t_1 and p_1 being the coordinates of the left point for the segment considered.

In the first step of the description procedure, a list of symbols is constructed by representing each segment by the appropriate shape. We call this list of symbols the "shape description." At this point, the shape description and the sketch correspond to each other pairwise.

In step 2, the shape description is modified by inserting the extrema. At the same time, a new list of segments is constructed by inserting in the list corresponding to the sketch, elements corresponding to the extrema. Those elements are not segments but points, and they are defined as

$$[log(t), log(p)] \qquad (6)$$

(where t and p are the coordinates of an extremum). This new list of segments is called the "segment description." The segment description corresponds pairwise to the shape description (the sketch does not any more).

In step 3, we group consecutive identical symbols of the shape description. In parallel, we merge the corresponding segments of the segment description. The new segment description contains element of two kinds:

1. Extrema: still defined as in the preceding section.
2. Segments: defined as

$$[slope, \ length, \ log(t_1), \ log(p_1), \ i_1, \ i_2] \quad (7)$$

where i_1 and i_2 mean that the segment was obtained by merging the segments of the sketch from index i_1 through i_2.

Therefore, at the end of the description, two attributes have been constructed from the sketch:

1. *The shape description*: A list of symbols that qualitatively describes the sketch.
2. *The segment description*: A list of segments and points that corresponds pairwise to the shape description. The segment description gives a global characterization of the sections of the sketch corresponding to the shapes, and also expresses the correspondence between the shape description and the sketch.

The segment description will be used for those two aspects. We mentioned earlier that qualitative information was not sufficient to perform model identification. As we will see later, in the matching procedure it will be necessary to check quantitative properties associated with a given shape. Some of those properties can be checked directly on the segment description, such as the length of a *plateau*, the level of a *valley*, the location of a *maximum*, etc. Others will require us to go back to the segments of the sketch, to determine properties such as the slope at the beginning of an *up*, etc.

Knowledge of the Interpretation Models

In the "Components of Model Identification" section, we explained that the knowledge of an interpretation model is divided into two distinct components. The first one is the representation of the information provided by type curves. For our purpose, it is a description of a typical model derivative in the language we have defined here. The second component represents quantitative properties of certain regimes.

Interpretation Models

An interpretation model is constituted by three different components (Gringarten 1982):

1. *Early time*: corresponding to the well and its surroundings.
2. *Middle time*: corresponding to the average reservoir behavior.
3. *Late time*: corresponding to the outer boundaries of the reservoir.

Several models have been developed for those three categories and we consider the most common cases:

1. Early time: well with wellbore storage and skin, vertically fractured well.
2. Middle time: infinite acting radial flow, pseudo-steady-state double porosity, transient double porosity.
3. Late time: Constant pressure fault, sealing fault, and any viable combinations of those.

By defining the possible models in the three categories, we implicitly define a set of interpretation models that can be obtained by combining them. For any of those models, we must be able to obtain the description of a typical derivative (or several) in the language introduced here. There are two possible approaches:

1. Make the knowledge explicit: list exhaustively all the models that can be obtained from the three categories above. Find their typical derivatives. Describe them in the language defined here.
2. Leave the knowledge implicit: describe a typical derivative for the component models only. Give the necessary knowledge to obtain from those the interpretation models.

The first approach would be cumbersome and subject to errors. Moreover, if the knowledge of the models were explicit, model identification would be a single step procedure. Solutions for given data would be determined by comparing at once *all* the data with the models. Instead, a human expert builds an interpretation model step by step, combining several component models that represent the behavior observed on different sections of the response. Therefore, the expert does not have an explicit knowledge of the interpretation models. Rather, the expert knows how to construct them. We present in the next section methods that we developed to give a computerized procedure the same type of knowledge—not of the interpretation models themselves but how to construct them. As in the case of human interpretation, this knowledge will be used to construct interpretation models that match some observed data.

Organization of the Knowledge

Because an interpretation model is the combination of several components, its derivative must be the combination of the derivatives of its components. Consequently, to describe the derivative of an interpretation model, it is sufficient to describe the derivatives of its components and to know the result of combining them. In this approach, there are three distinct knowledge bases required:

1. Knowledge of the component models: description of the derivative produced by all the models in early time, middle time, and late time categories.
2. Combination rules: rules defining the influence of combining the descriptions of component models.
3. Combination constraints: constraints defining what combination of the component models are viable.

In the next sections, we describe the methods used to provide those three types of knowledge.

Knowledge of the Component Models

We define three categories of models in a slightly different fashion from previously. In the first category, we consider basic interpretation models, constituted by a particular early time regime in a homogeneous infinite medium. The models in this category can then be made more complex by combining onto them the components of the two other categories. The second category corresponds to heterogeneous reservoir behaviors, and the third one to boundary effects. This organization was preferred to the one of the previous section because it

reduces the number of component models and therefore cuts down the number of possible combinations. For the moment we define only the description of one (or several) typical derivative(s) for each model in the language described in earlier sections. This description is only qualitative, and we will take the regime properties into account later. As for the data in the observation part of the analysis, the description is a list of shapes, and we call it the "shape description." It is possible for a given model to admit several shape descriptions.

Basic Models

1. Well with wellbore storage and skin in an infinite homogeneous reservoir:

   [*up, maximum, down, valley*] or [*up, plateau*].

2. Vertically fractured well in an infinite homogeneous reservoir:

   [*up, plateau*]

Heterogeneous Reservoir Behavior

1. Pseudosteady-state double porosity transition: [*down, minimum, up, plateau*]
2. Transient double porosity transition: [*up, plateau*]

The two descriptions listed above correspond to a heterogeneous transition followed by the infinite acting radial flow for the total system. The definition of these two models ensures that the combination of a model in this category with a basic model is still an interpretation model.

Boundary Effects

1. Sealing fault(s):

   [*up, plateau*]

   Depending on the height of the *plateau*, this description can correspond to either one sealing fault, or a wedge.
2. Constant pressure fault:
   [*down*]
3. Pseudosteady-state (closed reservoir):

   [*up*]

More complex models can be obtained by combination of those.

The sketch procedure described earlier may sometimes flatten extrema. For instance, a *maximum* on the data may be transformed into a short *plateau* by the sketch. To still be able to identify the adequate model, it is necessary to take this case into account in the description of models that exhibit extrema. For this reason, in the description of the wellbore storage and skin model (first basic model), we add

$$[up, \ plateau, \ down, \ valley]$$

and in the description of the pseudosteady-state double porosity model:

$$[down, \ valley, \ up, \ plateau]$$

Combination Rules

Given two models M_1 and M_2 with their respective shape descriptions $[s_1^1, \ldots, s_n^1]$ and $[s_1^2, \ldots, s_p^2]$, we want to determine the shape description for the model obtained by combining M_2 onto M_1, during the shape s_i^1. We express this description as

$$[s_1^1, \ldots, s_{i-1}^1, \ \mathrm{comb}(s_i^1, s_1^2), s_2^2, \ldots, s_p^2] \qquad (8)$$

where $\mathrm{comb}(s_i^1, s_1^2)$ represents the shapes obtained by combining s_1^2 onto s_i^1. Most of the time, this combination will leave both shapes intact and we refer to this situation as "nondestructive combination," expressed by the following rule.

$$shape_1 + shape_2 \rightarrow shape_1 + shape_2 \qquad (9)$$

The language we introduced earlier was constituted of both a vocabulary and a syntax. Although the descriptions produced with the rule above are guaranteed to use the right vocabulary, they may not be syntactically correct. For instance, if we combine the description for a sealing fault ([*up, plateau*]) with the first description of the wellbore storage and skin model ([*up, maximum, down, valley*]) during the *down*, we get

$$[up, \ maximum, \ down, \ up, \ plateau]$$

This description violates the syntax because a *minimum* should always occur between a *down* and an *up*. To avoid generating shape descriptions that are not syntactically correct, it is necessary to express the nondestructive combination rules for certain cases only. The only cases for which such a rule is useful are as follows.

Nondestructive Combination Rules

$$plateau + up \rightarrow plateau + up \qquad (10)$$

$$plateau + down \rightarrow plateau + down \qquad (11)$$

$$valley + up \rightarrow valley + up \qquad (12)$$

$$valley + down \rightarrow valley + down \qquad (13)$$

In other cases, the combination might modify the shapes and we refer to this second situation as "destructive combinations." The following rules are needed.

Destructive Combinations

$$valley + up \rightarrow minimum + up \qquad (14)$$

$$plateau + down \rightarrow maximum + down \qquad (15)$$

$$down + down \rightarrow down \qquad (16)$$

$$up + up \rightarrow up \qquad (17)$$

The first two rules are destructive because they modify the shape s_i^1 on which the combination is performed. The last two rules do not modify s_i^1. However, the section of the data corresponding to this shape is lengthened by the combination.

We give an example of how the combination rules can be used. We consider a system constituted by a well with wellbore storage and skin in an infinite homogeneous reservoir, described by [*up, maximum, down, valley*]. We search the effect of combining onto this model a sealing fault, described by [*up, plateau*]. If we combine the response of the sealing fault during the *valley*, we obtain two alternative descriptions.

Using the third nondestructive combination rule (12):

$$[up, \ maximum, \ down, \ valley, \ up, \ plateau]$$

and using the first destructive rule (14):

$$[up, \ maximum, \ down, \ minimum, \ up, \ plateau]$$

Those two descriptions define the possible responses when a sealing fault is added into the original infinite system.

Combination Constraints

Not all arbitrary combinations of the models listed earlier (basic models, heterogeneous reservoir behavior, boundary effects) are physically possible.

Therefore, we need to restrict the set of interpretation models that can be generated by expressing certain constraints. In our work, the following constraints were found to be necessary:

1. An interpretation model should contain one and only one basic model.
2. No heterogeneous reservoir behavior or boundary effect should occur prior to an early time regime. In other words, the first constituent of any interpretation model should always be a basic model.
3. An interpretation model should not contain more than one heterogeneous behavior.
4. An interpretation model should not contain more than two sealing fault effects.
5. Nothing should occur after pseudosteady state.
6. Nothing should occur after the effect of a constant pressure fault.

This list is not necessarily exhaustive. Other constraints may be required depending on the implementation of this approach.

Regime Description

In the previous section, we defined the necessary knowledge for describing qualitatively any possible interpretation model. As we emphasized earlier, it is necessary to take into account the quantitative properties of certain regimes. This information is all the more needed since the qualitative description alone may not always allow the distinction of different models.

To take this information into account, we first associate with the shape description of the models defined earlier a "regime description." The regime description is a list of regimes that simply gives a specific name to each symbol occurring in the shape description of a given model. For instance, for the first description of a well with wellbore storage and skin in an infinite homogeneous medium, to the list

$$[up, \; maximum, \; down, \; valley]$$

we associate the list

$$[ws, \; max \; ws, \; end \; ws, \; iarf]$$

where

ws stands for wellbore storage,
max ws for maximum after wellbore storage,

end ws for end of wellbore storage,
iarf for infinite acting radial flow.

In the process of constructing an interpretation model, the shapes corresponding to its different components may be altered (destructive combinations). It is therefore necessary to express what is the consequence (if any) on the corresponding regimes. The main two such consequences are given below with the associated combination rules for shapes.

$$valley + up \rightarrow minimum + up$$
$$r_1 + r_2 \rightarrow gt(r_1) + r_2 \tag{18}$$

where $gt(r_1)$ means that the level corresponding to the *minimum* is higher than the level that would have corresponded to the *valley*.

$$plateau + down \rightarrow maximum + down$$
$$r_1 + r_2 \rightarrow lt(r_1) + r_2 \tag{19}$$

where $lt(r_1)$ means that the level corresponding to the *maximum* is lower than the level that would have corresponded to the *plateau*.

Using those rules, we can now obtain for an interpretation model not only a shape description but also a regime description. If a regime corresponds to a particular section of a response, there may be some quantitative properties satisfied by this section. If we express those properties for the various regimes, then the regime description will implicitly give quantitative information about the models. We will describe in the next section how we actually express the regime properties.

Attributes for an Interpretation Model

A given model can have several descriptions. Any of those descriptions is constituted by a *shape description* and a *regime description* that correspond to each other pairwise. The shape description is a list of shapes that qualitatively describes the response of the model. The regime description is a list of regimes that implicitly gives quantitative information about the response.

Matching

We first review the results obtained in the sections on *"Observation"* and *"Knowledge of the Interpretation Models."* More precisely, we recall what

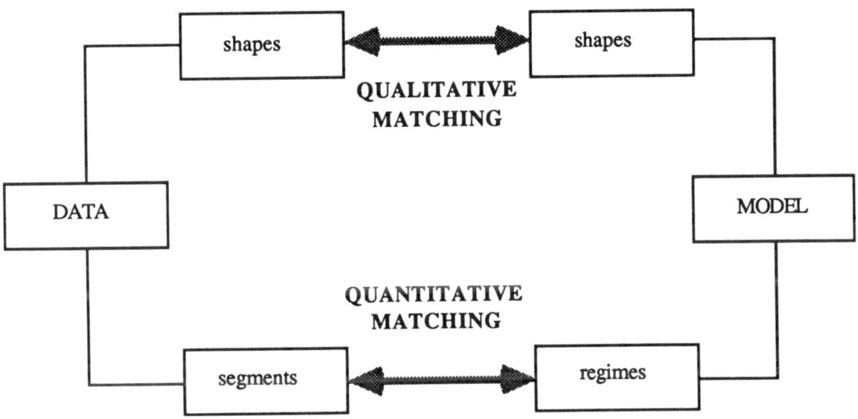

FIGURE 1.17. Components for matching.

descriptions are associated with a set of data and an interpretation model.

1. *Data*: The derivative of a given set of data is replaced by the *sketch*, from which are produced two different descriptions. The first, the *shape description*, qualitatively describes the main trends present on the derivative and is a list of shapes. The second, the *segment description*, is a list of segments that can be used to determine what section of the sketch corresponds to a particular symbol in the shape description. The segment description also gives global information about the sketch. The shape and segment descriptions correspond to each other, section by section.
2. *Model*: An interpretation model is characterized by a *shape description* (a list of shapes) and a *regime description* (a list of regimes). The regime description attributes a specific name to each element of the shape description. The two descriptions, shape and regime, correspond to each other, section by section.

For the moment, we consider matching under the following assumption. Each shape occurring in the data, (as derived from the sketch) can be interpreted with one of the available models. In other words, we assume that

1. The sketch has successfully recovered the true reservoir response from the original derivative.
2. The true reservoir response does not contain any anomalous behavior for which the models considered cannot account.

With this assumption, we can decompose matching in two distinct processes, as shown in Figure 1.17:

1. *Qualitative matching*: The shape description of the model and the data must be identical. Actually, it is sufficient that the shape description of the data could be obtained by truncating the shape description of the model.
2. *Quantitative matching*: The equality of the shape descriptions of the model and the data defines a relation between the segment description of the data and the regime description of the model. Since each segment of the segment description is associated with a particular section of the sketch, a correspondence is thus established between each regime of the model and a section of the sketch. This correspondence can be used to check that the model is not only qualitatively but also quantitatively admissible. To do this check we need to express for any regime that requires it, what properties of the corresponding section should be observed.

Quantitative Constraints

There are two types of constraints that need to be expressed, and we differentiate them by using the qualificatives "absolute" and "relative." An absolute constraint characterizes a specific regime. For instance the constraint imposed on a section when it corresponds to wellbore storage is absolute. It involves only the segments corresponding to this section. On the other hand, a relative constraint characterizes the relation between two different regimes. The doubling of value from infinite acting radial flow to the stabilization due to a single sealing fault is a relative constraint.

We emphasized earlier that a given regime usually corresponds to several segments of the sketch. It is necessary to keep this fact in mind when expressing the quantitative constraints. The thresholds added around the theoretical values of regime slopes, or level difference between regimes, are based on our experience with real well test examples. These threshold values may require further tuning.

1. *Absolute constraints*
   a. Among the segments of the sketch corresponding to wellbore storage, the segment with slope closest to 1 should have a slope at least equal to 0.7, provided that the initial *up* is at least 0.5 log cycles long.
   b. For wellbore storage and skin with low value of $C_D e^{2s}$ (no hump on the derivative), all the segments of the sketch corresponding to the initial *up* should have a slope less than 0.7.
   c. For a vertically fractured well, all the segments of the sketch corresponding to the initial *up* should have a slope smaller than 0.7.
   d. The segments of the sketch corresponding to the upward trend due to a sealing fault should not contain segments with slope greater than 0.7.
   e. The last segment of the sketch corresponding to pseudosteady state for a closed reservoir should have a slope greater than 0.7.
   In the previous section, we mentioned that the description of models containing extrema had to be augmented for the cases where those extrema had been flattened by the sketch. For instance, in the wellbore storage and skin model, to the ideal shape description [*up, maximum, down, valley*], we associated the second description [*up, plateau, down, valley*]. The *plateau* that we may recognize in this case should, however, be short. Therefore, we need to add an absolute constraint for this case:
   f. The segment of the segment description corresponding to the *plateau* in the wellbore storage and skin model should not have a length greater than 0.5 log cycles.

An equivalent constraint is imposed for the minimum of a pseudosteady-state double porosity transition.
   g. The segment of the segment description corresponding to the *valley* in a pseudosteady-state double porosity transition should not have a length greater than 0.5 log cycles.
2. *Relative constraints*
   a. In case of transient double porosity, the derivative value at the last point of the *up* due to the transition should not be more than twice the value at the bottom of the transition.
   b. For pseudosteady-state double porosity, the derivative value at the last point on the *up* due to the transition should not be higher than the value for the radial flow of the fissures (if it appeared on the response).
   c. There must be consistency between the different stabilization levels observed on the response, "radial flow," "less than radial flow," "greater than radial flow", etc.

From all those constraints we can see that checking for slope requires the algorithm to consider the segments of the sketch. On the other hand checking for length or level can be done on the segments of the segment description.

These constraints may at first seem simplistic. A given regime such as wellbore storage should in an ideal case correspond to a list of segments whose slopes are not only positive (or greater than 0.1), but also decreasing. We never check whether this is the case. Doing this check would amount to the consideration of a description of the curves at a lower level than allowed by the representation we used so far. By using a lower level representation, more information about the sketch could be taken into account. Because a real response seldom matches exactly the ideal behavior of the analytical model, some of this information would have to be transformed or ignored. Therefore, the assumption we made earlier that all the shapes in the data could be interpreted by the available models would not hold. To avoid for now dealing with too much information about the data, we use the simplest constraints that proved, based on our experience with real examples, to be sufficient for the discrimination of the models. Later, we will consider adding more information and propose methods for processing it.

The Matching Procedure

An expert, when looking at the derivative of given data, can recognize on this curve sections that are specific to particular models. The interpretation model is then given as the combination of the various components thus obtained (Gringarten, 1987). The order in which an expert recognizes the features is not obvious a priori. Does the expert follow the curve from early time to late time? Do some sections of the curve attract the expert's attention before others? We consider interpretation of the response in a linear fashion, from beginning to end. This procedure is natural because the descriptions of models and data are lists, and recursion is the adequate method for processing lists.

We define a "data element" as the section of the data corresponding to a particular element of the data description. A data element has two attributes, a shape and a segment. A given set of data is a list of data elements. Similarly, we define a "model element" as the section of a model corresponding to an element of the model description. A model element has two attributes, a shape and a regime. A given model is a list of model elements.

The procedure of constructing a model that matches a given set of data can be defined recursively by the following fact, and rule:

1. Fact: Any model matches a set of data with no element.
2. Rule: A model matches a nonempty set of data if
   a. the first element of the model matches the first element of the data, and
   b. the list constituted by the remaining elements of the models matches the list constituted by the remaining elements of the data.

With this recursive definition, we only need to define matching for elements:

1. A model element matches a data element if it matches this element both qualitatively and quantitatively.
2. A model element qualitatively matches a data element if their corresponding shapes are the same.
3. A model element quantitatively matches a data element if the constraints imposed by the regime of this model element are satisfied by the data (if there are no constraints, quantitative matching is obtained by default).

With those definitions, we give the exact procedure developed for matching. Using the knowledge of the models as we presented it in an earlier section, this procedure constructs interpretation models step by step (or element by element), looking at the data from left to right.

1. Find a basic model whose first element matches the first data element. Go to 2.
2. If there is no data element left, then stop. Otherwise go to 3.
3. Consider the next data element.
   a. There is no more model element:
      Find a component model allowed at this point (according to the combination constraints), whose first element matches the data element considered. Append this model onto the interpretation model constructed so far. Go to 2.
   b. Consider the next model element.
      If this element matches the data element considered, then go to 2. Otherwise, find a component model allowed at this point that, when combined onto the current model element, transforms this element to produce matching. Go to 2.

This procedure was implemented in Prolog in the model identification program described by Allain (1988). The method used by Prolog to solve a given problem is called "backtracking." Backtracking ensures that all the solutions are found, within the knowledge of the program's rule base.

At any time in the matching procedure, the interpretation model constructed matches the data until the element considered at that time. Complexity is added in the interpretation model only when there is an element of the model that does not match the data (either qualitatively or quantitatively). This is done by combining on the model a new component. No combination is performed until model and data disagree. For this reason, it is clear that the nondestructive combination rules are never used. However, in step 3a, the effect of appending a new component on the interpretation model is exactly equivalent to the effect of those rules.

Imposing the condition that combinations can be performed only when model and data disagree has several important advantages. First, it greatly reduces the search for appropriate interpretation models, and makes the matching procedure efficient. Also, it ensures that the models proposed are the simplest solutions to the interpretation. Actually this is not true without a restriction of the combination rule (17). This restriction will be presented and justified in a later section.

Unfortunately, there are some cases for which this matching procedure does not work properly. However, with the addition of new combination rules we can limit those cases, as we show in the next section.

Misses

If we try to match the response of a known interpretation model, then this same model should be proposed by the matching procedure. We show that this is not always true.

We consider matching analytical data defined by the response of a model M obtained as follows. M is constructed by combining two models M_1 (with shape description $[s_1^1, \ldots, s_n^1]$) and M_2 (with shape description $[s_1^2, \ldots, s_p^2]$) with a destructive combination. If M_2 is combined onto the shape s_i^1 of M_1, then M has the shape description given by

$$[s_1^1, \ldots, s_{i-1}^1, \text{comb}(s_i^1, s_1^2), s_2^2, \ldots, s_p^2]$$

Suppose that the first shape in $\text{comb}(s_i^1, s_1^2)$ is s_i^1. The only destructive combination rules that could ensure this are the last two rules, (16) and (17). For those two rules, (8) becomes:

$$[s_1^1, \ldots, s_{i-1}^1, s_i^1, s_2^2, \ldots, s_p^2] \qquad (20)$$

Until the $(i - 1)$th element of M, the model M_1 will be found admissible by the matching procedure. Now we look at the ith element. For this element, qualitative matching is obtained, since the shapes are equal. We pointed out in the "Knowledge of the Interpretation Models" section that even though the ith shape for M is the same as the ith shape for M_1, the corresponding segment is longer. The question now is whether a quantitative constraint applies to this segment, and whether it allows us to find a disagreement between the data and the model. Two cases must be considered:

1. If such a constraint applies and is not satisfied, then quantitative matching is not obtained. In this case, we will try to combine a new component onto the ith element of M_1. The combination of M_2 will provide a solution.

2. If no constraint applies, or if it is satisfied, then quantitative matching is obtained. No combination is performed and the matching procedure continues with the same model, M_1. When disagreement is found between the responses of M and M_1, it will be too late to perform the combination. The true solution will not be proposed.

We see that in the second situation, the matching procedure misses the true solution. We need to find out if this situation can occur at all, and what solution (if any) can be proposed in that case. As we mentioned earlier, the problem arises from the last two destructive combination rules (16) and (17). We study the possible effects of those rules in the next sections.

Rule (16): *down + down → down*

We consider matching analytical data defined as (part of) the response of a model obtained using the rule (16).

- M_1 is wellbore storage and skin with shape description
  [*up, maximum, down, valley*].
  M_2 is pseudosteady state double porosity with shape description
  [*down, minimum, up, plateau*].
  M has the shape description: [*up, maximum, down, minimum, up, plateau*].
  The data we consider are the complete response of M.

If we try to match those data, M_1 is found admissible until the *down* since no constraint applies to the corresponding segment. There is disagreement on the next shape (*minimum* versus *valley*), and it is too late to perform the combination.

We see that if we relaxed the constraint that combination should occur only at the location of disagreement between the data and the model, this problem could be avoided. However, this solution is computationally unacceptable. Moreover the procedure would in that case propose models whose complexity is not warranted by the data.

The other solution is to formulate ad hoc combination rules. The one we need in the case described here is

$$\textit{valley} + \textit{minimum} \rightarrow \textit{``nothing''} \qquad (21)$$

With this rule, the true solution will be found.

We look at a second example:

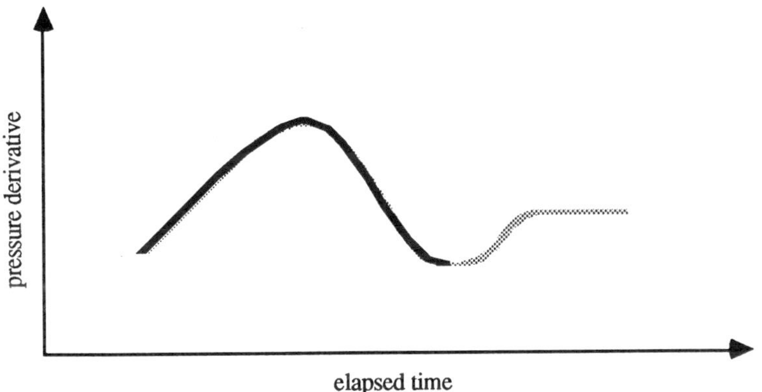

FIGURE 1.18. Truncated response for pseudosteady-state double porosity. Case 1.

- M_1 is the same model as before.
  M_2 is pseudosteady-state double porosity with shape description
  [*down, valley, up, plateau*].
  M has the shape description [*up, maximum, down, valley, up, plateau*].
  The data are the complete response of M.

This second description for M_2 (pseudosteady-state double porosity) was added in the "Knowledge of the Interpretation Models" section for cases where the *minimum* of the transition had been flattened by the sketch.

This time, M_1 will be found admissible by the matching procedure until the fifth shape (*up*). Again we need to add a rule allowing the algorithm to perform the right combination at this location. Unlike the previous rules, this one does not combine two shapes but a shape with a list of shapes:

$$up + [down, valley] \rightarrow [\ ] \qquad (22)$$

If we use this rule, then we need to go back and check the length of the *valley*, as dictated by the seventh absolute quantitative constraint.

The introduction of this new rule requires us to change the way we compute the combination of models, and more precisely the expression given by Eq. (8). We change this expression in the following way. We call $[s_1^1, \ldots, s_n^1]$, L^1 and $[s_1^2, \ldots, s_p^2]$, L^2 and consider a rule giving the effect of combining a sublist L_1^2 of L^2, with s_i^1. L_1^2 is equal to $[s_1^2, \ldots, s_k^2]$ for some $k \leq p$, and we call L_2^2 the list constituted by the remaining elements of L_2 (it can be empty). With those definitions, the combination of the two models considered is obtained by appending onto

$$[s_1^1, \ldots, s_{i-1}^1, \text{comb}(s_i^1, L_1^2)]$$

the list L_2^2.

For the two previous examples, the addition of new combination rules allows us to correct the behavior of the matching procedure. This is not always possible as illustrated by the third example:

- M_1, M_2 are the same as before.
  The data are defined as the truncation of the response of M after the *down*.
  The data have the shape description [*up, maximum, down*].

No disagreement will ever be found between the data and M_1. M_1 will be proposed as a possible solution. In reality, the effect of M_2 on the data may not be seen either. For instance, we show on Figure 1.18 a possible shape for the data derivative. This derivative is the truncation of the complete response of M shown in the shaded line. From the data only, there is no way to tell that the *down* actually resulted from a combination of M_1 with M_2. In that case, even an expert may propose M_1 as a possible solution.

There are some cases where the data exhibit a feature that would allow an expert to recognize that M_1 alone is not sufficient. We show such a case in Figure 1.19. The useful feature on the derivative in this figure is an inflexion. The description of this derivative is still [*up, maximum, down*] and as we pointed out earlier, the representation we use does not allow us to diagnose the presence of the inflexion. In this case the program will make a mistake that the expert would not make. Again, one way to cope with this problem would be to perform combinations even when the data and model agree. We

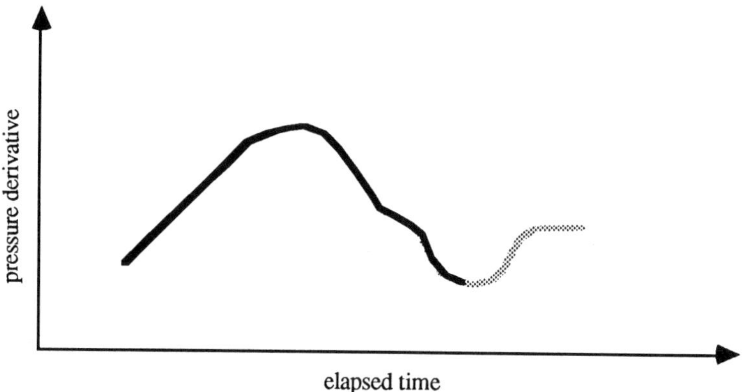

FIGURE 1.19. Truncated response for pseudosteady-state double porosity. Case 2.

would be sure in that case that the adequate model would be proposed. As we emphasized earlier, this solution is computationally unacceptable.

Rule (17): $up + up \rightarrow up$

The problems that arose in the previous section were due to the absence of quantitative constraint for a *down* and the fact that inflexions are always ignored. The situation is sometimes different for an *up*. We consider the two responses shown in Figure 1.20.

The first response on this figure is for a well with wellbore storage and skin in a reservoir with pseudosteady-state double porosity. The second one is for a similar system with a sealing fault added to it. Both responses have a similar shape description:

[*up, maximum, down, valley, up, plateau*]

For the first response, the only model proposed will be a well with wellbore storage and skin in a reservoir with pseudosteady state double porosity. For the second one, although the inflexion is ignored, disagreement will be found between this model and the data at the fifth shape (*up*). This is because the end of this *up* should not be higher than the level of radial flow (*valley*) (second relative constraint, listed earlier). Because of the disagreement, a combination will be performed. The only possible combination is to add a sealing fault effect, and therefore the right solution is proposed.

In the previous section we first noted that new combination rules could be added for some situations. Those rules corresponded to the combination of a *down* with the list [*down, minimum*]. The equivalent situation here is the combination of an

up with the list [*up, maximum*]. Such a combination could occur only for a change in wellbore storage. In that case, the description obtained for the data is the same as if there were no change in wellbore storage, even though the curve may show an inflexion in the initial *up*. Because the inflexion is ignored, no disagreement will be found between the data and a wellbore storage and skin model.

Restriction to the Rule (17)

We showed in the previous section an example where rule (17) helped correct a misinterpretation. This same rule is sometimes responsible for the generation of interpretation models that are too complex.

Suppose that in the matching procedure, we are looking at the ith element of the data and the model constructed so far is M_1, with shape description

$$[s_1^1, \ldots, s_n^1]$$

If there is a disagreement between M_1 and the data, we try to combine a new component on M_1. The only combination rules that we can use are the destructive combination rules, repeated below:

$$valley + up \rightarrow minimum + up \qquad (14)$$

$$plateau + down \rightarrow maximum + down \qquad (15)$$

$$down + down \rightarrow down \qquad (16)$$

$$up + up \rightarrow up \qquad (17)$$

Suppose that the ith shape of the data is a *minimum* and that the ith shape of the model is a *valley*. We

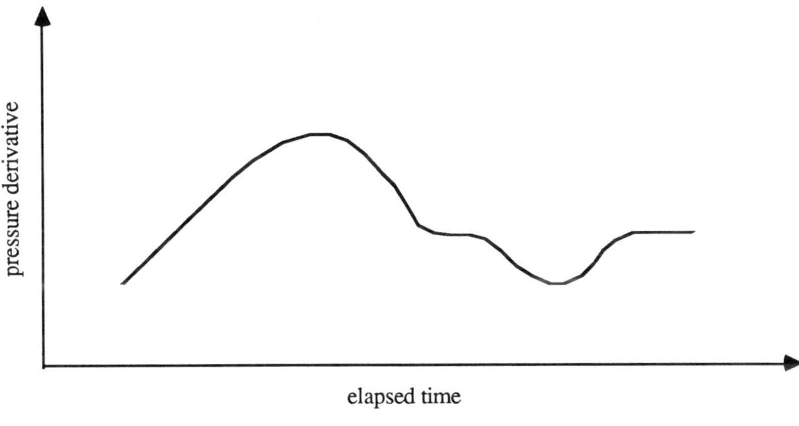

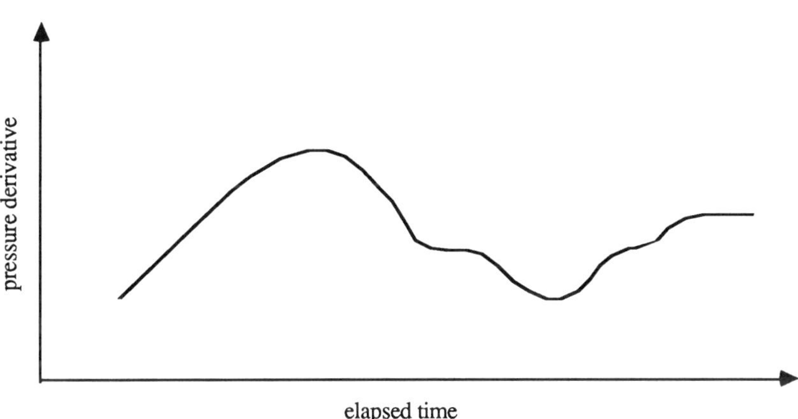

FIGURE 1.20. Two derivatives with the same shape description.

need to combine onto M_1 a model which shape description starts with an *up*. If

$$[up, s_2^2, \ldots, s_p^2]$$

is the shape description for M_2, then the shape description of the new interpretation model is given by

$$[s_1^1, \ldots, s_{i-1}^1, minimum, up, s_2^2, \ldots, s_p^2] \quad (23)$$

No quantitative constraint ever applies to a *minimum* and, therefore, we accept the new model and go to the next shape. The next shape of the data must be an *up* (because it is preceded by a *minimum*). The next shape of the model is also an *up* and qualitative matching is thus obtained. We call r_2 the regime associated with the initial *up* of M_2. If a quantitative constraint is imposed by r_2 on the

up and is not satisfied by the data, we try to combine a new component M_3 on the interpretation model. Because qualitative matching needs to be preserved the shape description for M_3 must start with an *up*:

$$[up, s_2^3, \ldots, s_q^3]$$

The rule that must be used for the combination is rule (17). This rule has no effect on the shape description of the interpretation model until the $(i + 1)$th shape (*up*). However, if r_3 is the regime associated with the initial *up* in M_3, r_2 is replaced by r_3 in the interpretation model. If the constraint imposed by r_3 is satisfied, the new interpretation model will be retained.

It is clear that the combination of M_1 directly with M_3 would have worked as well, and it will actually be proposed as a solution. Therefore, in that case, the matching procedure does not propose only the simplest solutions.

This situation occurs from the following sequence of combinations:

$$valley + up \rightarrow minimum + up$$
$$minimum + up + up \rightarrow minimum + up \quad (24)$$

When we perform the first combination, we check for quantitative matching on the *minimum*. If we checked on the *up* at that time, then we would reject the combination of M_2. Since we do not find any disagreement, we consider that this combination is admissible and go to the next shape. Unfortunately, when we find the disagreement, rule (17) allows us to completely substitute on the response the effect of M_2 by the effect of M_3. M_2, although completely "invisible" on the response, is still present in the interpretation model.

To cope with this problem, it is necessary to forbid the application of rule (17) in the situation described by Eq. (24). The only cases in which such a situation can occur are

1. M_1 is a transient double porosity transition. M_2 is a sealing fault effect.
2. M_1 is a transient double porosity transition. M_2 is pseudosteady state (closed reservoir).
3. M_1 is a sealing fault effect. M_3 is pseudosteady-state (closed reservoir).

Logically we expect that rule (16) could create similar problems to rule (17). The equivalent sequence of combinations to Eq. (24), for rule (16) is given by

$$plateau + down \rightarrow maximum + down$$
$$maximum + down + down \rightarrow maximum + down$$
$$(25)$$

A necessary condition for the application of this sequence is that the quantitative constraint on the *down* after the first combination is not satisfied. Since no constraint ever applies to a *down*, the sequence will never be used.

Summary

With a restriction to rule (17), the matching procedure is guaranteed to always propose the simplest models that represent the behavior observed on the data. Models are constructed starting with a basic model, and eventually combining new components onto this model. Combinations are performed only on elements of the model that disagree (qualitatively or quantitatively) with the data. This constraint allows us to greatly reduce the search space and makes the procedure very efficient. Although the models proposed are never too complex, they may sometimes be too simple. This problem usually arises when an inflexion on the data is the only indication available to determine the appropriate solutions. Two different approaches could be adopted to cope with this problem:

1. Relax the constraint that combinations should be performed only at elements of disagreement between data and model. This modification would guarantee that the solution set proposed would always contain the adequate interpretation models. In all cases, the procedure would also propose models that are too complex. In some cases, it would still propose models that are too simple. A method such as the one proposed by Watson et al. (1988) could be used to decide which are the appropriate models, after automated type curve matching has been performed. However, simply finding the models is computationally expensive because the search space is too large.
2. Take more information about the responses into account. If enough information is available, adequate solutions can be found in all cases. Adding this information does not require us to allow combinations at elements of agreement between data and model. Therefore
   a. the procedure would still be efficient.
   b. the models proposed would never be too complex.
   However, more information means more difficulties, because real data are never totally ideal. It is necessary to develop methods that allow the algorithm to ignore or modify some of the additional information. We will not present such methods here, although some examples can be found in Allain (1988).

For the moment, we simply add to the methods developed so far, the two new combination rules (21) and (22), and the restriction to rule (17). With those modifications, the matching procedure can in most cases propose the adequate interpretation models.

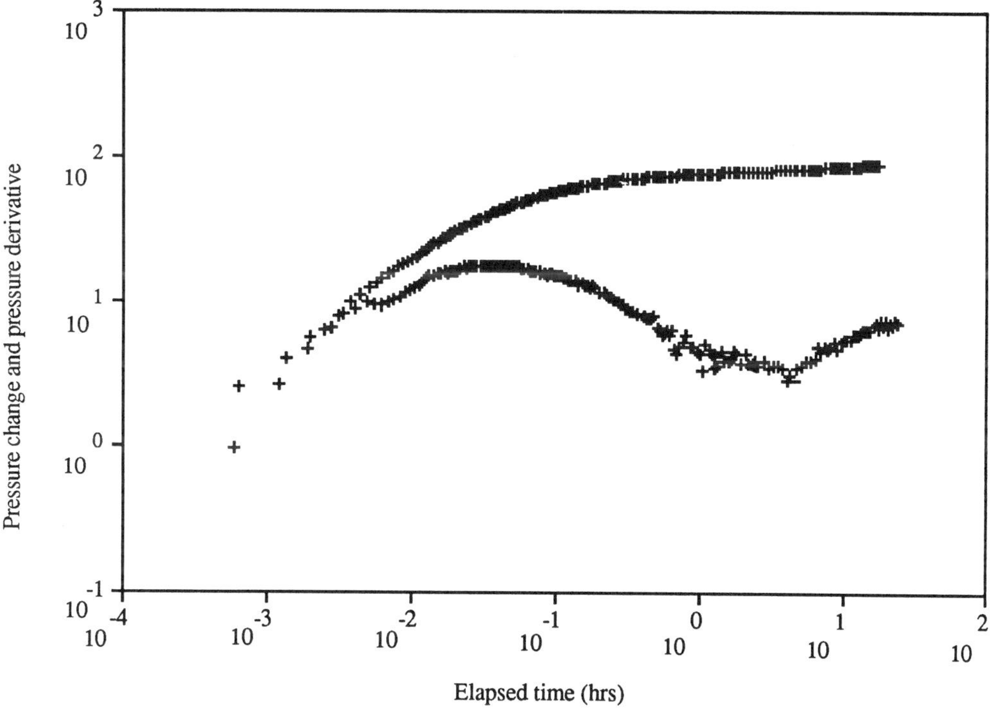

FIGURE 1.21. Example data. Flow period #4 (build-up).

Example

The use of a model identification program developed using the techniques described in earlier sections is illustrated on real data. The actual program was implemented in Prolog, and is described in more detail in Allain (1988). The test considered is a build-up test in which the well was flowed at three different rates before shut-in. This test is described by Bourdet et al. (1983).

Data

The pressure change was monitored during build-up. The data are plotted in Figure 1.21 versus elapsed time, along with the corresponding derivative.

As we emphasized earlier, the derivative is obtained with a central scheme using a differentiation interval of 0.2 log cycles. It is computed with respect to the superposition time function using the flow periods given in Table 1.1.

Observation

The first step involved in the observation, once the derivative has been computed, is the production of the sketch. The sketch obtained is shown in Figure 1.22, with the original derivative.

The description of this sketch in the language introduced earlier leads to the following shape description:

$$[up, plateau, down, minimum, up] \qquad (16)$$

(Also, the corresponding segment description is produced.)

A graphic equivalent of this description is shown in Figure 1.23, along with the sketch. We see that

TABLE 1.1. Flow periods

Test number	Duration (hours)	Rate (stb/d)
1	3.8	800
2	3.3	2500
3	16.45	830
4	18	0

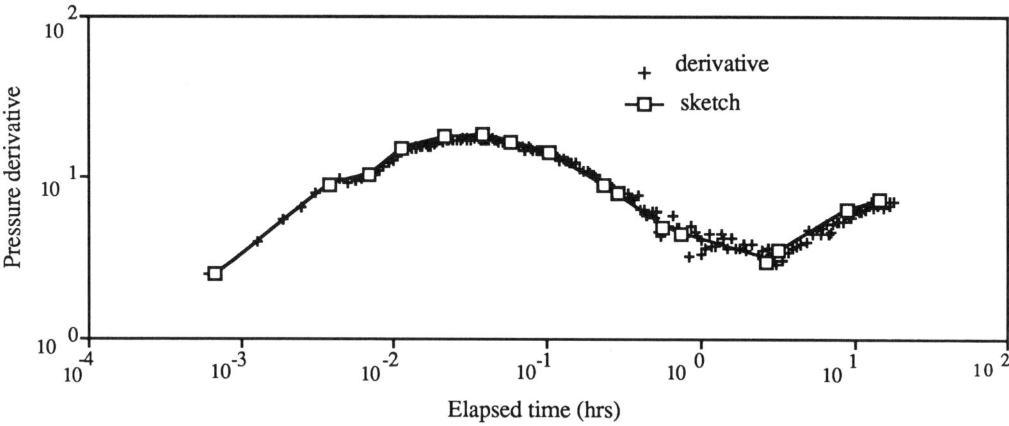

FIGURE 1.22. Example derivative and sketch.

the sketch has flattened the maximum of the hump occurring at early time. Also, the inflexion present in the initial *up*, which could be attributed to a change in wellbore storage, is ignored in the description. This is because the segment corresponding to the inflexion has a slope lower than 0.1.

Matching

There are three interpretation models proposed by the program:

1. A well with wellbore storage and skin in a reservoir with pseudosteady-state double porosity behavior.
2. A well with wellbore storage and skin in a reservoir with transient double porosity behavior.
3. A well with wellbore storage and skin in a reservoir with a sealing fault.

Those models were constructed by the matching procedure element by element, starting from the beginning of the data. We describe how this construction was actually realized.

First Element: up

To match this element, we need to pick one of the basic models. Any other choice is forbidden since one of the combination constraints imposes that the first component of any interpretation model be a basic model.

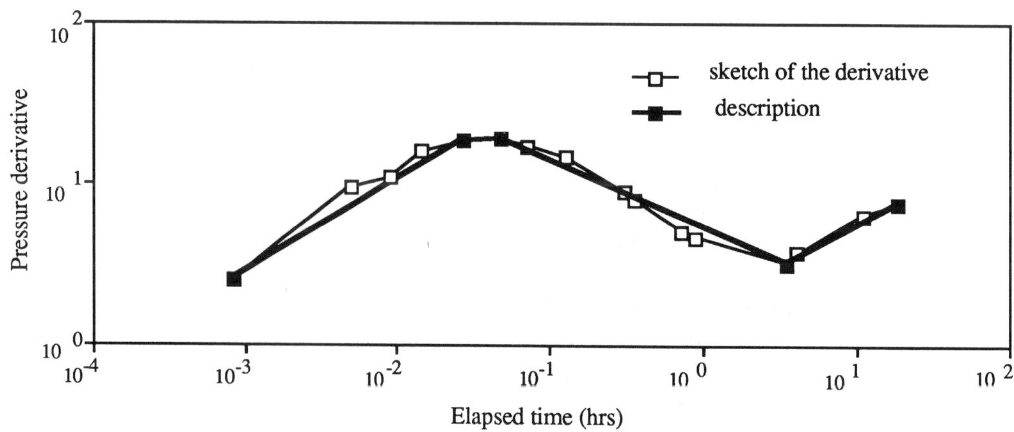

FIGURE 1.23. Description obtained from the sketch.

Qualitative Matching

There are four possible choices of model description that start with an *up*.

1. Wellbore storage and skin: [*up, maximum, down, valley*].
2. Wellbore storage and skin (flattened hump): [*up, plateau, down, valley*].
3. Wellbore storage and skin (low $C_D e^{2S}$): [*up, plateau*].
4. Vertically fractured well: [*up, plateau*].

Quantitative Matching

The constraint on the slope values of the segment corresponding to the *up* for all the descriptions above restricts the possible choices to

1. Wellbore storage and skin: [*up, maximum, down, valley*].
2. Wellbore storage and skin (flattened hump): [*up, plateau, down, valley*].

Second Element: plateau

Qualitative Matching

The first model proposed above disagrees with the data (its second element is a *maximum*). There is no combination that would allow us to correct this disagreement. The model is discarded.

The second model proposed still agrees with the data. It is now the only possible solution.

Quantitative Matching

The test made on the length of the *plateau* for the model retained succeeds. The only solution at this step is

1. Wellbore storage and skin (flattened hump): [*up, plateau, down, valley*].

Third Element: down

Qualitative Matching

There is agreement between data and model.

Quantitative Matching

Succeeds by default.

Fourth Element: minimum

Qualitative Matching

There is disagreement between the data and the model. On the fourth element of the model (i.e., *valley*), we try to combine a new component. To obtain qualitative matching with the data, this combination must replace the *valley* with a *minimum*. There are three different possibilities for the new interpretation model:

1. Wellbore storage and skin (flattened hump) in a reservoir with pseudosteady-state double porosity: [*up, plateau, down, minimum, up, plateau*]. [Using the combination rule (12)].
2. Wellbore storage and skin (flattened hump) in a reservoir with transient double porosity: [*up, plateau, down, minimum, up, plateau*]. [Using the combination rule (14).]
3. Wellbore storage and skin (flattened hump) in a reservoir with a sealing fault: [*up, plateau, down, minimum, up, plateau*]. [Using also the combination rule (14).]

Quantitative Matching

Succeeds by default since no constraint ever applies to a *minimum*.

Fifth Element: up

Qualitative Matching

The three models agree with the data.

Quantitative Matching

1. Since radial flow was not seen on the response, the constraint imposed on the derivative level at the end of the transition cannot be checked. Quantitative matching succeeds by default.
2. For the second model, we check the difference between the level of last point of the *up* and the level of the minimum. The test succeeds.
3. For the third model, we check the slope of the segments in the *up*. The test also succeeds.

At this step, the solution is complete. The three models are proposed as possible interpretation models.

Conclusions

In the artificial intelligence framework, we have developed methods to automate the model identification step of the well test interpretation procedure. Those methods consider model identification based on the pressure derivative. For each admissible interpretation model, they provide a first estimate of the parameter values, which can then be used for automated type curve matching analysis.

The first requirement to have a computer perform model identification is to provide this computer with the knowledge of the interpretation models. This knowledge was defined implicitly, in a form similar to the one used by an expert. In addition, we developed two different procedures, for observation and matching. The purpose of observation is to distinguish the true reservoir response from the noise on the data derivative, and then to describe this response in a form that can be compared to the models. Based on this description, the matching procedure constructs adequate interpretation models incrementally, considering the data from left to right. Models and data are represented symbolically as lists. Starting with the simplest model that matches the first data element, the matching procedure modifies this model only if it does not match all the data. This method guarantees that the models proposed are never too complex. However, true solutions can sometimes be missed. This situation occurs when a disagreement between the data and the model cannot be detected. In that case, the true solution and the model proposed cannot be distinguished, except by the fact that the later is more complex. Therefore, it would be possible to cope with this problem by relaxing the constraint that only the simplest models should be constructed. This modification enlarges the search space dramatically and makes the matching procedure computationally unacceptable.

Taking detailed information into account about the responses allows the algorithm to better distinguish the different models, and therefore to greatly reduce the number of cases where the true solution is missed. Because real data are never ideal, dealing with detailed information requires caution. The procedure requires methods for deciding whether this information is relevant, or should be ignored or transformed. We presented a tentative approach to this problem. The main idea behind this approach was that the description of a real response depends on the model considered to match this response. Comparison between the data and the model determines what part of the information about the data is relevant.

Future Work

Rating the Models

Certain sections of a model are characterized by quantitative properties, such as a particular slope value. Those properties were used to formulate quantitative constraints that determine if a model that qualitatively matches the data is admissible. The quantitative constraints are expressed by allowing some fixed maximum deviation around the expected theoretical value of regime slope or level difference between regimes. If the deviation allowed is too small, the adequate interpretation models may be missed. If it is too large, the matching procedure will propose models that are not adequate.

For those reasons, we believe that the quantitative properties of regimes should be used differently. Rather than determining if a model is admissible, they should be used to measure the *probability* that a model that qualitatively matches the data is admissible. For each regime property, a probability distribution would have to be chosen around the theoretical value.

A given model usually comprises several regimes with characteristic properties. Each regime will provide a probability value. The probability for the complete model could be obtained as the product of those values.

Interaction between Model Identification and Type Curve Matching

We have considered that the interpretation procedure was the succession of two different steps: model identification and type curve matching (Figure 1.24). The purpose of model identification is to propose the simplest interpretation models for given data. If a model that is a true solution cannot be distinguished from a simpler model, then the simpler model will be proposed and the true solution missed. Therefore, as emphasized earlier, the

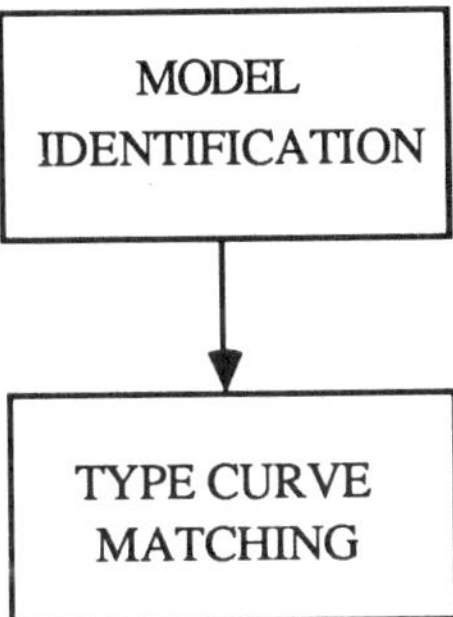

FIGURE 1.24. Decomposition of the interpretation procedure.

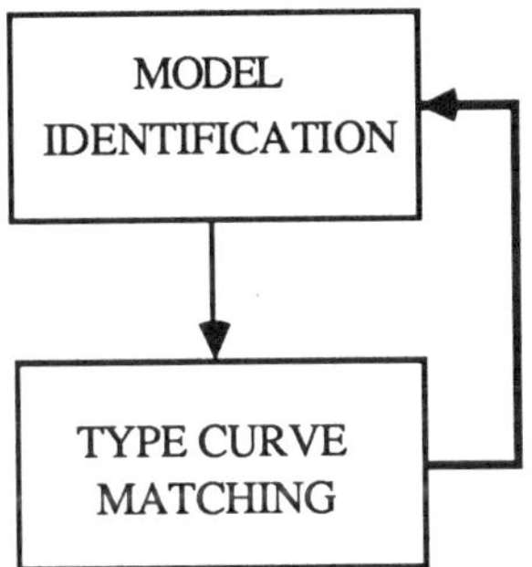

FIGURE 1.25. Revised decomposition of the interpretation procedure.

performance of a model identification program depends heavily on the information used about the models.

After choosing an adequate interpretation model, an expert will try to fit this model to the data. In some cases the expert may decide, on the basis of the best match obtained, that the model was too simple. In other words, the information used by an expert about the models does not always allow him/her to distinguish a given model from a simpler one. The information provided by type curve matching is in this case necessary to obtain the true solution.

Therefore, the decomposition of the interpretation procedure shown on Figure 1.24 is not totally correct. A feedback should be allowed from type curve matching to model identification (Figure 1.25).

In practice, the main problem is to find a criterion for deciding based on the best match obtained for the various models, whether those models are too simple. If such a criterion is found, then the following procedure can be adopted:

1. $n = 0$
2. $n = n + 1$
   Find the adequate interpretation models with n elements. Go to 3.
3. If there are no such models then go to 2. Otherwise go to 4.
4. Perform type curve matching. If the models are found too simple, then go to 2.

For model identification, it would be sufficient to modify the matching procedure developed in this study. This could be done very simply, replacing the constraint that only the simplest models should be proposed, by a constraint on the number of the components in a model.

References

Allain, O., 1987, An artificial intelligence approach to well test interpretation: MS Report, Stanford Univ.

Allain, O., 1988, An artificial intelligence approach to model identification in well test interpretation: Engineer Report, Stanford Univ.

Barua, J., Horne, R.N., Greenstadt, J.L., and Lopez, L., 1988, Improved estimation algorithms for automated type curve analysis of well tests: Soc. Petr. Engr., Formation Evaluation, 186–195.

Bourdet, D., Whittle, T.M., Douglas, A.A., and Pirard, Y.M., 1983, A new set of type curves simplifies well test analysis: World Oil 95–106.

Bourdet, D., Ayoub, J.A., and Pirard, Y.M., 1984, Use of pressure derivative in well test interpretation: paper SPE 12777 presented at the 1984 California Regional meeting of the Soc. Petr. Engr., Long Beach.

Clark, G., and Van Golf-Racht, T.D., 1984, Pressure derivative approach to well test analysis: A high-permeability North Sea reservoir example: paper SPE 12959 presented at the 1984 European Conference of the Soc. Petr. Engr., London.

Earlougher, R.C., 1977, Advances in well test analysis: Soc. Petr. Engr. Monograph, 5.

Genesereth, M.R, and Nilsson, N.J., 1987, Logical Foundations of Artificial Intelligence: Morgan Kaufmann, San Mateo, CA.

Gringarten, A.C., 1982, Interpretation of well tests in fissured reservoirs and multilayered reservoirs with double porosity behavior. Theory and practice: paper SPE 10044 presented at the International Petroleum Exhibition and Technical Symposium of the Soc. Petr. Engr., Bejing, 18–26 March.

Gringarten, A.C., 1987, Type curve analysis: What it can do and cannot do: J. Petr. Tech., 11–13.

Hochberg, J.E., 1957, Effect of the Gestalt revolution — The Cornell Symposium on Perception: Psychol. Rev. 64(2), 73–84.

Horn, B.P.K., 1986, Robot Vision: McGraw-Hill, New York.

Houzé, O., Horne, R.N., and Ramey, H.J., 1988, Infinite conductivity fracture in a reservoir with double porosity behavior: Soc. Petr. Engr. Formation Evaluation, 510–518.

Leeuwenberg, E., and Buffart, H., 1983, The perception of foreground and background as derived from structural information theory: Internal report, Department of Experimental Psychology, Univ. of Nijmegen, Netherlands.

Lowe, D.G., 1983, Perceptual Organization and Visual Recognition: Kluwer Academic Publishers, Boston.

McEdwards, D.G., 1981, Multiwell variable rate well test analysis: Soc. Petr. Engr. J. 21, 441–446.

Newell, A., and Simon, H.A., 1976, Computer science as empirical inquiry: Symbols and search: Commun. Assoc. Computing Machinery 19(3), 113–126.

Padmanabhan, L., and Woo, P.T., 1976, A new approach to parameter estimation in well testing: paper SPE 5741 presented at the Fourth Symposium on Reservoir Simulation of the Soc. Petr. Engr., Los Angeles, CA.

Proano, E.A., and Lilley, I.J., 1986, Derivative of pressure: Application to bounded reservoir interpretation: paper SPE 15861 presented at the European Conference of the Soc. Petr. Engr., London, England.

Ramey, H.J., 1976, Practical use of modern well test analysis: paper SPE 5878, presented at the California Regional Meeting of the Soc. Petr. Engr., Long Beach, CA.

Ramey, H.J., 1982, Pressure transient testing: J. Petr. Tech., Trans. AIME 273, 1407–1413.

Rosa, A.J., and Horne, R.N., 1983, Automated type curve matching in well test analysis using Laplace space determination of parameter gradients: paper SPE 12131 presented at the Soc. Petr. Engr. Annual Technical Conference and Exhibition, San Francisco, CA.

Stegemeier, G.L., and Matthews, C.S., 1958, A study of anomalous pressure build-up behaviors: Trans. AIME 213, 44–40.

Startzman, R.A., and Kuo, T.-B., 1986, A ruled-based system for well log correlation: paper SPE 15295 presented at the Symposium on Petroleum Industry Application of Microcomputers of the Soc. Petr. Engr., Silver Creek, CO.

Tsang, C.F., McEdwards, D.G., Narasimhan, T.N., and Witherspoon, P.A, 1977, Variable flow well test analysis by a computer assisted matching procedure: paper SPE 6547 presented at the 47th Soc. Petr. Engr. Annual California Regional Meeting, Bakersfield, CA.

Watson, A.T., Gatens, J.M., III, and Lane, H.S., 1988, Model selection for well test and production data analysis: Soc. Petr. Engr. Formation Evaluation 215–221.

Witkin, A.P., and Tenenbaum, J.M., 1983, On the role of structures in vision: in Human and Machine Vision, Beck, J., Hope, B., and Rosenfeld, A. (Eds.). Academic Press, New York, pp. 481–543.

Additional Reading

Brady, J.M., 1981, Computer Vision: North-Holland, Amsterdam.

Clocksin, W.F., and Mellish, C.S., 1987, Programming in Prolog: Springer-Verlag, Berlin.

Marr, D., 1982, Vision: Freeman and Co., San Francisco.

Winston, P.H., 1977, Artificial Intelligence: Addison-Wesley, Reading, MA.

2
Artificial Intelligence in Formation Evaluation

Tsai-Bao Kuo, Steven A. Wong, and Richard A. Startzman

Introduction

Artificial Intelligence

Artificial Intelligence (AI) is simply the use of reasoning processes that allow computers to make decisions in an apparently human fashion. In the early days of computing, the public viewed computers as electronic "brains" and assumed that somehow they had been endowed with human intelligence. This assumption vastly overestimated the ability of early computers to simulate reasoning. In fact, the overwhelming use of computers has been to replace and augment certain menial human tasks in the accounting and commercial areas requiring only a modicum of reasoning. Even in scientific and engineering applications, computers have been used almost solely as "number crunchers." Thus, they are programmed to perform tasks that their users find time and labor intensive.

It is interesting to note that the LISP programming language was developed in the 1950s, about the same time as FORTRAN and COBOL, the workhorses of scientific and commercial programming. LISP has become the preferred language of those who use the computer's ability to evaluate human reasoning. Nevertheless, few, if any, successful applications using LISP were apparent until recent years.

There are several reasons why the computer's use as a reasoning machine was delayed. One obvious reason is that virtually all development of software was dedicated to numerical applications. Many of these applications such as credit card and payroll processing represented basic functions of commercial terms. Savings and labor costs and time were apparent in commercial applications from the beginning. Another reason for the lag in "intelligent" computer applications concerns the difficulty in converting human reasoning to a set of computer instructions. Humans make decisions for a variety of reasons and their decisions often conflict even when all the pertinent facts are available. It is even hard to evaluate the reasoning process of a single individual because the human reasoning process, itself, is considerably complex.

A third reason for the delay concerns software available to process the intricate rules and chain of logic that we believe humans possess and use. Despite the power of languages such as LISP, programmers had difficulty developing computer instructions that would solve even simple problems involving only a small set of human-created rules. Software "shells" have helped overcome this programming difficulty. These shells allow users to input sets of rules and chains of logic in plain, everyday, language. They are called shells in the sense that they are empty until they are filled with sets of rules.

An Expert System is a set of computer instructions that uses AI methods to solve a specific, and often somewhat narrowly-defined, problem. Examples might include a problem to diagnose cardiovascular diseases or to recognize English language words in a sentence. Large numbers of expert systems are now developed using these shells.

What Is Formation Evaluation?

Jorden and Campbell (1984) defined formation evaluation as "the practice of determining the physical and chemical properties of rocks and their contained fluids."

From the start of the U.S. oil industry in 1859 to the use of the first wireline log in France in 1927 the primary formation evaluation method was the "drillers log." This was a written record of cuttings and fluids encountered by the drill bit. The first continuous, downhole wireline log measured the electrical resistivity of the rock and formation fluids. The electrical log was the only commercial type available until the early 1940s. Other types of logs that measured natural or induced radiation and acoustic properties were introduced later. By understanding the relationships between electrical, radiation, and acoustic responses of these logs and the physiochemical properties of the rock-fluid system, geologists and petroleum engineers could evaluate formations in ways not previously possible.

Another set of methods used in formation evaluation requires pressure and production data from direct well measurements. The mathematical foundation for these methods was established by Muskat (1937). Miller et al. (1950) and Horner (1951) later established a comprehensive theoretical foundation. Pressure and production methods are now used to help determine information regarding permeability, reservoir volume, heterogeneity, and average reservoir pressure among other parameters.

Why Do We Need AI in Formation Evaluation?

Formation evaluation, a broad field, includes both conventional well logging and pressure testing. Many problems in formation evaluation are amenable to numerical computations. For instance, estimates of formation permeability (pressure testing) as well as porosity and water saturation (logging) are derived numerically from physical measurements in wells.

Yet, in our view, most of the problems remaining to be solved in the area of formation evaluation involve the intelligent selection of rules derived from human expertise. Assume, for instance, that evaluating a certain geological formation requires that the water saturation be estimated accurately. In the formation evaluation design phase, the optimal measurement device(s) would have to be

selected. Some devices are better at measuring water saturation under a certain condition than other devices. The selection of the proper devices is, then, a matter of human knowledge, experience, and reasoning.

Once physical measurements (e.g., electrical resistivity, acoustic travel times) have been made then procedures must be established to make the final calculation of water saturation. This calculation requires that the geologic, geochemical, and petrophysical environment be considered even before a (usually) empirical water saturation equation is chosen. The quantitative water saturation is the end product of the application of number of mainly qualitative rules based on human knowledge, experience, and reasoning.

In conventional computer-based formation evaluation, the user selects the equations in advance. The rules are determined before run time and later applied during computations. An expert system, on the other hand, selects the rules and applies them during the run. The expert system therefore is more automatic and makes more efficient use of the true powers of the computer.

Expertise in Formation Evaluation

Formation evaluation is a specialized discipline within geoscience and petroleum engineering. The American Association of Petroleum Geologists (AAPG), The Society of Petroleum Engineers (SPE), and the Society of Professional Well Log Analysts (SPWLA) regularly provide professorial forums for the presentation and publication of professional papers on the subject of Formation Evaluation. The SPE's journal *Formation Evaluation* and the SPWLA's *The Log Analyst* are published regularly and contain peer-reviewed articles.

Many hundreds and, perhaps, thousands of individuals may be considered "expert" in the field of formation evaluation. They exist within oil-field service companies, the universities, and oil/gas firms. Recognition of expertise is probably best accomplished through peer selection. In other words, the workers in the field of formation evaluation know who the true experts are—at least within their circle of acquaintances.

Capturing this expertise, on the other hand, can be difficult. AI workers refer to expertise of a specialized field, such as formation evaluation, as

Domain Expertise. It is the job of the so-called "Knowledge Engineer" to transfer the knowledge, experience, and reasoning powers of the domain expert into useable computer code. As mentioned earlier, experts can and frequently do disagree. Resolution of disagreements may lead to better knowledge.

Well Log Interpretation

A problem associated with the computerized well log interpretation is that the conventional computerized methods cannot deal with varying geological environments. The real earth is so complicated that no formulas in formation evaluation are valid unconditionally. Numerical algorithms are efficient when a large sequence of formations with same rocks and fluids is to be analyzed; however, they perform poorly when "local" interpretation (without noting its adjacent intervals) is required.

WLAI System

Wu and Nyland (1986) proposed an AI-oriented method to emulate human expert's inference used in stratigraphic interpretation. They introduced algorithms consistent with the characteristics of the problem domain and developed a system called WLAI. This system can identify the formations, formation members, and sedimentary facies.

The human experts' rules considered in this system concern (1) contact recognition and (2) interval identification. The contacts correspond to rapidly changing geological environments and usually cause a rapidly changing log curve feature. To recognize the contacts, an iterative linear regression with variable break point algorithm (ILRV) is used. After the contacts are located, a well log becomes a set of intervals, each bounded by two contacts. Next, the system identifies the geological meaning of the intervals. It compares the test set to a pattern set, while each set represents one log. They use a system of symbols to represent lithological description. For example, a near shore marine silty and clayey sand is represented by "s3," an estuary sand with thin bed of clay is represented by "e2," a fluvial sand, silt, and clay with carbonaceous debris is represented by "f." In other words, a set of intervals is translated into a set of symbols such as "s3," "e2," and "f,"

which bear geological meanings. A predetermined set of interpretation is considered as a reference, or the pattern set.

A modified string-to-string matching algorithm (Liu and Fu, 1982) is then used to perform the comparison between the pattern set and the test set. The comparison process is essentially a procedure to match two strings of symbols. The similarity of two strings is measured by the sum of the costs for (1) inserting a pattern interval in the test set, (2) deleting a test interval from the test set, and (3) matching a test interval to a pattern interval. The best matching is the one with the least cost.

The result from one example shows that a total of 350 intervals were identified by computer, and about 86%, or 302 intervals, were consistent with geologists.

ELAS System

Conventional log analysis techniques have continuously been developed into many useful software programs. These programs attempt to offer many choices and require a large set of parameters to be selected by the user for the program to model a specific problem. They often become so complex and can be used efficiently by only a few experienced analysts. To provide the expertise to help an inexperienced person, Apté and Weiss (1985) developed a hybrid expert system to integrate existing well log analysis software programs with experiential knowledge about how and when to use the appropriate package. The system is called Expert Log Analysis System (ELAS). It is a rule-based advice system that can help the user in

1. understanding and controlling the log analysis program used,
2. advising on the appropriateness of using a method,
3. allowing the user to vary the assumptions and parameters used in different individual analyses,
4. monitoring and examining consistency between expected parameter values, and
5. providing interpretation of the results produced by the user's interaction with the program.

For example, an expert analyst usually has heuristics on the use of the Archie's equation for water saturation calculation. These heuristics suggest that this equation is appropriate for a given situation, or that other techniques should be performed if this method is used. Heuristics also suggest what parameters should be monitored and what interpretation should be made. It is these types of heuristics that are captured in the production rule and provide ELAS with its interpretive capabilities. Here is a typical interpretation rule:

IF WATER-FIT normal and WATER-
    QUALITY is (EXCELLENT or GOOD),
    and HYDROCARBON-FEET is more
    than 10
THEN Interpret: Indications of significant
    hydrocarbons.

The system also provides dynamic guidance to the user through a set of action-recommendation rules like the one below.

IF POROSITY done and S_w not done
THEN Advice: Compute S_w to determine water
    saturation and hydrocarbons in zone.

It is difficult for a user to keep track of the correctness and consistency of an analysis sequence that involves the use of several related methods. ELAS can help keep track of events by checking through a set of production rules. A typical example is

IF GAS expected, and
    Method C verifies Gas, and
    HYDROCARBON-FEET equal to 0
THEN Indicate: Hydrocarbons computation
    inconsistent with the amount of gas
    detected.

In general, the ELAS's production rules may be viewed as containing interpretive, consistency checking, and control knowledge that is organized around methods of analysis. In one sense, the power of well log analysis programs lies in mathematical methods, and the human expert's perspective of these methods is used to extract the necessary and sufficient set of parameters to control their usage.

INTELLOG

Data checking and parameter estimation are essential steps in well log analysis. To be able to use quantitative analysis programs, analysts are usually required to obtain all the data and parameters specified in those programs. If some data are not apparent to the analyst, one would need to rely on heuristic knowledge. However, this type of knowledge is difficult to integrate in quantitative analysis algorithms. To overcome this problem, Einstein and Edwards (1988) developed an expert system, called INTELLOG, to perform a series of data checking and parameter estimation before conducting quantitative analysis.

The data checking process that the INTELLOG performs includes (1) determining zone of interest, (2) normalizing input data, (3) performing environmental correction, (4) detecting non-porous lithology such as coal, anhydrite, gypsum, and salt, (5) evaluating the given data on the presence of trace elements, formation fluid, and lithology, and (6) examining the reliability of the logs.

In providing parameter estimation, the expert system (1) detects trace elements such as uranium, pyrite, and feldspar, (2) determines formation fluid, (3) estimates shale volume, and (4) predicts lithology using Hingle plots. The system then uses those estimated parameters to proceed quantitative analyses such as computation of shale volume, porosity, water saturation, and lithology.

They also compared the results produced by the expert system and various petrophysicists in over 100 cases. Although only six sample cases were presented, those wells are difficult to analyze because of the limited amount of information given. The results of their study show that the performance of the expert system is satisfactory and exceeds initial expectation.

LOGIX

Hoffman et al. (1988) successfully documented some 1600 petrophysical heuristic rules and developed a knowledge system that can

identify reservoir zones,
determine lithology,
determine the type of pore fluid,
calculate porosity, and
calculate hydrocarbon saturation.

The system also has capabilities to highlight inconsistencies, provide advice on how to reconcile the data set, and indicate alternative solutions. It contains expertise on evaluating siliciclastical and carbonate reservoirs in nine geological environments. A typical geological environment is described by 10–30 rock/porefill models. Each rock/porefill model contains approximately 10 conditions.

The reasoning mechanism used in the LOGIX is a hypothesis-and-test approach. It tests a number of rock/porefill models using all available information such as log responses, core measurements, well-test data, and geological descriptions of cuttings and side-wall samples. In general the information is represented as conditions of the petrophysical rules. The conclusion of a rule is the model hypothesized. The rule below is an example of this type of expression.

CONDITIONS	[MB,MD]
1. Gamma-ray log response is relatively low	[0.3, 0.5]
2. Calculated water saturation indicates hydrocarbon	[0.7, 0.3]
3. Separation between apparent density and neutron porosities is present	[0.5, 0.5]
4. Calculated porosity is greater than 0.08	[0.1, 0.7]
5. Bulk density is greater than [limestone density$\times(1-$porosity)] i.e., minimum density possible in gas bearing reservoir	[0.1, 0.7]

HYPOTHESIS: Gas-Bearing Limestone

Conditions in a rule can be the typical petrophysical evaluation methods. For example, in the rule above, the conditions 3, 4, and 5 can be obtained from a density-neutron cross-plot. Each condition is assigned a "measure of belief" (MB) when the condition is satisfied. On the other hand, if the condition is not satisfied, a "measure of disbelief" (MD) will be given. As a condition is tested, its MB and MD will be incorporated in the following way. When a condition is satisfied, then the MB will be calculated, but the MD remains the same, i.e.,

$$MB_{new} = MB_{old} + MB_{this\ condition} - (1.0 - MB_{old})$$
$$\times\ MB_{this\ condition}$$
$$MD_{new} = MD_{old}$$

When a condition is not satisfied, then the MB remains the same, and the MD will be calculated, i.e.,

$$MB_{new} = MB_{old}$$
$$MD_{new} = MD_{old} + MD_{this\ condition} -$$
$$(1.0 - MD_{old}) \times MD_{this\ condition}$$

The total belief in a certain hypothesis is expressed in terms of a confidence factor (CF), which is calculated from the final MB and MD, i.e., CF = $MB_{new} - MD_{new}$. If the CF is greater than 0.7, the analysis is considered of good quality. The hypothesis is accepted.

The system was expanded to include two knowledge bases for interpretation of micaceous sandstone reservoirs and vuggy carbonate reservoirs. It has been installed as a tool complementary to existing deterministic and statistical petrophysical interpretation systems.

Log Quality Control

One of the common problems in well log analysis is that logging data are usually imperfect. A novice analyst may overlook errors and anomalous occurrences encountered on well logs. While the analytical methods may be correct, interpretations and subsequent decisions may be inaccurate if the log quality problems are undetected. To overcome this problem, Warnken (1988) and Warnken et al. (1988) developed an advisory expert system for an end-user who is familiar with, but not an expert in, well logging operations to check the quality of logs.

When operating the expert system, the user is prompted through a brief interactive consultation session on the computer. A typical consultation session consists of 10 to 20 questions and answers for each log. The results from the consultation are presented as a list of conclusions that states the quality of the log in three major areas: (1) borehole, (2) tool, and (3) formation effects. Figure 2.1 shows the relations between log quality and the three major effects. Most of the conclusions are applicable to all logging environment and universal topic in log quality control. However, this system can be modified to include knowledge particularly applicable to a specific region or field.

The expert system was developed on a commercial expert system shell, which can handle a large number of heuristic rules for checking gamma ray, sonic, and dual induction logs. A backward chaining type of reasoning was applied to derive final conclusions from the rules in a fashion similar to a medical doctor's diagnosis procedure.

For example, in examining the quality of a sonic log, an end-user can check formation effects on the

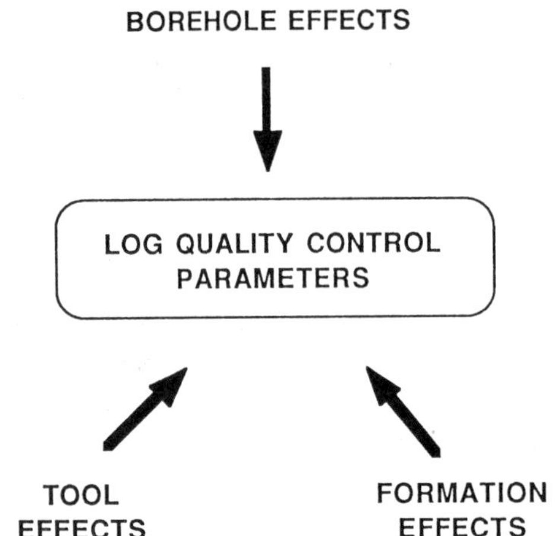

FIGURE 2.1. Major effects on log quality (Warnken, 1988).

log by going through the advice provided by the system and answering the associated questions. The items to be checked include

proper acoustic travel time at a known homogeneous formation,
agreement of the shale travel time on the offset log, if available, and the shale travel time on the log being checked,
porosity quality in connection with the porosity obtained from other porosity tool, and
on-depth with the primary log.

Tool effects are examined based on information obtained from

logging speed,
repeatability,
calibration before and after survey,
temperature and pressure ratings of the tool,
integrated time ticks,
log heading and scale,
cycle skipping, and
recorded travel time.

The following items are suggested to the user when the borehole effects are examined:

fluids in borehole,
washouts,
hole conditions indicated by caliper log, and
tension log.

This expert system can be further expanded to include quality control procedures for other logs. In current condition, it can be used as a complementary tool to any quantitative or qualitative analysis programs.

Well Log Correlation

Well log correlations are important to petroleum exploration and production practices because they provide the basic information required for reserve estimation and field development planning. However, a crucial problem is that correlations are extremely difficult to establish with certainty. The process of obtaining good correlations is always time consuming. Furthermore, correlations are often revived when new, supporting information is available. Therefore, the possibility of using computer programs to carry out the task has continuously been investigated.

Most computerized well log correlation methods described in the literature are curve matching procedures that are based on mathematical or statistical formulations. Typical examples are the methods using cross-correlation (Schwarzacher, 1964; Rudman and Lankston, 1973; Matuszak, 1972), Fourier transformation of the cross-correlation function (Rundman et al., 1975; Mann and Dowell, 1978; Kwon and Rudman, 1979), slotting procedures (Gordon and Reyment, 1979; Gordan, 1980), dynamic time warping (Hoyle, 1987), or other continuous functions (Neidell, 1969; Kerzner, 1983). Basically these methods perform point-to-point computation to obtain the similarity between the two logs at all possible positions of comparisons. These approaches can achieve good results for cases with simple geological settings or log traces obtained at very close distance, such as dipmeter curves; however, their applicability falls short in the cases with large thickness variation, missing sections due to faults or erosion, or facies change. Although some improved work has been attempted to incorporate stretching (e.g., Ghose, 1984) or zonation (e.g., Hawkins and Merrian, 1974) techniques in the curve matching procedures, complexity of the computation becomes a new problem.

Other types of approaches, such as the one proposed by Vincent et al. (1979), use information relating to the shape and structure of the log curves.

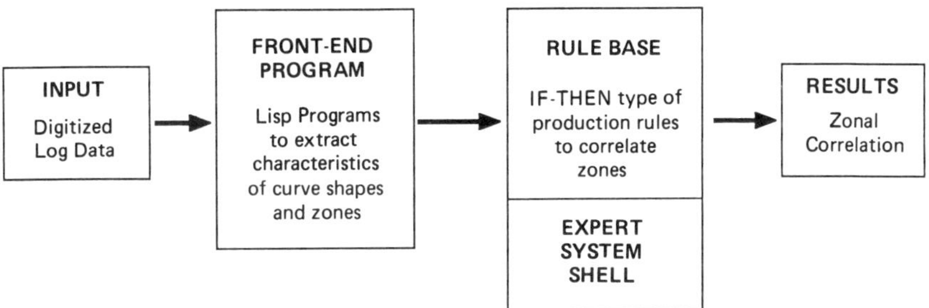

FIGURE 2.2. Well log correlation expert system (Kuo and Startzman, 1987. Reprinted with permission from Geobyte 2(2), pp. 30, 35. Copyright American Association of Petroleum Geologists).

Using this method, structured line patterns, instead of digital numbers, are compared, and the likelihood of the two patterns is measured. It takes the correlation process one step forward toward emulating the human's recognition process, and works fairly well in correlating dipmeter curves.

As the AI technology emerged in the 1980s, the use of AI to mimic human experts' correlation procedure was sought. Many researchers have attempted to apply these computer techniques to capture the knowledge and reasoning involved in the correlation process.

Startzman and Kuo Approach

Kuo (1986) used a symbolic computation technique and a rule-based approach to develop a system that performs correlation in a fashion parallel to human experts. This approach is based on a zone-to-zone correlation concept, and it

takes digitized log data as input and translates digital information into symbolic forms,
extracts shapes on the logs,
determines geological zones based on the shape characteristics,
characterizes zonal information, and
makes correlations using the zonal characteristics and a reasoning process.

The system, written in LISP Language, initially consists of four major parts: (1) Data Preparation Program, (2) Data Base, (3) Rule Base, and (4) Rule Interpreter. Later, it was reorganized into two modules (Kuo and Startzman, 1987), i.e., Front End Processor and Rule Base (see Figure 2.2.) The initial rule interpreter was replaced by a commercial expert system shell that became an independent module.

The Front End Processor performs functions such as digit-to-symbol translation, shape extraction, zone identification, and zonal attribute description. Figure 2.3 illustrates the application of these functions in a successive manner. It is the extracted, higher level information such as zonal attributes (instead of primitives or digital data) that is used in the correlation process.

The zonal attributes extracted from the log include (1) interval, (2) lithology, classified as shale or nonshale, (3) position of the zone, (4) general shape of the zone, (5) thickness, (6) average-amplitude, (7) name of the zone above, and (8) name of the zone below. For example, The Zone-A1 is the first zone in the well A and its attributes are represented as

(Zone-A1	Interval:	((−8478.0 71.72) (−8522.0 39.58))
	Lithology:	Nonshale
	Position:	Upper-section
	Shape:	Plateau-B
	Average-amplitude:	123.23
	Zone-above:	None
	Zone-below:	Zone-A2)

Thus, each zone is associated with a zone name and its eight attributes. These characteristics are stored properly in a temporary data base. The data base exists only when the system is preparing information to be used in the correlation rule base.

The Rule Base is a collection of IF–THEN type of production rules that are representations of experts' heuristics used to make zonal correlations between two logs. These rules are classified into three major sets: (1) attribute similarity rules, (2) correlation rules, and (3) confirmation rules. The

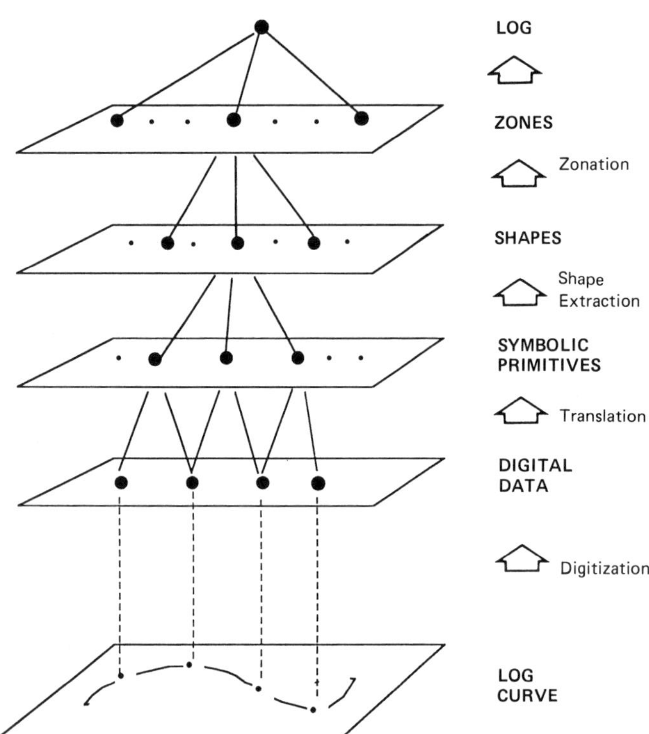

FIGURE 2.3. Schematic diagram of log characteristics extraction (Kuo, 1986).

attribute similarity rules take zonal attributes, such as position, shape, and thickness, of each zone and compare them to conclude an attribute similarity set for the two logs. The following is an example.

> IF The Position of Zone-A is at UPPER-SECTION, and
> The Position of Zone-B is at MIDDLE-SECTION.
> THEN The Relative Position of Zone-A and Zone-B is "CLOSE."

The precondition of applying this rule is that Zone-A and Zone-B are in two different wells. The identification of each zone pair is provided by a particular rule at the beginning of the correlation. It is designed to prevent correlations occurring between any two zones in the same well.

Next, the correlation rules use the attribute similarity information to conclude the quality of the match between two zones. The following is an example of the correlation rules.

> IF The Relative Position of Zone-A and Zone-B is "SAME," and
> The Shape Similarity of Zone-A and Zone-B is "SAME," and
> The Thickness-difference between Zone-A and Zone-B is "ALMOST-NO-DIFFERENCE."
> THEN The Correlation between Zone-A and Zone-B is "EXCELLENT" (10).

The quality of correlation, or degree of success, is represented both in a symbolic scale and a numerical scale. The symbolic scale consists of Excellent, Good, Fair, and Poor, which is equivalent to the numerical scale of 10–8, 7–5, 4–3, and 2–1 respectively.

The correlation rules attempt to establish the correlation of any two zones that are not in the same well. Therefore, possibilities to obtain several good matches for a single zone are very high. To eliminate the ambiguities, experts would examine the characteristics of the neighboring zones and use the thickness or lithology information of these zones to make final decisions. The confirmation rules are the coding of this type of reasoning. They are used to substantiate the final correlations.

Startzman and Kuo (1987a,b) showed that this method provides a conceptual basis leading to more realistic, expert-like correlation procedures. The correlations are made on geological zones, instead

FIGURE 2.4. Sample result of zonal correlation (Kuo and Startzman, 1987. Reprinted with permission from Geobyte 2(2), pp. 30–35. Copyright American Association of Petroleum Geologists).

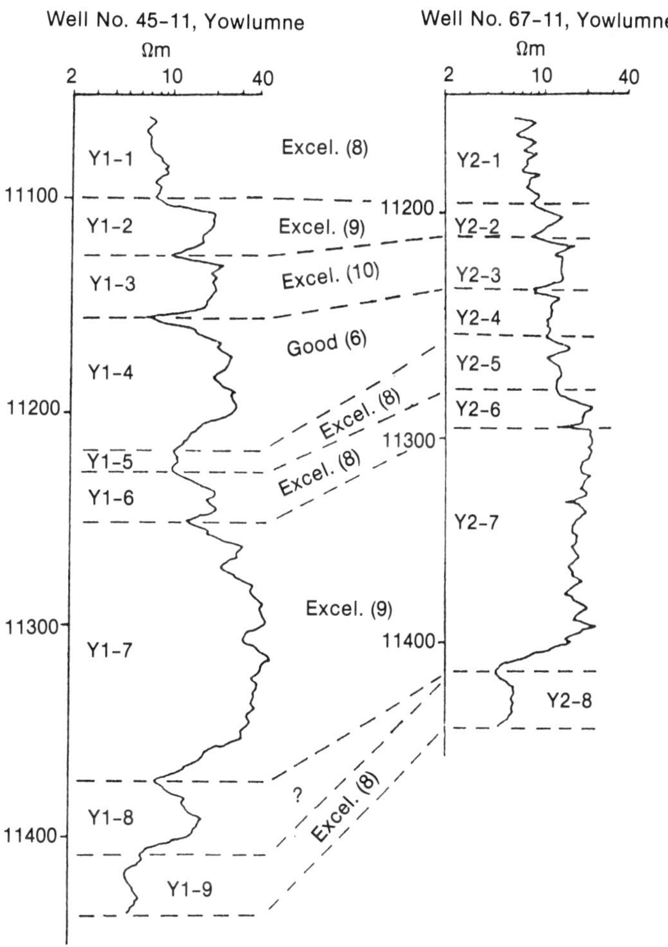

of discrete points or events. It appears that the correlation of thinning (or thickening) zones or missing zones can be treated more easily and accurately than other functional approaches. Figure 2.4 is a sample result of using this system to correlate two SP logs in a channel depositional environment which is considered a difficult case.

Other approaches have attempted to integrate expert systems with mathematical correlation algorithms. The experts' knowledge is coded to guide the implementation of dynamic correlation algorithms or used as constraints in curve matching procedures.

COREX System

Lineman et al. (1987) developed a system, called COREX, which includes a dynamic depth warping algorithm and an expert system. The expert system consists of a knowledge base and a data base. The knowledge base contains a priori knowledge of common depositional models and simple geological structures, and rules that allow this knowledge base to affect the data. The data base contains digital well logs, information about seismic ties, structure dip, and lithology.

In this system, the correlation is basically performed by the dynamic depth warping algorithm, which attempts to match two sequences of log readings (i.e., two well logs) in such a way that one sequence is transformed into the other. The transformation process involves (1) comparison, (2) editing, and (3) cost computing. The authors illustrated this process with an example of transforming the word "HILLIER," the test sequence, into the word "MILLER," a reference sequence. The process starts with comparison of characters at corresponding positions. For example, compare "H" to "M," "I" in the test sequence to "I" in the reference sequence,

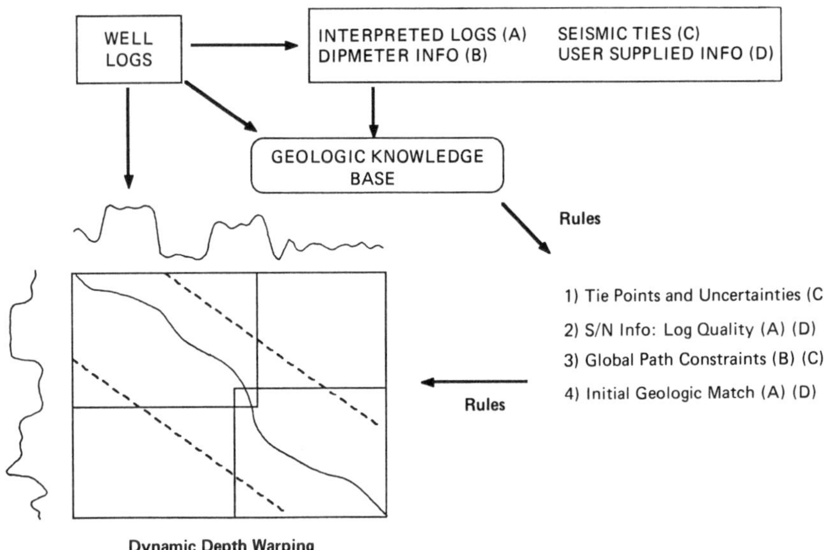

FIGURE 2.5. The COREX correlation system and example of warping path (Lineman et al., 1987).

etc. When the difference exists, an editing operation is applied to make two characters a match. The editing operation can be deleting, inserting, or changing a character, or a combination of any two or three. However, each operation is associated with a cost. For instance, deleting one character costs 1. When all the editing operations are completed, the total cost will be computed. Since many combinations of operations can all achieve same transformation, it is difficult to find the right one. The general rule is to choose the one with the least total cost. In other words, the path that involves the least editing operations will be chosen.

When this procedure is applied to the matching of log curves, the log reading at each depth point can be considered as a "character," and the entire sequence of log can be treated as a string of characters. The editing operations and cost computation are equally applicable to the matching of log curves. A depth warping algorithm calculates all possible matching costs between each point in the first log and all other points in the second log. A dynamic programming technique is used to recursively fill the cost matrix. Each point in the matrix representing the total cost of matching two sets of logs to that point.

The matching process can be improved by imposing constraints on the depth warping algorithm. The constraints are information resulted from applying the knowledge-based system to the data in the data base. With the constraints, the program reduces significantly the amount of calculation from the original, unconstrained problem, and enhanced greatly the chances of a geologically meaningful result (see Figure 2.5) (Lineman, 1986).

The following is a brief description of how this system is implemented.

At first, the system uses the matching algorithm described earlier to perform a large-scale lithology match. The result of the initial match is a series of tie points that limit the search space for the optimal warping path. The correlation of two logs is now becoming a task to correlate a number of smaller sections.

Second, the program runs through rules concerning the geologic structure of the particular section and uses seismic and dipmeter information to reduce the search space for the warping path. For instance, it may constrain the warping path to travel only in certain areas such as the lower half of the global cost matrix.

Finally, the system performs a point-to-point matching using dynamic programming. When all calculations are complete, the program uses a separate routine to trace back the minimum distance path through the cost matrix and draw tie lines between the match points.

The rules that produce constraints to the program falls into three sets: (1) lithologic or depositional environment rules, (2) geologic structure rules that translate structural information from

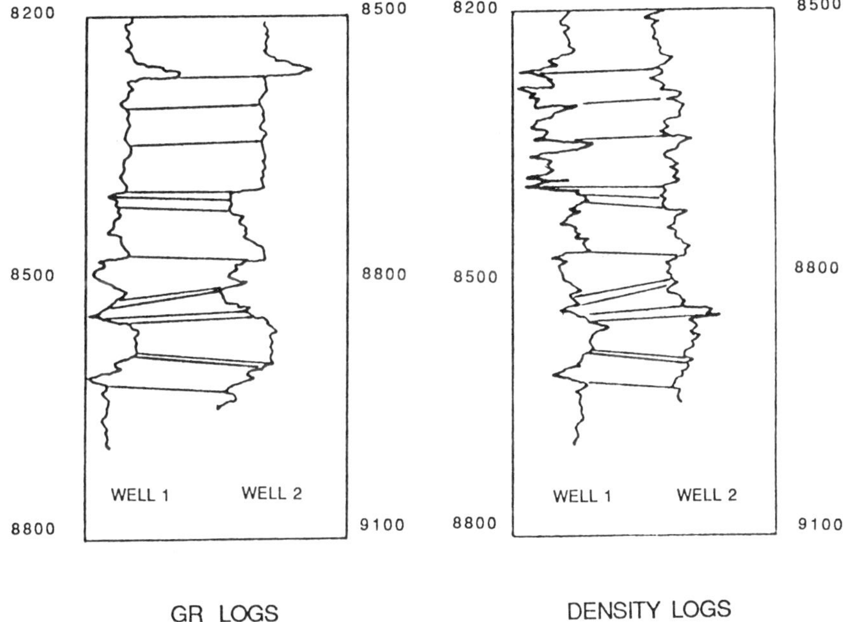

FIGURE 2.6. Sample result from COREX (Lineman et al., 1987).

seismic and dipmeter into dynamic programming constraints, and (3) local distance rules that modify the local distance matrix. The lithologic rules are usually applied in the first step, large-scale correlation, while the structure rules are used in the second steps of a correlation process. The local distance rules can be used to improve the correlation by modifying the local cost measurement scheme because of noise data.

This system has been applied to field examples from two widely separated oil provinces, and the results agree very well with correlations provided by experts. Figure 2.6 is an example of results from this system.

Olea and Davis Approach

Unlike the above hybrid system, a program by Olea and Davis (1986) containing production rules and correlation procedures was developed in one programming environment. The production rules are translated into FORTRAN procedures such that they can interact with the correlation algorithm, also in FORTRAN, without any difficulty. With this method, two logs are used per well, one for curve matching and the other for lithologic comparison.

The correlation is achieved by matching a fixed, short interval in one well against successive positions in a second well. The similarity between compared sections is measured using cross-correlation. The productions rules are restricted to assisting in the selection of parameters that control computation and in evaluating the final results. For example, the following two rules, which are extracted from human experts' practices, are used to control the matching.

Rule 1:   IF The degree of similarity between the traces is highest for lag **d**.
            THEN Lag **d** is the matching position.

Rule 2:   IF The shale content of two matched intervals is different.
            THEN Matching is infeasible.

This approach demonstrates that the use of multiple criteria for correlation seems to resemble closely the way experienced people correlate logs.

GEOLOGIX System

Rieuwerts (1989) discusses an interactive system for well log correlation. The system segments the logs into geologically meaningful units and infers

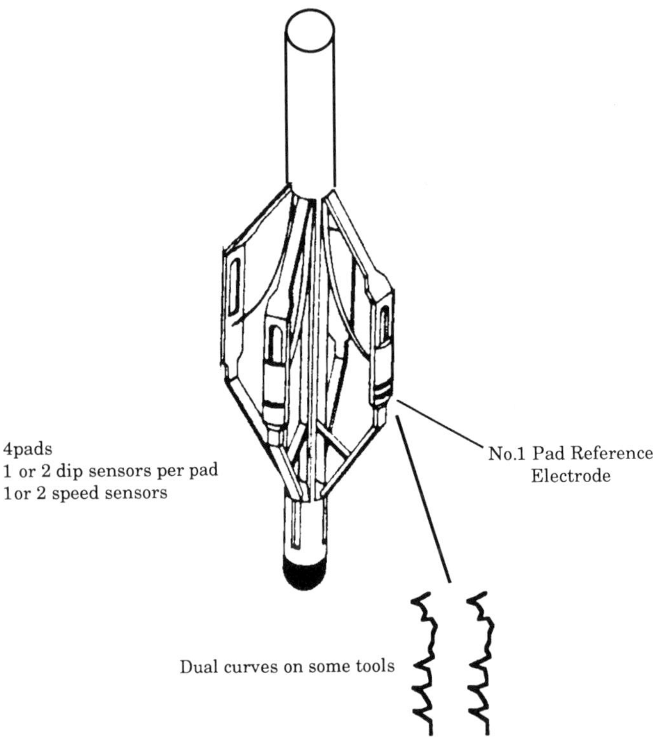

FIGURE 2.7. Dipmeter tool schematic from Schlumberger dipmeter interpretation fundamentals (1986).

lateral extent from its knowledge bases. Lithologies are determined first. Next, the system tries to establish a stratification network by focusing on thick shale layers. Third, the sands are classified. Finally, the rule-based modules are activated to help the interpreter with the correlation.

High-Resolution Logging

We will define high-resolution logging to refer to wireline logging devices that can resolve geologic events smaller than one foot. Devices in this category include the dipmeter, Formation MicroScanner (FMS), Borehole Televiewer (BHTV), and Electromagnetic Propagation Tool (EPT). Both the dipmeter and the FMS are pad devices that measure conductivity changes in the formation. The BHTV is a sonic device that uses high frequency pulses that reflect from the borehole surface or casing wall.

Pattern recognition and AI techniques have been applied to the dipmeter family of tools. Essentially two parts of the dipmeter processing chain have been studied: converting the raw data from the dipmeter family of tools to dip/azimuth information and interpretation of the dip patterns. We first review the basic design of the dipmeter tool family and show the motivation for automated processing/interpretation techniques.

Introduction to the Dipmeter Tool

Figure 2.7 shows a schematic of the dipmeter tool (Schlumberger, 1986). Each spring-loaded arm terminates in a section with the sensors called a pad. Each pad may have one or two dipmeter sensors or electrodes. Additionally, speed buttons or electrodes are located on pad 1 and pad 2. Each dipmeter sensor gives a one-dimensional correlation curve of conductivity changes as the tool traverses the subsurface.

The conductivity changes are caused by differences in the traversed formations. Since each pad has at least one button or sensor, multiple correlation curves are generated. In theory, a single formation or geologic event causes similar conductivity changes that can be traced from pad sensor to

FIGURE 2.8. Effect of planar features on the dipmeter tool.

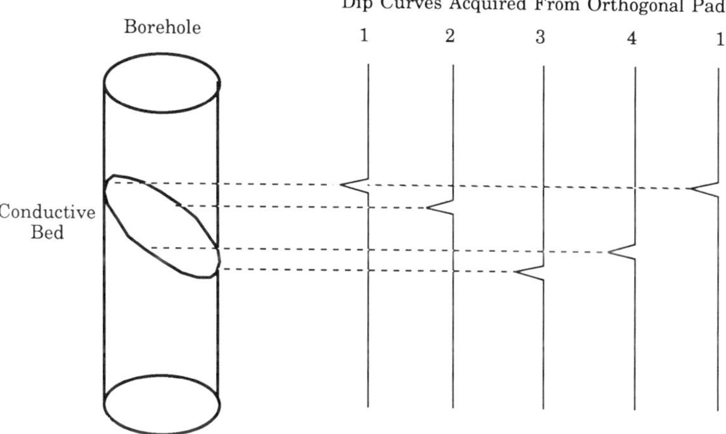

pad sensor. The conductivity changes are correlated by eye or a computer program. The correlations result in computed apparent dips. By correcting for tool orientation, true dips can be calculated and displayed. Figure 2.8 shows a schematic of a planar features intersecting a wellbore and its effect on the dipmeter sensors.

An important product of processing involves correlating the conductivity changes and plotting the results in the form of tadpoles. The tail of the tadpole points in the downward dip direction. The body of the tadpole indicates the magnitude of the dip. Figure 2.9 shows an example of a tadpole plot beside a sample gamma ray plot.

The tadpoles form patterns that can help in the identification of sedimentary structures. Sedimentary structures relate to the layering or stratification in sedimentary rocks (Berg, 1986). The stratification can result from sedimentary processes (primary sedimentary structures) or from changes shortly after deposition (secondary sedimentary structures). Samples of sedimentary structures include the shape, thickness, or fossil content of beds.

Researchers have devised various methods for classifying sedimentary structures based on the time of formation, the processes that created the structures, or the location of structures (e.g., Serra, 1985). Selley (1986) proposed a classification that combines the time of formation and the creating processes. McKee and Weir (1953) use layering thickness, splitting properties, stratification, and cross-stratification to describe primary sedimentary structures. Pettijohn and Potter

(1964) proposed a classification that depends on the location of features. Table 2.1 shows Pettijohn and Potter's classification scheme and its relation to well logging devices. The detection of sedimentary structures is important since they reflect energy conditions at the time of deposition.

Dipmeter Curve Matching

The first application of pattern recognition algorithms in dipmeter processing involved the semiautomatic computation of dip from the correlation curves. While human processing or optical processing of the dipmeter curves provides the best results, the process is time-consuming. Fixed interval correlation was the first algorithm used to

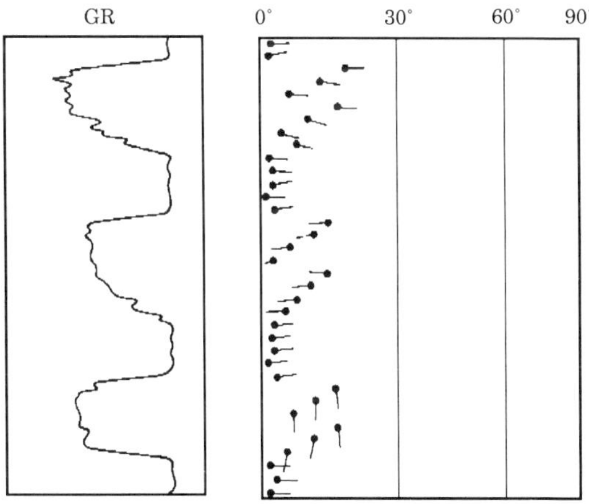

FIGURE 2.9. Tadpole plot.

TABLE 2.1. Classification of sedimentary structures and their relationship to well logging device[a]

Structural features			Appropriate tool
Bedding external	Vertical thickness	Apparent	All logs
		Real	High-resolution logs
	Shape	Lateral variations in sequence of beds	High-resolution logs
			All logs
	Boundaries	Abrupt	All logs
		Progressive	All logs
		Conformable	High-resolution logs
Bedding plane		Unconformable	High-resolution logs
	Lower boundaries	Physical origin	High-resolution logs
		Organic origin	High-resolution logs
	Upper boundaries	Physical origin	High-resolution logs
		Organic origin	High-resolution logs
Bedding internal	Massive		All logs
	Laminated		High-resolution logs
	Graded bedding		All logs
	Oriented internal		
	Growth structure		
Bedding	Physical origin	Load marks	High-resolution logs
		Fractures	High-resolution logs
		Faults	High-resolution logs
	Organic origin	Burrows	High-resolution logs
	Chemical origin	Stylolites	High-resolution logs

[a]From Pettijohn and Potter (1964).

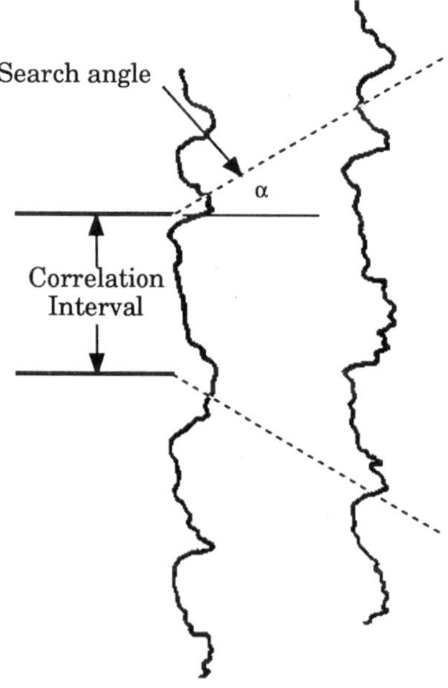

FIGURE 2.10. Fixed interval correlation.

correlate dipmeter curve automatically (Moran et al., 1961). Improvements to the fixed correlation method were described in Kemp (1980).

Figure 2.10 shows an example of fixed interval correlation. First, a fixed length of a base dipmeter curve is selected. A correlation coefficient indicating the degree of matching is calculated for all possible combinations of the reference interval and intervals on a second curve. The maximum of the correlation coefficient indicates the matching interval on the second curve. A maximum search angle is usually imposed to limit the possible correlations.

Certain checks can be performed to verify that the correlations are meaningful. The correlations can be checked for planarity, which implies that the correlations describe a planar feature. Nonplanarity implies improper correlation. Another check on the correlation process involves closure. Closure implies that a feature should correlate to itself. Figure 2.11 shows an example of closure. Nonclosure can also imply improper correlation.

Despite checks on the correlation process, the fixed interval method can calculate dips which exhibited excessive scatter not attributed to geological events. Hepp and Dumestre (1975) describe a program called CLUSTER, which uses overlapping correlation intervals and keeps statistically coherent correlations. Specifically, the correlations are taken three curves at a time. Eight different combinations are possible and result in eight different dip computations. Ideally, the dips should be the same. In practice, they tend to "cluster" together.

Nevertheless, fixed interval correlation has its problems. If a selected interval is too small, too many choices are available for correlation and correlation errors increase. If the selected interval is too large, the displacements calculated by correlation will be wrong due to averaging. For these and other reasons, Vincent et al. (1979) developed a symbolic pattern matching approach to dipmeter curve recognition. Their process consists of three phases: feature extraction, correlation, and dip computation. Feature extraction involves characterizing the dipmeter curves as in Figure 2.12. The digital curves are replaced by their descriptive counterparts. Each descriptive element also has numerical characteristics, which describe the shape of the symbol element. Some of the numerical characteristics include

the width of the peak
average
maximum
maximum minus average.

Again, a similarity coefficient is computed for each attempted correlation of elements. However, the correlations proceed according to a preset order

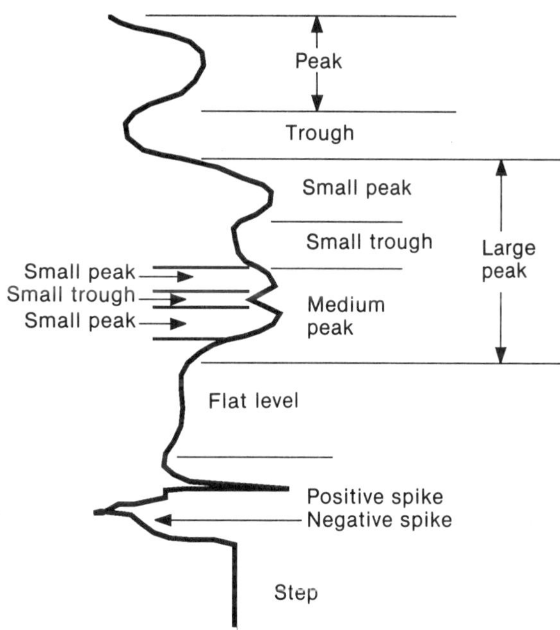

FIGURE 2.12. Feature extraction of the dipmeter curves (Copyright 1979, Society of Petroleum Engineers. Vincent, Ph., Gartner, J.E., and Attali, G.: "An Approach to Detailed Dip Determination Using Correlation by Pattern Recognition," *Journal of Petroleum Technology* (February 1979) (Figure 3).

or precedence (large troughs first, large peaks next, etc.). Accepted correlations are used to guide succeeding correlations. The correlation procedure operates on all four curves at once.

Kerzner (1983) devised an alternative approach to dipmeter curve correlation using a minimization procedure. Processing proceeds in three stages: preprocessing with an activity function and two stages of curve matching. Activity is defined as follows:

$$A = \sum_{i=\frac{-n}{2}}^{\frac{n}{2}} (x_i - \bar{x})^2$$

where x is the average over an interval of n points. A sample log and the corresponding activity function is shown in Figure 2.13. Note that the activity function acts like a smoothed derivative. The maximums of the activity curve are selected as correlation points. Some simple rules are used to control feasible correlations. The rule of noncrossing correla-

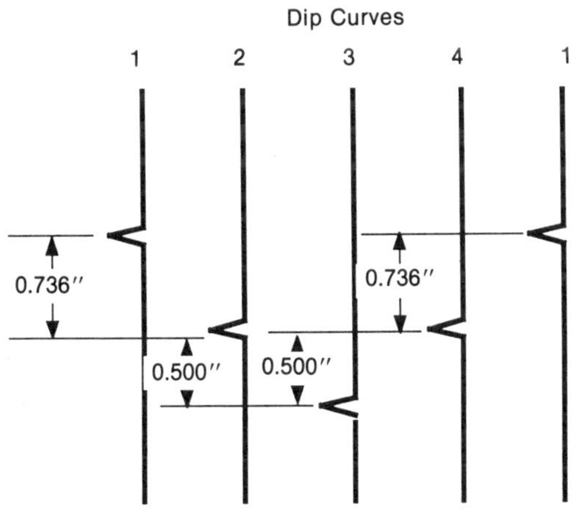

FIGURE 2.11. Closure.

FIGURE 2.13. Activity function (Copyright 1988, Society of Petroleum Engineers. Kerzner, M.G.: "A Rule-Based Approach to Dipmeter Processing." Paper, SPE 18128, presented at the 63rd. Annu. Tech. Conf. and Exhib., Houston, TX, October 2–5).

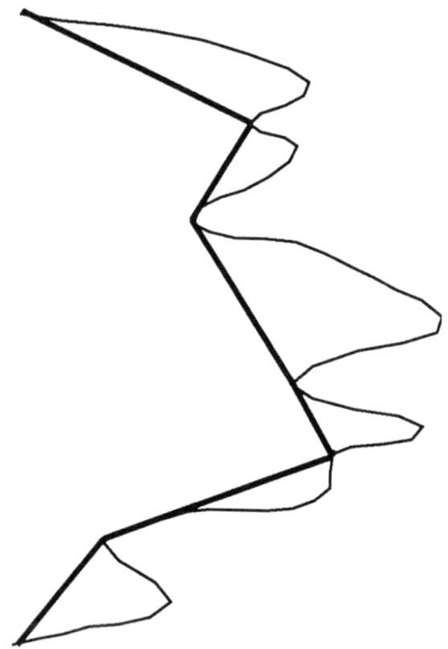

FIGURE 2.14. First stage of segmentation.

tions essentially indicates that the correlation lines between curves should be close to parallel.

In later work, Kerzner (1988) devised a rule-based approach to dipmeter curve correlation. The process consists of segmentation by successively connecting local minima on the correlation curves and correlating using the above optimization algorithm. Figure 2.14 shows how local minima are connected with straight lines. The local minima is found for the new curve as in Figure 2.15. Finally, the local minima of the second new curve is found as in Figure 2.16. This segmented curve is referred to as the base curve.

The rules to the curve correlation process serve to limit the search space of the optimization algorithm and direct the optimization. Examples of such rules are

Events must be greater than 0.25 ft and less than 6 ft in vertical extent.

If a 4 pad planar correlation combination is found, it is accepted.

If a 4 pad correlation cannot be found, change the base pad.

If an expected direction of correlation is established from previous correlations, it is used as a starting point for further correlations.

A radically different approach to dipmeter curve correlation is presented by Baldwin et al. (1989). Their study used neural network technology for pattern recognition of dipmeter and full wave sonic curve correlation. Readers interested in an advanced application of neural network technology are encouraged to read the original text.

Dipmeter Analysis

The second major application for pattern recognition and AI involves the interpretation of the dip patterns. We will concentrate on two different approaches to semiautomated dipmeter interpretation. The first system, the Dipmeter Advisor, represents the classical approach to knowledge-based system development. The second system, ESCAT, exemplifies an alternative approach to dipmeter interpretation by employing different plotting functions.

The Dipmeter Advisor System is a sophisticated collection of utilities and graphics designed to aid its user in the interpretation of dipmeter data. Its facilities include the ability to generate various plots and displays of dip data in conjunction with other well log-data, to suggest interpretations based on patterns in the dip data together with lithological and other information provided by the user, and to generate geologic cross sections from the data.

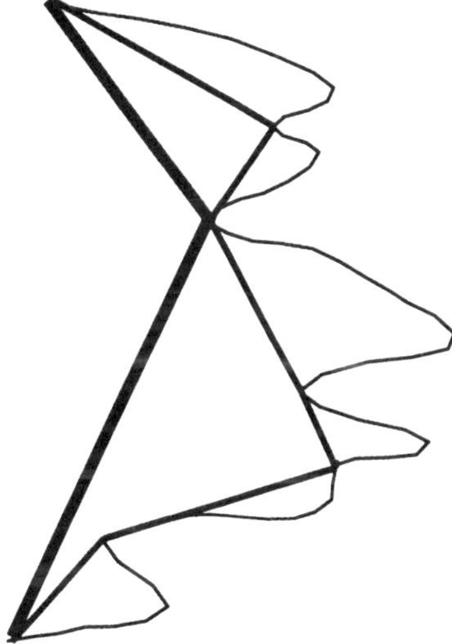

FIGURE 2.15. Second stage of segmentation.

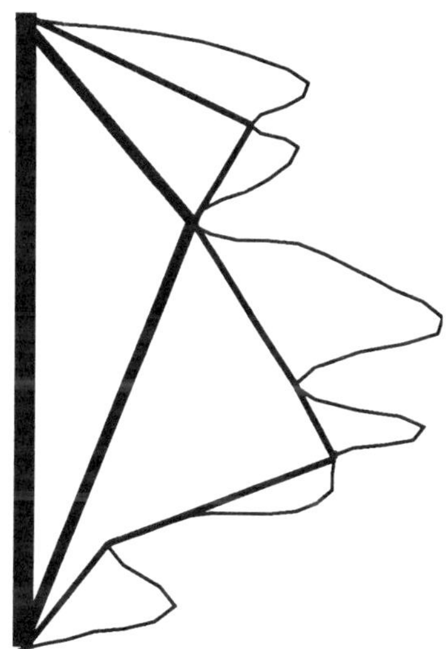

FIGURE 2.16. Third stage of segmentation.

The system consists of four parts: the graphic interface, a number of rule bases, a forward-chaining inference engine, and feature detection algorithms. Using the system involves a series of steps which lead to interpretation of the dipmeter data. The four major phases of processing are shown in Table 2.2.

The initial examination phase requires the interpreter to select the data he wishes to display and analyze. The system then checks the data for the possible preprocessing errors or tool malfunctions.

In structural dip analysis phase, the system finds zones where the tadpoles have similar magnitude and azimuth and asserts these zones as green patterns. The green patterns are merged and the system finds zones of constant structural dip. Structural dip is then removed before further processing.

Rules are applied to make a preliminary structural analysis of the data. Then symbolic pattern matching algorithms are applied to determine red, blue, and yellow patterns in the data. A final review of the data combines all information from the previous phases and confirms conclusions about structural features like faults.

The final stage of processing involves a lithology analysis with open hole data like the GR or SP to determine constant lithology zones. The depositional environment rules are then applied. Once again, the system applies symbolic pattern matching to find red, green, blue, and yellow patterns in areas of known depositional environments. Finally, rules are applied to conclude about stratigraphic features such as channels or bars.

Some of the experience gained from the Dipmeter Advisor (Hammock, 1988) include the caveats below:

A well designed interface is necessary for creating, editing, and debugging rules.

Close involvement of the experts whose skill the system is designed to imitate is crucial. The experts should be able to enter and test their own rules.

TABLE 2.2. Dipmeter advisor interpretation phases

Initial examination
Selection of data
Validity check
Structural dip analysis
Green pattern detection
Structural dip zone determination
Structural dip removal
Structural feature analysis
Preliminary structural analysis
Structural pattern detection
Final structural analysis
Stratigraphic feature analysis
Lithology determination
Depositional environment analysis
Stratigraphic pattern detection
Stratigraphic analysis

     T.-B. Kuo, S.A. Wong, and R.A. Startzman

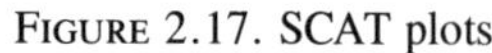

FIGURE 2.17. SCAT plots.

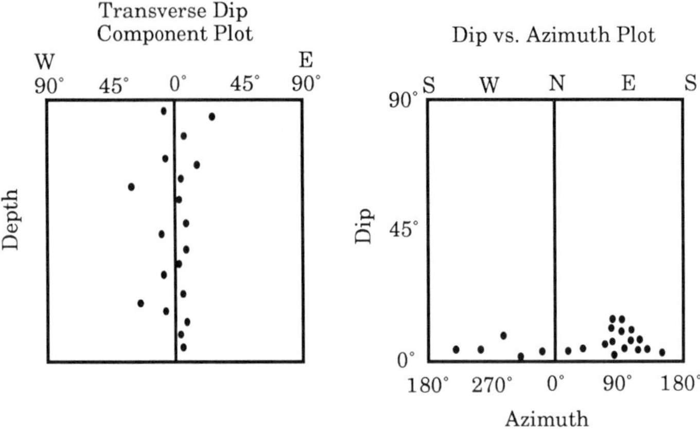

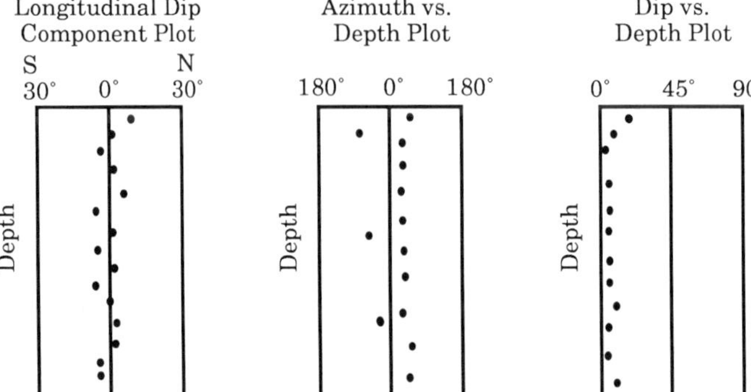

A clear representation of how results are to be used. For example, results from the Dipmeter Advisor system are intended to aid rather than replace human interpreters.

A straightforward method of representing conclusions drawn by the system, including the ability to query the system about its interpretations and to have it respond with text and pictures that explain its reasoning.

Non-order-dependent implementation of rules within a rule set is helpful.

An alternative approach to dipmeter interpretation involves Bengston's statistical curvature analysis techniques (SCAT) for analyzing dipmeter data (Bengston, 1988). Normally, we equate the tadpole plot with the representation of dipmeter data. SCAT used five different plots instead of the tadpole plot. The five plots consists of

transverse dip component plot
longitudinal dip component plot
azimuth vs. depth plot
dip vs. depth plot
azimuth vs. dip plot.

Figure 2.17 shows an example of the plots. The transverse dip direction refers to the direction of the cross section through the well that shows the most structural change. The longitudinal direction refers to the cross-section through the well that shows the least structural change. Interpretation of the five plots involves fitting internally consistent statistical trend lines to all five plots.

Thadani and Bengston (1988) developed an expert system called ESCAT (Essential SCAT), that was developed in the mid 1980s to assist in the interpretation of dipmeter data using the SCAT technique. The SCAT technique utilizes a combination of statistical and heuristic reasoning to synthesize structural and stratigraphic interpretation from dipmeter data. The SCAT heuristics are implemented using a back-chained rule-based paradigm. The statistical/ numerical component of the technique is implemented using standard FORTRAN procedures.

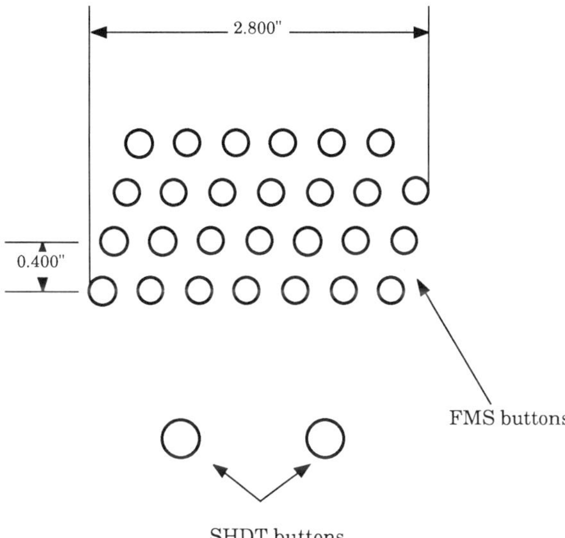

FIGURE 2.18. FMS 27 button configuration of pads from Schlumberger document on the Formation MicroScanner service (1986).

The ESCAT knowledge base currently consists of approximately 300 rules for structural/stratigraphic interpretation in extensional and compressional regimes. The system is configured so as to be highly interactive with the user—the user interface is entirely via graphic displays and menus. The heuristic rules "drive" the consultation with the user, prompting her about "what to look for" whenever appropriate. The entire system has been implemented in FORTRAN on a VAX 11/780 using a VAXstation graphics terminal. ESCAT has been in use at Standard Oil since 1985—several "tests" of the system with synthetical/real data confirm that interpretations produced with the aid of ESCAT compare favorably with those produced by expert interpreters.

Wellbore Imaging

Dipmeter interpretation is complicated by complex environments and must be augmented by whole core. However, core may not be available. Formation imaging provides an alternative way to get a continuous borehole representation. Two devices are currently available: conductivity (FMS) and sonic. The major applications for formation imaging include secondary porosity evaluation and thin bed evaluation.

Currently two types of logging devices are used to generate images of the surface of the wellbore. The Formation MicroScanner (FMS) is a resistiv-

ity tool with arrays of buttons each located on pads orthogonal to each other. Hence, the FMS gives a partial view of the wellbore surface. An electrical conductivity image is formed by an array of small electrodes that is in electrical contact with the formation (Ekstrom et al., 1986a). Porosity, formation fluid, rock textural characteristics, and borehole rugosity affect the quality of the FMS images. The FMS or conductivity imaging device is actually a modification of the current 8 button dipmeter tool. Figure 2.18 shows the pads for the dipmeter device augmented by the FMS buttons (Ekstrom et al., 1986b). The FMS comes in two configurations: 2 pads with 27 buttons apiece and 4 pads with 16 buttons apiece.

The Borehole Televiewer (BHTV) uses an ultrasonic transducer to send a short acoustic pulse out to the borehole or casing wall (Zemanek and Caldwell, 1969). The amplitude and transit time images can be made with varying frequency transducers. Higher frequencies give better resolution; lower frequencies are less sensitive to borehole rugosity. The BHTV gives complete borehole coverage of the borehole surface or casing wall. Similar factors affect the BHTV and the FMS.

High resolution images are necessary for detailed geologic interpretation. Both imaging devices (FMS and BHTV) have a spatial resolution less than 0.4 sec and a detection threshold of less than 0.04 sec (Georgi, 1985). The images provide qualitative information on the rock texture (Plumb and Luthi,

5	5	5
-3	0	-3
-3	-3	-3

Kirsch

1	1	1
1	-2	1
-1	-1	-1

Prewitt

1	1	1
0	0	0
-1	-1	-1

Robinson

1	2	1
0	0	0
-1	-2	-1

Sobel

FIGURE 2.19. Edge detectors.

1986). The cross-sectional geometry of the well can provide information on rock strength and contemporary stress directions (Zoback et al., 1985). Lithology and net sand counts can be made from the images (Hackbarth and Tepper, 1988). Fracture orientation can be obtained from the images (Laubach et al., 1988).

Pattern recognition and AI techniques can also be applied to wellbore images. Although few references to systems exist in the petroleum literature, a plethora of image understanding systems can be found in the pattern recognition and computer vision literature. We will cover three areas of interest: edge detection, pattern recognition, and knowledge-based systems.

Edge Detection

Intensity discontinuities tend to reflect object boundaries in images. Experiments with the human visual system support the notion that edges are extremely important in the recognition of an object. Edges are simply significant changes in intensity; edge detection involves finding these changes in intensity.

A good review of the literature can be found in Rosenfeld and Kak (1982). Ballard and Brown (1982) discuss three major types of edge detection classes: approximations to the gradient, parametric edge modeling, and template matching. Parametric edge models use optimization techniques to find parameters to fit particular image models. Because they involve much detail, the interested reader is encouraged to review parametric models

by Hueckel (1971), Nevatia (1978), and Abdou (1978). Template matching involves convolving a mask with an image. The convolved image reflects edge strengths. Figure 2.19 shows a few different types of edge templates that we use.

The templates are direction sensitive. The templates in Figure 2.19 are oriented to find edges in the north–south direction. The edge strength of an arbitrary pixel is simply a weighted average of the neighboring pixels. Different templates use different weights. If you want to find edges in a different direction, you must rotate the weights accordingly. A sample Kirsch edge calculation for an image is shown.

$$\begin{bmatrix} 5 & 5 & 5 \\ -3 & 0 & -3 \\ -3 & -3 & -3 \end{bmatrix} \times \begin{bmatrix} 1 & 4 & 3 \\ 2 & 3 & 1 \\ 2 & 3 & 3 \end{bmatrix} =$$

$$
\begin{aligned}
1.2.4 \quad &+ \quad (5\times4) \quad + \quad (5\times3) \quad + \\
(-3\times2) \quad &+ \quad (0\times3) \quad + \quad (-3\times1) \quad + \\
(-3\times2) \quad &+ \quad (0\times3) \quad + \quad (-3\times1) \quad = \quad 7
\end{aligned}
$$

The first matrix is the Kirsch weights. The second matrix represents the intensity values for a sample pixel and its neighbors in a sample image. The resultant value is the edge strength. Edge strengths are calculated for all pixels in the image. Figure 2.20 shows an example of an FMS image and the edge enhanced version.

Note that the weights are zero mean. That is, a region with constant gray scale values will have an edge strength of zero. Large edge strengths indicate a potential edge. Large negative edge strengths indicate a potential edge in the opposite direction of the preferred edge direction.

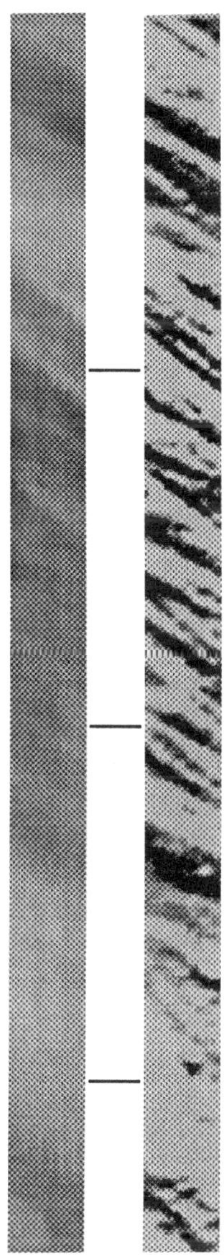

FIGURE 2.20. Edge enhanced FMS image.

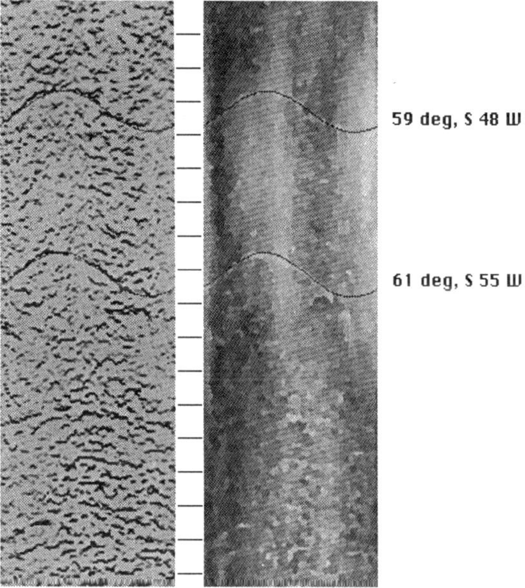

FIGURE 2.21. Hough transform results.

Wong et al. (1989) used edge detection as one of the first stages in pattern recognition of planar objects in wellbore images. First the images were processed with an edge preserving smoothing filter. Next, an edge template operator was used to highlight the edges of planar features. Finally, a pattern matching algorithm called the Hough transform which essentially converts the edges from image space to parameter space. Figure 2.21 shows an image that has been filtered, edge enhanced, and Hough transformed. The Hough transform finds one of the fractures in the middle of the BHTV image. The Hough transform has been used with success in various areas including tumor detection (e.g., Kimme et al., 1966), aerial photographic interpretation (e.g., Lantz et al., 1978), and quality control of wooden boards (e.g., Poelzleitner, 1986). A generalized Hough transform has been developed for cases where no simple analytical form can be obtained (Ballard and Brown, 1982). Gaps in data do not affect the Hough transform.

Texture Recognition

The recognition of image textures is of paramount importance in the identification of objects or regions of interest in digital images. Ways of quantifying texture have existed for over 30 years (Kaiser, 1955). Salient texture features must be extracted from an image prior to categorization by any pattern recognition technique.

Methodologies for the analysis of image textures fall into two broad categories: statistical and structural or syntactic. Statistical approaches to texture analysis include features based on the Fourier power spectrum, first-order statistics of gray level differences, and second-order gray scale statistics. All statistical methods require the definition of a metric that quantifies texture features. On the other hand, structural methods consider textures as repeated patterns that occur according to a set of previously defined placement rules. Examples of texture analysis can be found in Weszka et al. (1986) and Kjell and Dyer (1985).

Wong et al. (1989a) discuss how texture recognition could be used to find patterns in wellbore images. First, the target image texture is characterized by convolving edge and line templates with the image. Next, statistics on the convolved image are taken. Finding the target image texture in an arbitrary image involves comparing the database statistics with the statistics of the arbitrary image convolved with the same line and edge templates. Hence, image texture analysis methods could be used to find crossbedding or other patterns found in wellbore images if good training sets are provided.

Knowledge-Based Systems

Though wellbore images contain a wealth of information, the petroleum literature makes no reference to semiautomatic or automatic interpretation procedures for 2-D images other than the paper by Wong et al. (1989b). However, many image analysis articles have been written on image interpretation in the computing science and electrical engineering literature. The problem of image interpretation can be broken into two major parts: high and low level vision tasks. High level vision roughly corresponds to the cognitive or reasoning processing related to vision. Low level vision corresponds to the specialized and subconscious visualization ability. Researchers have generally approached the problem of low level vision by mimicking the eye through neural networks or inventing numerical/statistical approaches to pattern recognition and classification.

Because low level vision refers to capabilities that have not been documented, pattern recognition procedures have been developed as described above to circumvent the low level vision problem. However, high level vision can be modeled by using existing knowledge-based technology. Some of the image understanding systems outside of the petroleum industry based on KBS technology include programs by Davis and Hwang (1985) and Nagao and Matsuyama (1980) for aerial photography interpretation and Tucker's medical vision system (Tucker, 1984).

The processing chains and structures of image understanding systems are as varied as the applications themselves. We will summarize the aerial photography system of Nagao and Matsuyama since it parallels development of a wellbore imaging KBS system we are developing. They developed a symbolic knowledge-based system to interpret aerial photographs. They first use an edge preserving smoothing to remove small scale image patterns. Next, a simple region merging is performed. The regions of homogeneous intensity area characterized by elongatedness, area, color, and other attributes. Rules are then applied to interpret the image regions as vegetation, roads, buildings, or other man-made or natural objects.

Wong et al. (1989a, 1989b) developed a prototype knowledge based system called the Borehole Visualization Tool (BVT) to interpret wellbore images. The processing of data from digital data to symbolic primitives parallels the work of Nagao and Matsuyama. An edge preserving smoothing is used to remove small scale texture. Next, an edge detection and region merging scheme are used to form primitives. The primitives are characterized and placed a blackboard. Specialized rule bases work on the primitives and continually convert the data until geologic events such as fractures and faults are detected. Figure 2.22 shows a schematic of the system.

Well Test Analysis

Well test analysis is a task to find a model that adequately represents the physical situation existing in the wells and reservoirs being tested, and qualify the parameters that are incorporated in the model.

Erdle et al. (1986) pointed out major problems that a nonexpert might encounter in analyzing well test data:

the lack of a systematic approach to solving problems,
difficulty in judging the quality of the input data,
incomplete technical knowledge of available models and their underlying assumptions and limitations,
problems recognizing patterns and judging the goodness of fit between the model-generated data and the measured data,

FIGURE 2.22. Borehold visualization tool.

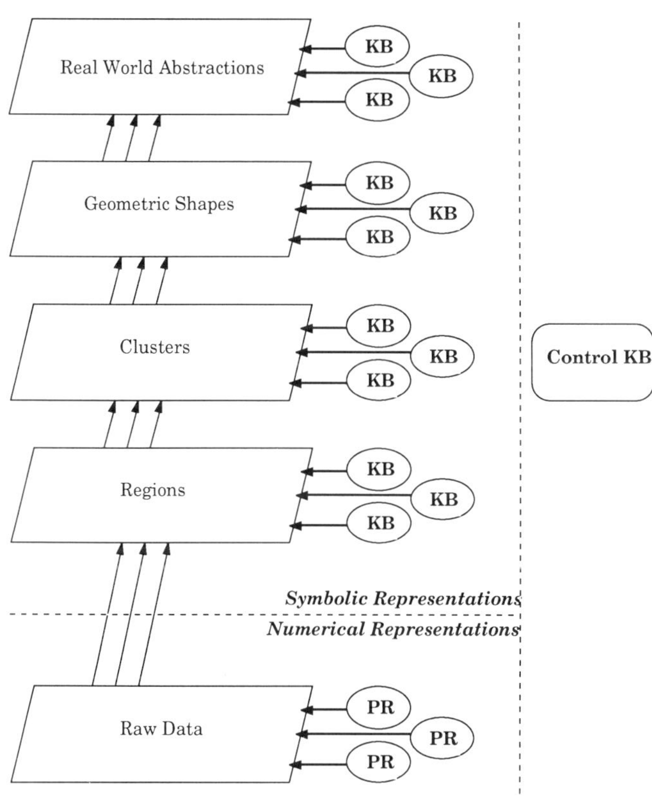

estimating the reliability of the interpretation, and difficulty formulating correct, meaningful conclusions from the interpretation results.

They suggested that the expert systems approach is ideally suited to minimizing the problems hindering successful interpretation. However, the expert systems should be used to increase the efficiency and accuracy instead of replacing the analysts.

Al-Kaabi et al. (1988) developed a knowledge base system that can seek interpretation models in a manner parallel to techniques used by human experts. The system is designed to interpret well test data using the descriptions of shapes that appear on the different plots of pressure test data. The major tasks that the knowledge base performs are to (1) recognize different flow regions on pressure derivative plots, and (2) identify applicable interpretation model.

Pressures recorded in a well test can reflect the characteristics of the flow during the test. The pressure derivative plot (i.e., pressure derivative vs. adjusted equivalent time) usually will reveal characteristics of different flow regions. A typical pressure derivative plot is a curve consisting of maxima, minima, stabilization, and upward or downward sections. Each section corresponds to a flow region and can be recognized easily. Figure 2.23 is a sample plot.

The process to recognize flow regions is essentially a process to segment a pressure derivative plot. A left-to-right visual inspection to the plot is performed and at the same time the distinct features such as maxima, minima etc. are recognized. The user of the system is guided to select a "marker" or the most distinctive feature that is usually a flat section (i.e., the stabilization region). Subsequently, the distinctive features preceding and succeeding the marker are recognized and segmented. As a result, the entire curve is divided into different flow regions (i.e., early time, middle time, and late time).

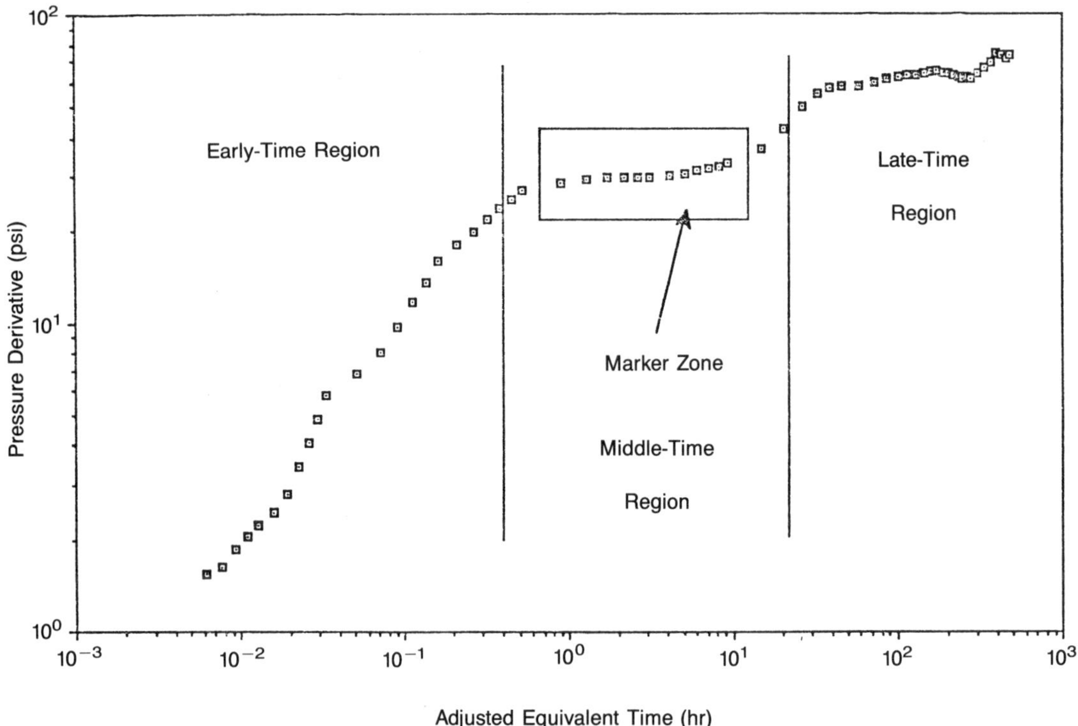

FIGURE 2.23. Example of a pressure derivative plot and major flow regions. (Copyright 1988, Society of Petroleum Engineers. Al-Kaabi, A.U., McVay, D.A., and Lee, W.J.: "Using an Expert System to Identify the Well Test Interpretation Model." Paper, SPE 18158, presented at the 63rd Annu. Tech. Conf. and Exhib., Houston, TX, October 2–5).

The recognition of flow regions is followed by a process to determine the interpretation model, which describes the wellbore condition, reservoir type, and the outer boundary condition of a test. First, the system determines the reservoir type: a homogeneous or heterogeneous reservoir, based on the characteristics of pressure derivative plot, semilog plot (i.e., a graph of pressure versus logarithm of the Horner time ratio), and log–log plot (i.e., a graph of logarithm of pressure change vs. logarithm of elapsed time). This is accomplished by firing rules such as the one below.

> IF The "minimum" section on the pressure
>   derivative plot is a U-shape;
>   The semilog plot shows evidence of two
>   straight line with possible slope dou-
>   bling;
>   The log-log plot indicates slope change.
> THEN The reservoir is heterogeneous; or

> The reservoir is homogeneous with
> possible fault.

In deriving the conclusions, a knowledge base architecture, called Blackboard, is used. The blackboard can be considered as a global memory space where the hypotheses to the solutions, intermediate solutions, the final solution, and the states of those solutions can be stored. Before the final solution is reached, all information on the blackboard can be modified according to the most recent conclusions. For instance, to identify the reservoir type, the system, by default, hypothesizes that the reservoir is homogeneous. This hypothesis (i.e., the current solution) is posted on the blackboard. Then, the system goes on to validate the posted solution by asking anticipated information from the user. An example is shown in Figure 2.24.

This expert system also identifies the types of the well bore (or inner boundary) condition and the pressure investigation limit (or outer boundary

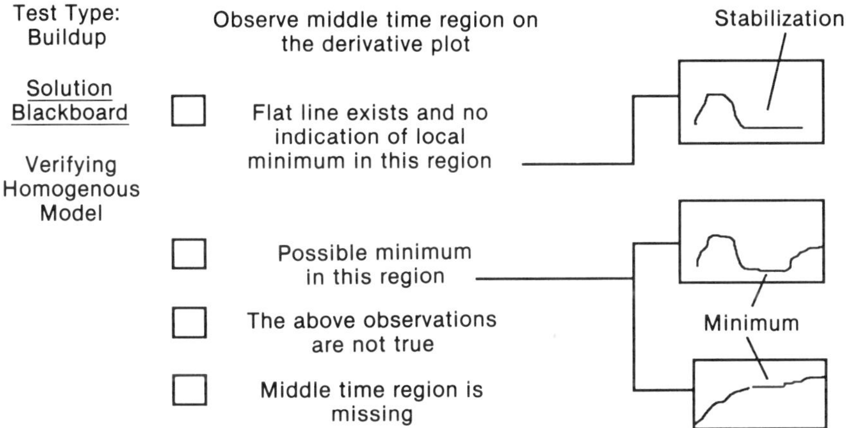

FIGURE 2.24. An example of a consultation window. (Copyright 1988, Society of Petroleum Engineers. Al-Kaabi, A.U., McVay, D.A., and Lee, W.J.: "Using An Expert System to Identify the Well Test Interpretation Model." Paper, SPE 18158, presented at the 63rd Annu. Tech. Conf. and Exhib., Houston, TX, October 2–5).

condition) associated with the reservoir. Since the pressure recorded in the early testing time is strongly affected by the conditions at or near the well bore, neglecting the existence of these conditions will easily result in erroneous reservoir parameters. The well bore conditions such as well bore storage, fractures in formation, skin effects caused by formation damage, or partial penetration can be identified from the characteristics of the pressure behavior in the early time region. A typical example to identify the well bore storage is to detect the unit slope straight line (in the early time region) on a log–log plot or a pressure derivative plot. If this characteristic is not found, the system will lead the user to check other observations.

In a similar manner, the system will describe the outer boundary condition of a pressure test and use heuristic rules, such as the one below, to conclude that the outer boundary is one of the three common conditions: infinite-acting, closed boundaries, and linear discontinuities.

IF Evidence of second straight line on the semilog plot with slope doubling, AND
The derivative plot shows an upward increase followed by stabilization, AND
The upward increase is preceded by stabilization,
THEN No flow boundaries (linear discontinuity) exist in a homogeneous system, OR

A heterogeneous system is infinite-acting.

This system has achieved an initial step to incorporate analytical solution methods developed in the past and human experts' qualitative interpretation procedures. It has further been expanded to include an automatic segmentation program to identify flow regions and a history matching module to verify the interpretation model determined by the expert system.

Allain and Horne (1988) also proposed an artificial intelligence method to identify a well test interpretation model. A detailed description of their method can be found in Chapter 1.

References

Abdou, I.E., 1978, Quantitative Methods of Edge Detection: USCIPI Report 830, Image Processing Institute, University of Southern California.

Al-Kaabi, A.U., McVay, D.A., and Lee, W.J., 1988, Using an expert system to identify the well test interpretation model: paper SPE 18158, presented at the 63rd Annu. Tech. Conf. Exhib. of the Soc. Petr. Eng., Houston, TX, October 2–5.

Allain, O.F., and Horne, R.N., 1988, Use of artificial intelligence for model identification and parameter estimation in well test interpretation: paper SPE18160, presented at the 63rd Annu. Tech. Conf. Exhib. Soc. Petr. Eng., Houston, TX, October 2–5.

Apté, C.V., and Weiss, S.M., 1985, An approach to expert control of interactive software systems: IEEE Transactions on Pattern Analysis and Machine Intelligence, PAMI-7, No. 5, 586–591.

Baldwin, J.L., Otte, D.N., and Bateman, R.M., 1989, Computer emulation of human mental processes: Application of neural network simulators to problems in well log pattern recognition: Proceedings of the Second Conference on Artificial Intelligence in Petroleum Exploration and Production, College Station, TX, May 17–19, 145–175.

Ballard, D.H., and Brown, C.M., 1982, Computer Vision: Prentice-Hall, Englewood Cliffs, NJ.

Bengston, C.A., 1988, Statistical curvature analysis techniques for structural interpretation of dipmeter data: Geobyte 3(2), 63–75.

Berg, R.R., 1986, Reservoir Sandstones, Prentice-Hall, Englewood Cliffs, NJ.

Davis, L.S., and Hwang, S.S., 1985, The SIGMA image understanding system: Proceedings of the Third Workshop on Computer Vision: Representation and Control, Bellaire, October 13–15, 19–26.

Einstein, E.E., and Edwards, K.W., 1988, Comparison of an expert system to human experts in well log analysis and interpretation: paper SPE 18129, presented at the 63rd Annu. Tech. Conf. Exhib. Soc. Petr. Eng., Houston, TX, October 2–5.

Ekstrom, M.P., Dahan, C.A., Chen, Min-Yi, Lloyd, P.M., and Rossi, D.J., 1986a, Formation imaging with microelectrical scanning arrays: Proc. 27th Soc. Prof. Well Log Anal. Logging Symp., Houston, TX, BB.

Erdle, J.C., Archer, D.A., Stiff, T.J., and Callihan, M.C., 1986, Well test engineering software with built-in expert advice: paper SPE15309, presented at the First Annu. Symp. Petr. Ind. Appl. Microcomps, Silvercreek, CO, June 18–20.

Georgi, D.T., 1985, Geometrical aspects of borehole televiewer images: Proc. 26th Soc. Prof. Well Log Anal. Logging Symp., Dallas, TX, O.

Ghose, B.K., 1984, STRETCH: A subroutine for stretching time series and its use in stratigraphic correlation: Comp. Geosci. 1(1), 137–147.

Gordan, A.D., 1980, SLOTSEQ: A Fortran IV program for comparing two sequences of observations: Comput. and Geosci. 6(1), 7–20.

Gordon, A.D., and Reyment, R.A., 1979, Slotting of borehole sequences: J. Int. Assoc. Math. Geol. 11(3), 309–327.

Hackbarth, C.J., and Tepper, B.J., 1988, Examination of BHTV, FMS, and SHDT images in very thinly bedded sands and shales: paper SPE 18118 presented at the 63rd Annu. Tech. Conf. Exhib., Houston, TX, October 2–5.

Hammock, D., 1988, Development of the dipmeter advisor system—What we learned: presented at the first Conf. A. I. Petr. Expl. Prod., Texas A&M University, College Station, TX.

Hawkins, D.W., and Merrian, D.F., 1974, Zonation of Multivariate Sequences of digitized geologic data: J. Int. Assoc. Math. Geol. 6(3), pp. 263–269.

Hepp, V. and Dumestre, A.C., 1975, CLUSTER—A method for selecting the most probable dip results from dipmeter surveys: paper SPE 5543 presented at the SPE Annu. Tech. Conf. Exhib., Dallas, September 28–October 1.

Hoffman, L.J.B., Hoogerbrugee, P.J., and Lomas, A.T., 1988, LOGIX—A knowledge-based system for petrophysical evaluation: Proc. Soc. Prof. Well Log Anal. 29th Annu. Logging Symp., June 5–8, San Antonio, TX, R.

Horner, D.R., 1951, Pressure build-up in wells: Proc., Third World Petroleum Congress, Leiden, II, 503.

Hoyle, I., 1987, Correlation of well logs using dynamic time warping: Ph.D. Dissertation, University of New England, Armidale, Australia.

Hueckel, M., 1971, An operator which locates edges in digitized pictures: J. ACM 18(1), 113–125.

Jorden, J.R., and Campbell, F.L., 1984, Logging I—Rock properties, borehole environment, mud and temperature logging: Monograph Vol. No. 9, Soc. Petr. Eng., Dallas, TX.

Kaiser, H., 1955, A quantification of textures on aerial photographs: Technical Note 121, Boston University Research Lab, Boston, MA.

Kemp, F., 1980, An algorithm for automatic dip computation: Comput. Geosci., 6, 193–209.

Kerzner, M.G, 1983, Formation dip determination—An artificial intelligence approach: Log Anal., 24(5) 10–22.

Kerzner, M.G., 1988, A rule-based approach to dipmeter processing: paper SPE 18128 presented at the SPE Annu. Tech. Conf. Exhib., Houston, TX, October 2–5.

Kimme, C., Ballard, D., and Sklansky, J., 1966, Finding circles by an array of accumulators, Part II: Commun. ACM 18, 120–122.

Kjell, B.P., and Dyer, C.R., 1985, Edge separation and orientation texture measures: Proc. IEEE Conf. Comput. Vision Pattern Recognition, San Francisco, CA, 306–311.

Kuo, T.-B., 1986, Well log correlation using artificial intelligence: Ph.D. Dissertation, Texas A&M University, College Station, TX.

Kuo, T.-B., and Startzman, R.A., 1987, Field-scale stratigraphic correlation using artificial intelligence: Geobyte 2(2), 30–35.

Kwon, B.D., and Rudman, A.J., 1979, Correlation of geologic logs with spectral methods: J. Int. Assoc. Math. Geol. **11**(4), 373–390.

Lantz, K.A., Brown, C.M., and Ballard, D.H., 1978, Model-driven vision using procedure description: Motivation and application to photointerpretation and medical diagnosis: Proc. 22nd Int. Symp. Soc. Photo-optical Instr. Eng., San Diego, CA.

Laubach, S.E., Baumgarder, R.W., Monson, E.R., Hunt, E., and Meador, K.J., 1988, Fracture detection in low-permeability reservoir sandstone: A comparison of BHTV and FMS logs to core: paper SPE 18119 presented at the 63rd Annu. Tech. Conf. Exhib., Houston, TX, October 2–5.

Lineman, D.J., 1986, An expert system for well-to-well log correlation, M.S. Thesis, Massachusetts Institute of Technology, Cambridge, MA.

Lineman, D.J., Mendelson, J.D. and Toksoz, M.N., 1987, Well to well log correlation using knowledge-based systems and dynamic depth warping: Proc. Soc. Prof. Well Log Anal. 28th Annu. Logging Symp., June 29–July 2, London, UU.

Liu, H.H. and Fu, K.S., 1982, A syntactic pattern recognition approach to seismic discrimination: Geoexploration **20**, 183–196.

Mann, C.J., and Dowell, T.P.L., Jr., 1978, Quantitative lithostratigraphic correlation of subsurface sequences: Comput. Geosci. **4**(4), 295–306.

Matuszak, D.R., 1972, Stratigraphic correlation of subsurface geologic data by computer: J. Int. Assoc. Math. Geol. **4**(3), 331–343.

McKee, E.D., and Weir, G.W., 1953, Terminology for stratification and cross-stratification in sedimentary rocks: Geol. Soc. Am. Bull. **64**, 381–390.

Miller, C.C., Dyes, A.B., and Hutchinson, C.A., Jr., 1950, The estimation of permeability and reservoir pressure from bottom hole pressure build-up characteristics: Transact., Am. Inst. Min. Metall. Petr. Eng. 189, 91–104.

Moran, J.H., Coufleau, M.A., Miller, G.K., and Timmons, J.P., 1961, Automatic computation of dipmeter logs digitally recorded on magnetic tapes: J. Petr. Tech. **14**(7), 771–782.

Muskat, M., 1937, Use of data on the build-up of bottom-hole pressures: Transact., Am. Inst. Min. Metallurg. Petr. Eng. **123**, 44–48.

Nagao, M., and Matsuyama, T., 1980, A Structural Analysis of Complex Aerial Photographs: Plenum Press, New York.

Neidell, N.S., 1969, Ambiguity functions and the concept of geological correlation: in Symposium on Computer Applications in Petroleum Exploration, D. Merriam (Ed.): Kansas Geol. Survey, Computer Contributions 40, pp. 19–29.

Nevatia, R., 1978, Evaluation of a simplified Hueckel edge-line detector: in Digital Image Processing, H.C. Andrews (Ed.): IEEE Computer Society, pp. 510–516.

Olea, R.A., and Davis, J.C., 1986, An artificial intelligence approach to lithostratigraphic correlation using geophysical well logs: paper SPE 15603 presented at the 61st Annu. Tech. Conf. Exhib., Soc. Petr. Eng., New Orleans, LA, October 5–8.

Pettijohn, F.J., and Potter, P.E., 1964, Atlas and Glossary of Primary Sedimentary Structures: Springer-Verlag, New York.

Plumb, R.A., and Luthi, S.M., 1986, Application of borehole images to geologic modeling of an eolian reservoir: paper SPE 15487 presented at the 61st Annu. Tech. Conf. Exhib., Soc. Petr. Eng., New Orleans, October 5–8.

Poelzleitner, W., 1986, A Hough transform method to segment images of wooden boards: Eighth Int. Conf. Pattern Recognition, Paris, October 27–31, 262–264.

Rieuwerts, H., 1989, GEOLOGIX, an interactive knowledge-based well correlation system: Proc. Second Conf. Artificial Intelligence Pet. Exp. Prod., College Station, TX, May 17–19, 177–186.

Rosenfeld, A., and Kak, A.C., 1982, Digital Picture Processing: Academic Press, San Diego, CA.

Rudman, A.J., and Lankston, R.W., 1973, Stratigraphic correlation of well logs by computer techniques: AAPG Bull. **57**(3), 577–588.

Rudman, A.J., Blakely, R.F., and Henderson, G.J., 1975, Frequency domain methods of stratigraphic correlation: Proc. Offshore Tech. Conf., Dallas, TX, 2, 265–277.

Schlumberger, Formation MicroScanner Service, Schlumberger Marketing Document 1986 M-089100 SMP-9100.

Schlumberger Document SMP-7002, 1986, Dipmeter Interpretation Fundamentals.

Schwarzacher, W., 1964, An application of statistical time-series analysis of a limestone-shale sequence: J. Geol. **72**(2), 195–213.

Selley, R.C., 1986, An Introduction to Sedimentology: Academic Press, London.

Serra, O., 1985, Sedimentary Environments from Wireline Logs: Schlumberger.

Startzman, R.A., and Kuo, T.-B., 1987a, An artificial intelligence approach to well log correlation: Log Anal. March–April, 175–183.

Startzman, R.A., and Kuo, T.-B., 1987b, A rule-based system for well log correlation: SPE Formation Evaluation, Sept. 311–319.

Thadani, S., and Bengston, A., 1988, ESCAT—An expert system for dipmeter log interpretation: presented at the Conf. A. I. Petr. Expl. Prod. College Station, TX, June 8–10, 6–7.

Tucker, L.W., 1984, Control strategy for an expert vision system using Quadtree refinement: Proc. Second Workshop on Computer Vision: Representation and Control, 214–218.

Vincent, Ph., Gartner, J.E., and Attali, G., 1979, An approach to detailed dip determination using correlation by pattern recognition: J. Petr. Tech. Feb. 232–240.

Warnken, D.K., 1988, An expert system advisor for well log quality control: Masters Thesis, Texas A&M University, College Station, TX.

Warnken, D.K., Startzman, R.A., and Kuo, T.-B., 1988, Well log quality control: presented at the Conf. Artificial Intelligence Petr. Exp. Prod., June 8–10, Texas A&M University, College Station, TX.

Weszka, J.S., Dyer, C.R., and Rosenfeld, A., 1986, A comparative study of texture measures for terrain classification: IEEE Transac. Syst., Man, Cybern. $6(4)$, 269–285.

Wong, S.A., Startzman, R.A., and Kuo, T.-B., 1989a, A knowledge-based approach to the interpretation of wellbore images: Proc. Second Conf. Artificial Intelligence Petr. Exp. Prod., College Station, TX, May 17–19,35–42.

Wong, S.A., Startzman, R.A., and Kuo, T.-B., 1989b, A new approach to the interpretation of wellbore images: paper SPE 19579 presented at the SPE Annu. Tech. Conf., San Antonio, TX, October 8–11.

Wong, S.A., Startzman, R.A., and Kuo, T.-B., 1989c, Enhancing Borehole Image Data on a High-Resolution PC: Paper SPE 19124 presented at the SPE Petroleum Computer Conference, San Antonio, TX, June 26–28.

Wong, S.A., Startzman, R.A., and Kuo, T.-B., 1989d, Texture Recognition in Images Using Energy Transforms: Technical Note ESL-89-4, Expert Systems Laboratory of the Crisman Institute for Petroleum Reservoir Management, College Station, TX.

Wu, X., and Nyland, E., 1986, Well log interpretation using artificial intelligence technique: Proc., Soc. Prof. Well Log Anal. 27th Annu. Logging Symp., Houston, TX, June 9–13, M.

Zemanek, J. and Caldwell, R.L., 1969, The borehole televiewer—A new logging concept for fracture location and other types of borehole inspection: J. Petr. Tech. June, 762–774.

Zoback, M.D., Moos, D., Mastin, L., and Anderson, R.N., 1985, Well bore breakouts and in situ stress: J. Geophys. Res. **90**, 5523–5530.

3
Intelligent Knowledge Based Systems and Seismic Interpretation

I. Williamson

An intelligent knowledge-based system (IKBS) consists of a knowledge base and an inference mechanism (plus an optional, possibly temporary, working database and a user interface). An IKBS may or may not be the same thing as an expert system. The IKBS has existed for some time but applications in geophysics are slow to appear. Seismic interpretation, which is almost entirely knowledge based, should yield readily to the IKBS approach, especially now that multitasking, dedicated interpretational workstations are becoming commonplace. An illustration is given using some sequence stratigraphy definitions and rules written in PROLOG for accessing the definitions.

Introduction

Intelligent knowledge based systems (IKBS) are one aspect of that subdiscipline of computer science which is loosely referred to as "artificial intelligence," an area of computer science that has existed since the late 1950s. It is difficult, if not impossible (and probably unnecessary), to find a consensus view of what exactly constitutes artificial intelligence (a task that assumes as a basic premise that we understand exactly what is meant by "natural" intelligence, no simple thing in itself). Because this is not the proper place for the pursuit of this philosophical and elusive goal, two definitions of what artificial intelligence may be are given:

1. "Artificial Intelligence is the study of how to make computers do things at which, at the moment, people are better" (Rich, 1983).

2. Artificial intelligence is the "field of research concerned with making machines do things that people consider to require intelligence. There is no clear boundary between psychology and Artificial Intelligence because the brain itself is a kind of machine" (Minsky, 1987).

Note that while these two definitions may overlap, to a considerable extent they are not coincident. For example, the first definition would include teaching machines to play golf, an activity at which humans are currently better than machines, but it could be argued—with apologies to golfers—that golfing is excluded by the second definition as it is not considered to require intelligence. Similarly, the first definition appears as if it may exclude psychology. On the basis of both of these definitions, however, the creation of a computer hardware and software combination that is able in some way to undertake or assist in seismic interpretation can be seen to be a task that falls within the realm of artificial intelligence.

Two basic artificial intelligence approaches to seismic interpretation are possible: (1) pattern recognition or image processing, and (2) the IKBS. Only the latter is pursued in detail here. Neither approach has received significant attention, although mentioned occasionally in print (Williamson, 1989). MYCIN, the most famous implementation of an expert system, may be consulted for advice on the diagnosis and treatment of bacterial infections of the blood. The next best known, the geological system PROSPECTOR, advises on the most favorable location for exploration for a fairly small range of ore minerals. EXPLOR, a second geological system, advises on hydrocarbon exploration in channel sandstones. IKBSs and expert systems are potentially useful both as consultation systems and as teaching tools; both uses are relevant in the context of seismic interpretation. Seismic interpretation is an area that should richly reward the IKBS approach because of its intensively knowledge-based nature. Unfortunately seismic interpretation suffers from the problem that much of its knowledge is empirical. In addition, very few published texts attempt to outline the fundamental rules of interpretation. One great advantage of a seismic interpretation knowledge base is that, while the state of the oil market has recently seen much seismic interpretation knowledge leave the industry permanently because of redundancy, an IKBS may retain all that is best of the experience, expertise and knowledge of former employees.

An IKBS may, in principle, be written in any language (EXPLOR is written in compiled BASIC), but in general the symbol-oriented languages, LISP and PROLOG, greatly facilitate implementation. As PROLOG, based on first order predicate logic, naturally incorporates the concept of querying a data (or knowledge) base, this is the more natural of the two languages for a seismic interpretation IKBS. Hardware for a seismic interpretation IKBS will almost inevitably include one of the several dedicated interpretation workstations now in use, possibly linked to a mainframe. The basic requirements here are for a great deal of high-speed permanent storage space (hard or optical disk, preferably extensively cached in several megabytes of fast RAM) and a fast CPU. These requirements are readily available so hardware is not the problem in implementing a seismic interpretation IKBS. Finally, as an example, some of the basic definitions of sequence stratigraphy are given as PROLOG clauses.

Seismic Interpretation Procedure

A seismic interpreter examines the data (the seismic section, velocity analysis, or whatever is available), either on paper or using an interactive interpretation workstation, for a pattern or a collection of patterns referred to as geophysical attributes, or attributes in short. These patterns are effects for which, unless the interpreter has confused signal with noise, there should be identifiable causes. The interpreter must then match the list of effects against a range of possible causes. If more than one cause can be identified, the interpreter needs a strategy for resolving the additional problem; this task will often involve a directed search for further attributes which in general the interpreter knows are associated with the list of tentatively identified geological causes. With the application of expertise, this second-level search will resolve the ambiguity and a conclusion may be reached.

Computer-Assisted Interpretation

Computer assistance with a variety of complex tasks is commonplace; it is, therefore, only natural that we should ask ourselves whether the computer may assist us with seismic interpretation. This proposed interpretational strategy may be implemented using a computer or workstation in one of the two artificial intelligence approaches referred to previously. The first approach is to make the entire procedure automatic where we rely on the machine to detect the patterns in the data and then determine, or interpret, their significance according to some predetermined criteria. Pattern recognition of some type is an approach that has been tackled by many researchers in artificial intelligence and which, in many contexts, is remarkably intractable; much has been borrowed from experimental and theoretical work in computer vision. It is far easier to program a machine to undertake, for example, Fourier or finite element analysis, tasks that humans find "hard," than it is to program a machine to recognize trees, pigs or unconformities on seismic sections, tasks that humans normally find trivial. However, the rate at which currently defined problems in this area are being tackled is bound to increase substantially in the future as various types of low-cost machines utilizing high-speed microprocessors wired together to permit a great deal of parallelism and simultaneity of processing appear. This type of approach in geophysics is currently experimental. However, simple pattern recognition has been in use for many years in the automatic interpretation of velocity analyses—velocity gathers, stacks, and spectra—although the final opinion has always been provided by a human interpreter. The only paper utilizing artificial intelligence techniques given at the 1989 E.A.E.G. meeting is concerned with this very area (Coppens, 1989). Because very little has appeared in the geophysical press no references are given here.

The Knowledge Based Approach

The second and simpler approach, the one discussed at some length here, should yield useful and valuable results much more quickly. In this approach, we rely on a human rather than a machine to do what people are extremely good at doing, and what is extremely difficult to program computers to do: that is, to recognize the patterns in the data. The user may then input these patterns or geophysical attributes (as text, menu choices or geophysical "icons" that have been "clicked" on with a mouse) to programs that compare them in both general terms and also for the specific problem with stored seismic interpretational data and geophysical (wire line log, for example) and geological knowledge derived from the user, from other seismic interpretation practitioners and experts, and from the geophysical and geological literature. Such a knowledge base may contain considerably more expertise than that possessed by an individual human expert and each (given that both are fully interactive) may learn from the other. This is the intelligent knowledge base system (IKBS) approach.

What Constitutes Knowledge?

Before a formal definition and description of an IKBS and expert systems are given, we need to define what exactly is knowledge. Definitions vary, but all seem to imply that knowledge consists of more than just facts. Facts, in the context of computing, are synonymous with data. Thus, a database consists simply of a collection of facts. Frost (1986) tells us that "knowledge is the symbolic representation of aspects of some named universe of discourse," and that data is "the symbolic representation of simple aspects of some named universe of discourse." This definition means in practice that the terms data and facts are effectively interchangeable and that, as Black (1986) says,

"'Knowledge' implies an imposing of organizing principles on these isolated facts to provide an understanding of the world which is predictive and embraces or explains the individual facts." These organizing principles may be better understood as rules, in the sense that the rules may be seen as generalizations which relate simple facts and enable us to infer from them new facts. Knowledge may thus be viewed as consisting of a combination of facts and rules.

What Constitutes an IKBS?

As the name suggests, an IKBS is the knowledge equivalent of a database. The basic requirements for an IKBS are a knowledge base and a means of drawing inferences from the knowledge (we may assume that there is also a means of communicating with the outside world—a user interface). This inference mechanism is sometimes referred to as an inference engine. In addition, we may wish to update the knowledge base from time to time, usually in one of two ways. Firstly, we may modify the knowledge base permanently—to add, delete, or change knowledge—as knowledge is dynamic. Secondly, we may temporarily store knowledge specific to the current usage of the IKBS, but which has no long-term or global significance and which therefore may be erased or otherwise discarded when that use terminates. An IKBS may also include elements for checking the correctness (in a semantic sense), the integrity and the consistency of its knowledge base. These are the essential elements of an IKBS.

An IKBS versus an Expert System

An expert system may or may not be more than an IKBS. The two terms are sometimes used as if they are synonymous. Some authors use the term "expert system" as if it means more than is implied by "IKBS." Whereas a basic IKBS draws inferences, reaches conclusions, and requests input from the user, an expert system should justify its line of reasoning and explain why it has requested a particular input (and perhaps also explain why a particular conclusion was not reached or offer any other relevant explanation). Thus, an expert system may in many ways mimic a human expert, a category to which many seismic interpreters would no doubt agree that they belonged. This paper is directed toward a simple IKBS rather than an expert system purely for pragmatic reasons—it is much simpler to illustrate the main aspects of an IKBS. The important elements of an expert system, written in PROLOG, are given in Hammond (1984). Hammond uses a dialect of PROLOG called micro-PROLOG (the dialect is, in fact, in a "subdialect," which runs under a front end called SIMPLE, which is itself a program written in, running under and distributed with micro-PROLOG).

Does Seismic Interpretation Need an IKBS Approach?

Does seismic interpretation need an IKBS approach? Almost inevitably the answer must be "yes and no." Seismic reflection data have been interpreted successfully for almost 70 years. The most obvious successes have been, and will probably continue to be, in the exploration for oil and natural gas but seismic data are increasingly being used to great effect by soft rock, structural, crustal and mining geologists among others. To that extent, an IKBS may seem irrelevant to many and they may well be right. However, the success of a seismic interpretation is often something that may be tested only approximately; in many cases no independent test is available and where such a test is available—say, a well—there is always a convincing explanation for any discrepancies that may come to light (the loggers, the geologists, the geochemists, the complex velocity distribution/ raypaths, the data processing, the poor signal-to-noise ratio, etc.). To this extent, seismic interpretation is not a science, but rather an art or, perhaps, a series of assertions. The reason that geophysicists have apparently found so much oil, for example, may well have more to do with luck and the fact that it becomes progressively more difficult to make stupid mistakes as a basin becomes more mature in exploration terms. Perhaps we should ask why so many dry holes are still being drilled despite the fact that the resolution of our data is at least an order of magnitude better than it was 30 or 40 years ago. Who really knows the nature of the lower crust to a resolution better than that apparently available

from seismic reflection data? What independent tests may we arrange for interpretations of deep seismic reflection data? To complicate matters, more and more non-geophysical geologists (and, it has to be admitted, some non-geological geophysicists) are interpreting seismic data and, for those who do not have a good grounding in both disciplines, it is easy to make mistakes.

This suggests that interpretations may be improved if the interpreter and the interpretation being produced may be constrained or calibrated in some way. In an ideal world this may best be done by an expert being permanently on call to offer help and advice, for example, the senior staff in a company. Where this is not possible, an IKBS, or expert system, with a broad and well-founded knowledge base offers a viable alternative. Indeed, the machine may surpass the expert because, given time, a knowledge base may contain the expertise of several experts.

Languages for Implementing an IKBS

An IKBS may, in principle at least, be implemented in any computer language. As with any computing problem, though, some languages are more appropriate than others. The EXPLOR expert system is written in compiled BASIC. This is not because this language is a particularly good medium for the implementation of artificial intelligence programs but because much of the input to EXPLOR is numerical and these data are resampled and contoured using polynomial trend surfaces. A scientific language is therefore an appropriate solution to this particular problem (no doubt FORTRAN, C or Pascal could have been used equally well if the authors of EXPLOR had so chosen). As we have seen earlier, an IKBS consists primarily of an inference mechanism and a knowledge base, so the problem of constructing an IKBS may be reduced to the two sub-problems of writing an inference mechanism and assembling the knowledge base.

The inference mechanism may be written in any language, but this task is greatly simplified or nonexistent if we use a language that is either designed for the implementation of inference mechanisms or is itself essentially an inference mechanism. This implies of course that we use one of the so-called artificial intelligence languages: specifically, either LISP or PROLOG. Both of these languages are extensible and share the use of the list as a fundamental data structure (in fact, the name LISP is an acronym for LISt Processing; PROLOG is derived from PROgramming in LOGic). Both are ideally suited for the manipulation of symbols and text rather than numbers. Other than this the two languages are rather different; which we choose to use will depend to a great extent upon personal taste (and, to a lesser extent, on geography—LISP is far more common in the United States whilst PROLOG is heavily used in Europe).

The debate about which is a better language for the implementation of an IKBS is one that will not be entered here. The interested reader should have no difficulty finding a plethora of books, tutorial, philosophical and pragmatic, about both languages. A useful start could be made with Baron (1988). What can be said is that PROLOG is easier to use and to read, does more or less all that LISP can do, if in sometimes different ways and uses far fewer brackets. The code that is included in the following therefore is written in PROLOG, and specifically in the Edinburgh syntax (see Clocksin and Mellish, 1984), which is the nearest approach there is to a standard dialect for the language; no guarantees are offered about the semantic correctness or bug-free nature of this code, though.

Seismic Knowledge Structure

Just as data have well-defined structures into which it may be fitted (sequential files, relational tables, etc.) there are also pre-defined structures into which knowledge may be fitted. These include production rules, Horn clauses, frames (also called slot and filler representations), semantic nets (as used by PROSPECTOR), various sorts of logic and others. Useful discussions may be found in Frost (1986) and Black (1986). It is too early to hazard a guess as to which will be the most relevant for the representation of seismic reflection knowledge, but it is likely that a number of different structures will be used simultaneously, particularly as no interpreter will ever need all seismic interpretation knowledge to be available at once.

The easiest structure to understand is the production rule; such a rule has two parts:

1. the condition or conditions (also called premises and antecedents), and
2. the conclusion or conclusions (or action or consequents).

A rule without conditions is a fact. We may summarize a production rule as:

    IF   condition(s)
THEN   conclusion(s)

as in

    IF A or   IF A and B or   IF W or X
THEN B   THEN C and D   THEN Y or Z

In order to simplify the knowledge structure so that it may be verified and maintained more easily it is preferable to split conjunctions (logical ANDs) of conclusions and disjunctions (logical ORs) of both conclusions and conditions, so that the second and third rules above would become:

    IF A and B   and   IF A and B
THEN C   THEN D

and

    IF W   IF X   IF W and   IF X
THEN Y   THEN Y   THEN Z   THEN Z

A geophysical example is

    IF the ITT is 67 µs/ft
        and the ITT is constant
        and the rock has flowed
THEN the lithology is halite

Seismic Interpretation Knowledge

There is no doubt that much seismic geophysical knowledge is not only empirical (and sometimes ad hoc) but also poorly documented. Most geophysical publications are technical and of little or no relevance to the interpreter (all of us forget some of the time, and some of us forget all of the time, that without an interpretation geophysics is worthless—it is only a means to an end, not an end in itself). Case histories are usually specific, with little of more general applicability. Geological knowledge, on the other hand, is generally well documented. Thus, as geological texts we find books such as Reading (1986) and Selley (1982) containing a wealth of information, while in the geophysical domain there are many books that promise a great deal by including the word "interpretation" in their title but that deliver very little. Notable exceptions to this are Badley (1985) and Jenyon and Fitch (1985).

The problem of seismic knowledge collection can be solved if we remember that knowledge is dynamic and almost always partial and that many things in geophysics and, to a different extent, geology can only be known approximately. Much seismic interpretation knowledge will need to be gathered from experts, some of whom fortunately teach short courses. What is much more abstruse is how the inevitable lacunae and approximations in the knowledge we have obtained affect the resulting interpretation.

An Example Using Sequence Analysis Knowledge

An example of an IKBS is constructed using some of the basic definitions given in Van Wagoner et al. (1987). The structure used is fairly simple and is adopted only for illustration; a "feature" is defined here as comprising a name followed by a list of attributes (PROLOG uses the square bracket— [ ]—to enclose lists). To avoid the lists becoming excessively long and to build the maximum flexibility into the knowledge base, many of the attributes for a given feature are themselves features, so some features are compound and some simple. A strategy for dealing with this complication is given later.

The first definition given is that for a sequence:

feature(sequence, [boundaries1,
                   systems__tract]).

This states that internally a sequence is comprised of system tracts and that it is delimited by type 1 boundaries (this designation of boundaries is used only for convenience as we will define other types of boundary later). We next give a definition for this boundary type:

feature(boundaries1, [unconformity,
                   correlative__conformity,
                   chronostratigraphic,
                   marine__flooding__surface,
                   overlying__strata__onlap]).

The definition for marine flooding surfaces is given below. In a fully interactive system, the IKBS should recognize that there are no definitions for "unconformities," "correlative conformities," and "overlying strata onlap" and query the user. This is easily accomplished, particularly if we include a list of all those things that the interpreter may be asked about the presence or absence of, as in, say ask__if__present (unconformities), and so on; the user responses may then be added to a temporary working database. We shall bear this in mind for the remainder of the definitions that follow but avoid this further complication here, and go on to define a systems tract:

feature(systems__tract, [parasequences,
                   parasequence__set]).

This is straightforward, so we deal with parasequences next:

feature(parasequences, [progradational,
                   shoal__upwards,
                   beds__relatively__
                   conformable,
                   beds__genetically__
                   related,
                   boundaries2]).

Parasequence boundaries have their own definition:

feature(boundaries2, [marine__flooding__surface,
                   correlative__conformity,
                   chronostratigraphic,
                   may__be__sequence__
                   boundary,
                   may__be__downlap__
                   surface]).

The definition of a marine flooding surface is rather long-winded and not necessarily easy to apply to a seismic section:

feature(marine__   [chronostratigraphic,
   flooding__surface,   water__depth__increases,
                   may__be__minor__
                   submarine__erosion,
                   no__subaerial__erosion,
                   no__basinward__facies__
                   shift,
                   may__be__nondeposition,
                   may__be__minor__hiatus,

```
                sequence__boundary__if__
                   onlapped,
                surface__planar,
                commonly__only__minor__
                   relief,
                may__correlate__coastal-
                   plain__surface,
                may__correlate__shelf__
                   surface]).
```

The following definitions should by now be self-explanatory:

```
feature(parasequence__set, [two__or__more__
                        related__para-
                        sequences,
                        parasequences,
                        stacking__pattern,
                        boundaries2,
                        boundaries3]).

feature(stacking__pattern, [progradation,
                        retrogradation,
                        aggradation]).
```

[*Note*: Stacking patterns are defined in part in Van Wagoner et al. (1987) by wireline log response—SP and resistivity in particular; this aspect is ignored here.]

```
feature(boundaries3,[separates__parasequence__
                     stacking__patterns,
                     may__be__sequence__
                     boundary,
                     may__be__system__tract__
                     boundary]).

feature(progradation,[basinwards__shift__in__
                      facies,
                      deposition__rate__exceeds__
                      accommodation]).

feature(retrogradation, [landward__shift__in__
                         facies,
                         deposition__rate__
                         less__than__
                         accommodation]).

feature(aggradation, [no__shift__in__facies,
                      deposition__rate__equals__
                      accommodation]).
```

The foregoing should illustrate the basic principles of the construction of a seismic interpretation knowledge base. We now need a rule that will determine if a feature is simple or compound—it is compound if one or more of its attributes is a feature that also appears in the knowledge base. We test for this as follows:

```
compound(Feature1):-
   feature(Feature1,Attributes1),
   feature(Feature2,[__]),
   not Feature1=Feature2
   in-list(Feature2,Attributes1).
```

The underscore in the second condition above indicates that we do not care what occupies this position. The negated condition is included to avoid any unnecessary checking. We do not need to write an explicit rule to test if a feature is simple, i.e., none of its attributes is represented elsewhere as features, because any feature that is not compound must by definition be simple. The rules for "in-list" are as follows [these are standard and appear in many texts on PROLOG—for example, they appear as "member-of" in Conlon (1985) and as "member" in Clocksin and Mellish (1984)]:

```
in-list(Feature,[Feature1__]).

in-list(Feature,[__1Tail]):-
   in-list(Feature,Tail).
```

These two rules say that a feature is a member of a list either if it is the same as the head (first member) of the list or if it is in the remainder of the list. Thus "in-list" recursively compares the feature with the first member of a list which gets progressively shorter as its first member is removed.

A Simple Example IKBS

Given that we now have an admittedly rather simple knowledge base (and knowledge structure) consisting of several facts and a couple of rules, how may we access it? How may we query and consult this information? We need to write a small number of PROLOG clauses that will do this for us. The first requirement is that we need to be able to tell the machine that we need advice. It should then respond by asking us to specify the type of query we wish to make and search the knowledge base for the answer. We may do this as follows:

```
advise:-
   ask(Query-type),
   give-query-answer(Query-type),
   advise.
```

(*Note*: Words beginning with a lower case letter are constants, words beginning with an upper case letter are variables, the symbol ":-" means "if," the comma separating conditions is to be read as "AND," and a disjunction of conditions may be shown by separating them with a semicolon, ";".)

The "advise" predicate is recursive in that, rather than terminate on completion, it calls itself. Thus we have created an advisory loop that will repeat forever. Later we will introduce a mechanism for breaking out of this if we desire. This form of recursion, where the recursive call is the last of a conjunction of conditions, is called "tail recursion"; in a well-behaved version of PROLOG, this type of recursion uses a constant amount of memory and hence can repeat indefinitely.

The clause for getting the query type is straightforward. We give the user only four options:

1. Describe the geophysical attributes that may generally be associated with a particular geological phenomenon. We thus wish to request a list of the attributes (seismic expression) of a particular feature (reef, halokinesis, system tract, etc.). This type of query is using the system in a purely reference capacity. In the examples given above, such a feature may be simple, in which case we will confine our program to listing the attributes stored for it in the knowledge base, or it may be compound, in which case we will need to list its "sub-features" and their attributes.

2. Suggest a possible interpretation. In this case, from a user-supplied list of seismic attributes, the IKBS will ascertain the feature or features that may give rise to them. This is a use that is more nearly what we generally mean by the term "seismic interpretation," the process whereby we arrive at an opinion as to the geological cause of a, usually, small number of geophysical effects. Geophysics is nothing if not ambiguous, of course (why else would we need geophysical interpreters?), so our observed collection of effects (what we are calling "attributes" here) may permit the identification of more than one cause. For example, we may record velocity pull-up at the base of a layer that shows local thickening immediately above this velocity effect. There is obviously a local lateral velocity anomaly in such circumstances. However, in the event of the lack of any other geophysical evidence we are left with a range of possible interpretations — reef, halite, channel, and so on. In such circumstances our IKBS should recognize that there is more than one possible explanation and list these. If we wish it to act more intelligently, however, it should be able to ask us if so-far unmentioned attributes, which are recorded as possibly being associated with our tentatively identified geological causes and which we may have no knowledge of ourselves or which may have (temporarily) slipped our memory, may be present in order to reduce the number of possible suggested interpretations.

The code listed here as an example of this type of approach is rather more trivial than this in that it will only suggest a possible cause (or causes) for observed effects if the user-supplied list of attributes exactly matches (down to the supplied number and order of attributes and the case of the letters used) the list stored in the knowledge base. Ideally the IKBS will recognize that the input list of attributes may only partially match a stored list, regardless of order and letter case. In such a situation there are two possible lines of action. First, the IKBS may suggest that we examine the data for attributes in its internal list but not in the input list. We may thus refine and focus better our interpretation (and perhaps improve our interpretational skills by learning a little more). Second, there may be attributes in our input list that the IKBS does not have in the stored list or lists that best match this input. In this case, a truly intelligent knowledge base will ask us if this discrepancy represents new knowledge that needs to be added to the knowledge base. This type of programming, where the program modifies itself (known in the logic programming context as "metalogical programming"), is very easy to accomplish in PROLOG. Thus, the program with a little help from the user, may learn from experience, an activity that we normally associate with intelligence.

3. List all those features known to the knowledge base. The IKBS should be able to inform us of the broad outlines of its contents so that we do not waste time formulating queries for which the system does not have an answer.

4. Exit. We will need to leave the advice session at some point, no matter how engrossing.

These four options are handled by the "ask" predicate:

```
ask(Query-type):-
  nl,
  write('Advice on Feature or Attributes?'),
  nl,
  write('Type F for Feature'),
  write('Type A for Attributes'),
  write('Type L for List of features'),
  write('Type E for Exit to operating system'),
  get(Query-type).
```

(*Note*: "nl," "write" and "get" are built-in predicates that produce a new line, send the term in brackets to the current output stream, e.g. the screen, and get the ASCII code of the next printing character from the current input stream — usually the keyboard — respectively.)

It can be seen that "get" returns one of four numbers: 70 if the input is "F," 65 for "A," 76 for "L," and 69 for "E." Note that it does not check for out of range or wrong case input. If the user input is "F," we print a request for the name of the feature then print its attributes:

```
give-query-answer(70):-
  nl,
  write('Type feature name, then full stop and
    CR'),
  read(Feature),
  feature(Feature,Attributes),
  write('The known attributes of the feature are:'),
  nl,
  print(Feature,Attributes).
```

(*Note*: "read" is a built-in predicate that reads the next term — terminated by a full stop and a non-printing character such as a carriage return — from the current input stream, i.e. usually the keyboard; "print" is a predicate we need to define here.)

We need to pass both the feature and its attributes to "print" because we need to test to see if the feature is compound before we print its attributes:

```
print(Feature,[   ]):-
  nl.

print(Feature,Attributes):-
  not compound(Feature),
  print(Attributes).

print(Feature,[Head/Tail]):-
  compound(Feature),
  write('   ',Head),
```

```
  feature(Head,Attributes),
  print(Head,Attributes).

print([   ]):-
  nl.

print([Head/Tail]):-
  write(Head),
  nl,
  print(Tail).
```

Of these five clauses for "print," the first gives a new line after we have exhausted all attributes for a compound feature, the second causes a list of attributes for a simple feature to be printed using the fourth and fifth clauses, the third recursively prints the attribute list for a compound feature, the fourth gives a new line when printing exhausts a simple list of attributes, and the last recursively prints a simple list. The clause for "compound" was given earlier.

When the user input is "A," for a list of attributes for a user-defined feature, the necessary actions as detailed here are relatively simple:

```
give-query-answer(65):-
  nl,
  write('Type the list of Attributes separated by
    commas'),
  write('   and terminated by a full stop and
    CR'),
  read(Attributes),
  feature(Feature,Attributes),
  nl,
  write('The feature with these attributes is:'),
  nl,
  write(Feature).
```

Obviously, this clause makes no allowance for features being compound. In addition, as has been said, this clause will succeed only if the user inputs the attributes in a fashion that matches exactly a stored set of attributes. The reader is invited to make modifications to correct for these deficiencies.

The clause that produces the desired output when the user requests a list of all features known about by the knowledge base is written in a way that allows us to use the predicate "append" — this is again a standard which appears in all the texts on PROLOG:

```
give-query-answer(76):-
  nl,
  feature(Feature,Attributes),
  append(Feature,Temporary-feature-list,Feature-
    list),
  print(Feature-list).

append([   ],List,List).

append([Head/Tail],Old-list,[Head/New-tail]):-
  append(Tail,Old-list,New-tail).
```

The last allowed user input here is "E" if the user wishes to exit—this calls the built-in predicate "halt." This fairly profound exit quits not only the program, but also PROLOG, and drops the user back into the operating system (or whatever the underlying software is):

```
give-query-answer(69):-
  halt.
```

This completes the simple IKBS.

Conclusions

Techniques common in artificial intelligence may be brought to bear on problems frequently experienced by the seismic interpreter. In addition, a seismic interpretation knowledge base may potentially contain the expertise of many interpreters. These are pursuits that are both possible and desirable. There are no great problems to be solved. All that remains is for someone to accumulate and code the knowledge—the body of facts and rules underlying seismic interpretation.

References

Badley, M.E., 1985, Practical Seismic Interpretation: Reidel, Dordrecht.

Baron, N.S., 1988, Computer Languages: Penguin, New York.

Black, W.J., 1986, Intelligent Knowledge Based Systems: Van Nostrand Reinhold, New York.

Clocksin, W.F., and Mellish, C.S., 1984, Programming in PROLOG: Springer-Verlag, Berlin.

Conlon, T., 1985, Learning Micro-PROLOG: Addison-Wesley, Reading, MA.

Coppens, F., 1989, An expert system for picking velocity spectra: in Technical Programme and Abstracts of Papers, 51st. Mtg. Eur. Assoc. Exp. Geoph.

Frost, R.A., 1986, Introduction to Knowledge Base Systems: Collins, London.

Hammond, P., 1984, micro-PROLOG for expert systems: in micro-PROLOG: Programming in Logic, K.L. Clark and F.G. McCabe (Eds.): Prentice-Hall, Englewood Cliffs, NJ.

Jenyon, M.K., and Fitch, A.A., 1985, Seismic Reflection Interpretation, 2nd rev. ed.: Gebruder Borntraeger, Stuttgart.

Minsky, M., 1987, The Society of Mind: Pan, London.

Reading, H., ed., 1986, Sedimentary Environments and Facies, 2nd. ed.: Blackwell Scientific Publications, Oxford.

Rich, E., 1983, Artificial Intelligence: McGraw-Hill, New York.

Selley, R.C., 1982, An introduction to Sedimentology, 2nd ed.: Academic Press, New York.

Van Wagoner, J.C., Mitchum, R.M., Jr., Posamentier, H.W., and Vail, P.R., 1987, Part 2: key definitions of sequence stratigraphy: in Atlas of Seismic Stratigraphy, Vol. 1, A.W. Bally (Ed.): Am. Assoc. Petr. Geol. Studies in Geology 27: Am. Assoc. Petr. Geol.

Williamson, I., 1989, Seismic interpretation: An egg in search of a chicken: in 75 Years of Progress in Oilfield Science and Technology, M.A. Ala, H. Hatamian, G.D. Hobson, M. King, and I. Williamson (Eds.): Balkema, Rotterdam.

4

An Expert System for the Design of Array Parameters for Onshore Seismic Surveys

Barrie K. Davis

Introduction

With the widespread acceptance and use of powerful, small, and inexpensive microcomputers in the oil industry it is surprising that the design of the arrays of geophones and shots used in seismic onshore shooting is sometimes still a back-of-the-envelope exercise. The importance of this initial planning stage in a survey cannot be overemphasized since despite the wonders of present-day processing techniques the old adage of "garbage-in, garbage-out" still applies. Even the stack array technique (Anstey, 1986) should give better results if the initial field record can be improved by using appropriate arrays for geophones and source (Vermeer 1990).

Array design is an ideal candidate for the use of a simple knowledge-based expert system. The process of designing arrays is deductive in that one starts from a series of facts concerning the equipment, field parameters, and target, and ends with a suggested configuration of geophones and source elements. Changes made in the input parameters will immediately affect the results, the examination of which may necessitate changes in input.

PROSEIS is a knowledge-based expert system that can be used on a personal computer in the office or in the field to help in the design of seismic arrays. The main parameters needed for the design are taken from a noise study made in the survey area together with the geophysical specification of the target formations, while extra information concerning the field geometry or logistics is requested at appropriate points.

The program is interactive so that changing any parameter immediately affects the results. There are copious help screens with explanations of terminology and methods of calculation that enable the program to be used as a tutorial in array design. For any array suggested by the program the user can see a response-plot on the screen, check the combined response of source and geophones, and obtain a hard copy on a graphics printer. When satisfied with the results, the explorationist can produce a report showing details of the survey together with copies of the individual and combined responses.

Special attention has been given to the user interface, so that an explorationist does not need to learn complicated commands or to be a computer expert to get the full benefit from the program.

Array Theory

A seismic array is a group of source or detector elements arranged on the surface of the ground in such a way as to attenuate coherent noise by acting as a wavelength filter (Savitt et al., 1958). In other words, in order to discriminate between the signal and the noise we use their different wavelengths as they appear to the array of geophones. Source and detector arrays are mathematically identical, and for simplicity we will only consider detector elements (geophones) and one frequency component.

The signal of interest, the reflected seismic wave, comes from below with its wavefront at a low angle to the surface and has a high apparent velocity along the array of geophones. If V is the seismic velocity at the surface, f is the frequency, and D is the angle between the wavefront and the surface, then the apparent velocity along the surface is $V/\sin(D)$ and the apparent wavelength as seen by the array is $V/(f\sin D)$. Clearly, when the reflected energy arrives in a vertical direction, the angle D is zero and the apparent velocity along the array is infinitely large since the seismic wave arrives at all geophones instantaneously. The wavelength is also infinitely large.

On the other hand, the coherent noise, ground-roll or air-wave, travels with velocity V_s horizontally along the surface so that the angle D is $90°$, and the apparent wavelength is equal to the actual wavelength which is given by V_s/f. Thus horizontally traveling waves generally have shorter wavelengths than the signal, and the array is designed so as to attenuate them.

To calculate the response of the array the seismic signal is represented by a sum of harmonic waves, and each of these components is considered separately (Smith, 1956). The wave will arrive at successive geophones with time lags that depend on the apparent wavelength and the geophone separation, and these will determine the degree to which the wave will be reinforced or canceled. The mathematics are simpler if only symmetrical arrays are considered since then only the cosine part needs to be summed, the sine parts cancel.

If the output of a geophone is given by $A\cos(wt)$, and the phase shift is p for geophone number r, the output of an array of n geophones will be given by

$$H(t) = \sum_{r=1}^{n} A(r) \cos[wt - p(r)]$$

Both the amplitude A and phase p of the output from the rth geophone are functions of r, the distance from the array center.

This expression can be evaluated and plotted by the following code fragment:

```
' Find total response if elements are in phase, i.e.,
' bunched.
total = 0                        ' initialize to 0
FOR r = 1 TO npops               ' add all weights
   total = total + weight(r)
NEXT r
' Calculate relative response and make a plot.
' xmax = number of x divisions required
' rfrac = fractional response of array
' f = scaling factor for x axis (to wavenumber)
' g = scaling factor for y axis (to decibels)
response=0                       ' initialize to 0
FOR x = 1 TO xmax                ' increment x by 1
   FOR r = 1 TO npops            ' add all responses
      response = response        COS(f ★ distance(r))
      + weight(r)
   NEXT
   rfrac = response/total        ' fractional response
   y = -g ★ LOG(rfrac ★
      rfrac)                     ' calculate y in db
   CALL DASHES(x,y,h)            ' plot a line to x,y
NEXT
```

The graphics function DASHES draws a dashed line from the previous point to the new point with coordinates x,y. The parameter h is passed in order to control the type of line, i.e., the number of pixels in and between the dashes. If $h = 0$ the line is continuous, and each time a response plot is added to a previous plot the value of h is incremented by 1 thus discriminating between responses.

The Expert System

The knowledge-based expert system has been one of the simplest developments in the world of "artificial intelligence" (James, 1988), and has been quick to catch the imagination of both software developers and computer users. Behind the hyperbole and exaggerations that often accompany the topic a knowledge-based expert system is simply a computer program that embodies facts, information, or data about a particular subject, and enables the user to reach conclusions based on a question-and-answer dialogue. The knowledge base can contain information gleaned from many papers and publications, from discussions between experts, and of course from many years of experience. The knowledge base encapsulates the specialist knowledge of many professionals in such a way as to make the data available to the nonexpert. In this accessibility lies power—but also danger.

An expert system should best be used by experts to extend their expertise. For example, one would not consider allowing a medical expert-system to be used by a neophyte to prescribe drugs, even though the machine could be programmed to print prescriptions, to yield an appropriate dose of penicillin, or even to give an antitetanus injection. Expert systems should ideally be considered as tools to be used by an expert to enhance performance, or, alternatively, as a learning tool by which a person can become more proficient in a subject.

A computer has no real intelligence. Calculating or reasoning better than a human being does not make a computer intelligent except in a very restricted sense, since everything the computer knows has been put there by human beings. The human has programmed the reasoning for it. A better less glamorous term than "artificial intelligence" might be "programmed reasoning." The same conceptual problem occurs when applying the terms "learn" and "think" to a computer, and perhaps the terms used should be "programmed learning" and "programmed thinking." The whole subject of trying to simulate human mental processes is surrounded by much jargon and mysticism, and in petroleum exploration one must never forget Pratt's adage (1952) "oil is found in the mind...."

We should, in fact, consider the relationship between artificial intelligence and real intelligence to be similar to the relationship between an artificial leg and a real leg, and think of an Expert System as a crutch that helps us walk in difficult or unknown terrain where without help we would be loath to venture.

Taking the above caveats into account the advantages over a consultation with human experts are clear: speed, availability, and cost.

Logical Structure

Expert systems are in practice based on the simple reasoning of a conditional statement that exists in all computer languages, the IF–THEN conditional. Thus after determining that we are going for a walk with the dog a British computer would ask:

Is it raining?

Our response could be processed as follows:

IF answer is "yes" THEN reply "Take an umbrella"
If answer is "no" THEN ask "Are there clouds?"

The intention being to continue the dialogue by asking questions about the clouds in order to determine if it might rain.

Of course an expert system which was programmed differently may have instead continued thus:

If answer is "no" THEN reply "Do not take an umbrella."

This latter expert system may simply have been badly programmed, or perhaps it was capable of learning from experience and would quiz us on our return, determine that we had been caught in a shower, and adjust appropriately the questions in the next session.

The structure of such a program is essentially a tree, where appropriate branches are chosen according to the user's answer to the computer's questions. The structure can be as simple or as complicated as is necessary to reach an appropriate result, with trunks, boughs, limbs, branches, twigs, shoots, sprouts, etc., together with jumps back from twig to bough for example. This tree with seven stages would have $2^7 = 128$ different sprouts if each stage gave only two choices. Three choices would give $3^7 = 2187$ possible outcomes, and larger examples where complicated jumps occur give rise to the appearance of artificial intelligence. Latest technological developments, such as the neural net, mimic the human brain by making such interconnections in the computer hardware rather than in the software, and with suitable programming may eventually enable better emulation of human intelligence.

Choice of Language

Since most computer languages are "problem oriented" in that each language is designed to be best for a certain type of application, the developer of an expert system must first make the choice of language. Thus a programmer may choose between several "artificial intelligence" languages, the best known being LISP and PROLOG with various implementations of each.

LISP is an acronym for LISt Processing, and was developed in the Massachusetts Institute of Technology in the early 1960s. A list is useful for storing data that must be regularly searched or sorted, and the LISP language is good for manipulating and comparing lists of any type of item and for determining whether particular items are present in each. LISP contains many functions that can be used to process lists. For instance a list 'A' may be defined as follows:

[SET 'A'(LINEAR TAPERED AREA_PATTERN)]

The function CAR then gives the first item on the list while CDR gives the truncated list. Thus the program CAR A would give (LINEAR) and CDR A gives (TAPERED AREA_PATTERN). One advantage of programming in LISP is that a function can be recursive, that is, the function may call itself.

The drawback of LISP on a PC for scientific work is that not all the mathematical and graphic functions necessary for creating a complete program are available. The language is text oriented, and would require a major effort to define all the functions needed to carry out all the tasks necessary for calculating array responses and plotting them on screen.

PROLOG is an acronym for PROgramming in LOGic and was designed in the early 1970s at the University of Aix-Marseilles specifically for use in the field of artificial intelligence. A PROLOG program consists of clauses rather than lists. An ordinary clause can be interpreted as a conditional or a procedure, so that

PATTERN(X) :- LINEAR(X), VIBROSEIS(X).

means "X is a pattern if it is linear and vibroseis," or "to prove X is a pattern show that it is linear and vibroseis."

Other types of clauses are the data clause and the query clause. Example of data clauses would be

METHOD(VIBROSEIS)
METHOD(TAPERED)

and a query clause could be

?-NOISE(X)

which would mean "find all X where X is noise."

After due consideration of the pros and cons of using LISP or PROLOG, I concluded that the disadvantages of lack of graphics and mathematics outweighed any advantages that might be gained from using the built-in logical operators. Most computer languages allow the building of similar logical facilities with greater or lesser effort, but constructing the missing mathematical and graphic functions could be a major undertaking. Add-on libraries for the personal computer versions of these languages that were investigated at the time were not sufficiently versatile to provide all the required services.

An alternative route would have been to use an expert-system shell into which the information could be entered, and in this way perhaps avoid writing from scratch the deductive or inferencing part of the program. Once again, at the time the program was researched in early 1987, the problem of mathematics and graphics appeared to be insoluble except for one or two expensive systems based on mainframe programs, which enabled the calling of subroutines written in other languages. Had I intended producing various expert systems in different subjects this may have been the way forward, but for the task in hand it would have been akin to using a sledgehammer to break a nut, with no certainty that at the end of the day the nut would break!

In the end the choice was simple. I would use the language that I knew best, and that was certainly suitable for the purpose.

Quickbasic

Language Description

PROSEIS is written in QuickBASIC, an integrated editor, debugger, and compiler from Microsoft, and many professional programmers will immediately stop reading this on finding that a version of BASIC was used! There is a continuing acrimonious dispute as to which language is best for general scientific programming, the protagonists usually being considered as C, Pascal, FORTRAN, and BASIC. BASIC is an acronym for Beginners All-purpose Symbolic Instruction Code, and perhaps it is this first word that repels the professionals. In fact BASIC has evolved a great deal since it was first developed for mainframe use at Dartmouth College in 1965, although in its latest reincarnation it has kept many of its advantages, especially its simplicity, its excellent facilities for handling text, and its capability of seeing immediate results. However, QuickBASIC is really a new language. It is a compiler rather than an interpreter, does not need line numbers, and is very modular and structured. Subroutines can be written in other languages and called from the QuickBASIC modules thus enabling the exploitation of features of Fortran, C, or Pascal if required. Libraries of functions or modules can be built from object files, used within the editing environment, and then linked into the final compiled program or loaded into memory to allow overlays to be used. QuickBASIC has an excellent debugger that enables the tracing of values of the variables to find and correct errors as the program executes. Recursion was introduced in version 4, and version 4.5 has a hypertext online help system that virtually makes obsolete the reference manual.

The language has two logical statements which are used to test conditions in order to make decisions:

> IF..THEN..ELSE branches to or executes different statements
> SELECT CASE conditionally executes different statements

A different construction, ON *j* GOSUB, is sometimes more convenient and precise for choosing between alternatives based on the passing of parameters between subroutines or modules. For instance, in the following program fragment the routine "array_type:" calls a subprogram named "hilight," that enables a choice to be made from a list of items on the screen by using the cursor keys or mouse. The choices are passed to this subprogram in a text array m$( ), where m$(0) is the title of the menu. The path followed by the program is then determined by *j*, the number of the chosen item that is passed back to the calling routine thus:

```
array_type:
    m$(0) =" Choose the type of array that you wish
        to see:"
    m$(1) =" Vibroseis"
    m$(2) =" Weighted elements at equal distances"
    m$(3) =" Equal elements at variable distances"
    m$(4) =" Area patterns (2D)"
    m$(5) =" HELP!"
    m$(6) =" Return to main menu"
    num_items = 6: j = 1
    CALL hilight (num_items, j, m$( ))
    ON j GOSUB vibroseis, tapered, variable_d,
area, array_help
    RETURN main_menu
```

One subroutine used several times in the program is "yes_or_no:." Thus in the following example the routine "land:" asks a question and receives its answer from the subroutine "yes_or_no:" which in turn calls "hilight" to get a value for *j*. This method is used for all routines that ask questions that can be answered simply by yes or no.

```
land:
    PRINT "Type of terrain:
    PRINT "Is the area flat with small elevation
        differences?"
    GOSUB yes_or_no
    ON j GOSUB flat, rugged, terrain_help
RETURN
yes_or_no:
    m$(0) = ""                              'no title
    m$(1) = " Yes"
    m$(2) = " No"
    m$(3) = " HELP!"
    m$(4) = " Go back"
    num_items = 4: j = 1
    CALL hilight (num_items, j, m$( ))
RETURN
```

A subprogram such as "hilight," which is used repeatedly, can be made part of the library of object files to avoid cluttering the main program.

In some cases the display of many choices all on one program line could be confusing, so the conditional statement SELECT CASE is used to control program flow according to the value of *j* as in the following example:

```
vibro_choice2:
    m$(0) = " Choose which type of vibroseis array
        to design ..."
    m$(1) = " noninteger—all sweeps at different
        positions:"
    m$(2) = " integer (A)—uniform, one sweep per
        position:"
    m$(3) = " integer (C)—nvibs sweeps, tapered at
        ends:"
    m$(4) = " design a nonvibroseis array:"
    m$(5) = " HELP!"
    m$(6) = " return to the main menu:"
    num_items = 5: j = 1
    CALL hilight (num_items, j, m$( );)
SELECT CASE j
    CASE 1
                        GOSUB noninteger
    CASE 2
                        GOSUB integer-A
    CASE 3
        .               GOSUB integer-C
    CASE 4
                        GOSUB other_array
    CASE 5
                        GOSUB array_help
    CASE ELSE
                        RETURN main_menu
    END SELECT
RETURN                              'to previous menu
```

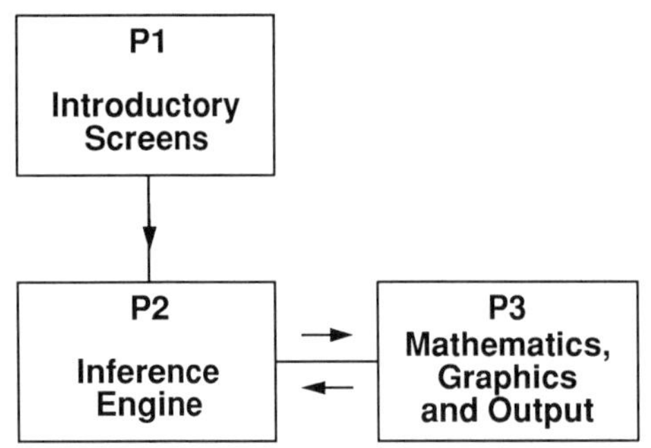

FIGURE 4.1. Schematic diagram showing the program comprising three modules which are swapped in and out of memory. P1 can call only P2, but P2 and P3 can call each other as required.

Graphics

The array responses are drawn in Hercules high-resolution monochrome graphics, a standard chosen because of the excellent screen resolution together with good quality plots obtainable on cheap dot-matrix printers. The Hercules screen resolution is 720×348 pixels as compared to the 640×200 of the color graphics adapter (CGA), 640×350 of the enhanced graphics adapter (EGA), and 640×480 of the video graphics array (VGA). Another advantage of using Hercules graphics was the availability from Laboratory Software Ltd. of HGRAPHICS, a library of extra graphics functions not available in QuickBASIC, which could be called from the main program. There are three alternative ways of using this library:

1. HGRAPHICS can be loaded into memory to allow the functions to be called when programming in the environment of the QuickBASIC editor.
2. The functions can be extracted from the library and permanently linked to the main program at the compilation stage.
3. The functions can be called at runtime from the library, which had been loaded into memory.

Type of terrain: Is the area flat, i.e. small elevation differences?

Type of terrain: Is the area flat, i.e. small elevation differences?

The effect of the ruggedness of the terrain depends on several factors only one of which is the elevation. Even rugged terrain will not effect a short spread if the surface velocity is high. The basic array design should not take into account one-off cases, although when working say in sand-dunes it would be wise to use as short an array as possible since many groups could be affected.

For more information on treating elevation problems and to calculate the maximum allowable elevation difference across an array please choose NO.

FIGURE 4.2. Menu allowing four choices. This help screen appeared after moving the highlight to HELP and pressing <RETURN>.

Use the up and down arrow keys to choose the data to enter,
When the desired choice is highlighted, type in the information.
< BS > erases and the < RET > key moves to the next entry.
When you have finished press the < ESC > key ...

Enter the following data for the shallowest horizon of interest:

Average velocity in meters/sec:
Two-way time in seconds:
Maximum dip in degrees (- = updip):
Highest signal frequency in Hz: 50
Maximum shot-geophone offset in m:

FIGURE 4.3. Screen showing a form to be completed in order to provide information for array design. Previous entries are remembered.

In Hercules graphics mode, two graphics and one text page can be employed, and HGRAPHICS enables the graphics screens to be saved on disk in compressed form and retrieved at will. In this way the screen could display a screen of graphics assistance without overwriting the other page on which the response plots were drawn, thus enabling several plots to be drawn on the same axes. Other useful functions enable the plotting of the screen in portrait or landscape mode on most popular types of 9-pin, 24-pin, and laser printers. Additional graphics functions replaced or enhanced the normal drawing functions available in QuickBASIC.

You have entered the following information:

Average velocity in metres/sec: 2000
Two-way time in seconds: 2
Maximum dip in degrees (- = updip):   10
Highest signal frequency in Hz: 50
Maximum shot-geophone offset in m: 1750

Then maximum effective array length N*D which will preserve signal
frequency is   34 metres,
The minimum effective array length N*D needed to cancel the longest
noise is   42 metres,
and the element spacing D needed to cancel the lowest noise velocity is
                5.6 metres.

The array necessary to cancel the longest wavelength noise will also
degrade shallow reflections at the far traces, but this can be corrected by
careful muting during processing. Check zero offset to see sensitivity to
dip. A uniform inline array could have 8 elements at a spacing of 5.6
metres to give an actual array length of 39.2 metres.

Do you wish to change any parameter before continuing [N] : ?

FIGURE 4.4. Details of a linear array which would attenuate the noise. The program warns that these array dimensions could also affect signal quality.

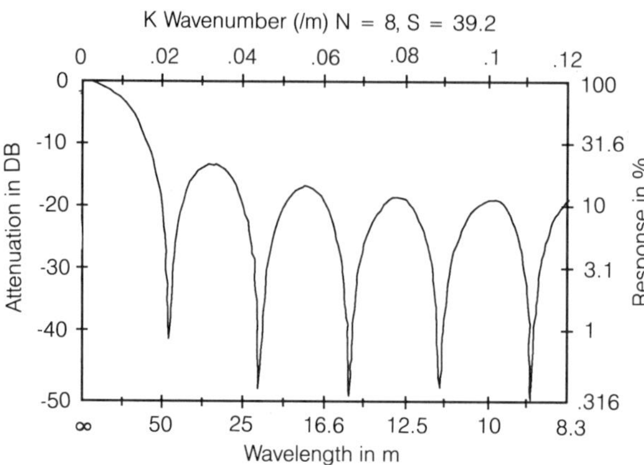

FIGURE 4.5. Plot of the response of the linear array suggested by the program as appropriate for the survey parameters as entered.

Structure of Program

The PROSEIS expert system is composed of three modules or overlays, so-called because each module when called overlays the previous one in the computer's memory. P1.EXE comprises the introductory screens together with various initializing routines. P2.EXE is the expert system proper, with its question-and-answer dialogues leading to the determination and calculation of the optimum array parameters. P3.EXE is the output unit which takes care of the response calculation together with the plotting of axes and responses on the screen or on paper. To enable the program to run on a machine with only 250K of RAM (random access memory) the modules were compiled to use the runtime library, which remains in memory together with only one module at a time (Figure 4.1).

P3.EXE was the first module to be developed as ARRAYS.EXE, a standalone program for calculating array responses for given field parameters. In ARRAYS the explorationist enters the type of array, number of elements, etc., and the program draws the response curve on the screen. The response can be calculated for vibrators, for dynamite, or for geophones, for uniform or tapered arrays, or for variable spacing, and different arrays can be combined to show the final plot. Reports can be generated showing response plots together with details of the survey, crew, and other information, or larger sized plots can be drawn without a heading if required to make, for instance, overhead transparencies. Any plot can be saved on disk and then annotated or edited using a special CAD program.

All the above facilities are available in PROSEIS except that no separate input is possible, all the parameters needed for drawing the response plots must be derived from the dialogue within P2, the expert system proper, the inference engine. The parameters are then passed to P3 when plotting is required.

Much of the P2 module is built around the simple multiple-choice menu "yes_or_no," shown in Figure 4.2. Selection is made by moving the highlight on a menu item by means of the up and down arrows or a mouse, and pressing the <RETURN> key or clicking on the left button of a mouse to confirm the choice. The program asks a question and the user gives the answer. If the user is unsure of the meaning of the question or wants more information then the HELP option can be chosen as in Figure 4.2. The GO BACK option allows changing previous choices in the light of the new question.

Another subroutine enables the explorationist to enter values to be used in the calculation of array parameters by completing a form. The user enters values in the appropriate place, moving from field to field either by pressing <RETURN>, using the cursor keys or moving the mouse, and then presses the <ESC> key or clicks on the right button of the mouse when satisfied. Note that values previous entered are remembered (Figure 4.3).

When an appropriate stage has been reached a plot of the array response can be shown on the screen. Several response curves can be plotted on the same axes and compared.

Examples of Results

After the user has answered the questions concerning the noise and target parameters, the details of a

Any Oil Company

Area _ _ _ _ _ _ _ _ _ _ _ _ Scotland
Prospect _ _ _ _ _ _ _ _ _ _ Loch Ness
License _ _ _ _ _ _ _ _ _ _ _ Q1000
Crew _ _ _ _ _ _ _ _ _ _ _ _ Davis Exploration
Date _ _ _ _ _ _ _ _ _ _ _ _ 9/8/76
Remarks _ _ _ _ _ _ _ _ _ Test Run

Source:
Vibroseis

Number of vibrators inline: 4
Distance between vibrators in meters: 6.83
Moveup between sweeps in meters: 8
Number of sweeps: 5

Total array length = 52.49 meters.

Array Response

X Wavenumber (/m) N = 20, S = 52.4

FIGURE 4.6. A page of a report generated by the program showing a heading together with the response of a suitable vibroseis array.

Any Oil Company

Area _ _ _ _ _ _ _ _ _ _ _ Scotland
Prospect _ _ _ _ _ _ _ _ _ Loch Ness
License _ _ _ _ _ _ _ _ _ _ Q1000
Crew _ _ _ _ _ _ _ _ _ _ _ _ Davis Exploration
Date _ _ _ _ _ _ _ _ _ 9/8/76
Remarks _ _ _ _ _ _ _ _ _ Geophone, Vibroseis and
Combined Response.

Comparison of Combined Array Responses

X Wavenumber (/m) N = 20, S = 52.4

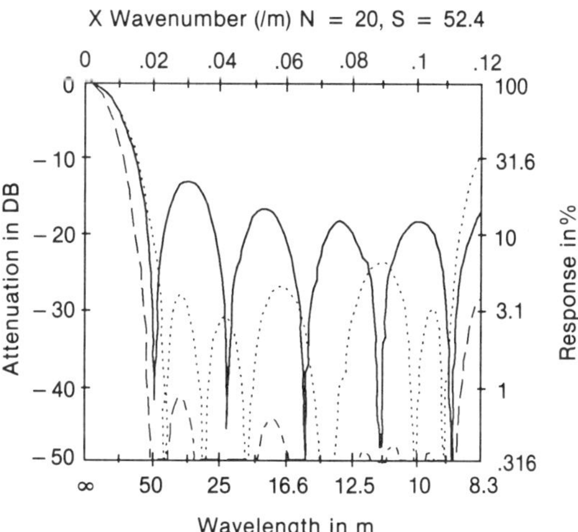

FIGURE 4.7. A page of a report generated by the program showing source and geophone array response together with the combined response.

linear array are shown, which will give the best noise attenuation. A warning is given if this length of array will attenuate the signal because of dip or differential moveout across the array, together with a suggestion that the explorationist may go back and alter parameters to see the effect on the array (Figure 4.4).

The response of the array can be seen on screen (Figure 4.5), or a different type of array (tapered, pattern, vibrator, etc.) can be designed to fit the same parameters.

When the explorationist is satisfied with the source and detector arrays, a report can be produced together with survey details showing the individual plots (Figure 4.6), and the combined response (Figure 4.7).

Hardware

The program runs on any MS-DOS microcomputer that has a Hercules or compatible graphics capability. Printed output can be obtained on any IBM or Epson compatible 9 or 24-pin graphics printer or HP laser printer. The program can be used in the field on a suitable battery-powered laptop computer, with printed output on a compatible printer.

Summary

The PROSEIS expert system allows an explorationist to determine the best seismic arrays for given field conditions. PROSEIS is extremely flexible, and can accommodate the principles of the stack array approach.

A battery-powered laptop computer and printer can allow calculations to be made in the field, thus providing the advantage of immediate modification of the field technique to suit local conditions.

Acknowledgments. Thanks are due to Enterprise Oil Exploration Ltd., for their support in the writing of this program, and to Dr. Ilan Davis for coding the mathematical and graphic subroutines. All trademarks mentioned are acknowledged.

References

Anstey, N.A., 1986, Whatever happened to ground roll?: The Leading Edge **5**(3), 40–45.

James, M., 1988, Expert systems demystified: Program Now, Adv. Program. J. **2**, 32–36.

Pratt, W.E., 1952, Towards a philosophy of oil finding: AAPG Bull. **36**, 2236.

Savitt, C.H., Brustad, J.T., and Sider, J., 1958, The moveout filter: Geophysics **23**, 1–25.

Smith, M.K., 1956, Noise analysis and multiple seismometer theory: Geophysics **21**, 337–360.

Vermeer, G.J.O., 1990, Discussion on ground-roll suppression by the stackarray, Geophysics **55**, 936–937.

5
An Expert System to Assist in Processing Vertical Seismic Profiles

Peter Crisi

The VSP Processing Advisor (VSPPA) provides general theory and methodology for processing vertical seismic profiles and offers specific advice for use with the VSEIS processing package.

The system development process included knowledge acquisition implementation and testing. Interviews with one expert, a VSP processing class, and background reading provided knowledge for the system. Development with the expert system building tool Personal Consultant Plus (PC+) resulted in a rule-based expert system with a backward-chaining inference engine. The system has not been tested extensively.

Design of this expert system includes one frame to obtain initial data, one frame to determine which process or processes to execute, and 13 frames to execute major steps in the processing sequence. The system contains 220 rules, 160 parameters, and 22 graphics, and requires 1.3 megabytes of space on an EXPLORER (Texas Instruments Inc.) computer.

The user is assumed to have little or no VSP processing experience. A knowledge map, tutorial, and user's manual are included to simplify understanding and using the system.

Introduction

An expert system (Waterman, 1986) is a specialized computer program that contains a high level of human expertise in a narrowly defined area. Expert systems and conventional computer programs differ in their application of knowledge and data. This information is analyzed by an expert system using an inferential process to apply the correct heuristics, i.e., rules of thumb. In contrast, a conventional computer program uses an algorithmic solution. The advantages of using an expert system are modularity and organization of rules and parameters in a problem domain.

The expert system development process has been termed knowledge engineering. A "knowledge engineer" must first define the problem, then gather background knowledge from publications in the problem domain. At this time he is ready to interview the domain expert or experts for strategies, rules of thumb, and domain rules. Next, he picks the expert system building tool best suited to the problem domain.

After the knowledge engineer encodes this knowledge, the expert critiques the system and recommends changes. When the expert is satisfied, the system is tested by future users and updated by the knowledge engineer. This process is outlined in Figure 5.1.

Vertical seismic profiles (VSPs) are similar to surface seismic surveys. A source is energized on the ground surface and the resulting vibrations are recorded either by a vertical- or a multiple-component geophone located in a borehole. The geophone is usually moved along the length of the borehole while the source is held at one position, but the source may also be moved. Processing the resulting data involves different procedures depending on the reason for running the VSP, the type of survey, the type of equipment, and other parameters.

Knowledge Acquisition

Before an expert system can be built, the "knowledge engineer" must understand the relevant decision processes and knowledge in the problem domain. Extensive reading and a VSP processing course provided the requisite background and supplemental information. Background readings minimized wasted time during interviews with the expert. Without questions focused on this knowledge base, the domain expert, like all experts, would naturally spend unnecessary time on topics of special and immediate interest to him. Asking the right questions required an understanding of the topic. This understanding came, in part, from readings. Further interviewing background came from a course in VSP processing. The course, which required actual processing of VSP data and was taken the same semester as much of the interviewing, provided the basic processing sequence on which interviews elaborated. The processing modules of the VSEIS (Spectrum Geophysical, 1986) package on the Gould computer and practical processing considerations were explained in the course.

Domain knowledge was acquired from Dr. Alfred Balch of the CSM Geophysics Department. The knowledge acquisition was performed as interviews with the expert. These interviews (about 60 hours) provided a methodology for extracting the important knowledge from a large quantity of information. They also helped create an efficient expert system structure including rules, parameters, and their dependencies.

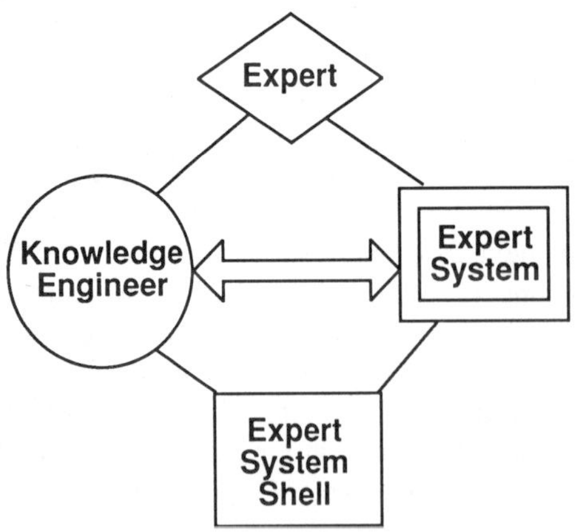

FIGURE 5.1. Expert system development process.

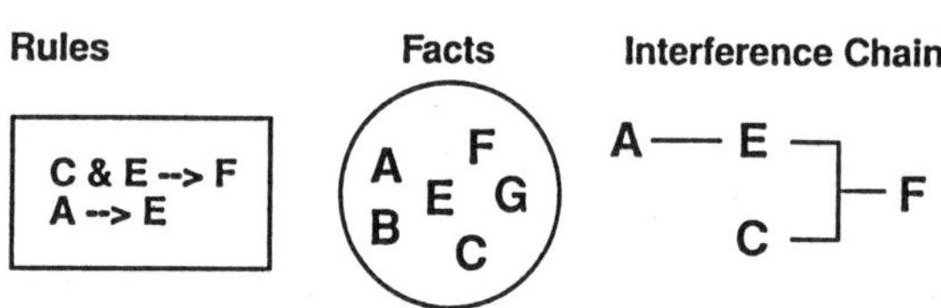

Check rule actions against initial facts
Rule action F not in initial facts
Need conditions C and E to determine F
Condition C in Initial Facts
Need condition A to determine action E
Condition A in initial facts
Execute rule A --> E and E and add E to facts
Execute rule C & E --> F and add F to facts

FIGURE 5.2. Backward chaining of rules.

The original system design goal was to give advice for processing VSP data for a tie to surface seismic data given a zero-offset source, a vertical-component downhole geophone, a land air gun source, and a nondeviated borehole. This basic system was expanded to include other options, but does contain limited advice for processing data recorded with a three-component downhole geophone.

Development Software

The VSPPA is a series of rules, parameters, and functions set up in a series of frames. An expert system development tool was used to create this structure and then to consult it. Without such a tool, writing the LISP code representing the expert system and the inference engine would be necessary. PERSONAL CONSULTANT PLUS (PC+) (Trademark of Texas Instruments Inc.) was the tool chosen for this expert system.

PC+ is a descendant of EMYCIN, the development tool created for the expert system MYCIN (Buchanan and Shortliffe, 1984). This expert system recommends drug types and dosages for patients with bacteremia, meningitis, or cystitis infections. MYCIN is a rule-based system implemented in LISP. The system employs a backward chaining inference engine as shown in Figure 5.2, can perform certainty calculations, and can provide explanations of the reasoning process used. An extended set of capabilities is found in PC+.

Although PC+ was chosen primarily for its availability, it worked well for developing the VSPPA. PC+ uses a rule-based knowledge representation with a default backward-chaining inference engine. In backward chaining, the system starts with a goal parameter and traces back from rule consequences (THEN) to rule antecedents (IF) until all of the traced rule antecedents can be satisfied, and the goal parameter is satisfied, or until all rules related to this parameter are tested. Backward chaining works well if the expert system's goals are clearly defined.

Forward chaining is also possible as in Figure 5.3. Forward chaining can be triggered in PC+ by flagging a rule as antecedent. In this case, the rule fires (executes THEN statement) immediately after all of its conditions (IF statement) are satisfied. Antecedent rules are passive; they fire only when all of their conditions are satisfied from user input or through backward chaining. Forward chaining works well as a primary inference method if multiple solutions are possible, requiring multiple inference chains.

For a particular goal, the same rules will be fired in forward and backward chaining. However, the rule testing sequence and inference chain will be different. With the exception of the VSPPA, all expert systems previously developed or currently under development by the Geophysics Department Artificial Intelligence Lab are coded on microcomputers. The VSPPA was developed on the Texas Instruments EXPLORER computer for four major reasons: memory, speed, ease of link to a mainframe, and expendability.

Of these four factors, memory and speed requirements were of immediate importance. The VSPPA, excluding graphics, fills about ⅓ of a megabyte of memory. Graphics take up another megabyte of memory. A system of this size running on a stand alone 8086-based microcomputer would be slow. The third consideration is the future ability to establish a high-speed mainframe link for interactive processing. Instead of giving advice to the user, the expert system can be modified to send system commands executing the desired processing modules in VSEIS. Of final consideration, the EXPLORER contains database and extended LISP capabilities that give the system added flexibility for future development.

Knowledge organization and control in PC-PLUS is handled by frames, parameters, and rules. Frames are the largest divisions of the expert system. They are organized in a tree structure, and

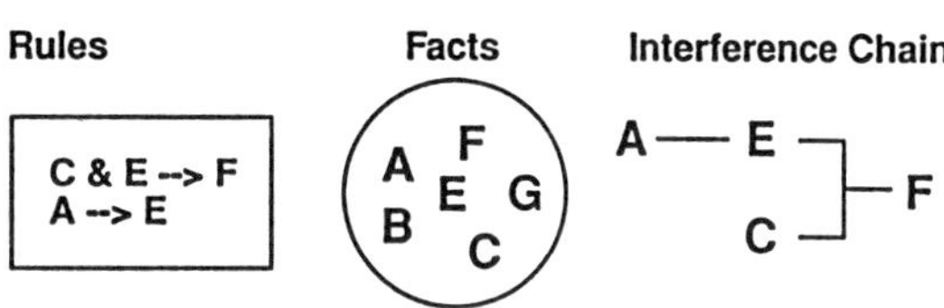

FIGURE 5.3. Forward chaining of rules.

each addresses a part of the problem defined by the knowledge base. Each frame contains a set of goals and prompts along with a translation. Initial data may also be obtained, and the conclusions displayed. Frames are flexible and expandable. Once the system creates and enters a frame, the necessary rules are applied to infer the value of each goal parameter. After the system finds the values for all goal parameters, it exits the frame.

Parameters store information used by the knowledge base. This information can be defined by the system user, or can be inferred by the system. Each parameter has a specific type and translation. A prompt, a help statement, certainty factors, and accepted values are a few of more than 20 properties available to further define a PC+ parameter.

Rules determine the solution of a problem. More specifically, rules can show how to infer a parameter's value, alter certainty factors, determine the logical relationship between parameters, and communicate information. Rules consist of one or more conditions (IF) and actions (THEN). Both IF and THEN statements can contain user-defined functions, parameters, and values.

Certainty factors are used in many expert systems, but not in the VSPPA. The primary reason for this structure is that the VSPPA is advice based. Users are guided by the expert system in making decisions involving uncertainty while processing their data. Since users are expected to know little about VSPs, they are asked about easily recognizable characteristics of their data. Advice for further processing is based on their answers. This advice may be to experiment with multiple processing techniques and to pick the best one by looking for specific results.

Execution control in the VSPPA is based on frame organization, ordering of goal parameters within frames, and ordering of rules using utilities. The first two controls specify the order of frames created and entered and the sequence of goals to satisfy within each frame. The third control specifies the order of rules to test for each goal. A rule with a higher utility will be tested first. This ordering allows testing of specialized cases before generalized cases, or common cases before rare cases.

Some consultation aids using PC+ are graphics, parameter helps, HOW's and WHY's, REVIEW, and SAVE PLAYBACK FILE. Graphics screens created on the EXPLORER can be attached to prompts, parameters, or rule conclusions. Each parameter may contain a HELP statement to assist the user in determining the parameter value. WHY tells users why they are being prompted for specific information. HOW tells users which parameters they have supplied and how the system has determined other parameter values. REVIEW displays the list of parameters prompted for so far in the consultation. If any responses are marked, the consultation will start over using unmarked responses and asking for values of marked parameters again. SAVE PLAYBACK FILE saves a file of parameter values for a partial or complete consultation. This file can be used to review, continue, or alter the consultation at a later date.

Knowledge Organization

In the VSPPA, each frame represents a fundamental step in the processing sequence. The top level frame obtains initial data required by lower level frames. The next lower level frame in this tree structure, as shown in Figure 5.4, determines which processing step or steps to execute. Each of the 13 frames on the lowest level gives advice. They advise on the basis of information from the two top-level frames.

In total, the system includes about 220 rules, 60 processing and 100 dummy parameters, 2 functions, and 22 graphic aids. The system requires about 320,000 bytes of storage space, excluding graphics. At the same time, system modular design makes information quickly accessible to the user. A complete listing of processing parameters is included in Appendix A, and the logical frames

and rule structures are included in Appendix B. The remainder of this section describes in detail the processing advice in each frame.

Initial Data

Many different types of VSP surveys exist. Additionally, many different formats may be used to record information about the survey, and this information may be incomplete. For these reasons, initial data requirements had to be flexible.

This expert system is designed so that the user need not provide parameter values unless processing is impossible without them or unless the user readily knows their values. If the user does not know a parameter value for which he is prompted, a HELP statement, a default value, or an UNKNOWN option are always available. Not all of the parameters will be used in a specific consultation. A few parameters are never used, but are available for future expansion of the expert system.

Values for a subset of the initial data parameters are requested in all consultations. These parameters are required initial data for the root frame and they likely will be used in processing any VSP data set. To know whether the user has severely limited processing time, observer's notes, a cement bond log, a caliper log, an acoustic log, a gamma–gamma log, and knowledge of bedding orientation and rock types ("geology") surrounding the borehole is important. The survey goal, source type, number of sources, and geophone type are also important parameters for any VSP survey.

Although not critical to processing the VSP, the well name, number of field tapes, and previous use of the VSPPA are also asked in initial data. If the user has not previously consulted the Processing Advisor, he is briefed on its operation. All of the goals in the root frame are designed to obtain additional initial data and to assign default parameter values if necessary. The first rule determines drilling information, borehole parameters, survey design, and recording equipment information if observer's notes are available. Next, if the expert system user knows the bedding orientation and rock types surrounding the borehole, he is asked whether dipping reflectors exist.

Many source types may be used in recording a VSP. Different sources will yield a variety of information about the survey. A set of rules determines

FIGURE 5.4. Frame tree of the expert system.

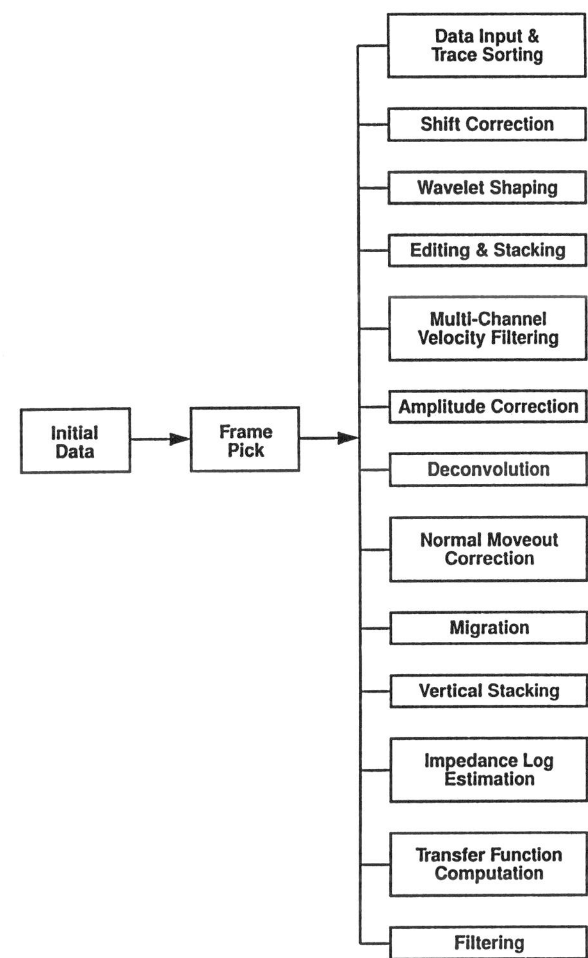

the number and size of sources, their orientation, and their impact along with their waveform and the location of multiple sources. If the source type is unknown, the default is a land air gun. Other defaults are a nondirectionally drilled well and a vertical component geophone. These defaults trigger messages warning that processed results may be wrong if the defaults do not accurately represent the data.

Another goal parameter gathers information if the user's survey goal is to tie his VSP data to surface seismic data. This information includes datum values, reference elevations, and correction velocities. Finally, if the data are in SEG-Y format, information is offered about processing using the VSEIS package on the Gould 9750 computer, and the next frame is called to determine which processing steps

to execute. If data tapes are not in SEG-Y format, VSEIS cannot read them. If necessary, the user is offered advice on how to determine the tape format using the Tape Format Advisor (Agena, 1988) and to convert the tapes to SEG-Y format.

Two functions defined in the initial data frame output all advice to a file for future reference by the user. Parameter values are not included in the printout, but all advice based on these values is included from this frame and all lower-level frames.

Processing Sequence

After entering initial data, the user is ready to consult one or more processing steps. If he does not pick any steps, the system will advise him on the appropriate sequence based on his initial data and

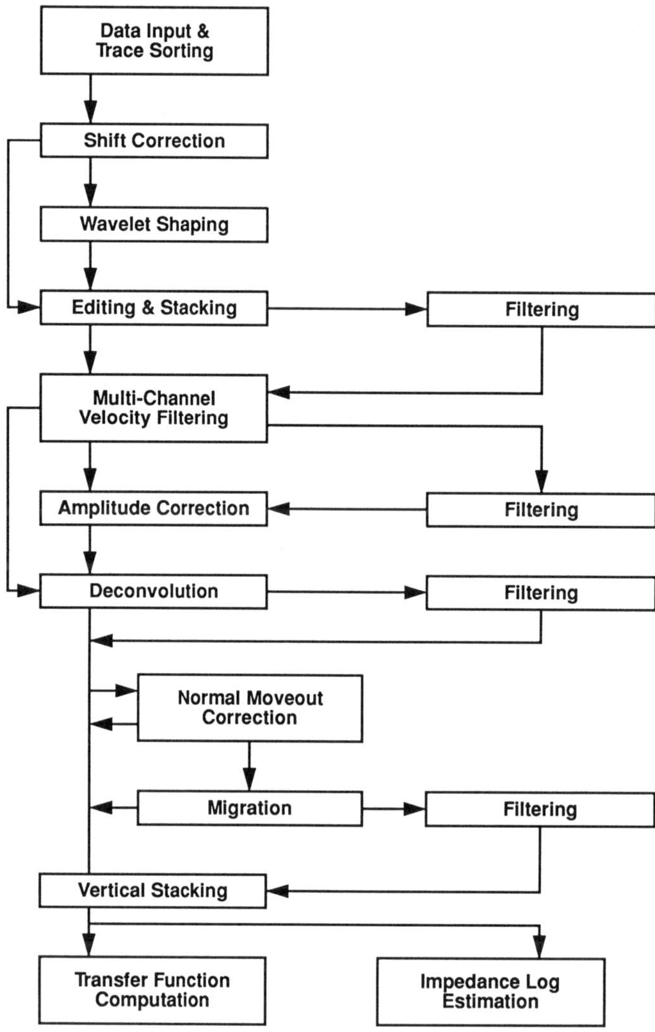

FIGURE 5.5. Consultation sequence for VSP Processing Advisor.

what he has already completed. This approach gives flexibility for the experienced user and guidance for the inexperienced user. The recommended processing sequence is shown in Figure 5.5. Filtering is advised in four places. Wavelet shaping and amplitude correction may be omitted in certain cases. All other steps have one occurrence.

If the user knows which processing steps he wants, then the appropriate frames are immediately entered. After consulting the original desired frames, the user is offered a chance to consult additional frames. This option saves retyping parameter values, which is necessary for a new consultation unless a REVIEW command is used or a playback file is recalled.

If users do not know the next processing step to consult, they are guided by the expert system. If no processing has been completed, data input and trace sorting is consulted. If data input and trace sorting has been completed, then a shift correction is applied. If a shift correction has been applied, the source type is not an explosive, and processing time is severely limited, then wavelet shaping is skipped and editing and stacking is performed. Otherwise, shot wavelet shaping is advised, after which stacking is performed. Following editing and stacking, filtering may be necessary to clean up the data before further processing. Upwave-downwave separation then extracts the source and reflected waveforms from the original data set.

Filtering is applied again after this step to remove tube waves and other coherent noise. Unless processing time is severely limited, the next step is a spherical divergence amplitude correction. Deconvolution follows either amplitude correction, or filtering if processing time is severely limited. Deconvolution boosts the high-frequency content of the data, so filtering is advisable again to limit the frequency range. After filtering, processing is complete for a zero-offset source if the processing goal is surface seismic evaluation, other, or unknown. If, after filtering zero-offset data, a surface seismic tie, stratigraphic trap exploration, prediction ahead of the drill bit, or geology near the well is desired, recommended processing is vertical stacking.

After filtering offset source data, a normal-moveout correction should be applied. Completion of this step leads to migration if reflections from dipping beds are anticipated. However, if no dipping beds are anticipated, processing is complete for a survey goal of surface seismic evaluation, other, or unknown, and vertical stacking is advised for any other survey goal. If migration was performed, filtering should follow to ensure clarity and coherency of events. After filtering, processing is complete for a survey goal of surface seismic evaluation, other, or unknown. Any other survey goal requires vertical stacking. Vertical stacking is the final step if the survey goal is a surface seismic tie or stratigraphic trap exploration. If the survey goal is to predict ahead of the drill bit, advice is offered for an impedance log estimation. And if the survey goal is to examine the geology near the well, a transfer function computation is advised. No further processing is recommended after a user has consulted the impedance log estimation or transfer function computation processing steps.

After determining which processing steps the user needs to consult, this intermediate level frame creates and enters the required lower level frames in the correct processing sequence.

Data Input and Trace Sorting

Before data can be processed, they must be read from field tapes into the computer. Next, the resulting files must be sorted into separate channels. These goals are handled by this frame. Also, advice is given to help derive unknown parameter

values that are required for processing. First, advice is offered for reading tapes into the computer. Then, warnings are given if the data format is unknown, and if the record length exceeds the VSEIS package limit.

If observer's notes are unavailable, the assumption is made that the user does not know the survey type, the number of recorded channels, or the data content of each channel. Advice is given to help the user derive values for these parameters from record headers. Once these parameters are known, the original data set is sorted by channel. Record headers and trace plots are obtained for the separated channels. Advice is then offered to correct the trace polarity on any reverse-polarity traces.

Shift Correction

The primary shift correction takes the data reorganized in step data input and trace sorting and corrects the downhole channel first arrivals to zero time. This correction represents the length of time between the time the recording is started and the time the source is energized. Since some source types may be energized below the ground or sea level datum, a second static correction trace shift to datum may be required. No advice is offered for time corrections to compensate for elevation changes.

The vertical-component data from which the shift correction is computed may contain refractions. These events may contain important information. However, they should be removed if they mask the desired downhole channel compressional wave first arrivals. Filtering to remove refractions, if necessary, is recommended before applying a shift correction. After checking for refractions, a static correction is applied to data recorded with a water gun, marine air gun, or explosive source. If data recorded with a vibroseis source has not been cross-correlated with the source waveform, advice is offered for cross-correlation at this time.

Finally, time breaks are picked to correct the downhole channel first arrivals to zero time. Time breaks are picked either interactively or manually from trace plots of the time break monitor channel. Once these picks are recorded, they are applied to the downhole data channel or channels. Following this time correction, advice is available for a spatial direct current bias correction to center each trace on its axis.

Editing and Stacking

Editing and stacking improve the downhole data quality dramatically by removing random noise. Additionally, they greatly reduce the size of the data set. Common practice in VSP surveys is to record test shots on levels moving down the borehole before recording data moving up from the bottom of the borehole. These test shots are discarded before stacking. If multiple shots were recorded on each level, selective stacking is the next step. Bad traces are eliminated by stacking only the good traces on each level. Three-component data are related differently from vertical-component data. Data from the three channels must be processed together to define events in three dimensions. As a result, if all traces are bad for one level on one channel, that level is discarded on all channels. After stacking, data files must be combined to yield one stacked trace for each level for each channel. The expected number of levels is computed from the shallowest level, the deepest level, and the level spacing. Missing levels must be interpolated, and extra stacked traces from test stacking and uneven level spacing must be deleted. Finally, if limited computer time is available, guidelines are given for resampling the stacked section.

Wavelet Shaping

This processing step compensates for source signature variations during the survey. If processing time is severely limited and the source signature exhibits minimal variation, this step may be omitted. If a source monitor channel was not recorded, this step must be omitted. Wavelet shaping is performed by arbitrarily picking a standard source monitor trace. A filter is computed to convert the source monitor trace for each other level to this standard trace. The filter for each level is then applied to the downhole trace or traces on the corresponding levels.

Upwave-Downwave Separation

VSP data have the unique advantage over surface seismic data in that the upgoing and downgoing wave fields can be separated and analyzed. Ideally, after wave field separation, the downwaves represent the source waveform, while the upwaves represent reflected events. This processing step and all following steps are markedly different for vertical-component and three-component data. Either a frequency–wavenumber or a median filter is used to extract the desired upgoing events from vertical component data. Median filtering takes longer since the first break arrival must be picked on each downhole trace. However, filtering accurately separates coherent downwaves from the original section. The frequency–wavenumber (f–k) filtering takes less time, but may remove some desired data with the downgoing wave train. Three-component data require triaxial data polarization to separate desired events. Further three-component processing employs some of the steps used for vertical-component data, but depends largely on the user's goals. The amplitude correction, deconvolution, vertical stacking, transfer function computation, and impedance log estimation processing steps were designed for vertical-component data.

Amplitude Correction

As a source waveform propagates through the earth, it incurs absorption and spherical divergence amplitude losses. After removal of field gain from the downhole data, two amplitude corrections are applied. Absorption is compensated for by normalization before separation of the upgoing and downgoing wave trains in upwave–downwave separation. Spherical divergence is corrected using a time-variant amplitude recovery curve.

Since these amplitude corrections must be applied at different points in the processing sequence, spherical divergence is the only correction applied in this frame. If automatic gain control is applied to the data, spherical divergence is not required. If processing time is severely limited, this processing step may be omitted.

Deconvolution

Deconvolution is the process of applying an inverse filter to remove the effects of the source signature, reverberations from a water layer or near-surface layers, and ghosts or other multiples. Other applications of deconvolution are to equalize all frequency components within a passband (whitening) to compensate for filtering actions that have strung out the wave train, or to convert one source waveform to another.

Wavelet shaping and either spiking-signature, predictive, or time-variant deconvolution can be performed from advice in this frame. If the user does not know which option to consult, advice is based on the user's goals and information about the data. If the user wishes to remove the effects of the source waveform and reverberations, then predictive, time- and space-variant, or spiking signature deconvolution is advised. Simple water bottom multiples or ghosts can be removed with a predictive deconvolution operator. A waveform changing rapidly in time or space indicates the need for time- and space-variant deconvolution. If neither of these cases exists, then spiking signature deconvolution is applied. This default is easy to apply and it accurately characterizes the downgoing waveform and multiples. If the user's goal is to shape the frequency response, then wavelet shaping deconvolution is advised. Among other uses, wavelet shaping deconvolution can be used to match the VSP waveform with the surface seismic waveform. Finally, the user is offered advice for writing his deconvolved data to magnetic tape. After filtering, no further processing is advised for zero-offset data if the survey goal is surface seismic evaluation, other, or unknown.

Normal-Moveout Correction

A normal-moveout (NMO) correction is applied to offset-source surveys to remove the variation of reflection arrival time due to variation in the shot-point-geophone offset distance. Before this correction can be applied, a depth or time model must be determined from a shotpoint log or other source. Advice is offered to write the NMO corrected section to magnetic tape. If no dipping reflectors are anticipated, processing is complete after this step for a survey goal of surface seismic evaluation, other, or unknown.

Migration

Dipping reflectors recorded with an offset source are moved to their true spatial positions using migration. Migration requires a geometry model and rms velocity model. This information is used to create reflection point images for the offset source VSP data, using linear raypaths and assuming horizontal bedding. After migration and filtering, no further processing is advised for the survey goals of surface seismic evaluation, other, or unknown.

Vertical Stacking

"Horizontal" stacking, or stacking on levels, was used to create a high-quality trace for each level. Now, traces from multiple levels can be stacked to produce an even clearer picture of the VSP waveform at the borehole. A vertically stacked trace can be used for a high-quality tie to surface seismic data or for detailed wavelet examination in stratigraphic trap exploration. Typically, a "sliding" corridor a few hundred milliseconds long starting just below the first arrival on each trace is stacked for traces representing a range of depth levels. With directionally drilled well data, a "sliding" corridor is not used. Instead, only the shallow levels are stacked using a corridor as long as the desired vertically stacked trace starting just below the first arrival on each trace. Quality of the stacked trace may suffer from using only shallow levels. However, obtaining the best possible composite downhole data trace representing the geology directly under the wellhead for a tie to surface seismic data at the wellhead is important. If a surface seismic tie is desired, a time shift may be required to match the vertically stacked VSP trace with surface seismic data at the wellhead. If the survey goal is a surface seismic tie or stratigraphic trap exploration, no further processing is advised beyond writing the trace to magnetic tape.

Transfer Function Computation

Attenuation changes as a function of frequency can be measured for zones of interest and can then be correlated with lithology. Use of a transfer function or reflection impulse response requires knowledge of the localized relationships between attenuation coefficients and sand percentage or sand/shale ratio. If the survey goal is to determine geology near the well, this is the last recommended processing step.

Impedance Log Estimation

A VSP-generated impedance log of target horizons can be correlated with a well log generated impedance log to predict impedances below the total depth of the well, ahead of the drill bit. Computation of an impedance log from borehole data requires a good acoustic log for velocity information and a good gamma–gamma log for density information. The product of velocity and density for each level produces a series of acoustic impedances. A series of reflection coefficients $R(n)$ can then be generated using the formula $R(n) = [Z(n) - Z(n + 1)]/[Z(n) + Z(n + 1)]$ where $Z(n)$ is the acoustic impedance of the nth level, $Z(n) = V(n) \times D(n)$, V is velocity, and D is density. This well log reflectivity series can be convolved with the VSP source waveform for direct correlation with VSP data at the well. Alternatively, acoustic impedances can be derived from VSP data for correlation with acoustic impedances from well log data. An inverse filter must be designed from a vertically stacked trace to return a reflectivity series from which acoustic impedances can be derived. The formula for $R(n)$ is used with $Z(1)$ from well log data to derive successive $Z(n)$ values. Impedances below total depth can be used to find information about geologic features and conditions that may change drilling tactics. If the survey goal is to predict ahead of the drill bit, this is the last recommended processing step.

Filtering

Background noise and undesirable coherent events can be removed with filtering. Five types of filters exist for different purposes: time-domain median, spatial median, band-pass, frequency–wavenumber (f–k), and mute (zero out undesirable samples). If the desired filter is unknown, a filtering goal is asked. The first option, mute, is used to remove high amplitude noise at the beginning or end of traces. The second possibility is to suppress random noise. A time-domain median filter may be used to eliminate high-frequency noise bursts. Ringing in the data may be caused by tool slip, and can be minimized by a band-pass filter. Coherent near-surface wave energy may appear to be dissipated. This effect appears to be a random noise source, but is caused by casing, and cannot be removed by filtering. Undesirable coherent events also exist. Tube waves and shear waves can often be removed from vertical-component data using a spatial-median or f–k filter. Noise from an alternating current source or a marine air gun bubble effect can be separated from the data with an f–k filter. A final option is to limit the frequency response of the data with band pass filtering. This goal is important in matching VSP data to surface seismic frequencies. Filtering can be performed at any time to enhance the data. However, filtering is recommended at three places in the data processing sequence: after stacking, after upwave-downwave separation, and after deconvolution. This concludes a cursory explanation of each processing step slanted toward the actual advice offered. The steps vary in size, but each covers a logical unit in the processing sequence. For this reason, steps are easily expandable in future development of this expert system. Each step covers about as much advice as can be used in one processing session.

Conclusions

Like any other expert system, the VSPPA contains only a portion of the expertise in its field. The system will never be "finished," however, it can always be expanded and improved. In its present stage of development, the VSPPA exhibits some noteworthy strengths and weaknesses. Originally, the system was designed to give advice for processing VSP data for a tie to surface seismic data given a zero-offset source, a vertical-component downhole geophone, a land air gun source, and a nondeviated borehole. Variations on these survey parameters were also considered, making this system applicable to the majority of VSP surveys. However, additional advice should be included for processing three-component data. Preprocessing steps would also be helpful in checking for erroneous and inconsistent input data.

Possibilities for future expansion include an interactive link between a mainframe and the EXPLORER computer. This link would initially allow simultaneous consulting with the VSPPA and processing with VSEIS would also reduce time wasted in converting advice from the expert system into data processing. Later, the Processing Advisor could be made an intermediary between the user and VSEIS; instead of offering advice, the Processing Advisor could execute the appropriate modules in VSEIS.

Although the VSPPA was designed for a variety of survey types, it was also designed for a variety of users. An experienced user can consult any processing steps he desires. On the other hand, the novice can either use a default value or obtain help deriving a value for any parameter. The target user has had minimal exposure to VSP data processing. One expert was consulted while developing this system. Additional experts should be consulted for different processing approaches. Their input will provide a scale on which to weigh the importance of specific advice and will also ensure that a complete set of the most important factors is used to determine advice for the system.

PC+ was well-suited to development of the VSPPA because the expert's knowledge could be described using an IF/THEN rule structure and specific goals. However, this development tool requires a rigid frame structure, and the possible processing sequences are limited by this structure. Additionally, rules are constrained in structure of their antecedents and consequences, while parameters are limited by assignment and use of their values.

Some of the extended functions in PC-PLUS did not work, so additional bookkeeping was necessary. For instance, if the observer's notes were available, the user was asked for additional parameter values. If no observer's notes were available, these parameters could not be used later in the processing sequence. An attempt to assign default values using extended functions failed, so two options existed. The chosen option was to confirm the existence of observer's notes before any additional parameter was used. The other option was to assign default parameter values using a LISP function.

If the user pressed carriage return without picking a parameter value, another problem with default values was exposed. In this case, the system assumes that the parameter value is unknown, and cannot make any inferences involving the parameter. Only after a value is entered can the system test it. As a result, the user is instructed to type in a value for every parameter, and is advised on the default values for unknown parameters. This problem is specific to PC+ versions 2.1 and older. Version 3.0 allows default parameter assignment when prompted parameter values are unknown.

This expert system's lack of certainty factors, its simple structure, and its requirement for external decisions indicate that the system could have been written in a conventional programming language (e.g., FORTRAN). However, writing the inference engine, and then creating and maintaining the system structure would have been difficult. For these reasons, an expert system development tool was utilized. PC+ had access to more memory, was faster, and had greater opportunities for expansion on the EXPLORER than on the TI PROFESSIONAL. When a link is established between the EXPLORER and the Gould 9750 computers, a user will be able to process data from the EXPLORER. The VSPPA can be expanded to call the required VSEIS modules, instead of telling the user which modules to call from a Gould 9750 terminal. Although PC-PLUS did not appear as robust on the EXPLORER as on the TI PROFESSIONAL, it had greater flexibility and capabilities. One consideration in the design of this system was that future modification and expansion should be simple. For this reason, the system was configured with one frame for obtaining initial data, one frame for deciding which processing steps to execute, and one frame for each major processing step. This design is simple in theory, but not in practice. The processing sequence is not always clearly defined. Different types of filtering can be applied at many points in the processing sequence. Amplitude corrections are applied in upwave–downwave separation, deconvolution, and amplitude correction frames. Wavelet shaping deconvolution can be applied in two different processing steps for two different reasons. Wavelet shaping and amplitude correction steps may be skipped if processing time is severely limited. Three-component data require a processing sequence only loosely comparable to that for vertical-component data.

Preliminary testing has been performed on the VSPPA. Students taking future VSP processing classes in the CSM Department of Geophysics will use the expert system as a tutorial. This testing will eventually reveal the weaknesses in advice offered by the system, and will highlight areas where additional advice is needed. The VSPPA, one of the first expert systems to assist in processing seismic data, applies to most VSP surveys, and provides detailed processing instructions. Advice is offered specifically for processing with the VSEIS package on the Gould 9750 computer. However, the system is flexible enough that advice could be easily substituted for another processing package on another computer. Although a prototype, the system contains the basic information for a complete VSP processing advisor. The era of the expert system is here, and VSP data processing may be changed forever.

References

Agena, W., 1988, "STAFOID," an expert system to identify seismic tape formats: Colorado School of Mines Publications, T-3561.

Balch, A.H., interviewed by P.A. Crisi, Colorado School of Mines Department of Geophysics, Golden, CO. spring semester 1987, fall and spring semesters 1988.

Buchanan, B., and Shortliffe, E., 1984, Use of MYCIN Inference Engine in Rule-Based Expert Systems: Addison-Wesley, Reading, MA, pp. 229–301.

Spectrum Geophysical Services, 1986, VSEIS processing system reference manual.

Texas Instruments, 1986, Personal Consultant^{Tm} plus for the Explorer^{Tm}.

Waterman, D.A., 1986, A Guide to Expert Systems. Addison-Wesley, Reading, MA, pp. 1–31.

Appendix A: VSPPA Parameters

Initial Data

ACOUSTIC LOG (Y/N): An acoustic log is required for velocities in computation of a synthetic seismogram.

CALIPER LOG (Y/N): A caliper log can be used to determine washout zones, and subsequent recording quality.

CASING HISTORY (Y/N): If the casing thickness and depth are known, degraded data might be traced to casing.

CEMENT BOND LOG (Y/N): A poor cement bond will yield poor coupling between the borehole and the downhole geophone, degrading data.

CHANNELS (DOWNHOLE-VERTICAL-COMPONENT, DOWNHOLE-VERTICAL-COMPONENT-GAIN-ADJUSTED DOWNHOLE-X-COMPONENT DOWNHOLE-X-COMPONENT-GAIN-ADJUSTED DOWNHOLE-Y-COMPONENT DOWNHOLE-Y-COMPONENT-GAIN-ADJUSTED BURIED-SOURCE-MONITOR WELLHEAD-MONITOR TIME-BREAK-MONITOR OTHER): Processing options depend in part on which channels were recorded.

CROSS-CORRELATED (Y/N): Vibroseis data must be cross-correlated to produce a Klauder wavelet before further processing.

DIPPING-BEDS (Y/N): If reflections are expected from anything other than nearly horizontal beds, migration will be performed.

DIREC-DRILL-WELL (Y/N): A directionally drilled well will require vertical stacking of levels near the surface for a surface seismic tie at the borehole.

ELEV-CORR-VEL (POS.#): The elevation correction velocity—typically the velocity of sound, 1100 ft/s—may be used in correlating a VSP with surface data.

EXPLOSIVE-DEPTH-LOG (Y/N): An explosive depth log can be used to apply a static correction to the data.

EXPLOSIVE-SIZE-LOG (Y/N): Explosive size may affect the shot waveform.

EXPLOSIVE-TYPE (DYNAMITE CHEMICAL OTHER UNKNOWN): Different explosive types may exhibit different shot characteristics.

FEET-METERS (FEET METERS): Distance units for the survey.

FILTER-HIGH-FREQ (POS.#): Frequencies up to 400 Hz may be recorded by equipment in a high-precision survey. The default cutoff is 200 Hz.

FILTER-LOW-FREQ (POS.#): Very little information is contained in frequencies under 5 Hz, the default instrument cutoff.

FORCE-LOG (Y/N): Weight drop force may affect the shot waveform.

GAMMA–GAMMA-LOG (Y/N): A gamma–gamma log is required for densities in computation of a synthetic seismogram.

GEOLOGY (Y/N): Knowledge of the bedding orientation and rock types of the area investigated with this VSP can help with processing.

GEOPH-HIGH-FREQ (POS.#): Frequencies up to 400 Hz may be recorded by geophones in a high-precision survey. The default cutoff is 200 Hz.

GEOPH-LOW-FREQ (POS.#): Very little information is contained in frequencies under 5 Hz, the default geophone cutoff.

GEOPHONE-REF-LEVEL (KELLY-BUSHING GROUND-SURFACE PLATFORM-DECK OTHER UNKNOWN): The geophone reference level is the VSP survey datum.

GEOPHONE-TYPE (X-Y-Z TRIAXIAL VERTICAL UNKNOWN): Vertical and three-component geophone orientations are used for different processing goals.

GUN-ARRAY (IN-LINE CROSS-LINE CIRCULAR VERTICAL UNKNOWN): A specific gun array may be chosen to yield destructive interference of noise.

GUN-CHAMBER-SIZE (Y/N): Gun chamber size will affect the source impulse and the resulting waveform.

GUN-DEPTH-LOG (Y/N): A marine static correction can be computed from a gun depth log and the water velocity.

GUN-PRESSURE-LOG (Y/N): Gun pressure will affect the source waveform.

HOLE-DEPTH (POS.#): The hole depth is used to project impedance log estimates below total depth.

LAND-MARINE (LAND MARINE UNKNOWN): A simple gapped deconvolution operator may be applied to remove water bottom multiples if a marine survey was run.

LIMITED-TIME (Y/N): The processing steps of WAVELET-SHAPING and AMPLITUDE-CORRECTION may be conditionally skipped if time is severely limited.

NUM-CHANNELS (POS.#): The number of recorded channels is found from either observer's notes or a plot of the original data.

NUM-FLD-TAPES (POS.#): Field tapes must be read into the computer before processing.

NUMBER-SOURCES (POS.#): Additional sources are desirable at a greater source-well offset.

OBSERVER'S-NOTES (Y/N): This information about the well and recording equipment may be helpful with processing.

P-S-SOURCE (COMPRESSIONAL SHEAR UNKNOWN): Vibroseis and land air guns may be compressional- or shear-wave sources.

PREVIOUS-USE (Y/N): Help is offered to first-time users of the VSPPA.

RECORD-LENGTH (POS.#): The data record length is checked against the maximum allowed VSEIS trace length.

RIG-HEADING (MULTILINE): Any information from the observer's notes that may be helpful to processing but has not been entered can be stored here.

SAMPLE-INTERVAL (POS.#): Data may be lost in resampling if the sampling interval is too sparse.

SEG-Y (Y/N): VSEIS cannot read data unless it is in trace-sequential, SEG-Y format.

SHOTPOINT-LOG (Y/N): A shotpoint log is used to determine the survey geometry for a normal-moveout correction.

SOURCE-SPACING (UNIFORM, NONUNIFORM, UNKNOWN): Sources may be spaced irregularly to yield destructive interference of noise.

SOURCE-TYPE (VIBRATOR, LAND AIR GUN, MARINE AIR GUN, WATER GUN, EXPLOSIVE, WEIGHT DROP, OTHER, UNKNOWN): Different source types have different shot waveform characteristics and require different processing procedures.

SURFACE-SOURCE (VIBRATOR, LAND AIR GUN, MARINE AIR GUN, WATER GUN, EXPLOSIVE, WEIGHT DROP, OTHER, UNKNOWN): The surface source type must be known for VSP correlation with surface seismic data.

SURVEY-DATE (SINGLE-LINE): The survey date is used for reference.

SURVEY-GOAL (SURFACE SEISMIC TIE, GEOLOGY NEAR WELL, PREDICT AHEAD OF BIT, SURFACE SEISMIC EVAL STRAT TRAP EXPLRN OTHER UNKNOWN): Different survey goals will require different information and different processing steps.

SURVEY-TYPE (ZERO OFFSET, FAR OFFSET, WALKAWAY): Investigation with a zero-offset source will yield information close to the borehole, while an offset source will give information farther from the borehole. Additional processing is required for offset sources.

SWEEP (UPSWEEP DOWNSWEEP UNKNOWN): The vibroseis sweep direction will be used along with the frequency limits and sweep length to define the sweep.

SWEEP-END-FREQ (POS.#): Frequency limits on the vibrator sweep will define frequency limits on the data. Limits of 0.5 to 200 Hz define the typical sweep range.

SWEEP-LENGTH (POS.#): A typical vibroseis sweep will last between 0.5 and 20 s.

SWEEP-SAVED (Y/N): The stored sweep can be cross correlated with the initial data to create a zero-phase Klauder wavelet section.

SWEEP-START-FREQ (POS.#): Frequency limits on the vibrator sweep will define frequency limits on the data.

TOOL-SLIP (Y/N UNKNOWN): Downhole geophone slip may be responsible for ringing in the data.

VSEIS (Y/N): Information is offered for new users of the VSEIS processing package.

WEATHERING-DEPTH (POS.#): The weathering depth can be used to determine whether the source was energized in a near-surface layer, and whether the velocity of this layer can be used to correct the source to ground elevation.

WEATHERING-VELOCITY (POS.#): The weathering velocity, typically in the range of 1500 to 2500 ft/s, is used to correct the source depth to ground elevation.

WELL-GROUND-ELEV (POS.#): Ground elevation at the well is used as a reference level for the survey.

WELL-LOCATION (SINGLE-LINE): The location of the well can be used to tie information from this well to other VSP and surface seismic surveys.

WELL-NAME (SINGLE-LINE): The well name is used for reference.

WELL-SEISMIC-DATUM (KELLY BUSHING, GROUND SURFACE, PLATFORM DECK, OTHER, UNKNOWN): The well seismic datum is the surface seismic survey datum.

Processing Steps

COMPLETION (NONE, FILTERING, DATA INPUT AND TRACE SORTING, SHIFT CORRECTION, WAVELET SHAPING, EDITING AND STACKING, MULTI-CHANNEL VELOCITY FILTERING, AMPLITUDE CORRECTION, DECONVOLUTION, NORMAL MOVEOUT CORRECTION, MIGRATION, VERTICAL STACKING, TRANSFER FUNCTION COMPUTATION, IMPEDANCE LOG ESTIMATION): Based on which processing steps are completed and initial data, the next applicable step will be executed.

NEXT-PROCESS (FILTERING,..., IMPEDANCE LOG ESTIMATION): A final chance is offered to execute additional processing steps using the input from this consultation.

PROCESS (I-DON'T-KNOW, FILTERING,..., IMPEDANCE LOG ESTIMATION): The user is given an opportunity to pick the processing steps to consult, or to ask advice on which steps to consult.

Data Input and Trace Sorting

INSRT-ADVICE (Y/N): Advice is offered for the processing step following INSRT.

REVERSE-POLARITY (Y/N): Traces recorded with reverse polarity are flipped.

TAPE-INPUT (Y/N): Advice is given for reading field tapes into data files.

Shift Correction

CONSTANT-PICK (Y/N): A constant time break correction is applied if first break picks are within half of a sampling interval.

DC-BIAS (Y/N): If necessary, a correction is made for direct current bias.

INTERACT-TERMINAL (Y/N): If a Tektronics terminal is available and the user wishes, he may interactively pick time breaks.

REFRACTIONS (Y/N): Refraction arrivals may exist in the vertical component. If so, they should be removed before a shift correction is applied.

SHIFT-ADVICE (Y/N): Advice is given for the processing step following SHIFT.

Wavelet Shaping

SHAPE-ADVICE (Y/N): Advice is given for the processing step following SHAPE.

VARIATION (Y/N): If noticeable variation exists in the source monitor channel, the source waveform is changing and must be standardized.

Editing and Stacking

DEEP-LEVEL (POS.#): The deepest recorded level is used to compute the total number of levels.

LEVEL-INTERVAL (POS.#): The distance between recording levels is used to compute the total number of levels.

MULTIPLE-SHOTS (Y/N): If more than one shot was recorded on each level, stacking will be required.

REPEATED-LEVELS (Y/N): Only one set of recordings on a level is needed, so repeated levels can be discarded.

SHALLOW-LEVELS (POS.#): The shallowest recorded level is used to compute the total number of levels.

STACK-ADVICE (Y/N): Advice is given for the processing step following STACK.

TEST-STACKING (Y/N): If stacking tests were performed, extra traces must be eliminated in the final stacked section.

UNEVEN-INTERVAL (Y/N): If more than one downhole geophone spacing interval was used in the survey, extra traces must be eliminated in the final section.

Multichannel Velocity Filtering

MCVF-ADVICE (Y/N): Advice is given for the processing step following MCVF.

SMED-FK-FILT (SPATIAL-MEDIAN, FREQUENCY-WAVENUMBER): Either an f–k or a spatial median filter can be applied.

Amplitude Correction

AMPCOR-ADVICE (Y/N): Advice is given for the processing step following AMPCOR.

BAD-TRACES (Y/N): If an explosive source was used, amplitude correction might not work.

SPHERICAL-DIVERGENCE (Y/N): A spherical divergence correction is applied separately to upwaves and downwaves to compensate for logarithmic amplitude decay.

Deconvolution

DECON-ADVICE (Y/N): Advice is given for the processing step following DECON.

DECON-DESIRED (SPIKING-SIGNATURE, TIME-SPACE-VARIANT, PREDICTIVE, WAVELET-SHAPING, UNKNOWN): The desired deconvolution is chosen.

DECON-GOAL (REMOVE-SOURCE-WAVEFORM-AND-REVERBS, SHAPE-AMPLITUDE-FRE-QUENCY-RESPONSE): If the desired deconvolution is unknown, the user's deconvolution goal will be used to recommend a type of deconvolution.

DECON-OUTPUT (Y/N): Advice is given to output the deconvolved section to magnetic tape.

TRANSMISSION-LOSS (Y/N): A time-and-space variant deconvolution filter is applied if the downwave changes rapidly in time or space.

Normal Moveout Correction

NMO-ADVICE (Y/N): Advice is given for the processing step following NMO.

NMO-OUTPUT (Y/N): Advice is given to output the NMO-corrected section to magnetic tape.

VEL-GEOM (Y/N): Velocity and geometry information is required for a normal moveout correction.

Migration

MIGRN-ADVICE (Y/N): Advice is given for the processing step following MIGRN.

MIGRN-NEEDED (Y/N): If an offset source is used, but no information about geology near the well is available, the need for migration can be determined.

Vertical Stacking

TIME-MATCH (Y/N): If the survey goal is to tie the VSP with surface seismic data, a correction may be required to match event times between the two data sets.

VSTACK-ADVICE (Y/N): Advice is given for the processing step following VSTACK.

VSTACK-OUTPUT (Y/N): Advice is given to output the vertically stacked trace to magnetic tape.

Impedance Log Estimation

IMPLE-ADVICE (Y/N): Advice is given for the processing step following IMPLE.

Transfer Function Computation

ATTEN-COEF (Y/N): If the attenuation coefficient K, has been locally correlated with sand percentage, then it is possible to determine the sand percentage in a section of a well from VSP data.

TRANFN-ADVICE (Y/N): Advice is given for processing following step TRANFN.

Filtering

AC-SOURCE (Y/N): Noise from an alternating current source can be removed with an f–k filter.

BUBBLES (Y/N): The bubble effect from a marine air gun can be removed with an f–k filter.

DISSIPATION (Y/N): Near-surface dissipation cannot be removed with filtering.

FILT-ADVICE (Y/N): Advice is given for the processing step following FILT.

FILTER-DESIRED (FREQUENCY-WAVENUMBER, BAND-PASS, SPATIAL-MEDIAN, TIME-DOMAIN-MEDIAN, MUTE, UNKNOWN): The desired filter is chosen.

FILTER-GOAL (ZERO-OUT-RANDOM-NOISE, REMOVE-RANDOM-NOISE, REMOVE-CO-HERENT-NOISE, REMOVE-SHEAR-WAVE-ENERGY, LIMIT-FREQUENCY-RESPONSE): If the desired filter is unknown, the user's filtering goal will be used to recommend a filter type.

HIGH-FREQ-BURSTS (Y/N): A time-domain median filter can be used to remove high-frequency bursts from data.

RINGING (Y/N): Band-pass filtering will suppress data ringing.

SMED-FK (SPATIAL-MEDIAN FREQUENCY-WAVENUMBER): Either an f–k or a spatial-median filter can be applied.

SURFACE-HIGH-FREQ (POS.#): VSP data frequencies can be band limited to match surface data frequencies if the VSP contains higher frequencies than the surface data.

SURFACE-LOW-FREQ (POS.#): VSP data frequencies can be band limited to match surface data frequencies if the VSP contains lower frequencies than the surface data.

TUBE-WAVES (Y/N): Either spatial-median or f–k filtering can be applied to remove tube waves.

Appendix B: User's Guide

This guide demonstrates how to start the VSPPA, how to use the expert system, and what to do with the advice offered. The VSPPA is available on the Colorado School of Mines Geophysics department EXPLORER computers. Before starting a consultation, gather as much information as possible about your VSP survey. The expert system needs your help to give the best possible advice.

Your next step is to log onto the EXPLORER and to start the expert system. Turn on the monitor and adjust the brightness and intensity controls to your satisfaction. Next, log onto the EXPLORER in a LISP Listener window. The current window type will be displayed in the lower left corner of the screen. If you are not in a Listener window, press the keys "SYSTEM" and "L" simultaneously and wait a few seconds. If nothing happens, press the keys "META," "CTRL," and "RUBOUT" on the left side of the keyboard, and the keys "META" and "CTRL" on the right side of the keyboard at the same time. This will reboot the EXPLORER.

Once the LISP Listener window is displayed, type (newuser) if you have not used the EXPLORER before, or (login 'YOUR-NAME) if you have, and wait a few seconds. Now, press the keys "SYSTEM" and "+." This will start PC+. Press key "F2" and pick the "CHANGE DIRECTORY" command by entering carriage return. Change the default path name from "lm:pcxkb" to "lm:pcrisi" and type carriage return. Then pick "VSPPA.KB#–––" and type carriage return again. The expert system will take a few minutes to load.

Now pick the "CONSULT" option to start a consultation. This expert system is designed with different modules corresponding to distinct steps in the VSP processing sequence. You will have the freedom to use any modules you desire, and the guidance to help you select the correct ones. A set of consultation aids is available to the user. Key "F1" will offer help in determining some of the system parameter values. Key "F2" will access options in the "Commands:" box at the bottom of the screen. All parameter values can be selected using the arrow keys and the keyboard. Use of the mouse to select options may result in a PC-PLUS error ending the consultation. Its use is not recommended. Instructions on how to answer parameter prompts are given at the bottom of the PC+ screen (above the "Commands:" box). When consulting the expert system, use the "F1" key if you are uncertain how to answer a question. Read the numbered procedures above the "Commands:" box for help entering parameter values. Be sure to answer every question. Options in the "Commands:" box are described below:

WHY: Lets the user ask why the system needs the information for which he is being prompted.

HOW: Tells the user how the system has determined parameter values other than those supplied in prompts.

REVIEW: Displays the list of parameters prompted for so far in the consultation, and allows modifications.

SAVE PLAYBACK FILE: Saves a record of a partial or complete consultation.

GET PLAYBACK FILE: Loads a partial or complete consultation record previously saved by a SAVE PLAYBACK FILE command.

PRINT CONCLUSIONS: Writes a record of the consultation and conclusions to a printer or disk file.

TRACE ON/OFF: Writes a copy of the rule tracing sequence to the screen, a printer, or a disk file. A record of advice given during your consultation is stored in file "lm:pcrisi;consult.txt" for future reference.

During the consultation, you may experience delays while writing to the "consult" file and while loading segments of the expert system. This activity is displayed on the flashing line at the bottom of your screen. Your "consult" file can be printed out and taken to the Gould computer for data processing. Terminals are located in the computer center at the north end of the Green Center. Use a FALCO terminal for interactive graphics plotting. The graphics mode on these terminals is entered and exited by pressing the keys "CTRL" and "F6" simultaneously. Field tapes should be checked in and all plots can be picked up at the computer center dispatcher's desk. Before processing VSP data, your directory must be tailored to run the VSEIS VSP processing package. First, make sure that you have access to enough disk space for a few files as large as a megabyte each. Next, add the command setenv path".: ~ /bin:/usr/local:/usr/ucb:/bin:/usr/bin:/usr/sup: /gplib "to your .login file. Also, create a file named "datadirec" with the line "/disk/directory" where "disk" is the name of the disk and "directory" is the name of the directory you plan to use. Before making these changes, check with the UNIX supervisor of the computing center or the seismic data processing supervisor to verify the current search path. A few characteristics of the VSEIS package are worth noting. All VSEIS processing is interactive. The processing suite is modular. Data files are stored on disk and data are handled diskfile to diskfile. Interactive plotting is possible using Tektronics-compatible terminals. The maximum number of traces in some modules is 600 and the maximum trace length is 6000 samples. Trace numbers are always resequenced in new files, so processed data will not retain the original trace numbering. The HELP module in VSEIS will list options and explanations. The command "vseis" will start the VSP processing package, and the command "exit" will get the user out of VSEIS. If additional information about the VSEIS module is desired, please consult the reference manual.

Appendix C: Knowledge Base Maintenance

Since this expert system is a prototype, it will be expanded and modified by users other than the knowledge engineer. These changes can be made quickly and simply in the VSPPA using PC+.

Any effort to modify the VSPPA should be done with a complete list of parameters (Appendix A) and a knowledge map in hand. These two items define the expert system. Changes should not conflict with the existing expert system structure. Design of the VSPPA was based on a survey recorded with a vertical-component downhole geophone and a zero-offset land air gun source. Additional work will be required to incorporate three-component data. Currently, one frame represents each major step in the processing sequence. Additions for three-component data and other desired options should fit into this existing framework.

The VSPPA contains 22 graphic objects and 2 functions for file output, along with 220 rules, 60 processing parameters, and 100 dummy parameters in the knowledge base frames. Each dummy parameter is denoted by a question mark following its name. All files used by the expert system are located in directory "lm:pcrisi;." The directory is large (1.3 megabytes), and only a few versions of the expert system should be saved at one time. To clean out the directory, use the "META-X CLEAN DIRECTORY" command in the ZMACS Editor on the Explorer™. If, during modifications, the PC-PLUS process receives an error, use the "TERM-0-S" command to enter the debugger, and then "ABORT" out of the error. To restart PC-PLUS after the error, enter the LISP Listener window using "SYSTEM-L," and then enter "(PC-PLUS)." Always save your revisions before leaving the Explorer.

6
Expert Systems for Seismic Interpretations and Validation of Simulated Stacking Velocity Functions

Kou-Yuan Huang

The expert systems studied are for (1) validation of simulated seismic stacking velocity functions and (2) for seismic interpretations.

Velocity Expert System (VELXPERT) is used to find knowledge representations and inference techniques that are appropriate for the seismic stacking velocity analysis problem domain and to use these representations and techniques to develop a prototype expert system for seismic stacking velocity validation that can detect several common errors and rate the quality of a given velocity function.

The Seismic Interpretation Expert System (SIES) is an attempt to apply expert system techniques in building a system to assist seismic interpreters. SIES is a backward-chaining system using symbolic pattern matching and streams of association lists with certainty factors to allow inexact reasoning. Rules in the knowledge base were built on seismic stratigraphy methods and seismic interpretation knowledge to determine subsurface features. Pattern recognition results can be used as facts. As a prototype, SIES illustrates how an expert system using pattern matching and streams can be used to make inferences and represent knowledge in a domain as complex as seismic interpretation.

Expert System for Validation of Simulated Seismic Stacking Velocity Functions

Introduction

Current seismic reflection data processing centers around the common depth point stacking technique to improve data quality (Mayne, 1962; Dobrin,

1976; Coffeen, 1978). To perform stacking the data must be corrected to compensate for variations in data collection geometry. This correction, called normal-moveout correction (NMO), is a nonlinear function of velocity (called Vnmo), distance, and time. Time and distance are known, but Vnmo must be determined. Vnmo determination has traditionally been done by a human expert who interprets time–velocity data. The interpretation results in a time–velocity function that is used to NMO correct the data. The interpretation process is a bottleneck in the seismic data processing stream. Several methods have been tried to automate the process (Taner and Koehler, 1969; Schneider and Backus, 1968; Garotta and Michon, 1967).

The technique presented here uses an expert system (VELXPERT), acting as a consultant, to test a given velocity function to determine its validity. Given several test cases the system was able to identify some common errors made by velocity interpreters. These results show the applicability and feasibility of expert system technology to the area of seismic velocity interpretation and its automation.

Figure 6.1 shows the major components of the VELXPERT system and how they are interconnected. Figure 6.2 shows the basic control flow of the system, and Figure 6.3 shows how data flow through the system. Inputs include the velocity function, some identification information about the gather, an area data base, and other area information if available. Area information could include almost any related information. Area information contains average stacking velocity curves for area and a set of constraining values (such as minimum and maximum velocities allowed). Other information that could be added to the area data base

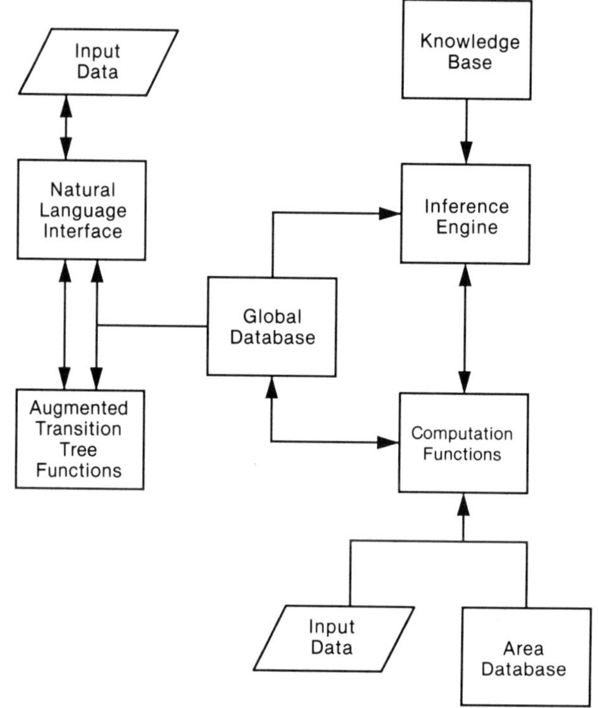

FIGURE 6.1. Components of VELXPERT system.

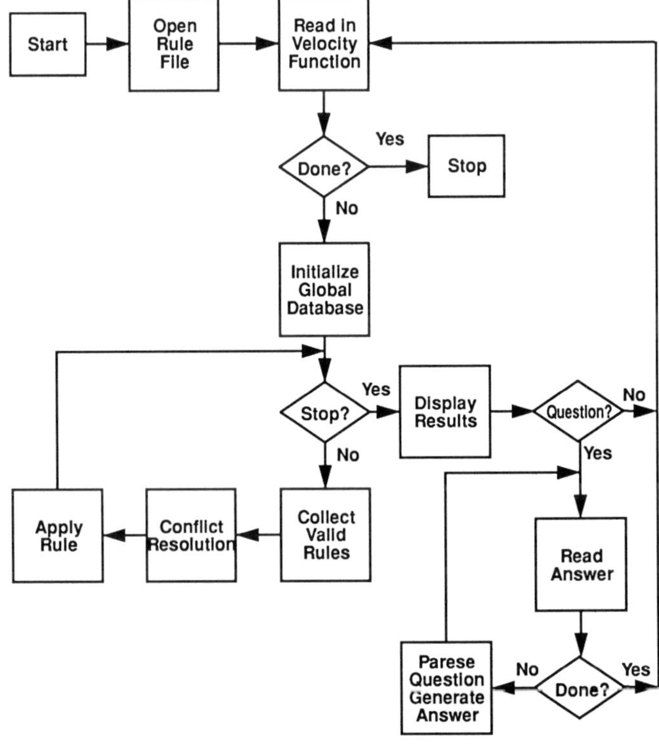

FIGURE 6.2. VELXPERT control flow diagram.

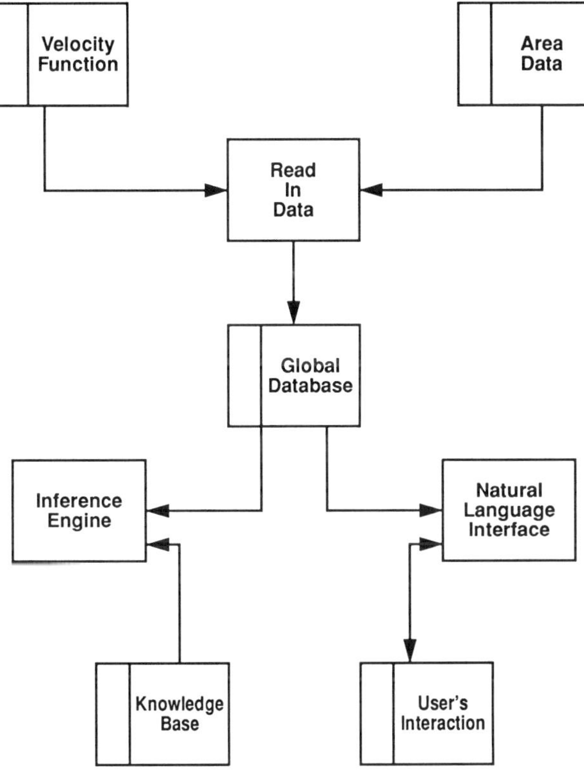

FIGURE 6.3. VELXPERT data flow diagram.

includes well logs, geochemical data, basin history data, etc.; anything that could be of use to the process. Other area information that can be used if appropriate includes sonic well logs and average velocity functions for specific lithologies. These inputs go through a preprocessing step to initialize the global data base. This information is used by the system to determine validity of the function and to answer questions about the function.

Knowledge Representation

The knowledge in the knowledge base is stored as rules, which are represented in an IF–THEN format. The production rules of VELXPERT system are listed in Appendix A. These rules are written by LISP type which is similar to Winston and Horn's (1984) rule-based expert system.

Inference Engine

The inference engine uses forward chaining and a combination of context limiting and specificity ordering as the control strategies (Winston, 1984) for choosing rules to be triggered and fired.

Currently the states GENERAL, LAND, MA-RINE, COMPUTE CERTAINTY, and STOP are implemented in the context limiting control. Initially the context is set to GENERAL by the preprocessing step. During this context the system checks for common errors that would make the velocity function invalid. These include invalid time NMO velocities (either too small or too large), invalid interval velocities, and several others. The next context to be invoked is the LAND or MARINE context depending on whether the data were collected in a land or marine environment. In this context further tests are made to determine if the function is invalid. In the MARINE context this includes testing for peg-leg and water bottom multiples. If the velocity function is rejected in any of these contexts the context is changed to STOP and the natural language interface takes over. If the function is still assumed to be valid, then the COMPUTE CERTAINTY context is invoked and the system tries to establish a confidence factor for the function based on heuristic measurements.

Natural Language Interface

Once a conclusion has been reached by the inference engine VELXPERT asks the user if they want to ask any questions concerning the conclusion. If the answer is yes then the natural language interface takes over and a question and answer mode is invoked that allows limited natural language communication between the user and VELXPERT.

VELXPERT uses augmented transition trees (ATTs) (Winston and Horn, 1984) to implement the parsing of the questions and the formation of answers. Currently five basic answer generation functions are available, one for each of the questions Why, How, Where, What, and Was. Was answers questions asking if a particular rule was used. Why answers questions about why an assertion was used. How answers questions about how a deduction was made. Where answers questions about where a particular variable value was found. What answers questions about what a particular variable's value is.

Heuristic Measurement Calculations

Once the system has weeded out the more obvious errors that fall into the fatal mistake category, then more subjective, heuristic knowledge is used to determine if the velocity function specified is valid or not and if valid to what degree.

The basic system used for computing a confidence factor in the COMPUTE CERTAINTY context is to assume that all heuristics computed have equal weight in determining the confidence factor placed on the velocity function. Several of the heuristic measurements made by the system compute their weights by measuring the closeness of the points in the velocity function to some other data.

The approach taken is to calculate closeness using a linear function that measures the distance of the two points being examined and normalizing to the range zero to one. These normalized distance measurements are then summed together and the sum is divided by the number of points being compared. This always gives a number between 0 and 1 that is used as the heuristic measurement of the validity of the function.

As an example, one of the rules wants to check that the difference between the Vnmo velocities found and the well average velocities are not too great. The procedure used is to calculate the difference between each point in the Vnmo velocity function and the well velocity function. This difference is then divided by the well velocity to give a percentage of difference between the two numbers. The percentage is then compared to the maximum allowed deviation from well velocities (5% in the current system). If the percentage is over the maximum allowed it is set to the maximum allowed. The percentage is then divided by the maximum allowed percentage and subtracted from one to normalize it in the range zero to one. These weights are added and divided by the number of points in the Vnmo velocity function. If the velocities were exactly the same we would get a weight of one. If all the velocities differed by more than 5% from the well velocities then we would get a weight of zero.

Once all the heuristic confidence factors are available the system will add them together and divide by the number of heuristics computed. This gives the final confidence factor for the function. Values greater than 0.9 (90%) are assigned an excellent confidence level, greater than 0.75 a good confidence level, greater than 0.6 a fair confidence level, greater than 0.45 a poor confidence level, and less than 0.45 are rejected by the system.

Testing

Test Data Generation

Two sets of tests were developed. The first set consists of test cases containing several common errors that would render the velocity function totally invalid. The second set consists of test cases that show the system using its heuristic measurements to demonstrate how the confidence factors vary with changes in the assumptions made about the function or changes in the function values.

The test data used here are synthetic velocity functions. Two basic geologic models were developed, both of which assume a horizontally bedded earth. One represents a typical Gulf Coast marine clastic section; the other is more interesting and represents a function from land data on the north slope of Alaska. The north slope model is more interesting because signs of permafrost and overpressuring are shown. Figure 6.4 contains the stacking velocity functions that would be used to NMO correct the gathers.

Gulf Coast Model			North Slope Model		
depth	Time (sec)	Velocity (ft/sec)	depth	Time (sec)	Velocity (ft/sec)
0	0.000	5,000	0	0.000	11,000
-500	0.200	5,000	-1000	0.185	10,800
-1000	0.393	5,325	-2000	0.382	10,516
-2000	0.690	6,099	-3000	0.586	10,277
-3000	0.969	6,348	-4000	0.776	10,327
-4000	1.231	6,635	-5000	0.961	10,418
-5000	1.480	6,882	-6000	1.141	10,530
-6000	1.710	7,158	-7000	1.314	10,678
-7000	1.927	7,418	-8000	1.482	10,824
-8000	2.133	7,670	-9000	1.646	10,970
-9000	2.329	7,916	-10000	1,804	11,124
-10000	2.509	8,190	-11000	1.961	11,267
-11000	2.682	8,453	-12000	2.113	11,410
-12000	2.850	8,694	-13000	2.273	11,490
-13000	3.013	8,921	-14000	2.437	11,538
-14000	3.172	9,140	-15000	2.655	11,365
-15000	3.326	9,354	-16000	2.836	11,241
-16000	3.476	9,558	-17000	2.979	11,485
-17000	3.622	9,759	-18000	3.117	11,635
-18000	3.765	9,953	-19000	3.251	11,793

FIGURE 6.4. Stacking velocity functions for geologic models.

The stacking velocity data were generated by developing a simple geologic model of the subsurface, performing nonzero offset ray-tracing on the model, and saving the time–velocity function computed from the model. The interval velocity and depth values used to generate the geologic models were taken from interval velocity functions shown in Sheriff and Geldart (1983). This stacking velocity data, when used to apply the NMO correction to the seismic gathers, give an ideal result (i.e., flattens the reflection curves) for each particular model. Next the functions were corrupted by errors. These corrupted functions are then input to VELXPERT to be tested.

Two area data bases were also generated, one for the marine Gulf Coast area and the other for the land north slope area. The values specified in these data bases are used by the VELXPERT rules to determine the validity of the velocity function. The area data bases contain a stacking velocity function that is the average of all stacking velocity functions in that area. The area data bases also contain values for area geophysical parameter extremes. These parameter extremes include information such as

- The minimum and maximum Vnmo velocities allowed

- The minimum and maximum interval velocities allowed
- Maximum amount of difference to allow between the area velocity function and the velocity function being tested.

For each area a root mean squared (rms) velocity function computed from a sonic log was also generated. This sonic log information and the stacking velocity function in the area data base were initially set equal to the ideal stacking velocity function expected from the models.

Finally, an average clastic rock velocity function was generated. The values for this function were set equal to the clastic velocity function found in the Gulf coast area.

Test Data Used

Test data used to determine if the function is to be rejected (set one tests) include

- Velocity function containing a water bottom multiple
- Velocity function containing a peg-leg multiple
- Velocity function containing an interbed multiple
- Velocity function containing a NMO velocity too low

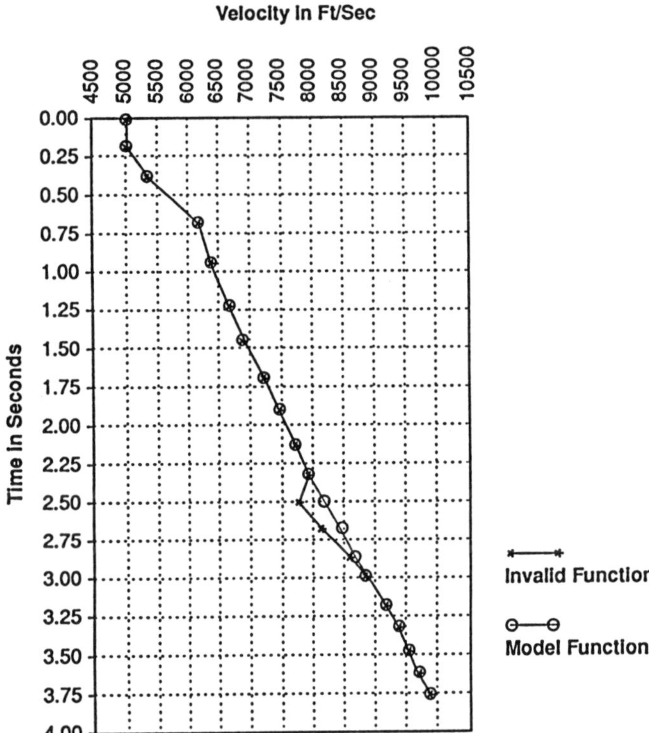

FIGURE 6.5. Velocity function containing peg-leg multiple Gulf Coast model.

- Velocity function containing a NMO velocity too high
- Velocity function containing times that are not increasing
- Velocity function containing times that are not increasing
- Velocity function containing an interval velocity too low (computed by Dix's equation)
- Velocity function containing an interval velocity too high (computed by Dix's equation)
- Velocity function containing a NMO velocity too high at too low a time
- Velocity function containing a NMO velocity too low at too high a time
- Velocity function containing an invalid slope between two time–velocity pairs
- Velocity function containing no shallow velocity inversion, in an area (north slope) that expects one.

Test data used to determine heuristic measurements on a function (set two tests) include

- Velocity function containing a velocity inversion up shallow when none is expected

- Velocity function containing a velocity inversion at depth when none is expected
- Velocity functions with abnormal time separations
- Velocity functions with abnormal velocity separations
- Velocity functions containing velocity inversions
- Perturbed area average stacking velocity functions
- Perturbed sonic velocity functions
- Perturbed average clastic velocity functions.

Experiments

Common Error Detection Results

Twelve tests cases were run to try to detect some common errors found in stacking velocity functions picked by a velocity interpreter. These test cases were designed to illustrate the rejection of a stacking velocity function due to general geophysical constraints. VELXPERT did find the error in each of the tests. Discussion of 4 test cases of the 12 cases follows.

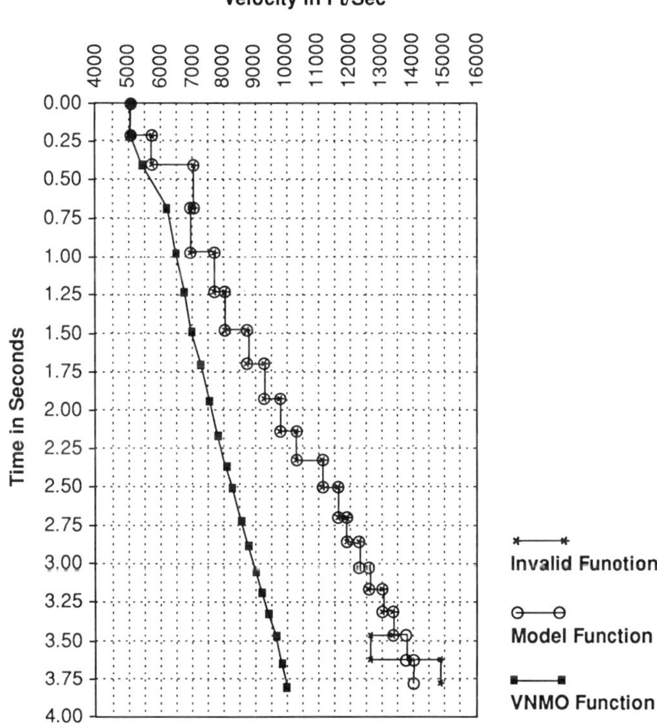

FIGURE 6.6. Velocity function generating an interval velocity too high Gulf Coast model.

Test case one is designed to illustrate the picking of a peg-leg water bottom multiple. This is a common error made by many velocity interpreters when the velocity of the primary events and the peg-leg are close together. Figure 6.5 shows the ideal Gulf Coast stacking velocity function and the invalid function containing the peg-leg water bottom multiple. The water bottom two-way time is 0.2 s and the first order peg-leg water bottom multiple is located at 2.529 s (being generated by the primary event located at 2.329 s) and has been included in the invalid function. VELXPERT correctly finds the error and declares the function invalid.

Test case two is designed to illustrate the picking of a velocity value that results in the calculation of an interval velocity (using Dix's equation in Dobrin, 1976) that is above the maximum interval velocity allowed for a particular area. The maximum interval velocity allowed for an area is determined by the expert and/or by statistical measurements of other functions and well data in this area. For these tests, the maximum interval velocity is allowed to 15,000 ft/s. This error, a common one, is made by velocity interpreters who pick a strong nonprimary event that looks very reasonable but generates an interval velocity that is not. Figure 6.6 shows the ideal Gulf Coast stacking velocity function, its interval velocity function, the invalid function containing the invalid pick, and its interval velocity function. The invalid pick has a velocity value of 9700 ft/s and is located at time 3.622 s. This pick and the one below it are used to calculate an interval velocity. The invalid interval velocity is 15 001 ft/s. VELXPERT correctly finds the error and declares the function invalid.

Test case three is designed to illustrate the picking of a velocity value that results in the slope between two adjacent picks, where the slope is too high. The minimum and maximum slopes allowed in an area are determined by the expert and/or by statistical measurements of other functions in this area. For these tests, the minimum slope is set to be 100 ft/s^2 and the maximum slope to be 3000 ft/s^2. The errors made in picking points in these areas are common involving the picking of strong nonprimary events. Figure 6.7 shows the ideal Gulf Coast stacking velocity function and the invalid function containing the invalid pick. The invalid pick has a velocity

K.-Y. Huang

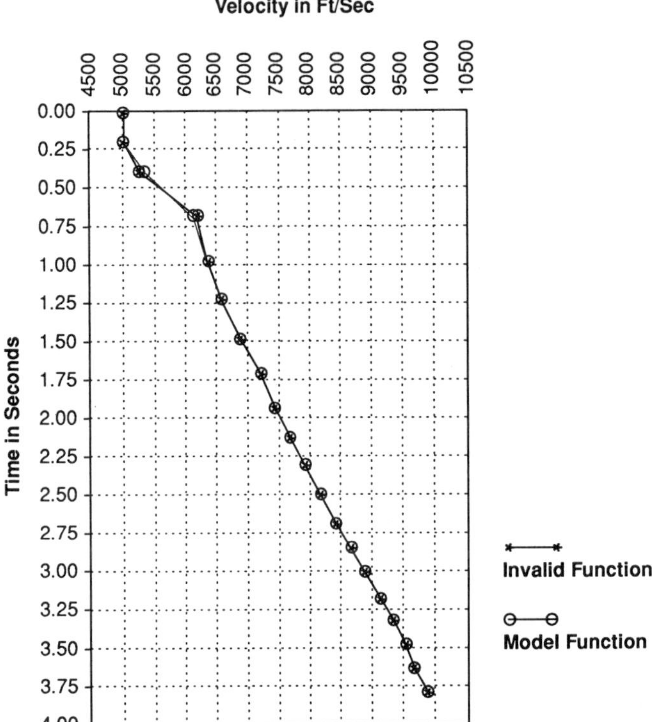

FIGURE 6.7. Velocity function generating a slope too high Gulf Coast model.

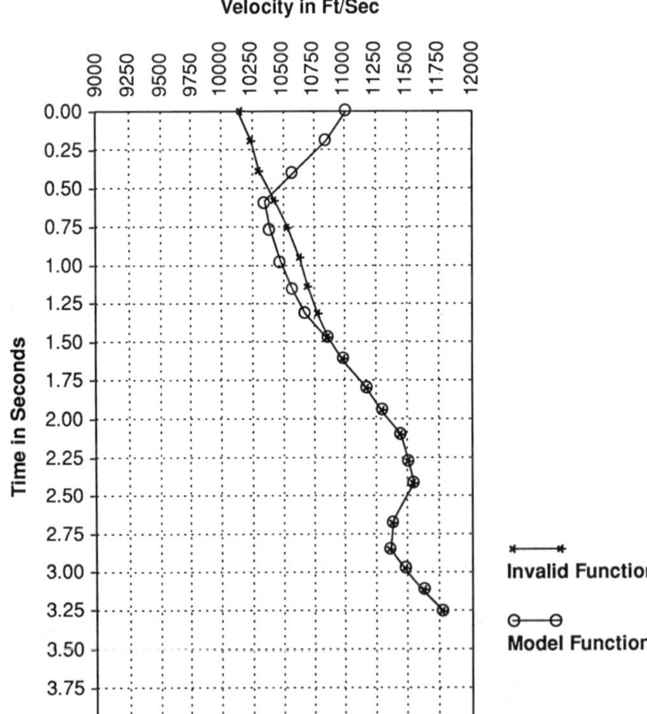

FIGURE 6.8. Velocity function containing no shallow velocity inversion north slope model.

FIGURE 6.9. Effect of failed heuristics on confidence factor.

Heuristic Measurement	Gulf Coast Confidence Factor	North Shope Confidence Factor
Ideal Function (all heuristics applied)	0.996 (excellent)	0.962 (excellent)
Remove sonic match	0.871 (good)	0.820 (good)
Remove Sonic and Clastic match	0.625 (fair)	0.820 (good)
Remove Sonic, Clastic and Area Average Match	0.500 (poor)	0.677 (fair)
Remove Time Separation Match	0.375 (unacceptable)	0.543 (poor)
Remove Velocity Separation Match	0.250 (unacceptable)	0.391 (unacceptable)
Add Velocity Inversions	0.125 (unacceptable)	0.286 (unacceptable)
Remove over-pressure Expected	0.063 (unacceptable)	0.143 (unacceptable)
Remove Shallow Inversions	0.033 (unacceptable)	0.060 (unacceptable)

value of 5260 ft/s and is located at time 0.393 s, resulting is a slope calculated from the pick and the pick above it of 3030 ft/s^2. VELXPERT correctly finds the error and declares the function invalid.

Test case four is designed to illustrate the picking of a velocity function that contains no velocity inversion in the shallow part of the function even though one is required. This test shows how information about an area can be used to require the function to conform to some given shape or pattern. This is a common error for an interpreter to make if he is not aware that a velocity inversion should occur due to permafrost or geological reasons (like a high velocity limestone on the surface over lower velocity clastic rock). Figure 6.8 shows the ideal north slope land stacking velocity function and the invalid function containing no velocity inversion shallow. VELXPERT correctly finds the error and declares the function invalid.

Heuristic Measurements Results

The results obtained from the heuristic measurement tests are best described by illustration of what the heuristics measure, how the confidence levels decrease as more and more velocity picks fail the heuristic measurements, and how many "bad" velocity picks are needed to degrade the confidence level chosen.

Figure 6.9 shows how the confidence factor for our test functions decrease as we allow each heuristic measurement to fail completely. In reality these heuristics would intertwine with each other and it would be difficult to find, for example, a function where the average velocity heuristic measurement was zero but the sonic measurement was one.

Figure 6.10 illustrates how one bad velocity pick affects the overall weight computed for that function. The figure shows the values for our test functions, which have 20 pairs of points per function. These results shown in Figures 6.9 and 6.10 can be expressed in another way as the number of "missed" heuristics allowed at each confidence level supported by the system. Figure 6.11 shows this representation for our 20 time–velocity pair functions.

Suggestions

The development of VELXPERT has shown that a part of the problem of stacking velocity validation

Heuristic Measurement	Gulf Coast Model	North Shope Model
Sonic Log Match	0.00625	0.00714
Clastic Rock Match	0.00625	0.00714
Area Average Match	0.00625	0.00714
Time Separation	0.00658	0.00752
Velocity Separation	0.00658	0.00752
Velocity Inversion	0.00658	0.00752
Shallow Inversion	0.06250	0.07143
Over-pressure expected	0.06250	0.07143

FIGURE 6.10. Effect of one bad velocity pick on confidence factor.

can be solved using expert system technology. More importantly the research has shown the representations and the implementation techniques used are viable.

Some areas deserving more study include

- The frame representational model (Winston, 1984)
- Testing with real data to document and improve the systems performance in one well-understood area.
- Incorporation of other informational sources (event files, adjacent gather's functions, seismic data, porosity data, depth data, age data, etc.)
- Addition of graphic interfaces
- Development of a robust set of heuristics, the determination of how these heuristics interact with one another and how they should be combined.

Seismic Interpretation Expert System (SIES)

Introduction

Seismic interpretation involves analysis of reflections on seismograms to recognize subsurface features that are usually associated with oil entrap-ments such as anticlines, faults, reefs, salt domes, pinchouts, and direct hydrocarbon indicators such as weak, bright, and flat spots. This step is most important in seismic processing because decisions about drilling locations are based on the results. However, the step is time-consuming and requires years of experience.

The Seismic Interpretation Expert System (SIES) applies expert system techniques in building a system to assist seismic interpreters.

Techniques used in the SIES are pattern matching and streams (Winston and Horn, 1984), backward inferencing using certainty factors (Winston, 1984; Shortliffe, 1976), augmented transition trees (Winston and Horn, 1984), and techniques to build the explanatory interface (Winston, 1984; Shortliffe, 1976). Backward-chaining was chosen because the goal is clear, i.e., looking for structures that contain oil or gas. Probability-based method is used in the propagating certainty factors.

The program database contains facts that are either provided by the user or derived from other facts.

The rule scanner is invoked at system initialization to detect syntax errors in the rules.

The natural language interface consists of augmented transition trees capable of parsing questions asked by the user.

	Number of Bad Comparisons Allowed (percentage of total comparisons) Using Given Heuristic Measurements		
Confidence Level	All	No Sonic	No Sonic or Clastic
Excellent	0 - 16 0% - 10%	0 - 14 0% - 10%	0 - 12 0% - 10%
Good	17 - 40 10% - 25%	15 - 35 10% - 25%	13 - 30 10% - 25%
Fair	41 - 64 25% - 40%	36 - 56 25% - 40%	31 - 48 25% - 40%
Poor	65 - 88 40% - 60%	57 - 77 40% - 60%	49 - 66 40% - 60%
Unacceptable	89 or more over 60%	78 or more over 60%	66 or more over 60%

FIGURE 6.11. Effect of bad comparisons on confidence level.

The explanatory interface has access to the knowledge base and keeps track of rules applied to answer "how" and "why" questions.

The knowledge base consists of a set of rules. Context-limiting and specific-ordering of the control strategies are used to design rules. A discussion of the construction of the rules follows.

SIES Rules

Basically, rules about seismic stratigraphy are derived from Mitchum et al. (1977) and Sangree and Widmier (1977) and are used by the inference system along with other geological information such as velocity to derive rock porosity information.

Rules about seismic features such as faults, anticlines, and how to detect and interpret them are from Mitchum et al. (1977a,b), Dobrin (1976), Bubb and Hatlelid (1977), and Coffeen (1984). Recognition of basic elements, such as anticlines, layers, areas with many or few reflections, and breaks in reflection continuity, provides first-level facts. From these, SIES derives intermediate facts and finally the goal. Hypothesis-supporting facts can be provided from seismic data and other geological information. Organization of the rules is illustrated in Figure 6.12. An arrow going from rule #x to rule #y means that #x's conclusion is one of #y's conditions. The complete list of rules is shown in Appendix B. These rules are written by LISP type, which is similar to Winston and Horn's (1984) rule-based expert system.

Control Strategies

Performance of a system relies heavily on its control strategy, i.e., the method to determine what rules to execute next. In this system, the context-limiting and the specificity-ordering strategies (Winston, 1984) are used.

In the context-limiting strategy, rules are divided into many subsets. Only one subset is active at a time. When selecting candidate rules, the system scans only the active subset rather than having to search the complete rule set. The pattern variable in the goal determines what subset is to be used. For example, if the goal is ((<layer) is porous), where < is the left delimiter of the variable layer, the subset for *layer* is activated.

SIES's rules are broken down into subsets as illustrated in Figure 6.13:

Rules about structures are rules 1, 2, 9, 11, 13, 14, 15, 15b, 16, 17, 18, 19, 20, 21, 22, 23, 30, 41, 42, and 43.

Rules about reflections are rules 3, 4, 5, 6, 7, and 8.

Rules about areas are rules 32, 33, 38, 39, and 40.

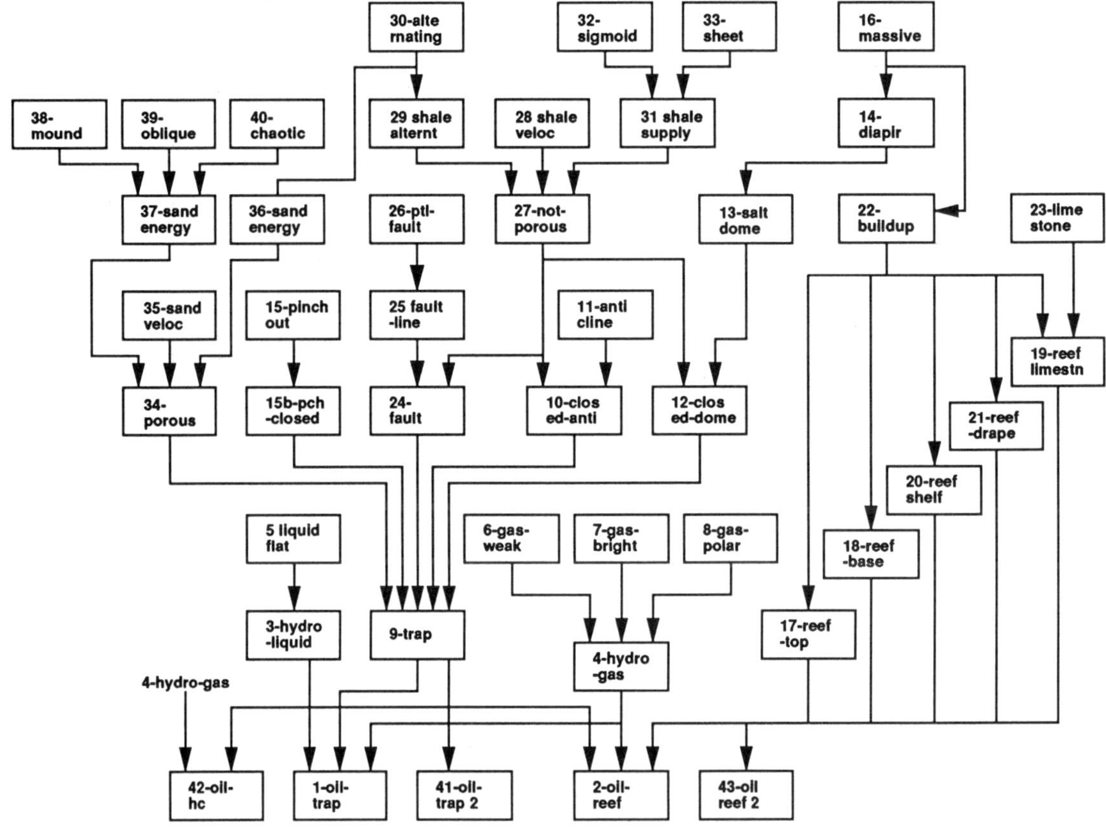

FIGURE 6.12. Organization of seismic interpretation expert system.

Rules about lines are rules 25 and 26.
Rules about layers are rules 10, 12, 24, 27, 28, 29, 31, 34, 35, 36, and 37.

The specificity-ordering strategy gives priority to rules that are more specific to the current situation. Those rules are evaluated first. If they are applicable, less specific rules will not be executed. For example, rule "1-oil-trap" is evaluated before rule "41-oil-trap2." "41-oil-trap2" will be evaluated only if "1-oil-trap" is not applicable.

```
(rule 1-oil-trap 0.95
if (((>structure) is a trap)
    ((>reflection) is an indication of hydrocarbon
     accumulation)
    ((<reflection) is part of (<structure)))
then ((<structure) contains oil or gas))
(rule 41-oil-trap2 0.8
if (((>structure) is a trap))
then ((<structure) contains oil or gas))
```

The following specificity relationships exist in the rule set:

Rule 1-oil-trap is more specific than rule 41-oil-trap.
Rule 1-oil-trap is more specific than rule 42-oil-hc.
Rule 2-oil-reef is more specific than rule 43-oil-reef 2.

Backward-Chaining System

The overall structure of the inference system is that of a backward-chaining system using streams of association lists with certainty factors to allow inexact reasoning.

A backward-chaining system works from goal to supporting facts, using desired conclusions to determine what facts to look for. Usually, a final goal cannot be deduced directly from supporting facts, but through many levels of intermediate results.

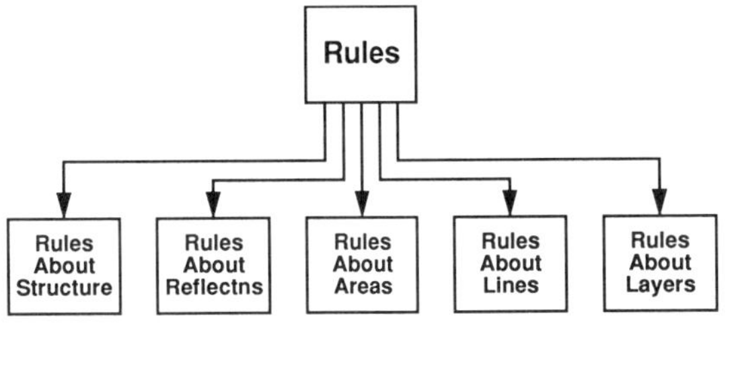

FIGURE 6.13. Rule subsets.

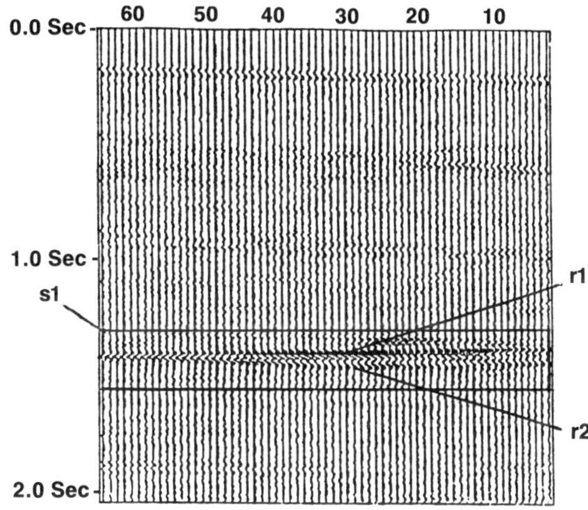

FIGURE 6.14. Seismogram at Mississippi Canyon.

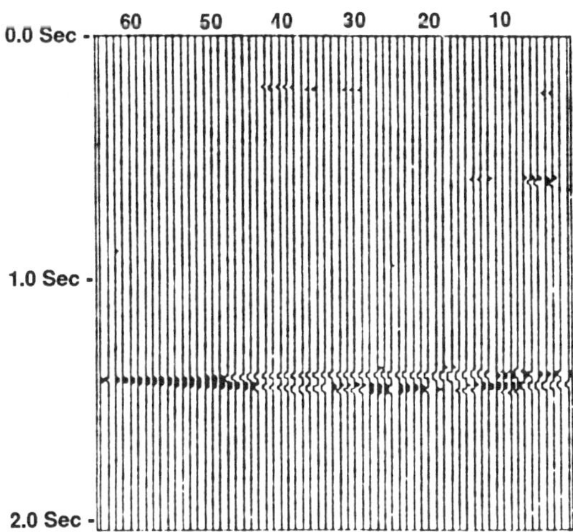

FIGURE 6.15. Tree classification result at Mississippi Canyon.

SIES's ultimate goal is to determine structures that contain oil or gas. In its symbolic representation, the goal is stated as follows:

((<: structure) contains oil or gas)

SIES first determines if the goal can be satisfied by existing facts. If no facts are found, the system searches for rules that can deduce the goal.

SIES's backward-chaining algorithm can be summarized in the following.
Calling sequence:

(backward goal initial-stream)

Algorithm 1:

1. If the goal can be satisfied by existing facts, return corresponding facts.
2. Select rules whose conclusions match the goal.
3. For each of such applicable rules do
   3.1. For each condition do
        3.1.1. Call "backward" recursively with the condition as the new goal. The initial stream passed to "backward" contains only variables that are present in the condition.
        3.1.2. Combine the current stream with that returned from the recursive call to "backward" to obtain the new current stream.
4. If no rules can be successfully applied, ask the user to provide supporting facts. Go to step 6.
5. Combine streams returned from the rules and build new facts using the conclusion pattern.
6. Return the result stream.

Experiments

Several real seismic data interpretations are conducted by SIES. The examples of Mississippi Canyon and High Island are shown as follows.

Mississippi Canyon

Figure 6.14 is a seismogram at Mississippi Canyon. Pattern classification result is used as facts in this experiment. Huang and Fu (1984) showed tree classification result in Figure 6.15. Reflection r1 has strong amplitude, low frequency

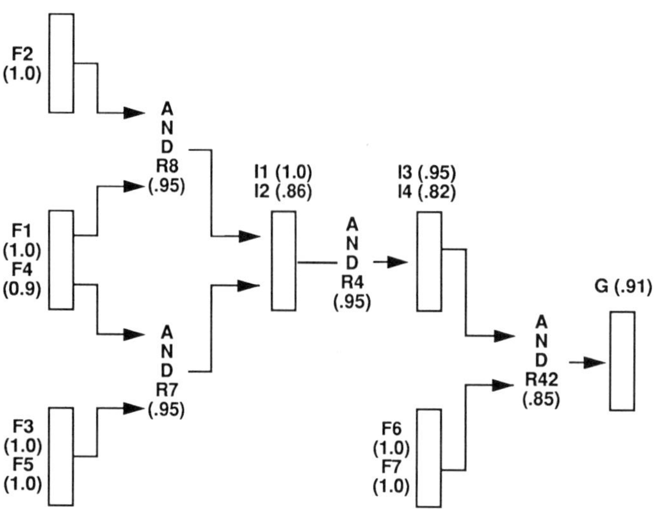

FIGURE 6.16. Inference graph of Mississippi Canyon example.

content, and polarity reversal. Reflection r2 has strong amplitude and low frequency.

The following facts are from pattern classification results and the seismic section:

F1: (r1 is unusually strong 1.0)
F2: (r1 has polarity reversal 1.0)
F3: (r1 has low frequency 1.0)
F4: (r2 is strong 0.9)
F5: (r2 has low frequency 1.0)
F6: (r1 is part of s1 1.0)
F7: (r2 is part of s1 1.0)

Figure 6.16 is the inference graph. Intermediate results and the goals are as follows:

I1: (r1 is an indication of gas 1.0)
I2: (r2 is an indication of gas 0.86)
I3: (r1 is an indication of hydrocarbon accumulation .95)
I4: (r2 is an indication of hydrocarbon accumulation .82)
G: (s1 contains oil or gas .91)

Figure 6.17 is a seismic section at High Island.

Similar to the Mississippi Canyon example, this example uses result from pattern classification. The tree classification result is shown in Figure 6.18 (Huang and Fu, 1984). Reflection r1 has strong amplitude, low frequency content, and

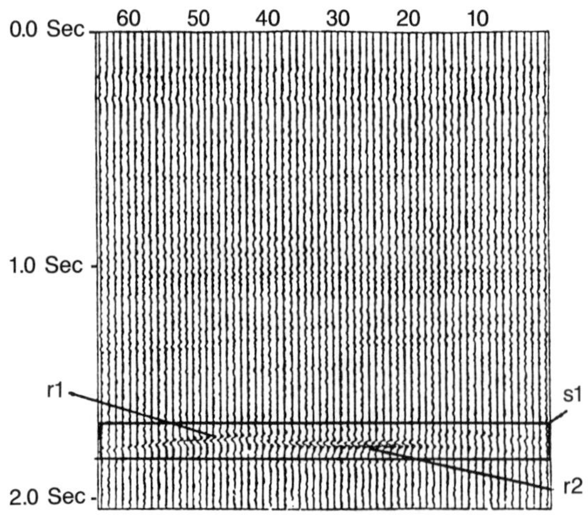

FIGURE 6.17. Seismogram at High Island.

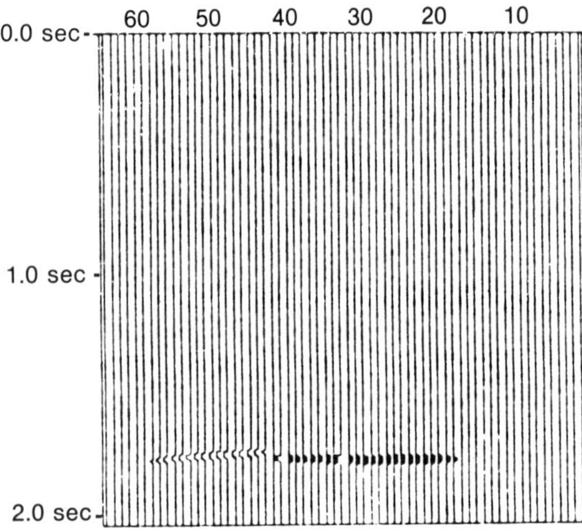

FIGURE 6.18. Tree classification result at High Island.

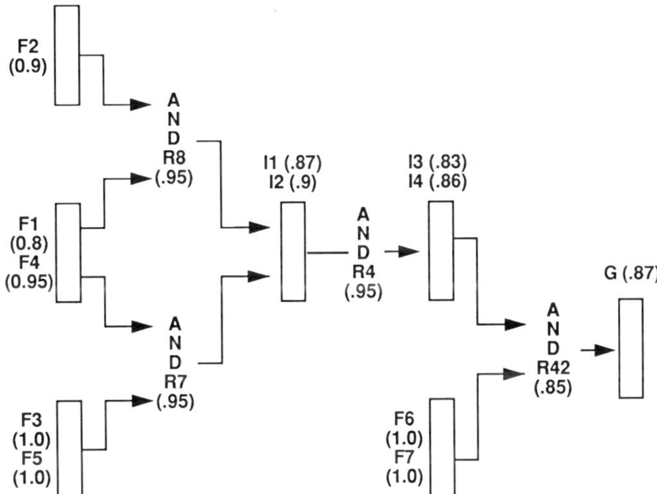

FIGURE 6.19. Inference graph of High Island example.

polarity reversal. Reflection r2 has strong amplitude and low frequency.

The inference graph shown in Figure 6.19 is similar to the Mississippi Canyon example. Facts from pattern classification result and the seismogram are as follows:

F1: (r1 is unusually strong 0.8)
F2: (r1 has polarity reversal 0.9)
F3: (r1 has low frequency 1.0)
F4: (r2 is strong 0.95)
F5: (r2 has low frequency 1.0)
F6: (r1 is part of s1 1.0)
F7: (r2 is part of s1 1.0)

Intermediate results and the goal are as follows:

I1: (r1 is an indication of gas .87)
I2: (r2 is an indication of gas 0.9)
I3: (r1 is an indication of hydrocarbon accumulation .83)
I4: (r2 is an indication of hydrocarbon accumulation .86)
G: (s1 contains oil or gas .87)

Conclusions

SIES, the Seismic Interpretation Expert System, is a backward-chaining system using pattern matching and streams as described in Winston and Horn (1984) and Winston (1984). The context-limiting control strategy greatly enhances system performance. The specificity-ordering strategy makes possible maximal utilization of known facts. Certainty factors allow probability-based inexact reasoning. The natural language interface uses augmented transition trees to parse user's inputs. The explanatory interface can answer "why" and "how" questions. The rule scanner verifies that rules are syntactically correct.

The problem area is seismic interpretation. The domain methods used are seismic stratigraphy and seismic section interpretation. Rules are written so that top-level facts are derivable from pictorial information in seismic sections and from other geological data. Those hypothesis-supporting facts can derive intermediate results, which in turn lead to other intermediate results and finally the goal. Pattern recognition results can be used as facts.

The economic return in applying artificial intelligence techniques in seismic interpretation is promising. Petroleum companies spend billions of dollars on exploration. A system such as the Seismic Interpretation Expert System could be of great value to oil companies by assisting interpreters in selecting the right prospects.

This system can be expanded by sufficient knowledge and to be used in complex real world applications.

Acknowledgments. The author thanks Kevin M. Barry of Teledyne Exploration for providing real seismic data.

References

Bubb, J.N., and Hatlelid, W.G., 1977, Seismic recognition of carbonate build-ups: in Seismic Stratigraphy—Applications to Hydrocarbon Exploration, C. Payton (Ed.): Am. Assn. Petr. Geol.

Coffeen, J.A., 1978, Seismic Exploration Fundamentals: PennWell Publ. Co., Tulsa, OK.

Coffeen, J.A., 1984, Interpreting Seismic Data: PennWell Publ. Co., Tulsa, OK.

Dobrin, M.B., 1976, Introduction to Geophysical Prospecting, 3rd ed.: McGraw-Hill, New York.

Garotta, R., and Michon, D., 1967, Continuous analysis of the velocity function and of the moveout corrections: Geophys. Prosp. 15, 584–597.

Huang, Kou-Yuan, and King-sun Fu, 1984, Detection of bright spots in seismic signal using tree classifiers: Geoexplor. 23, 121–145.

Mayne, W.A., 1962, Common reflection point horizontal data stacking: Geophysics 27, 927–938.

Mitchum, R.M., Jr., Vail, R.P., and Sangree, J.B., 1977a, Stratigraphic interpretation of seismic reflection patterns in depositional sequences: in Seismic Stratigraphy—Applications to Hydrocarbon Exploration, C. Payton (Ed.): Am. Assn. Petr. Geol.

Mitchum, R.M., Jr., Vail, R.P., and Thompson, S. III, 1977b, The depositional sequence as a basic unit for stratigraphic analysis: in Seismic Stratigraphy—Applications to Hydrocarbon Exploration, C. Payton (Ed.): Am. Assn. Petr. Geol.

Sangree, J.B., and Widmier, J.M., 1977, Seismic interpretation of clastic depositional facies: in Seismic Stratigraphy—Applications to Hydrocarbon Exploration, C. Payton (Ed.): Am. Assn. Petr. Geol.

Schneider, W.A., and Backus, M.M., 1968, Dynamic correlation analysis: Geophysics 33, 105–126.

Sheriff, R.E., and Geldart, L.P., 1983, Exploration Seismology 2: Data-Processing and Interpretation: Cambridge Univ. Press, Cambridge.

Shortliffe, E., 1976, Computer-based medical consultation, MYCIN: Elsevier, Amsterdam.

Taner, M.T., and Koehler, F., 1969, Velocity spectra—digital computer derivation and applications of velocity functions: Geophysics 34, 859–881.

Winston, P.H., 1984, Artificial Intelligence: Addison-Wesley, Reading, MA.

Winston, P.H., and Horn, B.K.P., 1984, LISP, 2nd ed.: Addison-Wesley, Reading, MA.

Appendix A: Velxpert Rules

```
( (general

   (rule1 (if  (have_not_checked velocity_extremes)

               (have_not_checked increasing_times)

               (have_not_checked time_velocity_extremes)

               (have_not_checked interval_velocity_extremes)

               (have_not_checked interbed_multiple)

               (have_not_checked time_velocity_slopes)

               (have_not_checked velocity_inversion)

               (have_not_checked sonic_log_available))

          (then (compute_velocity_extremes)))

   (rule2 (if  (have_not_checked increasing_times)

               (have_not_checked time_velocity_extremes)

               (have_not_checked interval_velocity_extremes)

               (have_not_checked interbed_multiple)

               (have_not_checked time_velocity_slopes)

               (have_not_checked velocity_inversion)

               (have_not_checked sonic_log_available)

               (have velocity_below_min))

          (then (set_state invalid)

                (set_step stop)))

   (rule3 (if  (have_not_checked increasing_times)

               (have_not_checked time_velocity_extremes)

               (have_not_checked interval_velocity_extremes)

               (have_not_checked interbed_multiple)

               (have_not_checked time_velocity_slopes)

               (have_not_checked velocity_inversion)

               (have_not_checked sonic_log_available)

               (have velocity_above_max))

          (then (set_state invalid)

                (set_step stop)))

   (rule4 (if  (have_not_checked increasing_times)

               (have_not_checked time_velocity_extremes)

               (have_not_checked interval_velocity_extremes)

               (have_not_checked interbed_multiple)

               (have_not_checked time_velocity_slopes)

               (have_not_checked velocity_inversion)
```

```
              (have_not_checked sonic_log_available))
         (then (check_times_not_increasing)))
  (rule5 (if   (have_not_checked time_velocity_extremes)
              (have_not_checked interval_velocity_extremes)
              (have_not_checked interbed_multiple)
              (have_not_checked time_velocity_slopes)
              (have_not_checked velocity_inversion)
              (have_not_checked sonic_log_available)
              (have times_not_increasing))
         (then (set_state invalid)
              (set_step stop)))
  (rule6 (if   (have_not_checked time_velocity_extremes)
              (have_not_checked interval_velocity_extremes)
              (have_not_checked interbed_multiple)
              (have_not_checked time_velocity_slopes)
              (have_not_checked velocity_inversion)
              (have_not_checked sonic_log_available))
         (then (check_time_velocity_extremes)))
  (rule7 (if   (have_not_checked interval_velocity_extremes)
              (have_not_checked interbed_multiple)
              (have_not_checked time_velocity_slopes)
              (have_not_checked sonic_log_available)
              (have_not_checked velocity_inversion)
              (have low_times_with_fast_velocities))
         (then (set_state invalid)
              (set_step stop)))
  (rule8 (if   (have_not_checked interval_velocity_extremes)
              (have_not_checked interbed_multiple)
              (have_not_checked time_velocity_slopes)
              (have_not_checked velocity_inversion)
              (have_not_checked sonic_log_available)
              (have high_times_with_slow_velocities))
         (then (set_state invalid)
              (set_step stop)))
  (rule9 (if   (have_not_checked interval_velocity_extremes)
              (have_not_checked interbed_multiple)
              (have_not_checked time_velocity_slopes)
              (have_not_checked velocity_inversion)
              (have_not_checked sonic_log_available))
         (then (check_interval_velocity_extremes)))
  (rule10 (if   (have_not_checked interbed_multiple)
              (have_not_checked time_velocity_slopes)
              (have_not_checked velocity_inversion)
              (have_not_checked sonic_log_available)
              (have interval_velocity_too_high))
         (then (set_state invalid)
              (set_step stop)))
  (rule11 (if   (have_not_checked interbed_multiple)
              (have_not_checked time_velocity_slopes)
              (have_not_checked velocity_inversion)
              (have_not_checked sonic_log_available)
              (have interval_velocity_too_low))
         (then (set_state invalid)
              (set_step stop)))
  (rule12 (if   (have_not_checked interbed_multiple)
              (have_not_checked time_velocity_slopes)
              (have_not_checked velocity_inversion)
              (have_not_checked sonic_log_available))
         (then (check_interbed_multiple)))
  (rule13 (if   (have_not_checked time_velocity_slopes)
              (have_not_checked velocity_inversion)
              (have_not_checked sonic_log_available)
              (have found_interbed_multiple))
         (then (set_state invalid)
              (set_step stop)))
  (rule14 (if   (have_not_checked time_velocity_slopes)
              (have_not_checked velocity_inversion)
              (have_not_checked sonic_log_available))
         (then (check_time_velocity_slopes)))
  (rule15 (if   (have_not_checked velocity_inversion)
              (have_not_checked sonic_log_available)
              (have invalid_time_velocity_slope))
         (then (set_state invalid)
              (set_step stop)))
  (rule16 (if   (have_not_checked velocity_inversion)
              (have_not_checked sonic_log_available))
         (then (check_velocity_inversion)))
  (rule17 (if   (have_not_checked sonic_log_available))
```

```
        (then (check_for_sonic_log)
              (set_step data_type))))
(land
  (rule18 (if  (equal area_location north_slope)
               (have_not velocity_inversion_shallow))
          (then (set_state invalid)
                (set_step stop)))
  (rule19 (if  (not_equal area_location north_slope)
               (have velocity_inversion_shallow))
          (then (setq shallow_inversion_weight .5)
                (set_step compute_certainty)))
  (rule20 (if  (equal step land))
          (then (set_step compute_certainty))))
(marine
  (rule21 (if  (have_not_checked multiples)
               (have_not_checked pegleg_multiples))
          (then (check_for_multiples)))
  (rule22 (if  (have_not_checked pegleg_multiples)
               (have found_multiple))
          (then (set_state invalid)
                (set_step stop)))
  (rule23 (if  (have_not_checked pegleg_multiples))
          (then (check_for_pegleg_multiples)))
  (rule24 (if  (have found_pegleg_multiple))
          (then (set_state invalid)
                (set_step stop)))
  (rule25 (if  (equal area_location gulf_coast)
               (have_not_checked clastic_velocity_ranges))
          (then (compare_clastic_vel_ranges)))
  (rule26 (if  (equal step marine))
          (then (set_step compute_certainty))))
(compute_certainty
  (rule27 (if  (have over_pressure_expected)
               (have velocity_inversion_deep)
               (have_not_checked over_pressure))
          (then (setq over_pressure_weight 1.0)
                (setq over_pressure t)))
  (rule28 (if  (have over_pressure_expected)
               (have_not velocity_inversion_deep)
               (have_not_checked over_pressure))
          (then (setq over_pressure_weight .5)
                (setq over_pressure t)))
  (rule29 (if  (have sonic_log)
               (have_not_checked sonic_velocity)
               (have_not_checked velocity_changes)
               (have_not_checked time_separation)
               (have_not_checked increasing_velocities)
               (have_not_checked area_average_velocity)
               (have_not_checked computed_weight))
          (then (compare_sonic_to_vel)))
  (rule30 (if  (have_not_checked velocity_changes)
               (have_not_checked time_separation)
               (have_not_checked increasing_velocities)
               (have_not_checked area_average_velocity)
               (have_not_checked computed_weight))
          (then (check_velocity_changes)))
  (rule31 (if  (have_not_checked time_separation)
               (have_not_checked increasing_velocities)
               (have_not_checked area_average_velocity)
               (have_not_checked computed_weight))
          (then (check_time_separation)))
  (rule32 (if  (have_not_checked increasing_velocities)
               (have_not_checked area_average_velocity)
               (have_not_checked computed_weight))
          (then (check_increasing_vels)))
  (rule33 (if  (have_not_checked area_average_velocity)
               (have_not_checked computed_weight))
          (then (compare_area_average_velocity)))
  (rule34 (if  (have_not_checked computed_weight))
          (then (compute_certainty)))
  (rule35 (if  (have computed_weight)
               (have computed_weight_too_low))
          (then (set_state invalid)
                (set_step stop)))
  (rule36 (if  (have computed_weight))
          (then (set_state valid)
                (set_step stop))))
```

Appendix B: SIES Rules

(rule 1-oil-trap 0.95

  if (((> structure) is a trap)

     ((> reflection) is an indication of hydrocarbon

       accumulation)

     ((< reflection) is part of (< structure)))

  then ((< structure) contains oil or gas))

(rule 2-oil-reef 0.95

  if (((> structure) is a reef)

     ((> reflection) is an indication of hydrocarbon

       accumulation)

     ((< reflection) is part of (< structure)))

  then ((< structure) contains oil or gas))

(rule 3-hydro-liquid 0.95

  if (((> reflection) represents a liquid surface))

  then ((< reflection) is an indication of hydrocarbon

       accumulation))

(rule 4-hydro-gas 0.95

  if (((> reflection) is an indication of gas))

  then ((< reflection) is an indication of hydrocarbon

       accumulation))

(rule 5-liquid-flat 0.95

  if (((> reflection) is unusually flat))

  then ((< reflection) represents a liquid surface))

(rule 6-gas-weak 0.95

  if (((> structure) is a reef)

     ((> reflection) is at the top of (< structure))

     ((< reflection) is unusually weak))

  then ((< reflection) is an indication of gas))

(rule 7-gas-bright 0.95

  if (((> reflection) is unusually strong)

     ((< reflection) has low frequency))

  then ((< reflection) is an indication of gas))

(rule 8-gas-polar 0.95

  if (((> reflection) is unusually strong)

     ((< reflection) has a polarity reversal))

  then ((< reflection) is an indication of gas))

(rule 9-trap 0.99

  if (((> layer) is a cap rock)

     ((> layer2) lies inside (< layer))

     ((< layer2) is porous)

     ((> structure) consists of (< layer) and

      (< layer2)))

  then ((< structure) is a trap))

(rule 10-closed-anti 0.99

  if (((> structure) is an anticline)

     ((> layer) is part of (< structure))

     ((< layer) is not porous))

  then ((< layer) is a cap rock))

(rule 11-anticline 0.95

  if (((> structure) consists of reflections which

      have locally high points that dip downward in

      all directions))

  then ((< structure) is an anticline))

(rule 12-closed-dome 0.99

  if (((> structure) is a salt dome)

     ((> layer) terminates updip against

     (< structure))

     ((< layer) is not porous))

  then ((< layer) is a cap rock))

(rule 13-salt-dome 0.99

  if (((> structure) is a diapir)

     ((< structure) contains salt))

  then ((< structure) is a salt dome))

(rule 14-diapir-trunc 0.9

  if (((> structure) is a massive body)

     ((> layer) terminates against (< structure) by

      structural truncation)

     ((< structure) has the shape of a column))

  then ((< structure) is a diapir))

(rule 15-pinchout 0.95

  if (((> layer) wedges out into (> layer2))

     ((< layer2) is not porous)

       (((< layer) is porous)

       ((> structure) consists of (< layer) and

         (< layer2)))

    then ((< structure) is a pinchout))

(rule 15b-pinch-closed 0.95

   if (((> structure) is a pinchout)

      ((< structure) wedges out updip))

    then ((< structure) is a trap))

(rule 16-massive 0.95

   if (((> structure) is an area with few

      reflections))

    then ((< structure) is a massive body))

(rule 17-reef-top 0.9

   if (((> structure) is a buildup)

      ((> reflection) is at the top of

        (< structure))

      ((< reflection) is strong))

    then ((< structure) is a reef))

(rule 18-reef-base 0.9

   if (((> structure) is a buildup)

      ((> reflection) is at the base of (< structure))

      ((< reflection) is strong))

    then ((< structure) is a reef))

(rule 19-reef-limestone 0.95

   if (((> structure) is a buildup)

      ((< structure) is made of limestone))

    then ((< structure) is a reef))

(rule 20-reef-shelf 0.9

   if (((> structure) is a buildup)

      ((< structure) was deposited on a shelf edge))

    then ((< structure) is a reef))

(rule 21-reef-drape 0.9

   if (((> structure) is a buildup)

      ((< structure) lies beneath a drape that dies

      out upward))

    then ((< structure) is a reef))

(rule 22-buildup 0.95

   if (((> structure) is a massive body)

      ((< structure) has the shape of a mound)

      ((> layer) terminates onlap against

      (< structure)))

    then ((< structure) is a buildup))

(rule 23-limestone 0.9

   if ((interval velocity of (> structure) calculated by

      the CDP method is about the same as that of

      limestone in the general area))

    then ((< structure) is made of limestone))

(rule 24-fault 0.95

   if (((> line) is a fault line)

      ((> layer) terminates updip against (< line))

      ((< layer) is not porous)

      ((< layer) is sealed at (< line)))

    then ((< layer) is a cap rock))

(rule 25-fault-line 0.95

   if (((> line) is a potential fault line)

      (reflections on two sides of (< line) can be

      correlated by section shifting))

    then ((< line) is a fault line))

(rule 26-potential-fault 0.95

   if (((> line) is a slanting line cutting across

      reflections))

    then ((< line) is a potential fault line))

(rule 27-not-porous 0.99

   if ((> layer) is shale)

    then ((< layer) is not porous))

(rule 28-shale-velocity 0.9

   if ((interval velocity of (> layer) calculated by

      the CDP method is about the same as that of

      shale in the general area))

    then ((< layer) is shale))

(rule 29-shale-alternate 0.7

   if (((> structure) contains alternating sand and

      shale layers)

      ((> layer) is part of (< structure)))

    then ((< layer) is shale))

(rule 30-alternating 0.95

   if (((> structure) contains many strong

      reflections))

then ((< structure) contains alternating sand and

shale layers))

(rule 31-shale-supply 0.9

if (((> area) has indication of low sediment supply)

((> layer) is part of (< area)))

then ((< layer) is shale))

(rule 32-energy-sigmoid 0.8

if (((> area) primarily consists of sigmoid facies))

then ((< area) has indication of low sediment

supply))

(rule 33-energy-sheet 0.8

if (((> area) primarily consists of sheet facies))

then ((< area) has indication of low sediment supply))

(rule 34-porous 0.99

if ((> layer) is sandstone)

then ((< layer) is porous))

(rule 35-sand-velocity 0.9

if ((interval velocity of (> layer) calculated by the

CDP method is about the same as that of sand in

the general area))

then ((< layer) is sandstone))

(rule 36-sand-alternate 0.8

if (((> structure) contains alternating sand and

shale layers)

((> layer) is part of (< structure)))

then ((< layer) is sandstone))

(rule 37-sand-supply 0.9

if (((> area) has indication of high sediment

supply)

((> layer) is part of (< area)))

then ((< layer) is sandstone))

(rule 38-energy-mound 0.8

if (((> area) primarily consists of mound facies))

then ((< area) has indication of high sediment

supply))

(rule 39-energy-oblique 0.8

if (((> area) primarily consists of oblique facies))

then ((< area) has indication of high sediment

supply))

(rule 40-energy-chaotic 0.8

if (((> area) primarily consists of chaotic facies))

then ((< area) has indication of high sediment

supply))

(rule 41-oil-trap2 0.8

if (((> structure) is a trap))

then ((< structure) contains oil or gas))

(rule 42-oil-hc 0.85

if (((> reflection) is an indication of hydrocarbon

accumulation)

((< reflection) is part of (> structure)))

then ((< structure) contains oil or gas))

(rule 43-oil-reef2 0.85

if (((> structure) is a reef))

then ((< structure) contains oil or gas))

7
Pattern Recognition to Seismic Exploration

Kou-Yuan Huang

Decision-theoretic and syntactic pattern recognition techniques are employed to detect the physical anomalies (bright spots) and to recognize the structural seismic patterns in two-dimensional seismograms. Here, decision-theoretic methods include Bayes classification, linear and quadratic classifications, tree classification, partitioning-method and tree classification, and sequential classification. The generated features are envelope, instantaneous frequency, polarity, moments, and contrasts from cooccurrence matrix. A hierarchical system is proposed for seismic syntactic pattern recognition. Syntactic methods include error-correcting finite state automaton, picture description language (PDL), and tree automaton. Experiments using simulated and real seismograms are presented. The recognition results are quite encouraging.

Introduction

The use of pattern recognition techniques in seismic exploration is relatively new. Bois (1980, 1983), Hagen (1981), Huang et al. (1981), Huang and Fu (1982, 1983, 1984, 1985a,b, 1987a,b), Huang and Sheen (1986), and de Figueiredo (1982) presented the pattern recognition approach to seismic exploration data. Publications by Huang and Fu (1982), Chen (1985), and Aminzadeh (1987) were the highlights of using pattern recognition for seismic exploration data.

In the early 1970s, two major oil companies had been successful in predicting the occurrence of offshore gas from seismic reflection data (Dobrin, 1976). The predictions were based on the anomalies that would be expected in the amplitudes of reflections. The high amplitude portion of the reflection is referred to as a bright spot, which is the indicator of gas accumulation (Dobrin, 1976). There are many hydrocarbon indicators derived from seismic data (Dobrin, 1976; Payton, 1977). The major physical indicators are high amplitude due to high reflection coefficient, low frequency content due to high frequency attenuation, and polarity reversal due to negative reflection coefficient at the reflection of the gas sand zone. The interpretation of seismograms by experience is a tedious and subjective task. Furthermore, subtle changes in the nature of the reflected signals often cannot be seen by visual analysis. From a pattern recognition point of view, these physical anomalies can be recognized as one class, bright spot, and can be detected by using decision-theoretic pattern recognition methods.

In the seismic interpretation and recognition of seismic patterns, structural information is also very important. A hierarchical system is proposed for seismic syntactic pattern recognition. The error-correcting finite-state automaton, picture description language, and tree automaton are employed to recognize structural seismic patterns.

Nonparametric Method: Linear Classification of Ricker Wavelets

Two major seismic anomalies are high amplitude and low frequency. Analytic signal analysis can be applied to extract such anomalies (Huang and Fu,

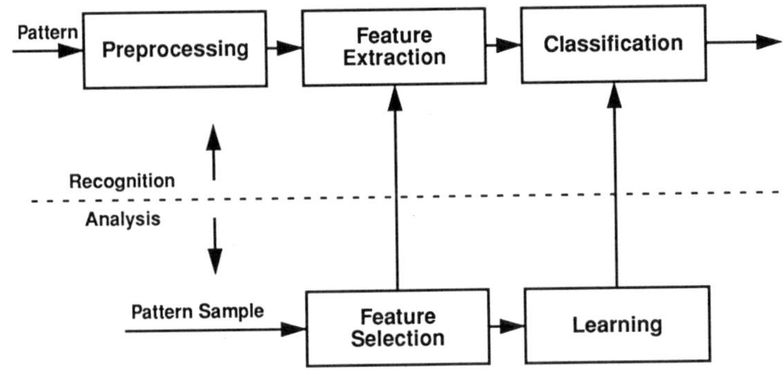

FIGURE 7.1. Block diagram of a decision-theoretical pattern recognition system.

1984; Farnback, 1975; Taner et al., 1979; Robertson and Nogami, 1984). Envelope describes the outer shape of the signal. Instantaneous frequency extracts the internal property of the signal. Here, envelope and instantaneous frequency are extracted as the two features used in classification. A parametric method (Bayes decision rule) and two nonparametric methods (linear and quadratic classifications) are used as the classification methods.

Zero-phase Ricker wavelets are usually used in the simulation of complex trace analysis. Huang and Fu (1984) pointed out that the major part of the energy of wavelets in real seismic data is close to the central part of zero-phase Ricker wavelets Also, the central part of zero-phase Ricker wavelets is close to the cosine waveform. The instantaneous frequency of the $\cos 2\pi f_0 t$ is equal to f_0. So zero-phase Ricker wavelets are tested in the experiments. In this paper, a 20-Hz zero-phase Ricker wavelet is used to simulate the physical anomalies of high amplitude and low frequency content, and a 30-Hz zero-phase Ricker wavelet is used to simulate the reflection wavelet without physical anomalies in the seismogram.

At first, Ricker wavelets are classified in a seismic trace. Then the simulated seismograms using Ricker wavelets are classified trace by trace. To analyze the seismic patterns, a decision-theoretic pattern recognition system is presented in Figure 7.1.

In a simulated seismic trace (Figure 7.2a), 20- and 30-Hz Ricker wavelets and Gaussian white band 10–60 Hz random noise are generated. The three pattern classes can be defined as

C_1 : $r(t) = n(t)$ Gaussian noise
C_2 : $r(t) = 20$ Hz Ricker wavelet $+ n(t)$
C_3 : $r(t) = 30$ Hz Ricker wavelet $+ n(t)$

Using analytic signal analysis, envelope and instantaneous frequency are extracted from the signal itself and used as the two features in the decision-theoretic pattern recognition system. Here the envelope value is multiplied by 200 to have the same magnitude order of instantaneous frequency. The scattering diagram of two features is shown in Figure 7.2b.

A supervised linear classification technique is used to analyze Figure 2b. A modified fixed-increment training procedure (Huang and Fu, 1987a,b) is employed to design the classifiers. In this seismic trace, the linear classification result is shown in Figure 7.2c.

Linear Classification for the Detection of Seismic Anomalies

The one-trace pattern recognition techniques can be directly applied to two-dimensional (2-D) reflection seismograms to classify seismic anomalies (bright spots) because the processing is trace-by-trace. The primary reflection synthetic seismogram is generated in Figure 7.3b. The geologic model from Dobrin (1976) is shown in Figure 7.3a. Two pattern classes are considered in the simulated seismogram. One class is the zero-phase 20-Hz Ricker wavelet at the boundary of shale and gas sand zones (bright spot), i.e., at the top of the trap.

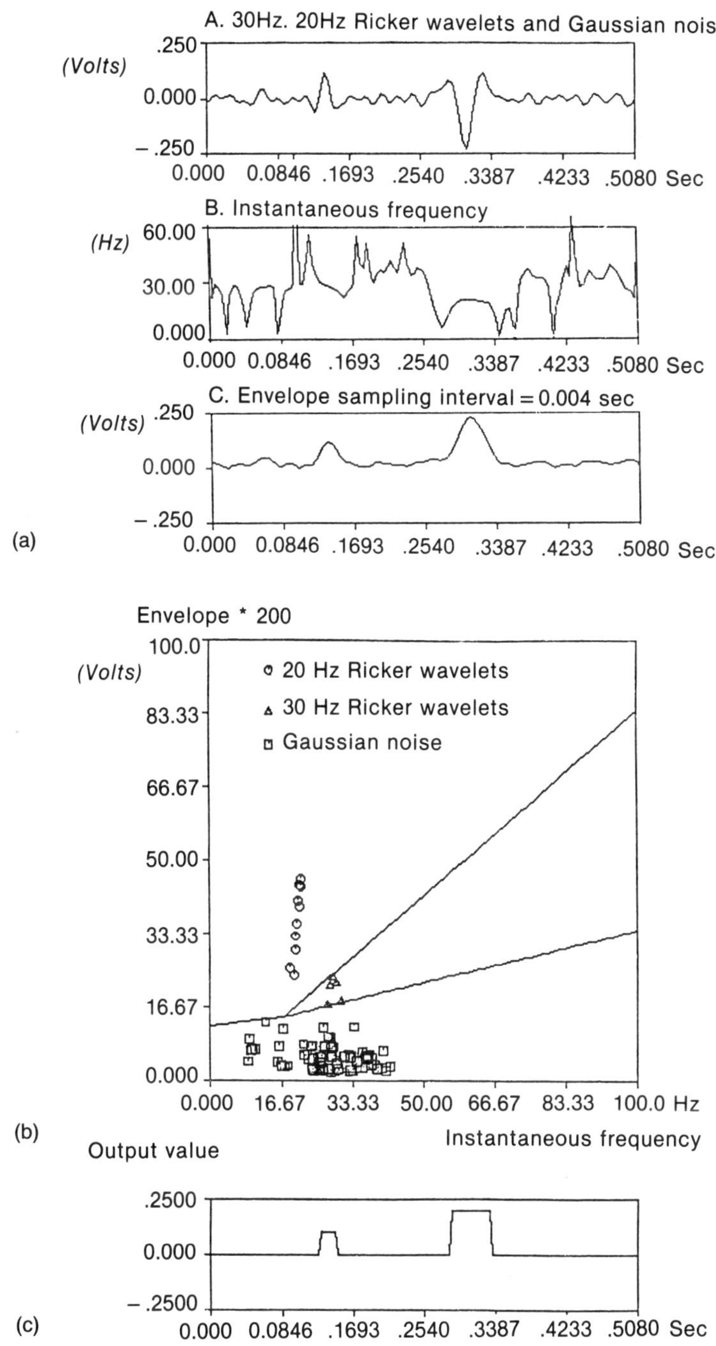

FIGURE 7.2. (a) Signal, its instantaneous frequency, and envelope. (b) Linear classifier. (c) Linear classification result of Ricker wavelets.

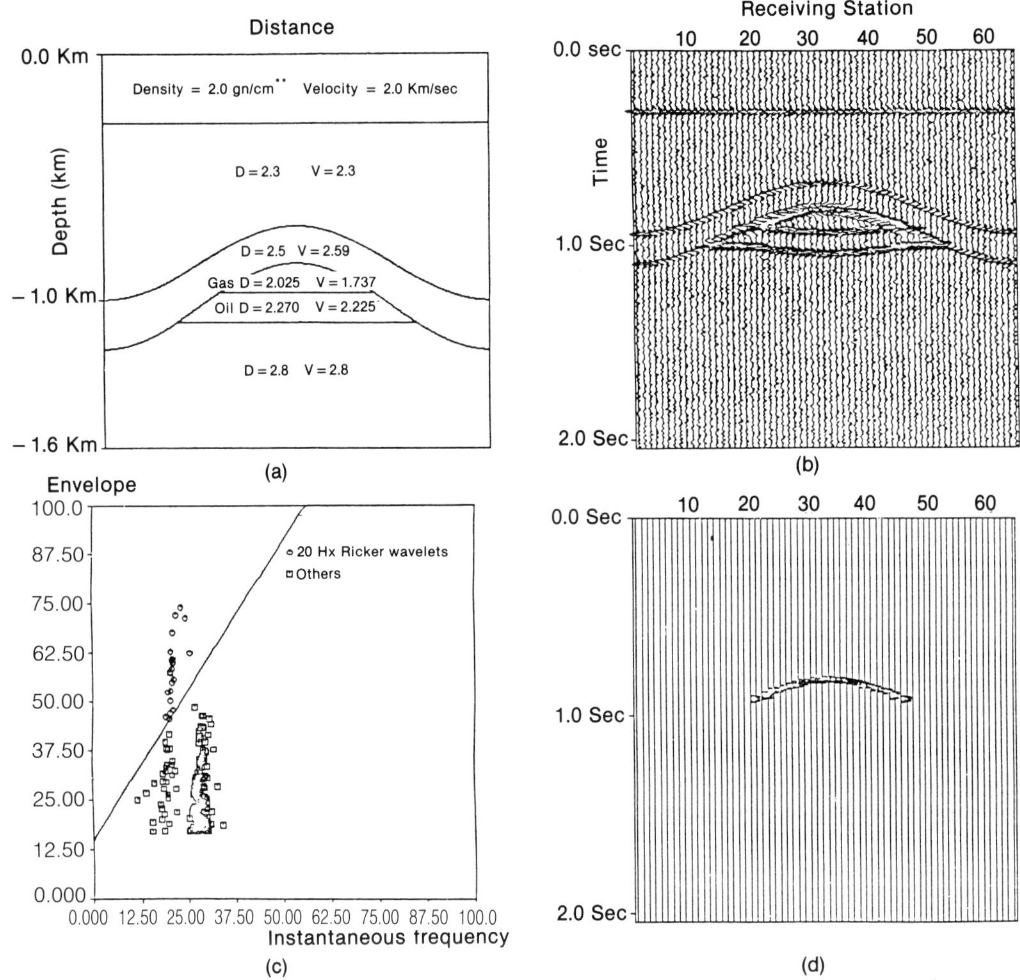

FIGURE 7.3. (a) Geological structure. (b) Synthetic seismogram of bright spot. (c) Feature distribution of training samples and linear classifier. (d) Linear classification result of bright spot.

The zero-phase 20-Hz Ricker wavelet has high amplitude proportional to the high-reflection coefficient, phase reversal and low frequency content due to high frequency attenuation in the gas sand zone. The other class is the nonbright spot, the zero-phase 30-Hz Ricker wavelet at the other layer boundary reflection and Gaussian white band 10–60 Hz random noise corresponding to the random noise collected in the field. There are 64 traces in the seismogram. Every trace has 512 points. The sampling interval is 0.004 s.

Training traces are equally selected from the seismogram for easy selection and convenience. The training traces are the 4th, 12th, 20th, 28th, 36th, 44th, 52nd, and 60th traces. The feature distribution of the training samples is shown in Figure 7.3c. Envelope is multiplied by 200 in order to have the same magnitude as instantaneous frequency. A linear classifier is used. The modified fixed-increment training procedure is applied. The result of using the linear classifier is shown in Figure 7.3d. Compared with the original seismogram in Figure 7.3b, the classifier appears to be quite good in the detection of physical anomalies.

Linear Classification in Thin-Bed Seismogram

The time to pass through the gas sand zone is less than the half cycle of 20-Hz Ricker wavelet, and the gas sand zone is called "thin-bed." In the exper-

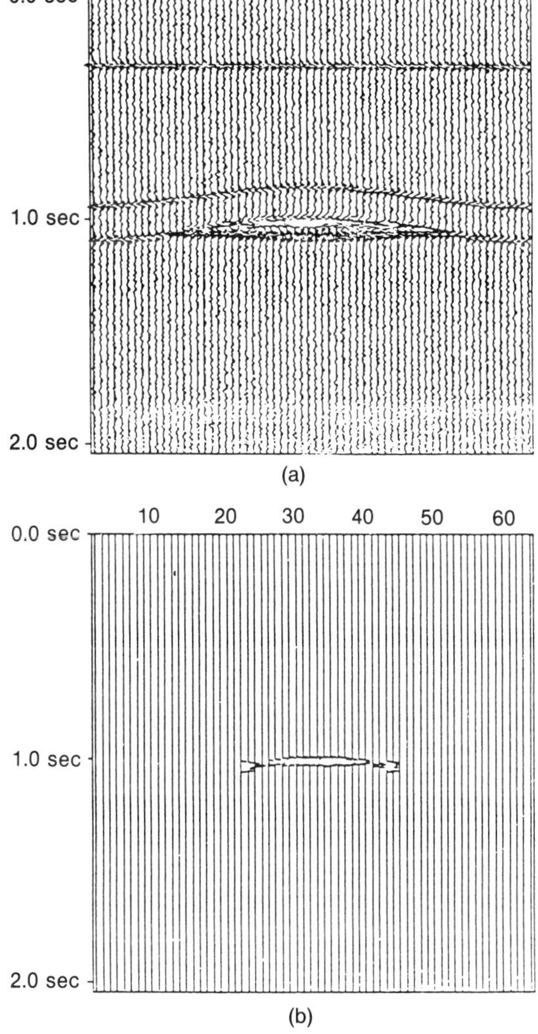

(a)

(b)

FIGURE 7.4. (a) Synthetic seismogram of bright spot (thickness of gas sand zone is two-thirds quarter wavelength). (b) Linear classification result of bright spots (two-thirds quarter wavelength at gas sand zone).

iments, several thin-bed seismograms are tested. One example is shown. A simulated seismogram for a model in which the thickness of the two-thirds quarter wavelength in gas sand zone is shown in Figure 7.4a. From Figure 7.4a, although the 20-Hz Ricker wavelet at the top of the gas sand zone is mixed with the 30-Hz Ricker wavelet at the bottom of the gas sand zone, the 20-Hz Ricker wavelet is dominant and can overcome the interference, i.e.,

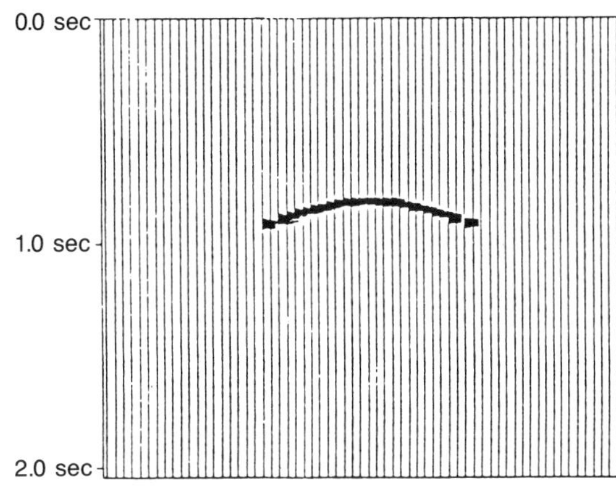

FIGURE 7.5. Quadratic classification result of bright spot.

the physical properties are preserved. Compared with the original simulated seismogram in Figure 7.4a, the classification result shown in Figure 7.4b still seems to be good.

Quadratic Classification in Seismograms

The general type of quadratic decision boundary between two classes (Fu, 1982) is

$$D(X) = \sum_{k=1}^{N} w_{kk}x_k^2 + \sum_{j=1}^{N-1} \sum_{k=j+1}^{N} w_{jk}x_jx_k$$
$$+ \sum_{j=1}^{N} w_jx_j + w_{L+1}$$

where x_j is the feature, w_j is the weighting, and N is the number of features. To simplify the problem, only the parabolic function is calculated here. If we apply a parabolic function and modified fixed-increment training procedure to get the decision boundary in the two-feature space of Figure 7.3b, the parabolic decision boundary is

$$D(x,y) = ax^2 + bx + cy + d$$
$$= 188.03x^2 + 10589.4x - 11877.86y$$
$$+ 1960.0 + 0 = 0$$

where x represents instantaneous frequency and y represents envelope. The result of classifying Figure 7.3b is shown in Figure 7.5.

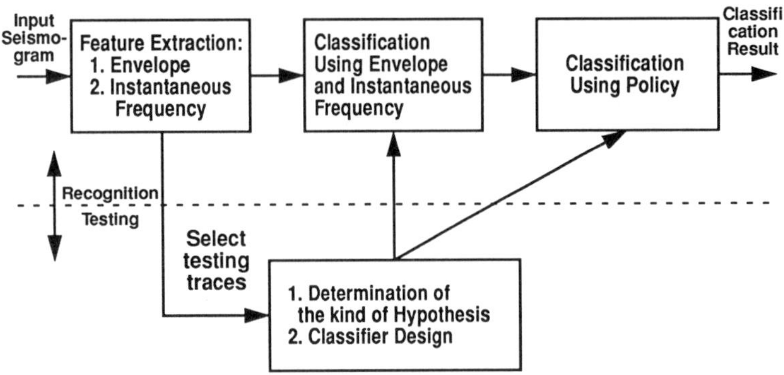

FIGURE 7.6. Block diagram of a tree classification system.

Parametric Method: Bayes Classification of Ricker Wavelets

The envelopes and instantaneous frequencies of the 20- and 30-Hz Ricker wavelets in Figure 7.2b can be shown to be Gaussian distributions by χ^2 tests. The feature distribution of each class is calculated as the multivariate Gaussian. The Bayes decision rule (Fu, 1982) is applied and the Bayes classifiers are determined from

$$P(w_i)p(X/w_i) = P(w_j)p(X/w_j)$$

where $P(w_i)$ is the a priori probability for class w_i and $p(X/w_i)$ is the multivariate probability density function for class w_i or

$$\log \frac{P(w_i)}{P(w_j)} - \frac{1}{2} \log \frac{|K_i|}{|K_j|} - \frac{1}{2} [(X-M_i)^T K_i^{-1}(X-M_i)$$

$$- (X-M_j)^T K_j^{-1}(X-M_j)] = 0 \qquad i,j=1,2,3,$$

$$i \neq j, \text{ and } P(w_1)=P(w_2)=P(w_3)=\tfrac{1}{3}$$

The classifiers are quadratic because of the different covariance matrices.

From Figure 7.2b, the decision boundaries are

$$g_{12}(x,y) = 5.546x^2 - 0.037y^2 - 1.106xy$$
$$- 188.0595x + 18.425y + 1604.572 = 0$$

$$g_{13}(x,y) = 0.403x^2 - 0.014y^2 - 0.0093xy$$
$$- 23.418x - 2.169y + 372.102 = 0$$

$$g_{23}(x,y) = 5.143x^2 - 0.023y^2 - 1.0967xy$$
$$- 164.64x + 20.591y + 1232.47 = 0$$

where x represents instantaneous frequency and y represents envelope.

Tree Classification

The major physical indicators of bright spots (physical anomalies) are high amplitude, low frequency, and polarity reversal.

From feature selection techniques (Huang and Fu, 1987b) tree classification can be proved as the optimal method in the decision-theoretic approach. The block diagram of a tree classification system is shown in Figure 7.6. Tree classification is easy to design and computationally efficient. Three kinds of hypotheses of bright spots are presented:

1. The bright spot has high amplitude, low frequency content, and polarity reversal. Envelope, instantaneous frequency, and polarity are used in the tree classifier.
2. The bright spot has high amplitude and low frequency content. For this case, only the envelope and instantaneous frequency are used in the tree classifier.
3. The bright spot has high amplitude and polarity reversal. Here, only envelope and polarity are used in the tree classifier. For the input seismogram, some testing traces are selected from the seismograms, a nonsupervised clustering analysis is performed. The bright spots are detected if one of the three hypotheses is satisfied. After determining the tree classifiers from the testing traces, the whole seismogram will be processed.

Tree Classifier Design

From feature selection, instantaneous frequency is the best one feature after envelope thresholding. So envelope is used as the first level and instan-

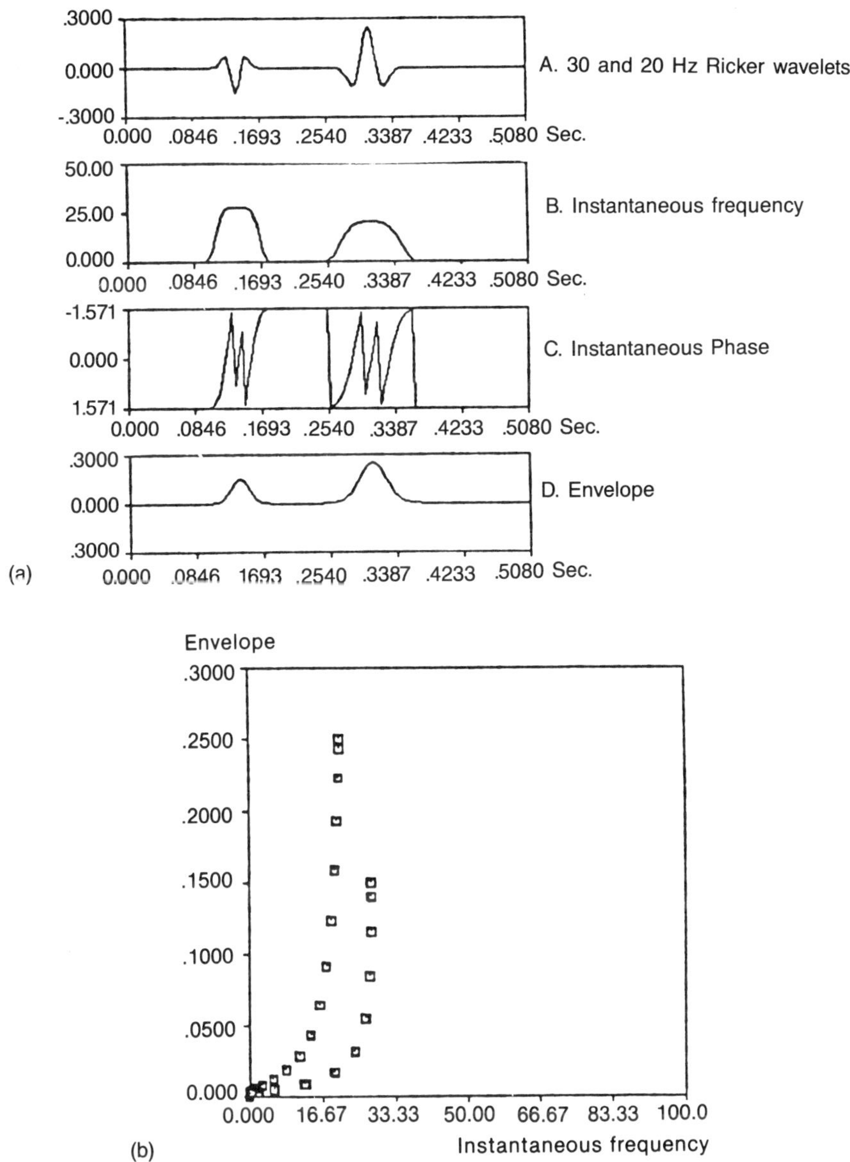

FIGURE 7.7. (a) Ricker wavelets (30 and 20 Hz) and their analytic signal representations. (b) Feature distribution of a seismic trace.

taneous frequency is used as the second level in the tree classification. The purpose of envelope thresholding is to separate signal and noise. The envelope of Gaussian noise is proven as Rayleigh distribution (Huang and Fu, 1984). From scatter diagram in Figure 7.7b, the envelope of a zero-phase Ricker wavelet is approximately, uniformly distributed between one-third of the maximum envelope and the maximum envelope. In the classification of Ricker wavelets plus Gaussian noise (Huang and Fu, 1984), the threshold of envelope is determined between Rayleigh (from Gaussian noise) and uniform distribution (from Ricker wavelet). Usually the Bayes decision rule is used. For high signal-to-noise ratio, the threshold of envelope is determined by selecting the maximum

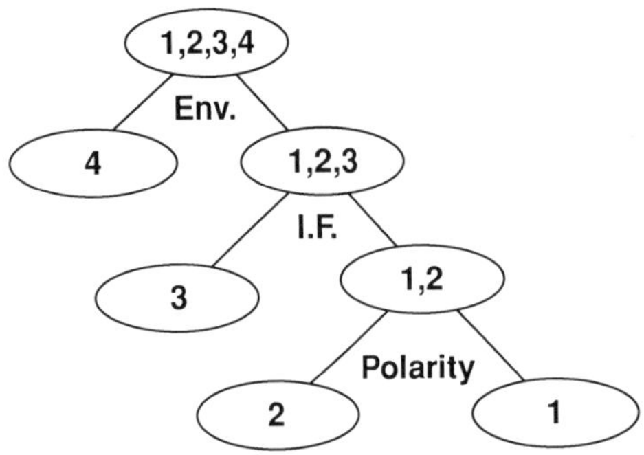

FIGURE 7.8. Tree classifier of the first hypothesis.

between one-third maximum envelope of signal and 3σ of Gaussian noise. For the data in the scatter diagram above the envelope threshold, the threshold of instantaneous frequency is determined by inspection if data are separable or by a K-mean clustering analysis and maximum pseudo F-

statistics (PFS) value (Vogel and Wong, 1979). The tree classifier of the first hypothesis is shown in Figure 7.8.

Tree Classification in Real-Data Experiment

A relative amplitude seismogram at Mississippi Canyon is processed. In Figure 7.9a, at 0.2 and 1.4 s, the dominant wavelets are approximately the minimum-phase wavelets. Testing traces are selected on the 4th, 12th, 20th, 28th, 36th, 44th, 52nd, and 60th traces. The scatter diagram of envelope and instantaneous frequency is shown in Figure 7.9b. Above envelope threshold 0.08, most of the data in the scatter diagram are centered in the low frequency part. So by inspection, $K=2$ clusters are selected. The threshold of instantaneous frequency is 26.69 Hz. Polarity classification

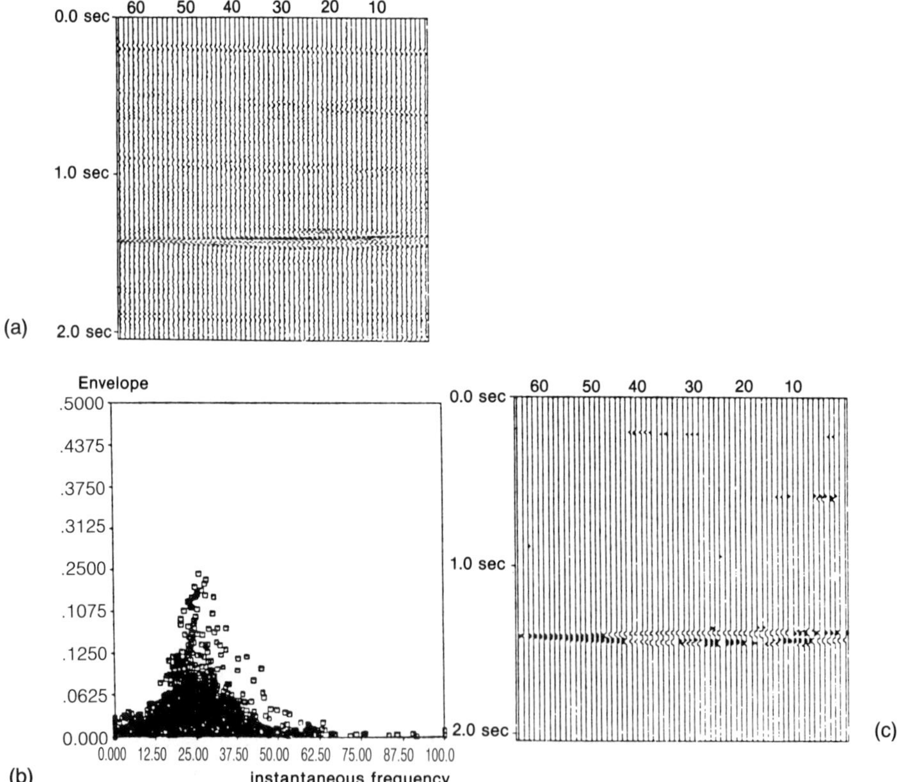

FIGURE 7.9. (a) Relative amplitude seismogram at Mississippi Canyon (negative on the right). (b) Feature distribution of testing traces. (c) Tree classification result of bright spots.

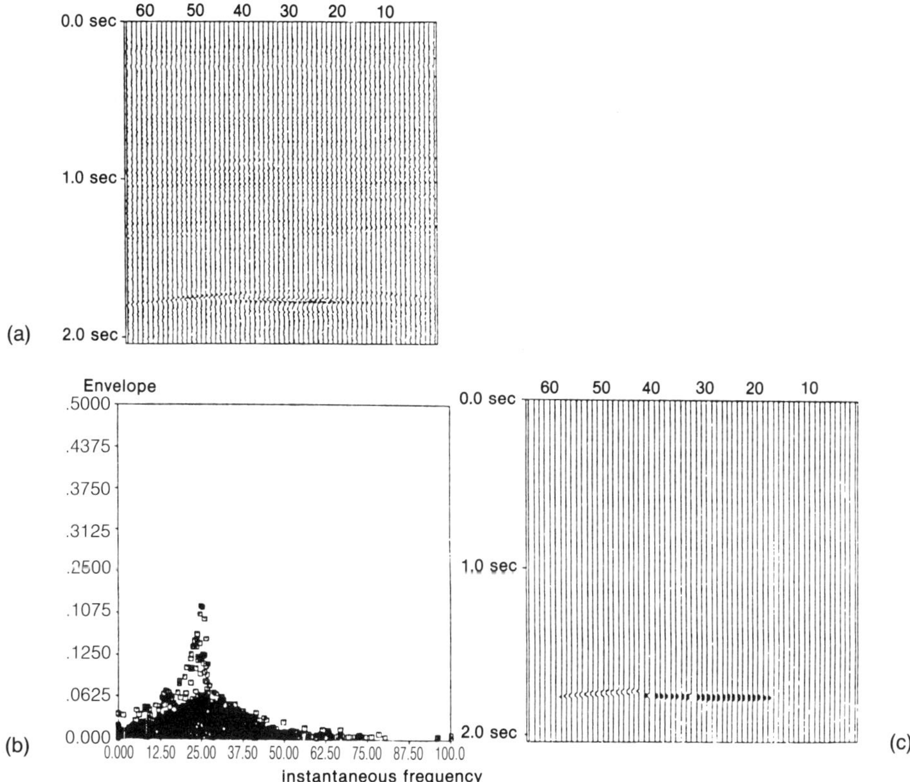

FIGURE 7.10. (a) Relative amplitude seismogram at High Island (negative on the right). (b) Feature distribution of testing traces. (c) Tree classification result of bright spots.

is after envelope and instantaneous frequency. The tree classifier designed is the first kind of hypothesis. The detection of bright spot is shown in the curve leftward white portion of Figure 7.9c, approximately at the center of 1.4 s. We can predict that the gas sand zone is approximately at 1.4 s with negative polarity. The left-hand side and the black portion of Figure 7.9c at 1.4 s is the result of using envelope and instantaneous frequency with positive polarity and comes from the interference effect of thin bed.

Also, the seismogram at High Island is processed. In Figure 7.10a, at 1.7 s, the dominant wavelets are the zero-phase wavelets. From the scattering diagram of envelope and instantaneous frequency in the testing traces, the low-frequency content is not significant (Figure 7.10b). The tree classifier is the third hypothesis. Envelope and polarity are selected as the features in the tree classification. The classification result is shown in Figure 7.10c. The bright spot is at 1.7 s, which is the most probable to accumuate gas. At the middle part of the layer of 1.7 s, the reflection of the positive polarity is the most probable to accumulate oil and water. Without using the classification technique, the bright spot may be detected at the middle part of the layer of 1.7 s from visual inspection.

Partitioning Method and Tree Classification

There are many attenuation effects in a seismogram, with the quality factor Q being the dominant one. Frequency attenuation effect may appear in a seismogram. A bright spot has important indicators of high amplitude, low frequency content, and phase reversal. But reflections from other layers in a seismogram probably also contain some of these

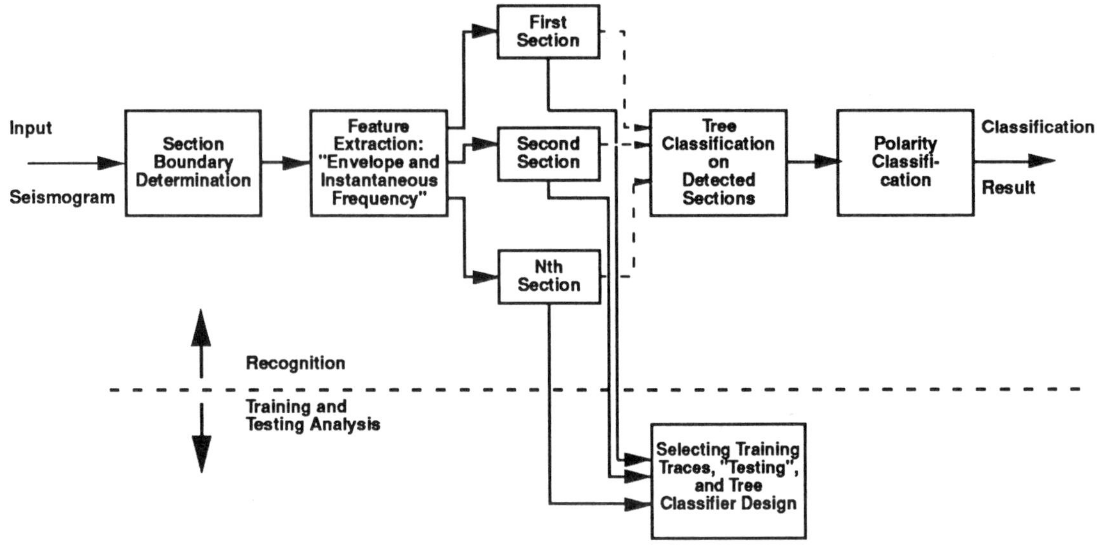

FIGURE 7.11. Block diagram of partitioning method and tree classifier system.

indicators, especially the low frequency content. So a "partitioning method" is proposed. The basic idea is that bright spots having low frequency content are only true in a small section of a seismogram. A block diagram of the partitioning method and tree classification system is shown in Figure 7.11. A seismogram is first partitioned into small sections. For each section, three hypotheses and their corresponding tree classification techniques are used to detect bright spots. The major advantage of the partitioning method is that the overlapping distributions of the envelope and instantaneous frequency in different sections can be separated, i.e., the distribution of envelope and instantaneous frequency from one seismic section is not disturbed by those of other seismic sections. The design of the tree classifier in one seismic section is therefore easier than that for processing the whole seismogram. An experiment on a real seismogram at Mississippi Canyon is presented.

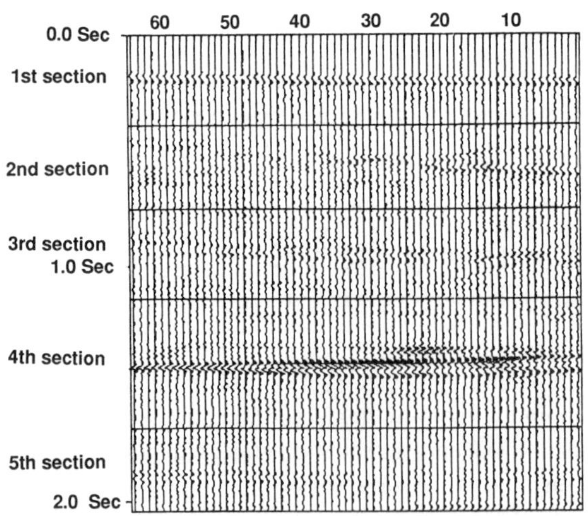

FIGURE 7.12. Relative amplitude seismogram at Mississippi Canyon and partitioning boundaries (five sections).

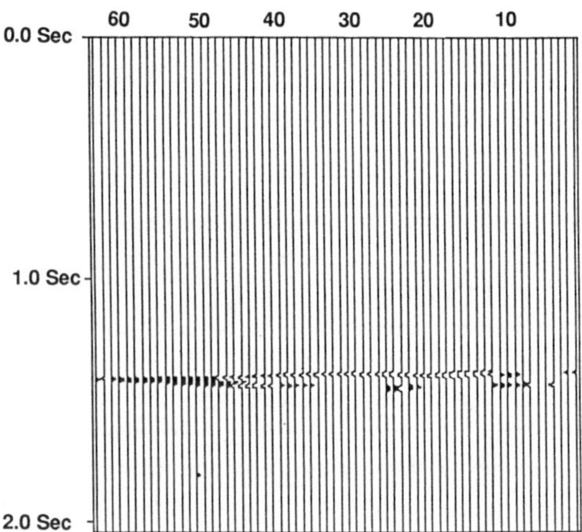

FIGURE 7.13. Tree classification result of bright spot (partitioning method and tree classification).

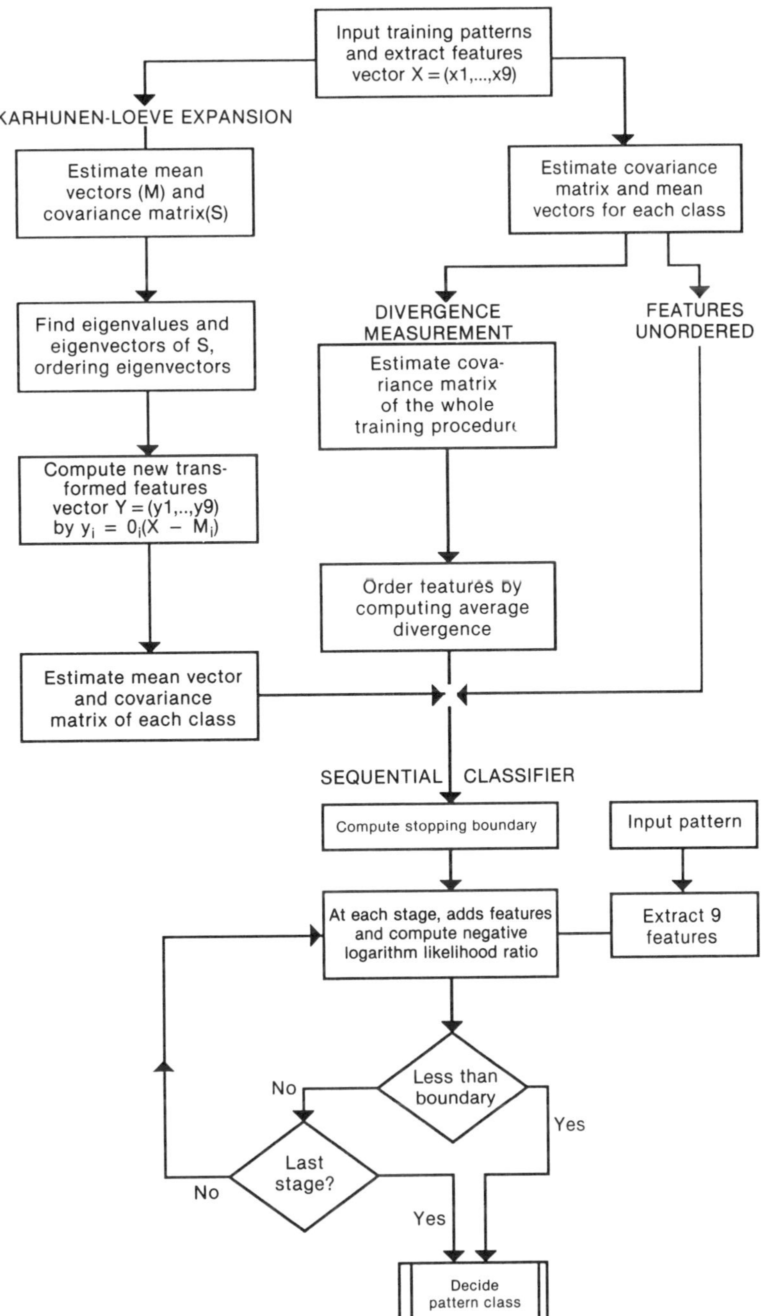

FIGURE 7.14. System of pattern recognition.

Results by Using the Partitioning Method and Tree Classification

Using partitioning algorithm, we can get five sections in Figure 7.12. Three hypotheses and their corresponding tree classifiers are tested on each section. Testing traces are the 4th, 12th, 20th, 28th, 36th, 44th, 52th, and 60th traces. From the feature distribution of envelope and instantaneous frequency in the testing traces of each section, only the 4th section returns the bright spot information. The first hypothesis of tree classification is applied in the experiment. After classifying the 64 traces in the 4th section, the complete classification result is

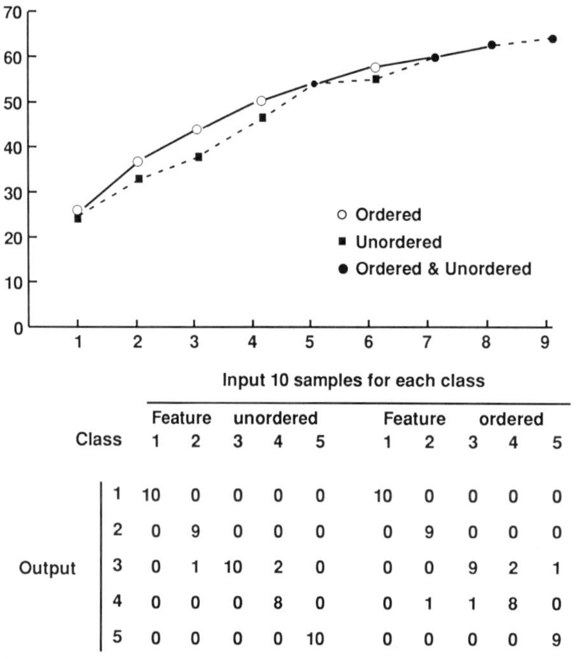

Input 10 samples for each class

		Feature	unordered				Feature	ordered			
Class		1	2	3	4	5	1	2	3	4	5
Output	1	10	0	0	0	0	10	0	0	0	0
	2	0	9	0	0	0	0	9	0	0	0
	3	0	1	10	2	0	0	0	9	2	1
	4	0	0	0	8	0	0	1	1	8	0
	5	0	0	0	0	10	0	0	0	0	9
Average no. of stages		2.9	7.0	5.9	7.3	4.9	3.8	6.0	5.6	6.5	5.3
Accuracy (%)		100	90	100	80	100	100	90	90	80	90

(b)

FIGURE 7.15. (a) Average divergence versus number of feature for 50 seismic training patterns. (b) Recognition results on 50 input seismic patterns (unordered and ordered cases).

shown in Figure 7.13. In Figure 7.13 using the partitioning method, the bright spot (physical anomalies) is clearly located at 1.4 s only. Figure 7.13 has a significant improvement on the classification result that is quite important in seismic interpretation.

Sequential Classification

Sequential classification technique is employed on seismic pattern recognition. The system of the sequential classification is shown in Figure 7.14. Nine features are generated through the extraction of moments, which are invariant of translation, slanting, stretch or squeezing, size, and mirroring. After features are extracted, a statistical sequential test (Fu, 1968) is applied on the seismic pattern recognition. Pattern is recognized by adding one feature at each stage and compared negative logarithm likelihood ratio to stopping boundary until the target class is determined. From the simulated experiment, 90% of the correct recognition is achieved.

Given a 2-D function $f(x,y)$, for digital image, we define the moment of order $(p + q)$ as

$$M_{pq} = \sum_x \sum_y x^p y^q f(x,y)$$

Let $(\bar{x},\bar{y})$ be the centroid of object, the central moment is

$$U_{pq} = \sum_x \sum_y x_0^p y_0^q f(x,y)$$

$$x_0 = x - \bar{x}, \qquad y_0 = y - \bar{y}$$

where

$$\bar{x} = \frac{M_{10}}{M_{00}}, \quad \bar{y} = \frac{M_{01}}{M_{00}}$$

The steps to extract nine invariant features are as follows:

1. Thin seismic patterns.
2. Compute M_{00}, M_{01}, M_{10}, M_{20}, and M_{02}, find centroid $(\bar{x}, \bar{y})$.
3. Compute U_{20} and U_{02} in terms of M_{00}, M_{01}, M_{10}, M_{02}, and M_{20}.
4. Compute standard deviations σ_x and σ_y.
5. Normalize coordinates, $x' = (x - \bar{x})/\sigma_x$ and $y' = (y - \bar{y})/\sigma_y$.
6. Compute $a = \sum x'y' / \sum y'^2$ and $x'' = \frac{(x' - ay')}{(1 - a^2)}$, $y'' = y'$.

TABLE 7.1a. Ordered eigenvalues and the cumulative of 50 seismic training patterns.

	1	2	3	4	5	6	7	8	9
Eigenvalue	0.637	0.689	0.044	0.016	0.011	0.006	0.003	0.003	0.002
Cumulative	0.787	0.896	0.950	0.970	0.983	0.990	0.994	0.998	1.000

TABLE 7.1b. Recognition results of sequential classification associated with KL expansion method on 50 seismic input patterns.

		Input class			
Output class	1	2	3	4	5
1	10	0	0	0	0
2	0	10	0	0	0
3	0	0	8	3	0
4	0	0	2	7	0
5	0	0	0	0	10
Average no. of stages	1.0	3.2	2.4	4.9	3.1
Accuracy (%)	100	100	80	70	100

7. Compute invariant moments of order 3 and 4,
$$M_{pq}'' = \sum \sum x''^p y''^q f(x'', y'')/M_{00}.$$
8. For seismic patterns compute $V_{pq} = M_{pq}''$ for even p and $V_{pq} = |M_{pq}''|$ for odd p.

Nine features are extracted, V_{30}, V_{21}, V_{12}, V_{03}, V_{40}, V_{31}, V_{22}, V_{13}, and V_{04}, which are the moments invariant of thickness, translation, size, slanting, stretching or squeezing, and mirroring.

Experiment

There are five classes of seismic patterns based on the reflection data included in the experiments.

Class 1: Pinch-out seismic pattern
Class 2: Flat-spot seismic pattern
Class 3: Sealevel fall seismic pattern
Class 4: Sealevel rise seismic pattern
Class 5: Bright-spot seismic pattern.

Fifty training patterns (10 samples for each class) are generated and 50 input seismic patterns (10 samples for each class) are tested.

Experiment 1: Sequential Classification Using Divergence

The average divergence versus the number of features when features are ordered and unordered is shown in Figure 7.15a. The first few feature measurements under the ordered case provide the effec-tive correct classification. Figure 7.15b shows the results of recognizing 50 input seismic patterns when features are ordered and unordered.

Experiment 2: Sequential Classification Using Karhunen–Loeve Expansion

Table 7.1a shows the ordering eigenvalues. It is noted that the variances of the first two principal components cover 90% of the total variances, which implies that the first two features can effec-tively achieve the correct classification. The rec-ognition results in Table 7.1b show the effective-ness of using this technique.

TABLE 7.2. The 12 features used in the experiments.

Feature	Displacement	Direction (degrees)
1	1	90
2	2	90
3	4	90
4	1	45
5	2	45
6	4	45
7	1	0
8	2	0
9	4	0
10	1	−45
11	2	−45
12	4	−45

```
..87....87.......78......22......233...............35.5..........
87....77........6.....2237......55................765..2.........33
.....6.......673..5236....2333.............888...........3356..
..3657653..55.........223................77........33167......
66.......33.........2231................2222........33..........6
.....2233......2335...........................11.........36657.
....3........33........22.2....5555...22..11.......7536......
.667......6522.......333....3335.........22........5...5......
5.....335........2223................2.2.........356.......567
.....22........112..........8.....53..........55...23....33...
...55.......532....667..7777.....66.........666...............
.33......225......5...77.....98.....66...77...........532...3
5.....633.......................999..777.........6..6...636.
..2355.................667....888.......6363....53.35........
33.............7.666..55...7778.........'.5....6653........133..
......5522..........33.............533..6.............122...55
..2233....555365..............322536.............122..........
11.......6223...3.........5552...........55......21...221........
.....3555........3....555..........123......33............5336
..333..................1...333.......22..............3....
23........12..33..2333.....3.3333.....23..267....56...1253....5
...............5...33.33.............25...8878..325......53.
.2....32....12..33.112..3........23552.....................3...
3.3....................253553.............363.36...........
...556..3......333..312535...............636.5...6..3.652.756..
11..253..3....3...12..3.6..3536....35..36...3........6...5...73
..366.....36.........5...55.......5..35....................
62..........353535..3..........353.............632657.........3
......535........35..........56....8..73..8.83.......57683223.
56...5..........6.....567.8565..2366.53..32.6........5........
.3367.732......3.23.32...3....................3.......3
....2351.23................35.76..8...326.....65....23.353...3.
.237....6..23.3.55.35..76....6..32.233...23.................
6.................65..23..................535.2..........32.
.535........566............567.............73...5.2........56..2
.........33.........3..........35.........767........22.36.3.....
......535........77......56.......65657.............2..3......
6................5326...3...........5.....676867...278..77..
....5.32.......7.........3..5..665.......32......6.2...83....
...5.......................3..5.........565........3........7

        Features :  2.0191 1.9457 2.0243 1.8093 1.9474 1.9374
                    1.2936 1.6553 1.7834 1.9595 1.8947 2.0042
```

FIGURE 7.16. Chaotic reflections and its 12 features.

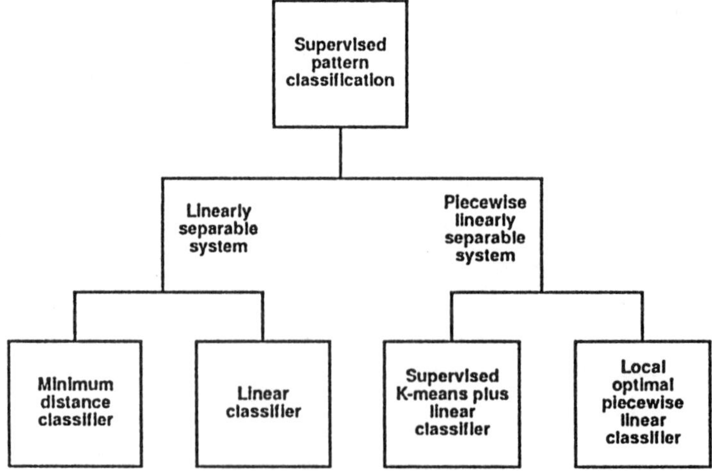

FIGURE 7.17. Supervised pattern classification.

FIGURE 7.18. Direct-wave, reflection, and refraction patterns of one-shot seismogram.

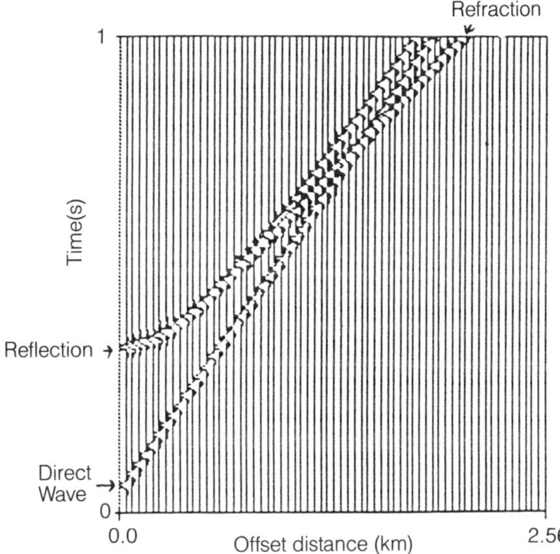

Contrasts and Cooccurrence Matrix in Seismic Pattern Inspection

Don et al. (1984) had successfully inspected the metal surface using pattern classification techniques. Due to the similarity between the seismic data images and the metal surface images, several pattern classification techniques are applied to inspect the seismic patterns.

Based on the cooccurrence matrix (Don et al., 1984), 12 effective features, contrasts, are extracted and used to the seismic pattern recognition. Table 7.2 shows the 12 features, the contrasts from the cooccurrence matrix. Figure 7.16 shows the chaotic reflections. Eight analyzed seismic patterns are chaotic reflections, few relections, many reflections, bright-spot, flat-spot, pinch-out, gradual sealevel fall, and gradual sealevel rise. Four patterns for each class are used as the training patterns and four patterns for each class are used as the testing patterns. The classification techniques include four clustering algorithms, two linear classifiers, and two piecewise linear classifiers. Figure 7.17 shows the pattern classification techniques used here.

Experiment

The experiment is to work on all the 8 classes (32 training samples and 32 test samples). For the minimum distance classifier, four test samples are misclassified. This yields 87.5% correct rate (28 out of 32). Linear classifier results in 84.4% correct rate (27 out of 32). Supervised K-means plus linear classifier yields 78.1% correct rate (25 out of 32). The local optimal piecewise linear classifier achieves 84.4%. The results are quite good in the seismic pattern inspection.

Hierarchical System, Finite-State Automaton, and Hough Transformation

In a seismogram, there are many seismic patterns. In a common source (one-shot) seismogram (Figure 7.18), traveltime curves of the direct wave and refracted wave patterns are straight lines and the travel time curve of the reflected wave pattern is hyperbolic. Direct wave and reflected wave patterns have some severe interferences beyond some distance from the source. In the stacked seismogram (Figure 7.19), there are diffraction curves and a continuous straight line reflection. Huang et al. (1985) showed the usefulness of the Hough transformation for detection of straight lines and hyperbolic curves in a seismogram. However, the detection results showed some interferences existed among nearby patterns in the common source seismogram.

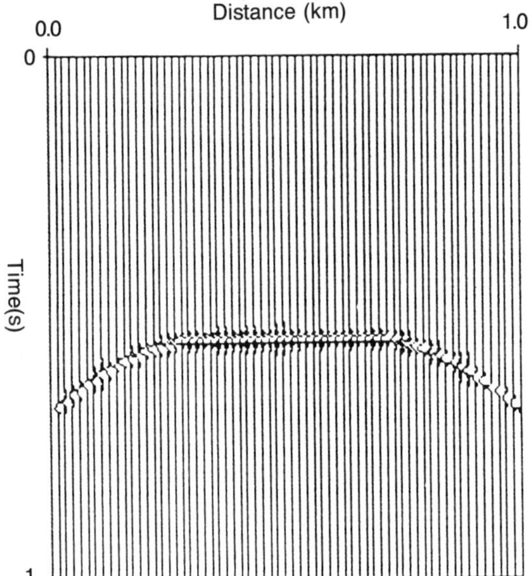

FIGURE 7.19. Horizontal reflector and diffraction patterns of stacked seismogram.

To detect different and varied types of patterns in a seismogram, a hierarchical syntactic pattern recognition and Hough transformation system is developed to improve the pattern detection. This is done by decomposing the complex patterns into uniform patterns consisting of similar properties, and further decomposing the uniform patterns into each type of pattern. Figure 7.20 shows the hierarchical pattern recognition schema for composite seismic patterns. A syntactic pattern recognition system (Fu, 1982) is used in hierarchical detection. Figure 7.21 shows the block diagram of a syntactic

pattern recognition system. After each single pattern is detected by using syntactic pattern recognition, a Hough transformation (Hough, 1962; Huang et al., 1985; Illingworth and Kittler, 1988) is used in the reconstruction of the seismic patterns, pattern by pattern. Figure 7.22 shows the block diagram of the hierarchical syntactic pattern recognition and Hough transformation system.

The Hough transformation part of the system includes envelope generation, thresholding, Hough transformation, and parameter determination (Huang et al., 1985). Envelope describes the outer shape of the wavelet (Huang and Fu, 1984). A Hough transformation transforms each image element in the picture space (T–X) into elements in parameter space. In the line detection using Hough transformation, the line equation of the direct wave and refracted wave patterns in the T–X space is $\rho = x_i^*$ $\cos \theta + t_i^* \sin \theta$. The line in the T–X plane corresponds to one point in the ρ–θ plane. Every point (x_i, t_i) in the picture space maps one line of ρ–θ in the parameter space. N points (x_i, t_i) of the same line map N curves of ρ–θ in the parameter space. The intersection of N curves in the parameter space is one point (ρ, θ), which is the parameter of the line equation $\rho = x_i^* \cos \theta + t_i^* \sin \theta$. Hough transformation is also used to detect hyperblic patterns. (Huang et al., 1985). A visual inspection method, local peak detection method, maximum peak detection method, and clustering algorithm determine the parameters. The clustering algorithm includes a K-mean algorithm with bottom-up hierarchical algorithm and pseudo F-statistics (PFS) value method (Vogel and Wong, 1979). Parameters corre-

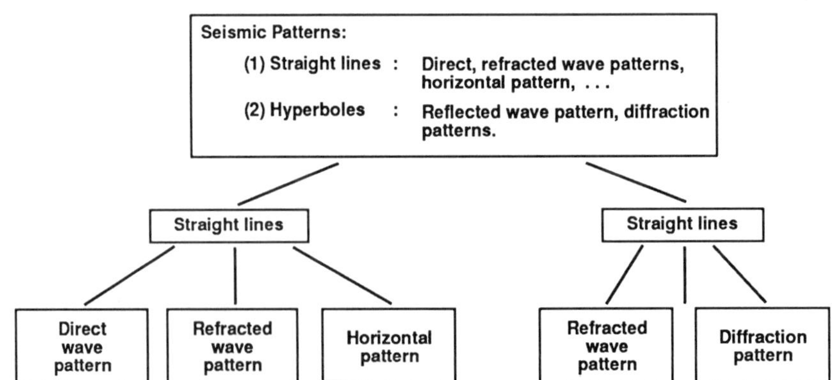

FIGURE 7.20. Hierarchical pattern recognition for composite seismic patterns.

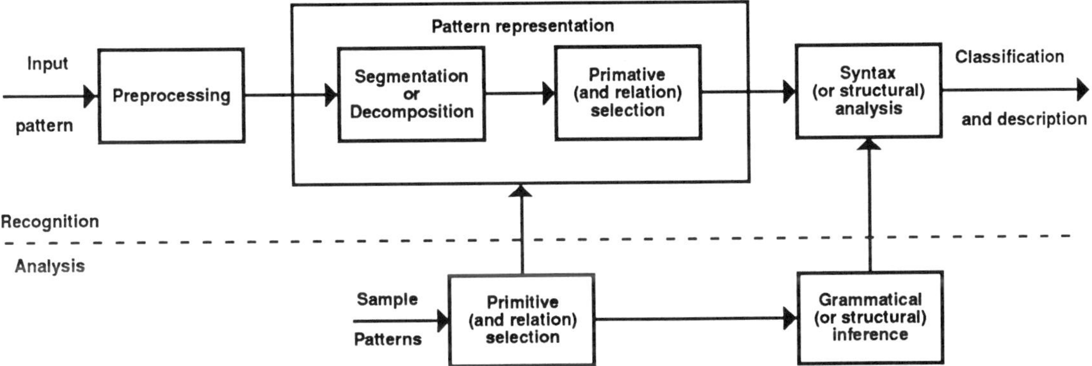

FIGURE 7.21. Block diagram of a syntactic pattern recognition system. *Syntactic Pattern Recognition and Applications,* (K.S. Fu, © 1982, pp. 9, 98. Adapted by permission of Prentice-Hall, Inc. Englewood Cliffs, NJ 07632).

sponding to the detected patterns are transformed back to their pattern equation. Then, the patterns are detected in the picture space.

The system of syntactic pattern recognition includes envelope generation, linking process in the seismogram, segmentation, primitive recognition, grammatical inference, and syntax analysis. Linking processing extracts seismic patterns using a linking algorithm based on the pattern growing technique and a function approximation algorithm. Primitives are assigned by amplitude-dependent encoding and a grammar is inferred by K-tail finite-state inference.

Syntax analysis is performed by an error-correcting finite-state automaton. Finally, the seismic patterns are automatically recognized and reconstructed.

Horizon Linking Processing and Segmentation

The linking process technique (Huang et al., 1990) extracts the seismic horizon patterns from the seismogram. First, each waveform of each trace is detected by scanning the seismogram in a vertical direction from left to right. Second, these detected

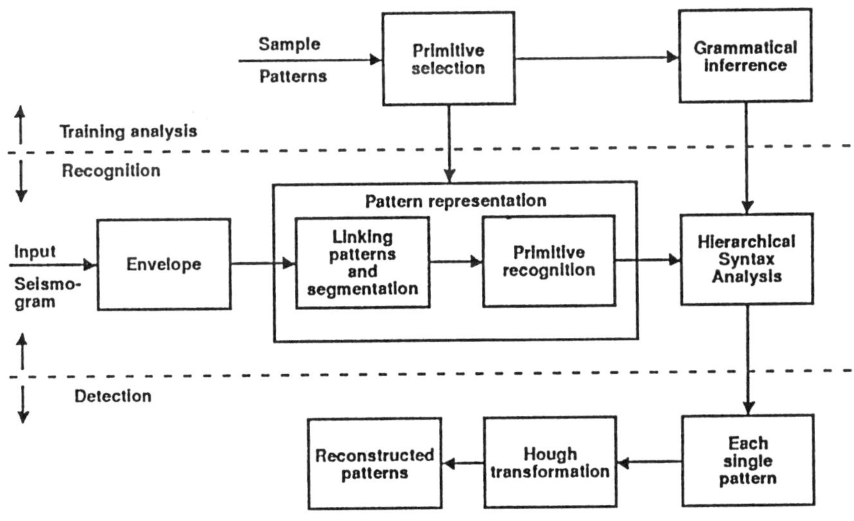

FIGURE 7.22. Block diagram of hierarchical syntactic pattern recognition and Hough transformation system.

waveforms are linked as skeletons according to their geometric properties.

A pattern growing algorithm in linking processing automatically extracts patterns using the principle of branch and bound search (Winston, 1984) and function approximation. Initially, the algorithm creates a new skeleton when the peak of the detected waveform is in the starting trace. The assignment of each new peak point (x_j, y_j) on the next trace to one of the existing skeletons or to a new skeleton is based on the calculated difference between the estimated peak location at the current trace from the previous existing skeleton and the current analyzed peak point. The current new peak is assigned to one of these existing paths if the difference is within the threshold of the function approximation. Otherwise, a new path with this new peak point is created. Using this linking algorithm, all the input peak points are linked as several skeletons.

Primitive Recognition

Next, each linked pattern is segmented. Every four sampling intervals (traces) comprise one segment. Each segment is assigned to a primitive, so each linked pattern can be represented as a sentence, i.e., a string of primitives. The amplitude-dependent encoding method (Huang and Fu, 1985a,b) is used in the assignment of primitives. Average slope is used in the encoding.

Training Patterns

Training patterns to be used for grammatical inference are generated for a common source seismogram using different apparent velocities for direct wave patterns, refracted wave patterns (straight

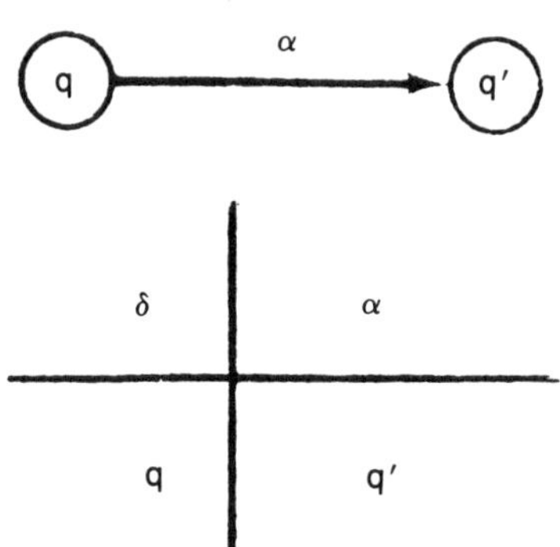

FIGURE 7.24. (Top:) Graphic representation of $\delta(q,a) = q'$. (Bottom:) A state transition table of $\delta(q,a) = q'$.

lines), and reflected wave patterns (hyperbolas). Training patterns for a stacked seismogram with different velocities in horizontal patterns (straight lines) and diffraction patterns (hyperbolas) are also generated.

In addition to the whole training patterns, a training pattern can be a portion of a whole pattern, which is formed by the intersection of many patterns. The intersected patterns may be discontinued because of interference. All these patterns are included in the training pattern set. By using segmentation and primitive recognition, all these training patterns are transformed into strings of primitives. The length of segment at the end of each pattern may not be enough, so truncation is considered. A finite-state grammar can be inferred from a set of training strings in each class.

Grammatical Inference

To describe the structural information about the class of patterns under study, a systematic approach to the design of syntactic pattern recognition systems is needed. The use of algorithms capable of obtaining a grammar from a set of patterns that are representative of some training patterns is required. This system is called grammatical inference.

A finite-state grammar G (Fu, 1982) is a quadtuple, $G=(V_N, V_T, P, S)$, in which V_N is the finite set of nonterminal (nonprimitive) symbols, V_T is the finite set of terminal primitive) symbols, $V_N \cup V_T = V$, $V_N \cap V_T = \phi$. $S \in V_N$ is the starting symbol of a sentence, P is a finite set of rewriting rules (production) or procedures denoted by $A \rightarrow aB$, or

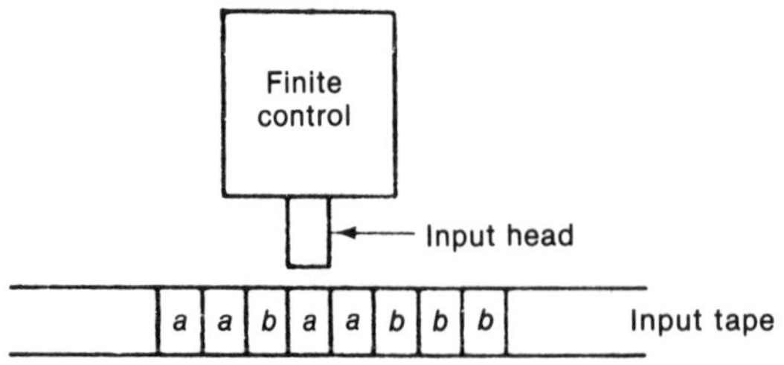

FIGURE 7.23. Finite-state automaton.

$A \rightarrow a$, where A, $B \in V_N$, and $a \in V_T$. $L(G)$ is the language generated by G.

Example 1

Consider a finite-state grammar $G = (V_N, V_T, P, S)$, where $V_N = \{S, A\}$, $V_T = \{a, b\}$, and P:

$$S \rightarrow aA$$
$$A \rightarrow aA$$
$$A \rightarrow b$$

Then the language generated by the grammar G is $L(G) = \{a^n b \,|\, n = 1, 2, \cdots\}$.

The K-tail finite-state grammatical inference algorithm (Fu, 1982) is used to infer grammars from a set of training patterns. The algorithm reduces the number of derived grammars. It finds the canonical grammar first and then merges the states that are K-tail equivalent. The algorithm is adjustable. The value of K controls the size of the inferred grammar.

A set of training sentences for each class is used to generate the source grammar and each input sample can be recognized as one member of a particular class.

Finite-State Error Correcting Parsing

Due to distortion and noise problems, the syntax analysis uses finite-state error correcting parsing. Finite-state error correcting parsing tries to correct an input sentence by looking for and removing errors in the sentence. Three types of errors can occur. These are insertion errors, substitution errors, and deletion errors. In this study, all segments are continuously sampled without insertion error, therefore, only substitution and deletion errors are considered. The finite state grammar is expanded to include only the substitution and deletion errors. The original production forms of a finite state grammar are

$$A \rightarrow aB$$
or
$$A \rightarrow a$$

The production forms added to account for substitution errors are stated as follows:

$$A \rightarrow bB \qquad a \neq b$$
or
$$A \rightarrow b, \qquad a \neq b$$

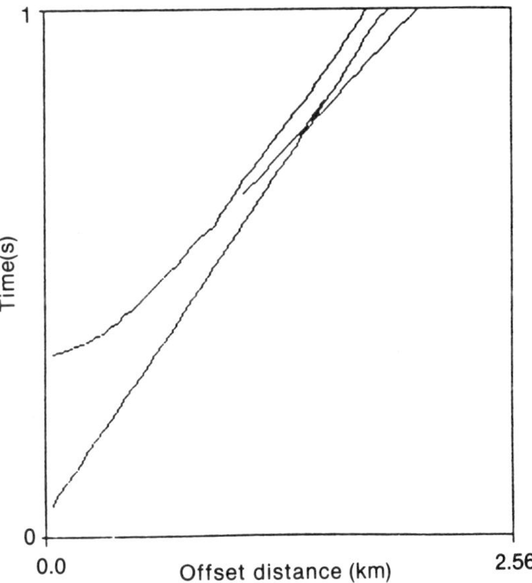

FIGURE 7.25. Skeleton of the one-shot seismogram.

The production forms added to account for deletion errors are stated as follows:

$$A \rightarrow \lambda B$$
or
$$A \rightarrow \lambda$$

where λ is the empty terminal (empty string).

Input pattern strings are analyzed by a finite-state automaton (Fu, 1982) that can accept only languages defined by finite-state grammars. Finite-state automaton is the simplest recognizer (recognition device) to recognize the strings (sentences) of the language.

A deterministic finite-state automaton A is a quintuple

$$A = (\Sigma, Q, \delta, q_0, F)$$

where Σ is a finite set of input symbols, Q is a finite set of states, δ is a mapping of $Q \times \Sigma$ into Q (next state function), $q_0, \in Q$ is the initial state, and $F \subseteq Q$ is the set of final states, $T(A)$ is the language accepted by A.

A convenient representation of a finite-state automaton is given in Figure 7.23. The final control, in one of the states in Q, reads symbols from an input tape in a sequential manner from left to right. Initially, the finite control is in state q_0 and is scanning the leftmost symbol of a string in Σ^*,

TABLE 7.3a. Training patterns for velocity data of the one-shot seismogram.

	Velocity 1st layer (m/s)	Velocity 2nd layer (m/s)
(1)	1900	2500
(2)	1900	2700
(3)	1900	2900
(4)	2100	2700
(5)	2100	2900
(6)	2100	3100
(7)	2300	2900
(8)	2300	3100
(9)	2300	3300

TABLE 7.3b. Training strings for the one-shot seismogram.

Direct wave pattern	Refracted wave pattern	Reflected wave pattern
	cc	
	ccc	
dd	cccc	
ddd	ccccc	
dddd	cccccc	
ddddd	ccccccc	
ddddddddd	cccccccc	
dddddddddd	ccccccccc	abbcccdddd
dddddddddddd	cccccccccc	abbcccddddd
ddddddddddddd	ccccccccccc	abbcccdcddd
dddddddddddddd	cccccccccccc	abbcccccccccccc

which appears on the input tale. Σ^* denotes the set containing all strings over Σ including λ, the empty string. The interpretation of

$$\delta(q,a) = q', \ q,q' \in Q \ a \in \Sigma$$

is that the automaton A, in state q and scanning the input symbol a, goes to state q' and the input head moves one square to the right. A convenient way to represent the mapping δ is by the use of a state transition diagram or a state transition diagram or a state transition table. The state transition diagram and table corresponding to $\delta(q,a) = q'$ are shown in Figure 7.24 (top) and 7.24 (bottom). The mapping δ can be extended from an input symbol to a string of input symbols by defining

$$\delta(q,\lambda) = q, \ \delta(q,xa) = \delta[\delta(q,x),a], \ x \in \Sigma^* \text{ and } a \in \Sigma$$

Thus, the interpretation of $\delta(q,x) = q'$ is that the automaton A, starting in state q and scanning

through the string x on the input tape, will be in state q' and the input head moves right from the portion of the input tape containing x. A string or a sequence x is said to be accepted by A if

$$\delta(q_0,x) = p$$

for some $p \in F$ (a set of final states). The set of strings accepted by A is defined as

$$T(A) = \{x \mid \delta(q_0,x) \in F\}$$

The theorem that transforms a finite-state grammar to a finite-state automaton (Fu, 1982) is used.

Let $G = (V_N, V_T, P, S)$ be a finite-state grammar. Then there exists a finite-state automaton $A = (\Sigma, Q, \delta, q_0, F)$ with $T(A) = L(G)$ exists.

A deterministic and reduced number of states of finite-state automaton is used to parse the input strings and is derived from the nondeterministic finite-state automaton (Fu, 1982). The input linked strings are analyzed by this deterministic finite-state automaton machine; therefore, the input patterns are automatically detected by using the computer.

Common Source Simulated Seismogram Results

Figure 7.25 is the skeleton of the common source (one-short) seismogram (Figure 7.18) produced by the linking processing algorithm. Each skeleton is encoded as a sentence.

When the hierarchical pattern recognition system is applied to the common source seismogram, lines and hyperbolic curves are first separated using the deterministic finite-state automaton derived from the first two terminals of the straight line training patterns (cc, dd) and the hyperbolic curve pattern (ab). Here, cc is equivalent to a refraction pattern with slope between 12 and 17 over 8 traces, dd is equivalent to a direct wave pattern with slope between 17 and 25 over 8 traces. After separating patterns of different types, each pattern can be further separated from other patterns of the same type based on the class recognizers generated by the training patterns.

Table 7.3a shows the nine velocity combinations used in generating nine training patterns of each type of pattern in a common source seismogram. Using the primitive recognition method, Table 7.3b lists all possible training strings. These

TABLE 7.4a. *K*-tail nonterminal table for reflected-wave patterns.

$K = 14$
U1 = {abbccccccccccc, abbccccdcddd, abbcccddddd, abbcccdddd}
U2 = {bbccccccccccc, bbccccdcddd, bbcccddddd, bbcccdddd}
U3 = {bcccccccccccc, bccccdcddd, bcccddddd, bcccdddd}
U4 = {ccccccccccc, ccccdcddd, cccddddd, cccdddd}
U5 = {ccccccccccc, cccdcddd, ccdddddd, ccdddd}
U6 = {ccccccccc, ccdcddd, cddddd, cdddd}
U7 = {ccccccc, cdcddd, ddddd, dddd}
U8 = {ccccccc, dcddd}
U9 = {dddd, ddd}
U10 = {cccccc}
U11 = {cddd}
U12 = {ddd, dd}
U13 = {ccccc}
U14 = {ddd}
U15 = {dd, d}
U16 = {cccc}
U17 = {dd}
U18 = {d}
U19 = {ccc}
U20 = {cc}
U21 = {c}

TABLE 7.4b. *K*-tail grammer table for reflected-wave patterns.

$K = 14$
$G = (V_N, V_T, P, S)$
V_N = {U1, U2, U3, U4, U5, U6, U7, U8, U9, U10, U11, U12, U13, U14, U15, U16, U17, U18, U19, U20, U21}
V_T = {a, b, c, d, e, f}
S = U1
P:
U1→a U2
U2→b U3
U3→b U4
U4→c U5
U5→c U6
U6→c U7
U7→c U8
U7→d U9
U8→c U10
U8→d U11
U9→d U12
U10→c U13
U11→c U14
U12→d U15
U13→c U16
U14→d U17
U15→d U18
U15→d
U16→c U19
U17→d U18
U18→d
U19→c U20
U20→c U21
U21→c

TABLE 7.4c. Expanded grammer table for reflected-wave patterns.

$K = 14$
$G = (V_N, V_T, P, S)$
V_N = {U1, U2, U3, U4, U5, U6, U7, U8, U9, U10, U11, U12, U13, U14, U15, U16, U17, U18, U19, U20, U21}
V_T = {a, b, c, d, e, f}
S = U1
P:

U1→a U2	U7→e U9
U2→b U3	U8→c U10
U3→b U4	U8→d U11
U4→b U5	U9→d U12
U4→c U5	U10→c U13
U4→d U5	U11→c U14
U5→b U6	U12→d U15
U5→c U6	U13→c U16
U5→d U6	U14→c U17
U6→b U7	U15→d U18
U6→c U7	U15→d
U6→d U7	U16→c U19
U7→b U8	U17→d U18
U7→c U8	U18→d
U7→d U8	U19→c U20
U7→c U9	U20→c U21
U7→d U9	U21→c

TABLE 7.4d. Deterministic finite-state automaton for reflected wave patterns ($K = 14$).

	a	b	c	d	e	f	Final state
1	2						0
2		3					0
3		4					0
4		5	5	5			0
5		6	6	6			0
6		7	7	7			0
7		8	9	9	10		0
8			11	12			0
9			11	13			0
10				14			0
11			15				0
12			16				0
13			16	17			0
14				17			0
15			18				0
16				19			0
17				20			0
18			21				0
19				22			0
20				25			1
21			23				0
22				25			0
23			24				0
24			25				0
25							1

TABLE 7.5a. Input extracted strings from the one-shot seismogram.

cc
cccc
abbcccdddd
dddddddddd

TABLE 7.5b. Results of the one-shot seismogram.

Input strings	Type	Results
cc	Line	Line 2
cccc	Line	Line 2
abbcccdddd	Hyperbolic curve	Hyperbolic curve
dddddddddd	Line	Line 1

include complete strings and partial strings caused by the intersection of two patterns. Deletion errors at the end of each pattern are included. In the grammatical inference step, different K-values are tried. For $K = 14$, Table 7.4a shows the set of nonterminals used to infer the finite-state grammar for reflected wave patterns of the common source seismogram. The inferred finite-state grammar is shown in Table 7.4b. Table 7.4c is the expanded grammar with substitution errors included. The deterministic finite-state automaton is derived in Table 7.4d. Similar procedures are performed for direct wave patterns and refracted wave patterns.

The input extracted strings of the common source seismogram (Figure 7.18) are shown in Table 7.5a. Four linked skeleton strings are listed. The results are shown in Table 7.5b. The deterministic finite-state automata are constructed to analyze the input strings and the detected patterns reconstructed by using the Hough transformation. Figure 7.26 shows all the reconstruction of the detected patterns of the common source seismogram.

Stacked Simulated Seismogram Results

Figure 7.19 shows the input patterns of the stacked seismogram. By using linking processing to separate the hyperbolic curve patterns from straight line patterns, we obtain the skeletons of the stacked seismogram shown in Figure 7.27. Each skeleton is encoded as a sentence.

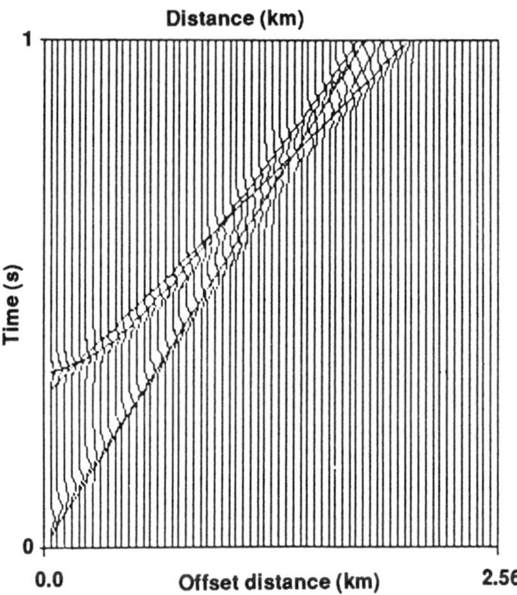

FIGURE 7.26. Summing patterns for the one-shot seismogram.

The first three symbols of each pattern are used to separate linear and hyperbolic curve patterns of the stacked seismogram, then each individual class recognition process is performed.

Table 7.6a lists the nine training patterns with different velocity combinations. Table 7.6b lists all sets of training strings including any deletion errors at the end of patterns. Similar to the above procedures, the set of nonterminals, the inferred finite-state grammars, the expanded grammars, and the finite-state automata are performed for horizontal patterns, the left diffraction patterns, and the right diffraction patterns. Table 7.7a is the input extracted strings for the stacked seismogram. Table 7.7b shows the results for the seismic patterns detected in the stacked seismogram. Summing up all the detected patterns results in Figure 7.28, which shows all patterns in the stacked seismogram.

Discussion

Hierarchical system is quite important in the recognition of seismic patterns. The syntactic pattern recognition technique has a great potential and may be applicable to complex 2-D and three-

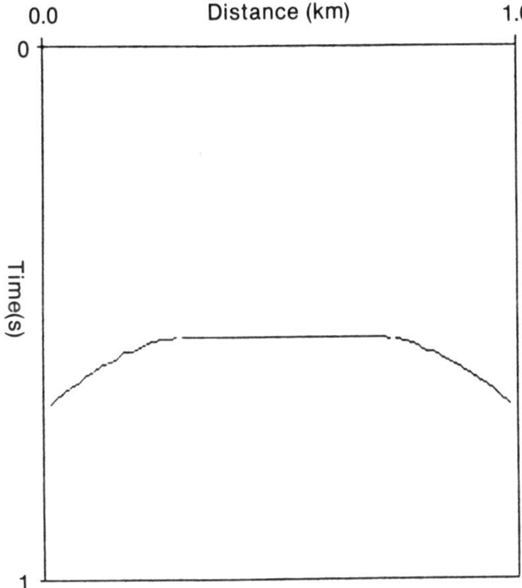

FIGURE 7.27. Skeleton of the stacked seismogram.

dimensional (3-D) seismic patterns caused by complex geology and velocity distribution.

Picture Description Language for Recognition of Seismic Patterns

In the interpretation and recognition of seismic patterns, structural information is very important. This study presents a structurally based analysis using Picture Description Language (PDL) to recognize seismic patterns.

In the area of picture description language, Shaw (1969) presented a formal linguistic description schema for computer processing systems. PDL had been applied in character recognition (Fu, 1982) and spark chamber photograph classification (Shaw, 1969). However, the analysis using PDL in the recognition of seismic patterns had not been used.

In PDL, a picture primitive is labeled at two distinguished points, a tail and a head. A primitive can be linked or concatenated to other primitives only at its tail and/or head. Because only two points of possible concatenations are defined, a primitive can be represented as a labeled directed edge of a graph, pointing from its tail to its head node. PDL is a linear string language; a picture may be expressed as concatenations among primitives according to four binary concatenation operators $\{+, -, \times, *\}$ and a unary operator $\{\simeq\}$. The four operators and their descriptions are shown in Figure 7.29.

TABLE 7.6a. Training patterns for velocity data of a stacked seismogram.

	Velocity 1st layer (m/s)	Velocity 2nd layer (m/s)
(1)	1900	2500
(2)	1900	2700
(3)	1900	2900
(4)	2100	2700
(5)	2100	2900
(6)	2100	3100
(7)	2300	2900
(8)	2300	3100
(9)	2300	3300

TABLE 7.6b. Training strings for a stacked seismogram.

Horizontal reflector patterns	Diffraction patterns (1)	Diffraction patterns (2)
	eef	
	eff	
	deef	
cccccc	deff	aab
ccccccc	eeff	aabb

TABLE 7.7a. Input extracted strings from a stacked seismogram.

eeff
aabb
cccccccd

TABLE 7.7b. Results of a stacked seismogram.

Input strings	Type	Results
eeff	Hyperbolic curve	Hyperbolic 1 curve
aabb	Hyperbolic curve	Hyperbolic 2 curve
cccccccd	Line	Line

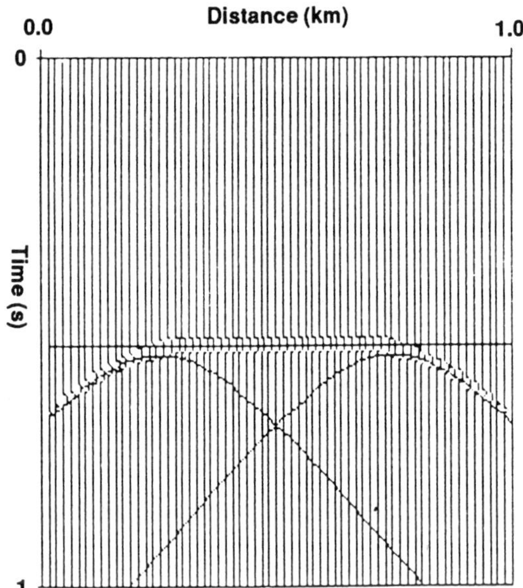

FIGURE 7.28. Summing patterns for the stacked seismogram.

The recognition system first maps a pattern to a string through the following stages:

1. Preprocessing—a mechanism to remove the superfluous points while maintaining the skeletal structure of the pattern.
2. Segmentation and primitive extraction—a process to segment a pattern into a set of segments and represent each segments with primitives.
3. String representation—a process to integrate the primitive coded segments by PDL using the directed graph concept.

After the string representations are obtained, the clustering algorithms are used to maximize the adequacy of the representations of a partitions in classes or clusters. The results have demonstrated the consistency and accuracy of the pattern recognitions using PDL.

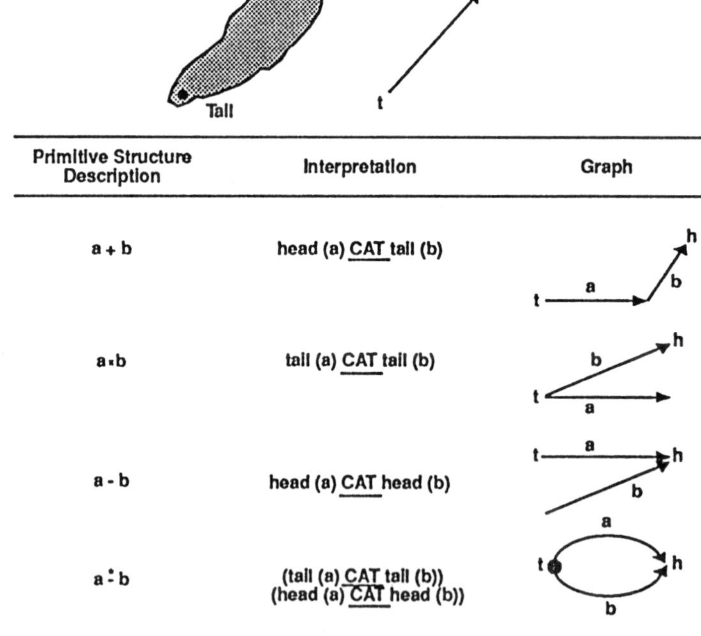

FIGURE 7.29. PDL primitive and operations. (K.S. Fu, *Syntactic Pattern Recognition and Applications*, © 1982, pp. 9, 98. Adapted by permission of Prentice-Hall, Inc. Englewood Cliffs, NJ 07632).

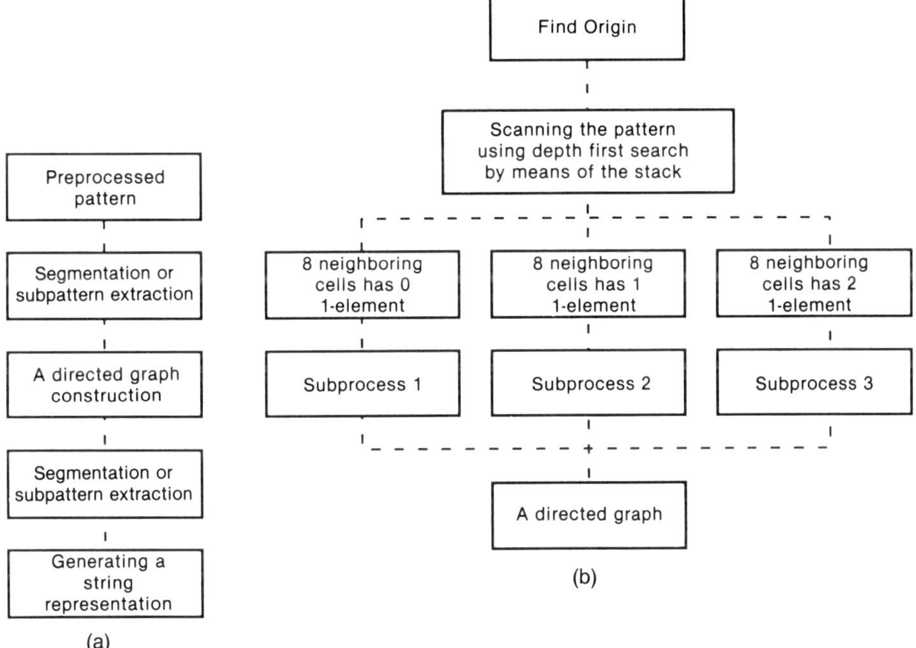

FIGURE 7.30. (a) Flow chart to transform pattern to string representation. (b) Flow chart to construct directed graph.

Pattern Representation Process

After the preprocessing stage, the input image passes through pattern representation. The preprocessed seismogram is a continuous line pattern on 64 × 64 grids.

Data Structure

The PDL pattern structure is implemented as a directed graph. The pointer-based multilinked list data structure is used for the directed graph.

Primitive Recognition and PDL Representations

The size-dependent primitive and size-independent primitive recognition systems are considered. For size-dependent primitive recognition system, each three symbols are encoded as one symbol. For size-independent primitive recognition system, eight-direction Freeman's chain-code is used.

A complete process of transforming a 2-D pattern to a one-dimensional (1-D) string representation in PDL for analysis is shown in Figure 7.30.

Results of PDL Method

In the supervised pattern recognition, a set of training seismic patterns is used as prototypes, representing different classes. For the classification and description of an unknown pattern, the input pattern can be matched against the prototypes.

A total of nine training seismic patterns are used as prototypes. The constructed directed training patterns are shown in Figure 7.31a–i. The weighted Levenshtein distances between the input and each of the nine training patterns are calculated. Nearest-neighbor classification rule is used in the assignment of input pattern.

The results of applying these processes to each input seismic pattern, i.e., bright-spot, pinch-out, flat-spot, gradual sealevel fall, and gradual sealevel rise seismic patterns are presented. The input seismic patterns are shown in Figure 7.32a–g.

The results of pattern recognition using Levenshtein distance and nearest-neighbor classification rule are shown in Table 7.8a. for size-dependent representation and in Table 7.8b for size-independent representation.

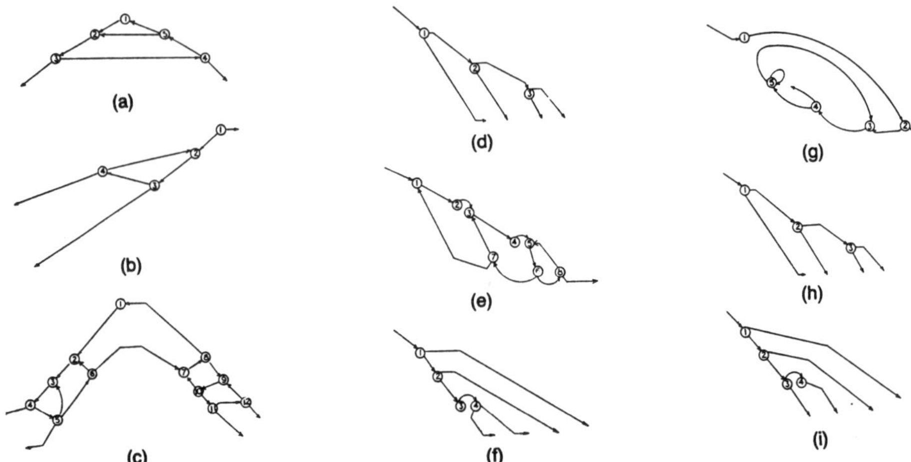

FIGURE 7.31. Directed graph of training patterns. (a) Bright-spot, (b) pinch-out, (c) flat-spot, (d) gradual sealevel fall, (e) gradual sealevel fall, (f) gradual sealevel rise, (g) gradual sealevel rise, (h) gradual sealevel fall, and (i) gradual sealevel rise.

A Tree Automaton System for Recognition of Seismic Patterns

Seismic patterns have two kinds of properties: (1) physical anomalies and (2) structural information. Decision-theoretic (statistical) pattern recognition can handle the properties of physical anomalies. In the seismic interpretation and recognition of seismic patterns, structural information is very important. Tree grammar and automaton was applied in bubble chamber photographs, LANDSAT data, fingerprint and English characters recognitions (Fu, 1982), but has not yet been applied in seismic patterns.

The block diagram of tree automaton system proposed in this study is shown in Figure 7.33.

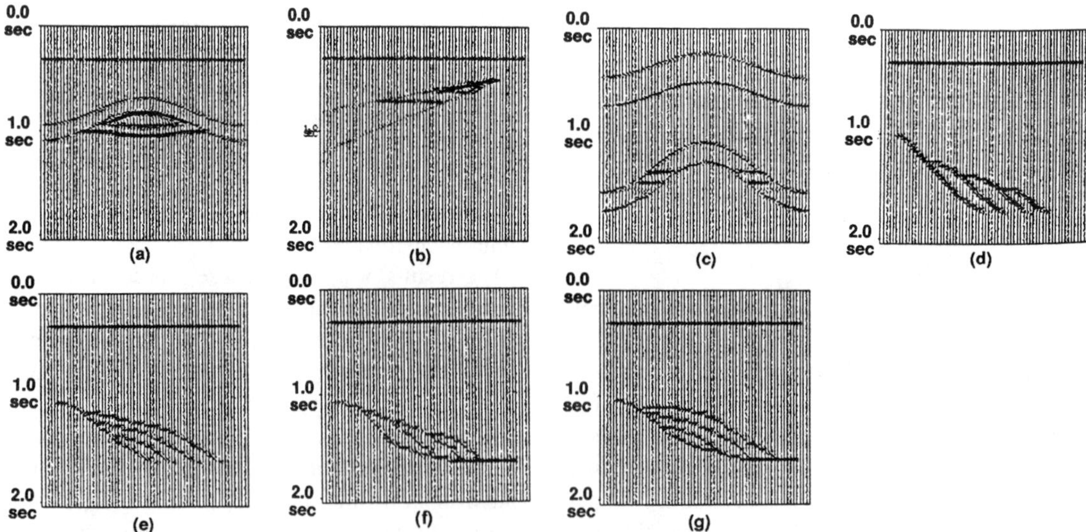

FIGURE 7.32. Synthetic seismic patterns. (a) Bright-spot, (b) pinch-out, (c) flat-spot, (d) gradual sealevel fall, (e) sealevel rise, (f) gradual sealevel fall, and (g) sealevel rise.

TABLE 7.8a. Clustering result of seismograms for size-dependent representation.

Input pattern	Figure 7.31a Bright-spot	Figure 7.31b Pinch-out	Figure 7.31c Flat-spot	Figure 7.31d Gradual sealevel fall (A)	Figure 7.31e Gradual sealevel rise (A)	Figure 7.31f Gradual sealevel fall (B)	Figure 7.31g Gradual sealevel rise (B)	Figure 7.31h Gradual sealevel fall (C)	Figure 7.31i Gradual sealevel rise (C)	Class result
Figure 7.32a	5	28	45	38	36	35	40	39	34	Bright–spot
Figure 7.32b	26	3	46	35	37	35	40	35	36	Pinch-out
Figure 7.32c	47	49	5	62	52	57	57	65	60	Flat-spot
Figure 7.32d	38	32	59	6	34	22	34	13	19	Gradual sealevel fall (A)
Figure 7.32e	37	36	49	35	3	31	30	35	32	Gradual sealevel rise (A)
Figure 7.32f	35	32	51	24	28	6	31	8	24	Gradual sealevel fall (B) or (C)
Figure 7.32g	41	41	56	36	29	32	5	37	32	Gradual sealevel rise (B)

TABLE 7.8b. Clustering result of seismograms for size-independent representation.

Input pattern	Figure 7.31a	Figure 7.31b	Figure 7.31c	Figure 7.31d	Figure 7.31e	Figure 7.31f	Figure 7.31g	Figure 7.31h	Figure 7.31i	Class result
Figure 7.32a	0	12	45	18	24	21	17	18	19	Bright-spot
Figure 7.32b	12	0	54	15	30	23	18	16	19	Pinch-out
Figure 7.32c	45	54	0	51	38	48	47	52	48	Flat-spot
Figure 7.32d	18	15	51	0	28	17	18	12	12	Gradual sealevel fall (A)
Figure 7.32e	24	30	38	28	0	24	22	27	25	Gradual sealevel rise (A)
Figure 7.32f	21	23	48	17	24	0	21	12	9	Gradual sealevel fall (B)
Figure 7.32g	17	18	47	18	22	21	0	19	18	Gradual sealevel rise (B)

Seven synthetic seismograms are used as input to the system. The synthetic seismograms of bright-spot, pinch-out, flat-spot, gradual sealevel fall, and sealevel rise patterns are generated in Figure 7.32a–g, the same as in PDL.

To construct the tree representation, envelope seismogram generation, thinning processing, pattern segmentation, primitive recognition, tree representation construction, and binary form tree construction techniques are used. Two thinning algorithms are used to extract the skeleton of the input seismogram. Pattern segmentation and primitive recognition are done by analyzing geometric properties. A breadth-first tree search algorithm is used to construct the minimum-height tree representation of the seismic pattern. The binary form tree representation is constructed (Fu, 1982) for the generalized error-correcting tree automaton (GECTA).

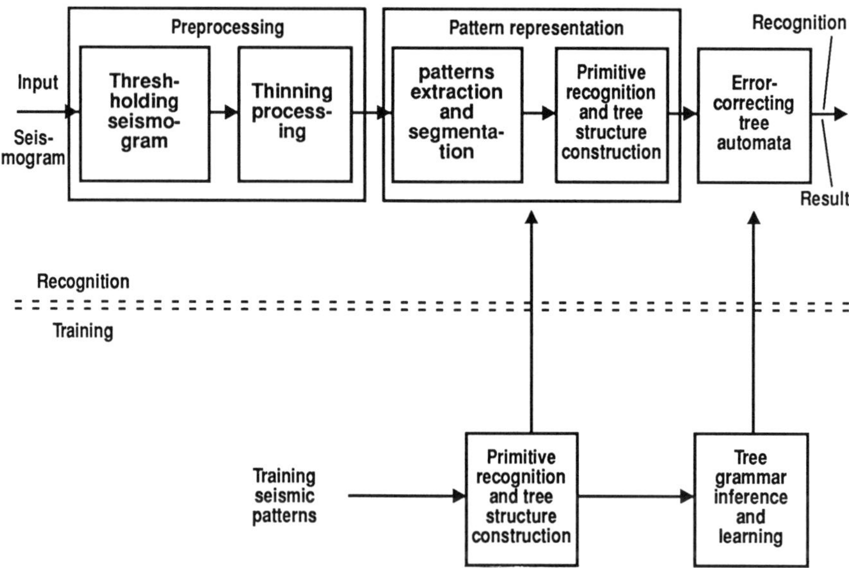

FIGURE 7.33. Block diagram of a seismic tree automaton system.

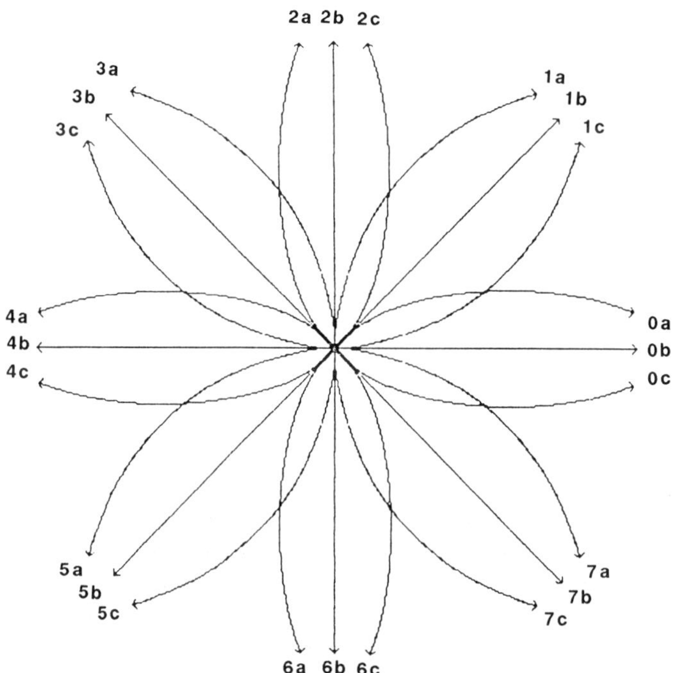

FIGURE 7.34. Twenty-four primitives.

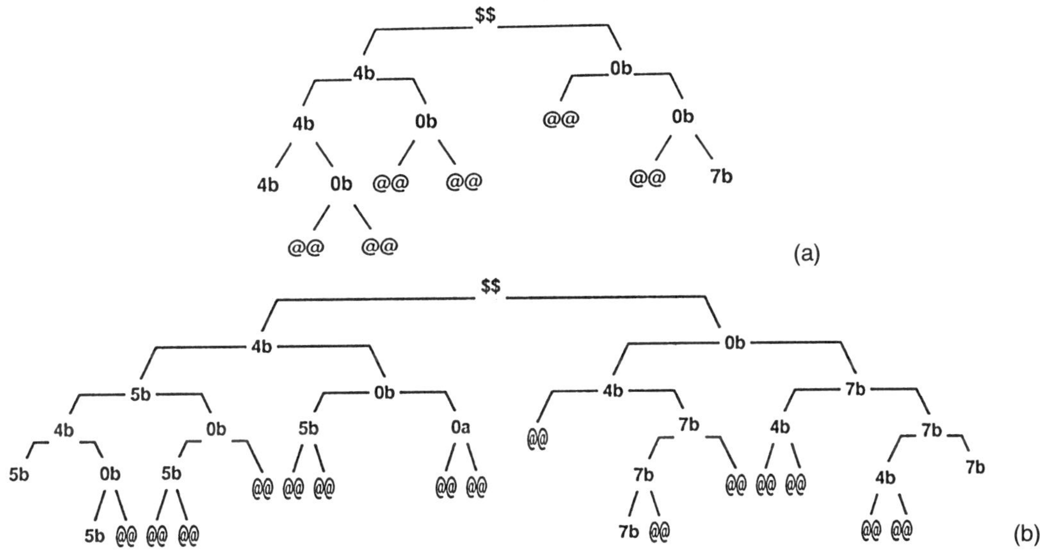

FIGURE 7.35. (a) Tree representation of bright-spot pattern. (b) Tree representation of flat-spot.

Expansive tree grammar inference technique is used to transform the training patterns into tree grammars. In cases of gradual sealevel fall and sealevel rise patterns, repetitions occur in some portions of the patterns. A learning from near-example procedure is proposed to automate the extension of the tree grammar.

Using both tree grammar and tree representations of seismic training patterns, error-correcting tree automata (Fu, 1982) can be constructed and can recognize and classify the input seismic patterns in the recognition process. The error-correcting tree automaton is a backward procedure of constructing a tree-like transition table from the leaves to the root of the tree to be recognized.

Primitive Recognition

Primitives are the basic components of a pattern. Each segment is assigned to a primitive. Twenty-four primitives are used in the system and they are shown in Figure 7.34. In this system, the scaling of the pattern is independent.

A pattern is conceptually converted into a directed graph in the tree representation construction. In Figure 7.34, the first character of the primitive is determined by matching the basic direction of the segment with the eight directions used in Freeman's chain code. The second character of the primitive is determined by the shape, i.e., straight line or curve, of the segment.

Tree Representation and Primitive Recognition of Training Patterns

Here, a scanning algorithm is proposed. The scanning is from left to right and top to bottom. In the algorithm, breadth-first tree expansion algorithm is used to construct tree structure. Nine training patterns are generated, the same as in PDL. The tree representations of bright spot and flat spot are shown in Figure 7.35.

A pattern identification will be assigned to each pattern and the pattern name will be stored in a file along with this identification. This identification will be used as the prefix of the nonterminal symbols of inferred production rules and for learning patterns to designate the pattern from which some learning can be achieved. In our examples, the pattern identifications are assigned to training patterns as follows:

Pattern identification	Pattern name
A1	Bright-spot
A2	Pinch-out
A3	Flat-spot
A4	Gradual sealevel fall
A5	Gradual sealevel rise
A6	Gradual sealevel fall
A7	Gradual sealevel rise
A8	Gradual sealevel fall
A9	Gradual sealevel rise

Construction of Expanded Tree Grammars

For training patterns, it is necessary to transform them to the tree grammar. Expansive tree grammar inference technique is applied in this step. The syntax errors, deletion, substitution, and insertion errors (stretch, branch, and split) are considered in this study. Binary tree grammar construction is used. Expanded tree grammar is needed to construct the GECTA (Fu, 1982).

Experimental Results

Minimum-Distance Structure Preserved Error-Correcting Tree Automaton

Substitution error is only considered. The parsing results of input patterns, Figure 7.32a–g using structure preserved error-correcting tree automaton (SPECTA) is shown in Table 7.9. Minimum-distance classification rule is used.

Maximum-Likelihood Structure Preserved Error-Correcting Tree Automaton

When the probability distribution of patterns and the deformation probabilities on each terminal symbol are available, error-correcting parsing based on maximum-likelihood criterion may provide a better recognition performance. Substitution error is considered.

TABLE 7.9. Summary of recognition results of minimum-likelihood SPECTA.

Input parameters	Distance — Starting states									Recognition results	Parsing time (min)
	A100	A200	A300	A400	A500	A600	A700	A800	A900		
Figure 7.32a bright-spot	0.2	2.4	–	–	–	–	–	–	–	A1 bright-spot	3.2
Figure 7.32b pinch-out	–	–	–	–	–	–	–	–	–	–	3.0
Figure 7.32c flat-spot	–	–	2.1	–	–	–	–	–	–	A3 flat-spot	5.5
Figure 7.32d gradual sealevel fall	–	–	–	–	–	–	–	0.2	–	A8 gradual sealevel fall	2.0
Figure 7.32e gradual sealevel rise	–	–	–	–	–	–	–	–	0.2	A9 gradual sealevel rise	2.2
Figure 7.32f gradual sealevel fall	–	–	–	–	–	0.5	–	–	–	A6 gradual sealevel fall	3.5
Figure 7.32g gradual sealevel rise	–	–	–	–	–	–	0.2	–	–	A7 gradual sealevel rise	3.7

TABLE 7.10. Summary of recognition results of maximum-likelihood SPECTA.

Input parameters	Distance									Recognition results	Parsing time (min)
	Starting states										
	A100	A200	A300	A400	A500	A600	A700	A800	A900		
Figure 7.32a bright-spot	4.8E-4	–	–	–	–	–	–	–	–	A1 bright-spot	2.3
Figure 7.32b pinch-out	–	–	–	–	–	–	–	–	–	–	2.0
Figure 7.32c flat-spot	–	–	1.0E-11	–	–	–	–	–	–	A3 flat-spot	3.1
Figure 7.32d gradual sealevel fall	–	–	–	–	–	–	–	6.1E-4	–	A8 gradual sealevel fall	2.1
Figure 7.32e gradual sealevel rise	–	–	–	–	–	–	–	–	6.1E-4	A9 gradual sealevel rise	2.2
Figure 7.32f gradual sealevel fall	–	–	–	–	–	1.4E-5	–	–	–	A6 gradual sealevel fall	2.4
Figure 7.32g gradual sealevel rise	–	–	–	–	–	–	6.1E-5	–	–	A7 gradual sealevel rise	2.6

TABLE 7.11. Summary of recognition results of GECTA.

Input parameters	Distance									Recognition results	Parsing time (min)
	Starting states										
	A100	A200	A300	A400	A500	A600	A700	A800	A900		
Figure 7.32a bright-spot	0.2	2.4	26.6	11.0	11.6	13.4	13.9	10.4	11.5	A1 bright-spot	57
Figure 7.32b pinch-out	5.8	3.6	26.6	13.0	12.2	13.4	13.3	13.0	12.1	A2 pinch-out	54
Figure 7.32c flat-spot	15.4	17.0	2.1	26.0	25.6	17.8	16.5	24.4	29.7	A3 flat-spot	136
Figure 7.32d gradual sedalevel fall	12.0	13.4	33.8	1.0	5.2	14.1	16.2	0.2	8.1	A8 gradual sealevel fall	24
Figure 7.32e gradual sealevel rise	11.7	12.4	33.3	3.0	1.3	14.7	12.4	4.0	0.2	A9 gradual sealevel fall	17
Figure 7.32f gradual sealevel fall	8.4	10.0	30.4	12.6	12.4	0.5	13.7	11.8	11.7	A6 gradual sealevel fall	58
Figure 7.32g gradual sealevel rise	9.1	9.2	25.9	11.6	11.2	8.7	0.2	13.6	11.2	A7 gradual sealevel rise	64

The parsing results of input pattern of Figure 7.32 using maximum-likelihood SPECTA is shown in Table 7.10.

Generalized Error-Correcting Tree Automaton

Consider five types of error transformation. The recognition results of input pattern of Figure 7.32 is shown in Table 7.11 and recognized as the bright spot with minimum distance.

Discussion

For input pinch-out pattern, the tree representation is distorted, so the pattern cannot be recognized by the minimum-distance SPECTA and maximum-likelihood SPECTA. From the recognition results, the tree approach of the high-dimensional grammar for the recognition of seismic patterns is quite encouraging and important, and can improve the automatic seismic pattern interpretation.

Acknowledgments. The author thanks Kevin M. Barry of Teledyne Exploration for providing relative-amplitude seismic data.

References

Aminzadeh, F., Ed., 1987, Handbook of Geophysical Exploration: Section I. Seismic Exploration, **20**, Pattern Recognition & Image Processing, Geophysical Press, London.

Bois, P., 1980, Autoregressive pattern recognition applied to the delimitation of oil and gas reservoirs: Geophys. Prosp. **28**, 572–591.

Bois, P., 1983, Some application of pattern recognition to oil and gas exploration: Inst. Electr. Electron. Eng., Trans. Geosci. Remote Sensing, **GE-21**, 416–426.

Chen, C.H., Ed., 1985, Pattern Recognition, **18**, Pergamon Press, Oxford.

deFigueiredo, R.J.P., 1982, Pattern recognition approach to exploration: in Concepts and Techniques in Oil and Gas Exploration, K.C. Jain, and R.J.P. deFigueiredo, Soc. Expl. Geophys.

Dobrin, M.B., 1976, Introduction to Geophysical Prospecting, 3rd ed., Chapter 10. McGraw-Hill, New York.

Don, Hou-Son, Fu, King-sun, Liu, C.R., and Lin, Wei-Chung, 1984, Metal surface inspection using image processing techniques: Inst. Electr. Electron. Eng. Trans., System, Man, Cybernet. **SMC-14**, 139–146.

Farnback, J.S., 1975, The complex envelope in seismic signal analysis: Bull. Seismol. Soc. Am. **65**, 951–962.

Fu, K.S., 1968, Sequential Methods in Pattern Recognition and Machine Learning: Academic Press, New York.

Fu, K.S., 1982, Syntactic Pattern Recognition and Applications: Prentice-Hall, Englewood Cliffs, NJ.

Hagen, D.C., 1981, The application of principal components analysis to seismic data sets: Proc. 2nd Int. Symp. Comput. Aided Seismic Anal. Discrimination, North Dartmouth, 98–109.

Hough, P.V.C., 1962, Method and means for recognizing complex patterns: U.S. Paten 3,069,654.

Huang, K.-Y., 1990, Branch and bound search for automatic linking process of seismic horizons: pattern recognition, **23**, 657–667.

Huang, K.-Y., McGillem, C.D., and Anuta, P.E., 1981, Detection of bright spots using pattern recognition techniques: presented at the 51st Annu. Int. Mtg., Soc. Expl. Geophys.

Huang, K.-Y., and Fu, K.S., 1982, Decision-theoretic pattern recognition for the classification of Ricker wavelets and the detection of bright spots: presented at the 52nd Annu. Int. Mtg., Soc. Expl. Geophys.

Huang, K.-Y., and Fu, K.S., 1983, Detection of bright spots in seismic signal using pattern recognition techniques: TR-EE83-35, Purdue Univ.

Huang, K.-Y., and Fu, K.S., 1984, Detection of bright spots in seismic signal using tree classifiers: Geoexploration **23**, 121–145.

Huang, K.-Y., and Fu, K.S., 1985a, Syntactic pattern recognition for the classification of Ricker wavelets: Geophysics **50**, 1548–1555.

Huang, K.-Y., and Fu, K.S., 1985b, Syntactic pattern recognition for the recognition of bright spots: Pattern Recognition **18**, 421–428.

Huang, K.-Y., and Fu, K.S., 1987a, Decision-theoretic approach for classification of Ricker wavelets and detection of seismic anomalies: Inst. Electr. Electron. Eng., Geosci. Remote Sensing **GE-25**(2), 118–123.

Huang, K.-Y., and Fu, K.S., 1987b, Detection of seismic bright spots using pattern recognition techniques: in Handbook of Geophysical Exploration: Section I. Seismic Exploration, **20**, Pattern Recognition and Image Processing, F. Aminzadeh (Ed.): Geophysical Press, London.

Huang, K.-Y., and Sheen, T.H., 1986, A tree automaton system of syntactic pattern recognition for the recognition of seismic patterns: presented at the 56th Annu. Int. Mtg., Soc. Expl. Geophys.

Huang, K.-Y., Fu, K.S., Cheng, S.W., and Sheen T.H., 1985a, Image processing of seismogram: (A) Hough transformation for the detection of seismic patterns; (B) Thinning processing in the seismogram: Pattern Recognition **18**, 429–440.

Illingworth, J. and Kittler, J., 1988, A survey of the Hough transform: Comput. Vision, Graphics, Image Process. **44**, 87–116.

Payton, C.E., Ed., 1977, Seismic Stratigraphy — Application to Hydrocarbon Exploration: Am. Assn. Petr. Geol. Mem. **16**.

Robertson, J.D., and Nogami, H.H., 1984, Complex seismic trace analysis of thin beds: Geophysics **49**, 344–352.

Shaw, A.C., 1969, A formal picture description scheme as a basis for picture processing system: Inform. Control **14**, 9–52.

Taner, M.T., Koehler, F., and Sheriff, R.E., 1979, Complex, seismic trace analysis: Geophysics **44**, 1041–1063.

Vogel, M.A., and Wong, A.K.C., 1979, PFS clustering method: IEEE Trans. Patt. Anal. Mach. Intel. **PAMI-1**(3), 237–245.

Winston, P.H., 1984, Artificial Intelligence: Addison-Wesley, Reading, MA.

8
Pattern Recognition for Marine Seismic Exploration

Ferial El-Hawary

Underwater seismic exploration using acoustic arrays aims at determining the structure of the underwater layered media in terms of properties that are important to the exploration task. Identifying hydrocarbon formations is a significant aim of the process. Signal interpretation and layer identification have been somewhat subjective and practiced by experienced marine geologists who have relied on the interpretation of acoustic gray scale graphic records. Thus identifying subsurface structures is a process that involves a great deal of human judgement, and knowledge. The interpretation problem is an imaging problem involving acquisition, processing, and pattern recognition (or interpretation). An expert system approach is eminently suited for this purpose to allow reasoning under dynamic uncertainties and to enable inference of information on the structure of the underwater layered media.

An overview of problems in underwater seismic image analysis and interpretation is given. Acquisition and processing of signals due to reflections and multiple reflections from the seabed and the underlying media are discussed. Modeling and compensating for dynamic heave component present in the signals, using conventional and parallel Kalman filtering, are discussed. Feature extraction algorithms involved in the interpretation process including delay and amplitude parameter estimation using a combined cross-correlation-minimum variance filter, linearized recursive estimation, and an event enhancement filter are covered. The inclusion of attenuation in the interpretation process as a potentially useful feature is dealt with.

A great deal of interest exists in expert system implementations in diverse application areas, including the application to underwater acoustics and geophysical problems. Some of these developments are reviewed, and ingredients of a knowledge-based interpretation system for marine seismic image analysis and interpretation are discussed.

Introduction

The marine seismic exploration method using acoustic arrays involves many challenging problems where the object is to determine the structure of the underwater layered media in terms of significant geometry and material properties that are required in a given exploration task. Identifying hydrocarbon formations is one important goal of the marine seismic process. The process involves the stages of image (or data) acquisition, processing, and pattern recognition (or interpretation). Classical acoustic techniques have been major tools for marine layer identification and classification of sediments over the recent past.

In marine subsurface mapping applications, firing pulses from marine energy sources are used to impart energy to the media underwater. High pressure air suddenly introduced into the water column creates a pressure pulse, which usually suffers from air bubbles. The pressure waveform is not periodic, but may be approximated by a damped sinusoidal wave. In practice, the optimum waveform is obtained after the firing pulse has traveled for about 50 m in the water column. The wave hits the seabed and propagates through the ocean floor

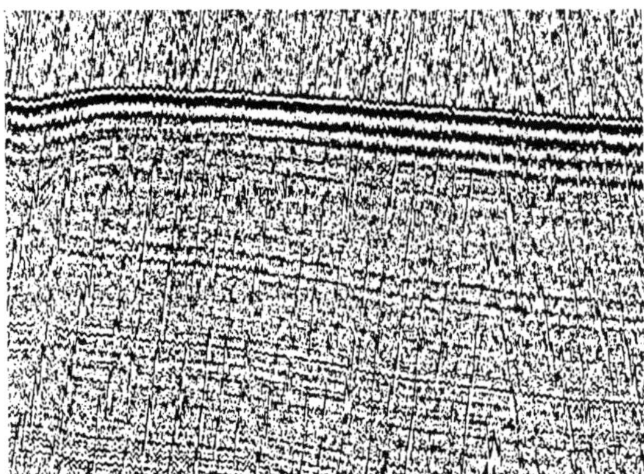

FIGURE 8.1. Typical underwater layers' image prior to processing.

and underlying structures, undergoing multiple transmissions and reflections at the layers' boundaries due to material properties changes. As a result of the reflections encountered, echo pressure signals are returned to the surface. These signals contain much information about the nature of the layers traversed. Arrays of pressure sensors (hydrophones) contained in cables receive the reflection traces, which are then converted to digital form, and transmitted down the cable to the recording instruments on board the ship or a remotely controlled vehicle.

The data collected by the array of sensors are essentially a raw image (or map) of the underwater structure. Figure 8.1 shows a typical underwater layers' image prior to processing. The acquired raw image is corrupted by noise, with a major component due to the ship's dynamics, coupled to the towed body (fish) through the towing cable, and the hydrodynamics of the towed body, which cause vertical motions of the source and sensor. These components have the effect of a varying acoustic wave travel path to the sea floor and to the subbottom reflectors between successive pings of the source. The motion's effects (commonly called heave effects) appear on the reflection records along the ship's track as additional undulations of the sea floor and of the subbottom reflectors.

Compensating for the heave component in the received signals is an important preprocessing task for improving displays of the raw and filtered reflection data, for extracting media parameters such as reflection coefficients and reflector depths, and in general for trace feature extraction. Partial

removal of these effects can be done by use of estimates of the vertical source motions from hydrostatic pressure and motion sensors to delay or advance the pulse firing instants relative to a clock pulse reference. This removal is done so that the source-bottom–sensor pulse travel time corresponds to that of a source and sensor located at a constant depth relative to the mean water surface level. Modeling the heave phenomenon for a given application is a main task in the marine seismic method, as a prelude to the application of Kalman based filters to identify and compensate for the heave component in the raw image. The heave motion is also of interest in buoy wave data analysis (Severance, 1975).

Following compensation for heave effects, the filtered image is then processed to classify and identify the subsurface layers using a number of interpretation procedures that rely on a selection of trace attributes. The processing aims at extracting required information about the properties of the media underwater. The parameters of the media are related to characteristic points on the received arrays, which are prefiltered to compensate for source and sensor heave components. The filtered image is then processed to find important trace attributes such as amplitude, attenuation, and delay parameters in each individual record. Parallel and array processing techniques can play a significant role in the processing phase.

The area of seismic signal modeling has been covered extensively in the literature. Mendel (1986) gives a tutorial introduction to the subject area. A classical treatment that provides excellent discussions of the theoretical foundations of the estimation process is found in Robinson (1967). Robinson and Durrani (1986) is a more recent contribution to the area commonly referred to as seismic deconvolution. A state space-based treatment can be found in Mendel (1983). An approach to extracting the subsurface features using a system theoretic treatment can be found in El-Hawary (1985).

The final step toward locating oil and gas fields and sediment classification is the data interpretation phase. This phase is not purely a geological problem. The correct mapping of the subsurface interfaces is important to identify horizons. All relevant geologic information such as faults and oil wells are included in the mapping. The geologic history of the region is an important part of the interpretation process. The interpretation is based on experience and knowledge of the region. Layer identification and signal interpretation procedures have been to a large extent subjective and practiced by very few able and experienced marine geologists who have relied primarily on the interpretation of acoustic gray scale graphic records. Current advances allow interpreters to use computer aids that are essentially enhanced graphic displays. Identifying subsurface structures on this basis is a process that involves a great deal of human judgment and knowledge.

A wealth of recent literature documenting progress of expert system implementations in diverse application areas exists. Most expert systems are rule based. This is valuable in medical applications, resulting in many recent contributions such as Bonamini et al. (1981). Moderate attention has been given to applying pattern recognition principles to reflection seismology over the past decade. Mathieu and Rice (1969) present results to distinguish between sand and shale horizons using linear discriminant analysis. This technique has also been used by Khattri and Gir (1976) and Khattri et al. (1979) to classify sand–shale interbedding sequences. Clustering techniques are employed by Bois (1980, 1981) to delimit reservoir boundaries. Huang and Fu (1984) use instantaneous trace values and a tree-based classifier to detect bright spots. Hagen (1981) uses instantaneous frequency measures to discriminate between porous and nonporous media. Love and Simaan (1985) deal with the problem by segmenting stacked data into zones of common signal character using texture-based image processing and pattern recognition techniques. More recently, the application to underwater acoustics and geophysical problems has been the subject of interest as can be found in Chen (1985), Hassab and Chen (1985), and Justice et al. (1985).

Marine seismic interpretation is a labor-intensive task. The opportunity exists, however, to go further and incorporate expertise in the computer in the form of a knowledge-based system. The problem of underwater seismic signal interpretation is particularly well suited to the application of artificial intelligence methods to incorporate as much human experience into the computer and to provide an assist to the human analyst. The ultimate purpose of constructing an expert system for marine seismic interpretation is to allow reasoning under dynamic uncertainties and to enable inference of information on the structure of the underwater layered media.

Requirements of Marine Seismic Interpretation System

Developing an integrated knowledge-based marine seismic interpretation system is of interest as an aid to the human operator. Central to such a system is a number of ingredients for acquired image filtering and subsequent processing for the purpose of parameter estimation, attribute extraction, pattern recognition, and rule-based interpretation. The components of this system can be classified broadly as those related to the preprocessing for heave compensation stage, and subsequent activities involved in extracting distinctive features of the underwater layers in preparation for the interpretation stage.

The object of the preprocessing stage is to eliminate the portion of the received signal that is attributable to heave effects in as much as possible. In Figure 8.2, the raw image received by the sensor is shown in the left side for 500 m of track. In the right side of the figure, five individual records are shown in expanded form to illustrate the heave effect evident by the nonalignment of the first peaks representing the water–sediment interface reflection. This stage is unique to the marine seismic method as compared to the land-based seismic exploration task. Pattern recognition methodologies applied at this stage can play an important role in producing an effective compensation strategy.

Extracting distinctive features of the underwater layered media is carried out using the received signals subsequent to heave compensation. The analyst is able to draw on a variety of possible trace features to assist in the interpretation process. These features can be either mathematically or visually based. The mathematically derivable features can be classified in terms of the domain used for representing the received signals. Time domain-based features include seismic event related values such as peak amplitudes and delay times as well as zero crossings and polarities. If the time series representing the seismic trace is modeled as an autoregressive (AR) process, then the coefficients of the resulting difference equations can be employed as

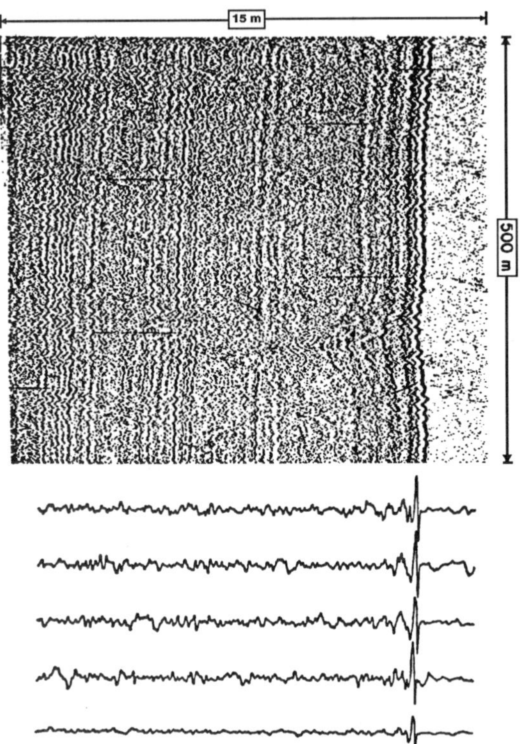

FIGURE 8.2. Raw underwater image showing heave effects.

potentially useful features. A frequency domain representation of the signals in terms of the power spectrum allows extraction of some feature frequencies and their corresponding amplitudes as candidates for the interpretation process. The autocorrelation function of the signal offers yet another set of possible features such as its peak values and the corresponding delay times.

Some mathematically derivable features are useful as the basis for arriving at significant physically based quantities that characterize the properties of the layered media underwater. For example, as discussed later, the velocity of sound propagation in the medium traversed and the density of each layer can be inferred from the acoustic impedance profile obtained from reflection coefficients related to amplitude and delay parameters of the seismic trace. The wavelet distortion attributed to attenuation effects can provide the basis for finding the attenuation rates in the various media traversed by the wave and hence another set of candidate features is made available for the interpretation stage.

The eventual outcome of the interpretation stage is a decision on the nature of the layers being probed, which provides insight into the potential of the area for hydrocarbon trapping. This stage involves the selection of the appropriate features or attributes combined with the expert's knowledge of the general geological, geophysical, and other experiences to arrive at a conclusion as to the potential of the area considered. A knowledge-based system can provide a tremendous assist in this regard.

Preprocessing for Heave Compensation

Compensation for source heave is a processing activity that distinguishes the marine seismic method from its land-based counterpart. The preprocessing requirement using Kalman filtering involves two steps. First, a model of the heave phenomenon is obtained on the basis of available data. The literature in marine hydrodynamics, such as Bhattacharya (1978), Price and Bishops (1974), and McCormick (1973), provides details of modeling the heave hydrodynamics. On the basis of the frequency response of the heave record, a model for the heave dynamics that is consistent with those found in the area of hydrodynamics can be assumed. The type of model as well as its order are important considerations (El-Hawary and Richards, 1987). In early applications such as El-Hawary (1982) a linear second-order model of the phenomenon was found satisfactory for certain field data. In this step the estimation of the model parameters is an important aspect. The resulting estimation software uses an iterative procedure (see El-Hawary, 1987a).

The heave model provides the basis for formulating the heave extraction problem as one of Kalman filtering in the theory of optimal linear estimation. The design of the Kalman filter enables reducing the residual heave effects, i.e., for delaying and advancing the recording trigger on successive firings so as to effect a smoothing or removal of such undulations. The filtering can be applied to postexperiment data records, or preferably in a real-time mode during acquisition of the reflection responses. The Kalman filter applications are well known for some different physical situations as reported in Gelb (1974) and Meditch (1969).

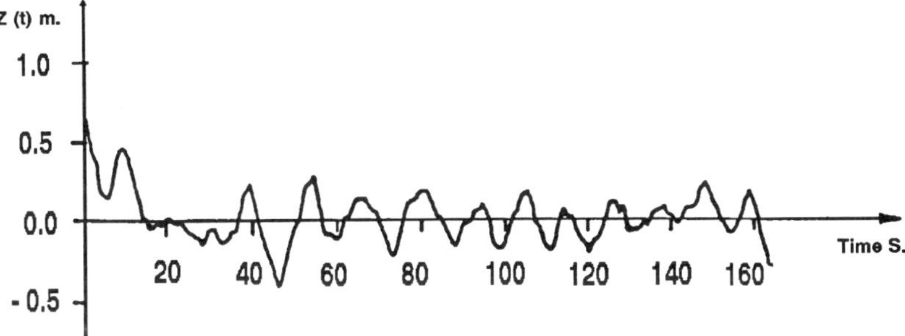

FIGURE 8.3. Time domain heave waveform.

Heave component extraction details are presented in El-Hawary and Vetter (1981).

Heave Motion Modeling

Typically, the record $z(t)$ is available as shown in Figure 8.3, to model the heave motion and exhibits a somewhat periodic random component. The process of obtaining $z(t)$ is documented in El-Hawary (1982). Frequency response methods have been used successfully in many fields of application including the task of identifying the model of the phenomenon of source heave. The Fourier transform $Z(f)$ of the record $z(t)$ gives the frequency response of the heave dynamics. This response is assumed to be due to a purely random excitation (white noise) caused by current and wave effects on the towed body, towed cable, and ship. To avoid aliasing, the Fourier transform of the record with a given time spacing rate is obtained through an appropriate time scaling that corresponds to a sampling frequency that is much higher than the anticipated frequency components in the record.

It is postulated that the heave process is represented by a continuous linear time invariant (LTI) dynamic system. The simplest model in this class is the second-order model that has been utilized in a number of practical applications. From an accuracy point of view, higher order models can be expected to provide less error in modeling the process. We assume that the heave process is modeled using the following transfer function:

$$H(s) = K \frac{N(s)}{D(s)} \tag{1}$$

The system's gain is denoted by K. The numerator function $N(s)$ has M zeros z_i, and is therefore given by

$$N(s) = (s - z_1)(s - z_2) \cdots (s - z_M) \tag{2a}$$

The denominator function has N poles p_i, and is given by

$$D(s) = (s - p_1)(s - p_2) \cdots (s - p_N) \tag{2b}$$

The heave process model estimation task resolves into finding the optimal values of poles and zeros using least-squares model parameter estimation for a given model order. Ideally, an attempt should be made to determine a priori the optimal model order, but this process is quite involved in practice (El-Hawary and Richards, 1987). The process is ideally suited for a knowledge-based approach.

Kalman Filtering Application

Once a model and its parameters have been identified, the process of Kalman filtering is performed. In the early implementation conventional Kalman filtering (El-Hawary, 1982) is used satisfactorily in a majority of cases. The heave motion model of Eq. (1) can be written in discrete state space form as

$$x(k + 1) = \phi(k + 1, k)x(k) + \Gamma(k + 1, k)w(k) \tag{3}$$

The state transition matrix $\phi(k + 1, k)$ and the matrix $\Gamma(k + 1, k)$ are constants and therefore

$$x(k + 1) = \phi x(k) + \Gamma w(k) \tag{4}$$

The input sequence $w(k)$ is assumed to be a Gaussian white sequence with zero mean and a covariance matrix $Q(k)$, being positive semi-definite. The initial state is assumed to be a Gaussian random vector with zero mean and known covariance matrix $P(0)$. Further $w(k)$ is assumed to be independent of $x(0)$. The record of heave

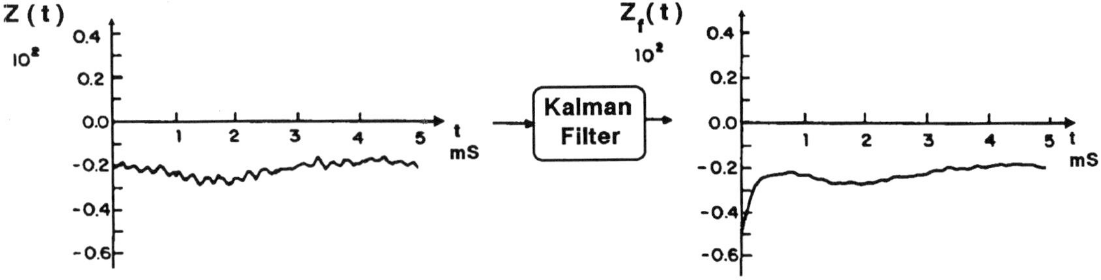

FIGURE 8.4. Effect of Kalman filtering on heave component.

component is assumed to be the basis for the measurement model given by

$$z(k + 1) = Hx(k + 1) + v(k + 1) \qquad (5)$$

The measurement error sequence v is assumed to be Gaussian with zero mean and a covariance matrix $R(k)$.

Assume that measurements $z(1), z(2), \ldots, z(j)$ are available, from which we like to estimate $x(k)$, denoted by $x(k|j)$. In filtering $j=k$, and therefore $x(k|k)$ is to be determined. The standard predictor-corrector form of a Kalman filter is used.

In the predictor stage a prediction of the state is obtained based on the previous optimal estimate

$$x_k(-) = \phi_{k-1}x_{k-1}(+) \qquad (6)$$

In addition, the error covariance matrix is obtained as

$$P_k(-) = \phi_{k-1}P_k(+)\phi_{k-1}^T \qquad (7)$$

In the corrector stage an updated state estimate is obtained

$$x_k(+) = x_k(-) + K_k[y_k - H_kx_k(-)] \qquad (8)$$

In addition, an update of the covariance matrix is obtained as

$$P_k(+) = (I - K_kH_k)P_k(-) \qquad (9)$$

Here K is the Kalman gain matrix given by

$$K_k = P_k(-)H_k^T[H_kP_k(-)H_k^T + R_k]^{-1} \qquad (10)$$

In Figure 8.4, the input to the Kalman filter and the resulting heave component are shown, to illustrate the filtering effects. Figure 8.5 shows time domain and power spectrum representations of a typical field case with seabed profile including the heave effect, the heave component, and finally the seabed profile compensated for heave effects.

Parallel Kalman Filtering

The real time application of the standard Kalman filter is limited by the filter's computational complexity. It is evident that faster implementations are desirable and parallel versions of Kalman filtering are required. Kalman filtering is a sequential process that evaluates the predictor equations prior to the computation involving the corrector equations. This coupling introduces computational delays, which can be avoided by using a parallel Kalman filter based on decoupling the predictor and corrector stages by forcing the corrector to lag the predictor by one discrete step. The resulting filter is called the parallel or [decoupled] Kalman filter. The filter equations are given by the following predictor–corrector form:

In the predictor stage a prediction of the state is given by

$$x_{k+1}(-) = \phi_k[\phi_{k-1}x_{k-1}(+)] \qquad (11)$$

This is a modified version of standard Kalman Filtering process

$$x_{k+1}(-) = \phi_k[x_k(+)] \qquad (12)$$

In addition the error covariance matrix is

$$P_{k+1}(-) = \psi_kP_{k-1}(+)\psi_k^T \qquad (13)$$

where

$$\psi_k = \phi_k\phi_{k-1} \qquad (14)$$

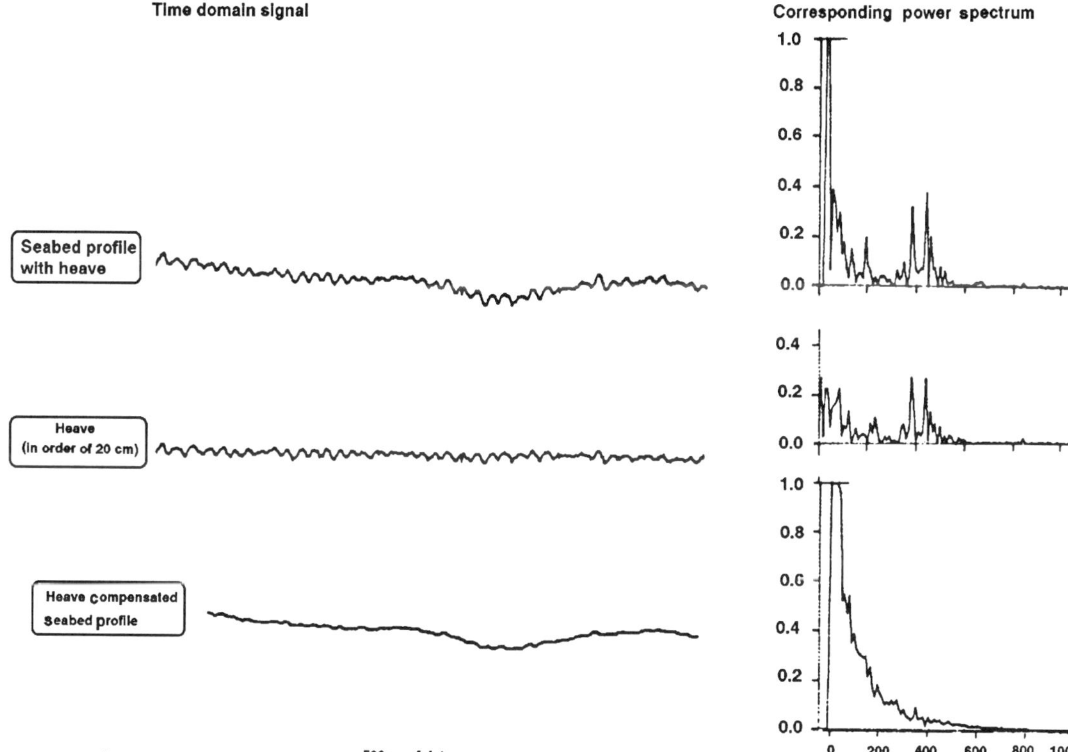

FIGURE 8.5. Typical time domain and spectral representations of seabed profile prior to heave removal, heave component, and compensated seabed profile.

The preceding equations are a modified version of standard Kalman filter.

In the corrector stage an updated state estimate is given by

$$x_k(+) = x_k(-) + K_k[y_k - H_k\hat{x}_k(-)] \quad (15)$$

In addition an update of the covariance matrix is

$$P_k(+) = (I - K_kH_k)P_k(-) \quad (16)$$

The Kalman gain matrix is given by

$$K_k = P_k(-)H_k^T[H_kP_k(-)H_k^T + R_k]^{-1} \quad (17)$$

The predictor is now decoupled from the corrector.

From a computational point of view, it is convenient to define

$$F_{11} = I - K_kH_k \quad (18)$$

and

$$F_{12} = K_k \quad (19)$$

which allows defining

$$F_k^T = [F_{11}^T \mid F_{12}^T] \quad (20)$$

$$S_k^T = [x_k^T(-) \mid Z_k^T] \quad (21)$$

A compact form for the optimal estimate is given by

$$x_k(+) = F_kS_k \quad (22)$$

The covariance matrix is now given by

$$P_k(+) = F_{11}P_k(-) \quad (23)$$

The computations can be carried out concurrently using two processors. Recent contributions in signal processing, computer architecture, and VLSI design show that systolic array processing techniques are extremely useful for designing special purpose devices to solve problems in linear algebra and system analysis (Kung, 1985; Travassos, 1985). The motivation of systolic architecture is that for special purpose hardware a systematic means of design is required and, therefore, a methodology for mapping high level computation into hardware structures. Systolic array architectures can be designed to take advantage of the parallel Kalman filter. Kalman filtering theory and parallel Kalman filters are applied to the exploration problem using a second-order model in recent work (El-Hawary and Ravindranath, 1986). Particular emphasis is given to the multireceiver case as an important application of array processing methodology. Parallel Kalman filtering is designed to take advantage of systolic array implementation.

Issues in Preprocessing

A number of important issues that are crucial to the success of a heave compensation strategy should be mentioned at this point. The first issue concerns the choice of model order to adequately represent the heave motion encountered while acquiring the data being interpreted. It seems sensible to suggest that a number of features of the power spectrum of heave record can be used to guide the model order selection process. The second issue relates to the appropriate choice of the covariance matrices Q and R involved in the Kalman filtering procedure. Yet another issue relates to the actual process of finding the optimal estimates of the model parameters, which were stated to be the result of an iterative solution procedure. Once again the selection of an appropriate initial guess appears to be an important factor in the success of such a scheme. Most of the issues discussed here can be treated by an experienced analyst and therefore present excellent candidates for knowledge-based system implementations.

Time Domain Attribute Extraction for Lossless Media

Presently, some aspects of trace feature extraction from the reflected signal waveforms subsequent to heave compensation via time domain representations are discussed. The acoustic pressure wave travels through the subsurface layered media and undergoes multiple reflections as it impinges on boundaries between successive layers. The received signal at the sensors (hydrophones) contains replicas of the original source signal that are corrupted by noise components due to various sources and modulated by the transmission media.

Assuming a linear lossless model of the wave propagation one can express the received signal $y(t)$ as the sum of N_L terms, each consisting of a version of the source signal $x(t)$ delayed by an accumulative two-way travel time τ_i with an amplitude scale factor a_i as well as an additive noise term. The amplitude scale factors are related to the layers' reflection coefficients and the delay times are related to layer depth and the velocity of sound propagation in the given media. Thus

$$y(t) = \sum_{i=1}^{N_L} \alpha_i x(t - \tau_i) + v(t) \tag{24}$$

According to de Figueiredo and Shaw (1987), the model Eq. (24) is called a Tauberian approximation of the actual trace. In essence, the required information about the media is extracted in a time domain-based approach using peak and delay parameter detection techniques performed with the aid of (24). Figure 8.6 shows in graphic form how the Tauberian model arises as a natural consequence of wave propagation through media with varying properties such as the acoustic impedance. Performing the required information extraction task may appear to be straightforward. Many detection procedures are available, and we review three schemes in this section.

Sequential Correlation-Based Detection

The estimation of delay and amplitude parameters can be carried out effectively using cross-correlation processing combined with minimum variance analysis (El-Hawary, 1985). Two steps are involved in this procedure.

The first step uses the fact that a signal such as $x(t)$, when cross-correlated with the signal $y(t)$, which is the sum of delayed replicas of $x(t)$ plus a random waveform will yield a cross-correlation function exhibiting large peaks at the values of the respective time delays. Peak and delay times are estimated using the cross-correlation between source and received signals. The result is a cross-correlation function with peaks at the unknown delay times. A direct search for the extrema results in an unrefined estimate of the delay parameters. In

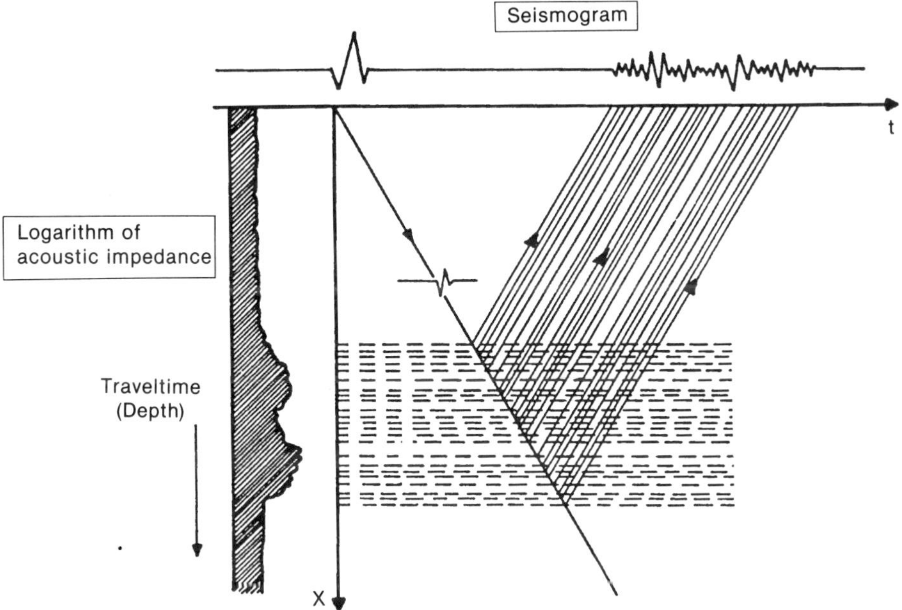

FIGURE 8.6. Elements of a marine seismic trace.

this step it is important that a threshold be established on the amplitude of the extrema considered. In practical implementations past experience is relied on in a trial and error procedure to set a threshold that is appropriate for the area surveyed. A database of prior knowledge is an important component of an automated implementation of this fast procedure.

The second step refines the estimates using minimum variance information. For this purpose, the problem resolves to finding an optimal parameter estimate of the amplitude scale factors with optimality defined in the minimum of error variance sense. The optimal estimates of the amplitude scale factors are given by

$$\hat{a}_i = \frac{[(\sigma_v^2/\sigma_{a_i}^2)\,\bar{a}_i + \phi_{xy}(\hat{\tau}_i)]}{[(\sigma_v^2/\sigma_{a_i}^2) + \phi_{xx}(0)]} \qquad (25)$$

Note that prior knowledge of the statistics of the noise process $v(t)$ is needed. This knowledge can also be achieved through the use of a matched filter implementation, where the impulse response of the filter is a reverse time replica of the source waveform. The matched filter simply maximizes the signal-to-noise ratio at the filter output.

Without a priori knowledge an estimate of peak amplitude is given by the ratio of cross-correlation of source and receiver signals to the autocorrelation of the source signal. Thus

$$\hat{a}_i = \frac{\phi_{xy}(\hat{\tau}_i)}{\phi_{xx}(0)} \qquad (26)$$

Linearized Delay Model Detection

In this procedure (El-Hawary and Vetter, 1980), it is assumed that each delayed replica of the source signal is approximated by a first order Taylor expansion about an estimate of each the time delay.

$$\Delta\tau_i = \hat{\tau}_i - \tau_i \qquad (27)$$

A further assumption is made that a rough estimate of the delays involved is available. The delayed replica is then represented by the values of the source signal and its derivative both evaluated at the estimate of the delay.

$$x(t - \tau_i) = x(t - \hat{\tau}_i) - \Delta\tau_i x'(t - \hat{\tau}_i) \qquad (28)$$

As a result an approximation to the Tauberian expansion can be written as

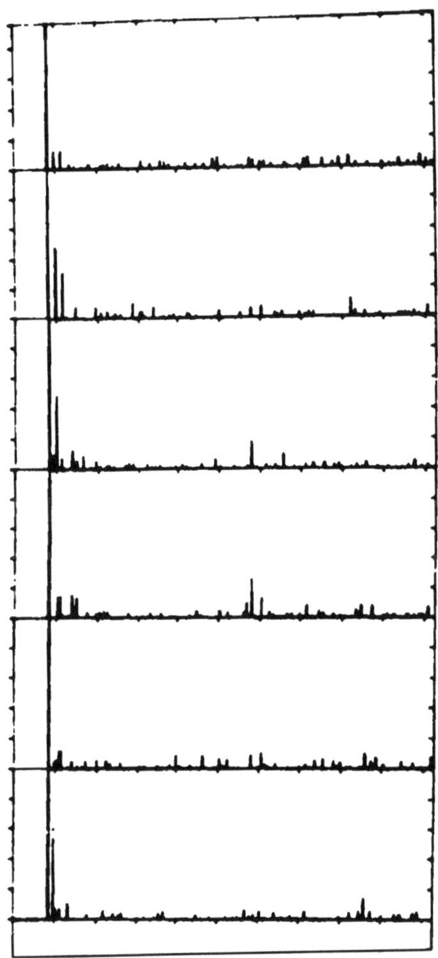

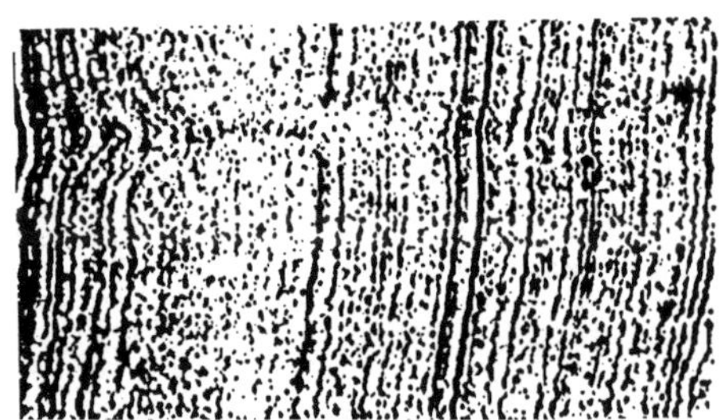

Figure 8.7. Typical enhanced underwater layers' image.

$$y(t) = \sum_{i=1}^{N_L} a_i x(t - \hat{\tau}_i) + b_i x'(t - \hat{\tau}_i) + v(t) \quad (29)$$

The new variables b_i are given by the products

$$b_i = a_i \Lambda \tau_i \quad (30)$$

In discrete form Eq. (29) is written in the familiar linear measurement model given by

$$Y = HX + V \quad (31)$$

The problem, therefore, is one of linear parameter estimation, which is then solved efficiently using recursive algorithms.

It is important to realize that the choice of the estimate of the delays plays a central role in determining the success of this procedure. Aids in choosing these values include the requirement of consistency between successive records and in some instances (especially in the initial processing phase) it may be necessary to use results of a sequential correlation detector.

Event Enhancement Filtering

Event enhancement filtering is a heuristic based procedure (El-Hawary, 1982) that has been applied to field data successfully for enhancing image reflections from subsurface layered media. The effect of amplitude parameters is amplified by a squaring process. The positive and negative parts of the reflected image play a main role in event enhancement. The filtering process is essentially one of prewhitening through differentiation, followed by a shifting and multiplication to sharpen the signal.

Note that if the noise term in the Tauberian expression is neglected, a more direct approach can be employed to extract the amplitude scale factors and delays. This method, suggested by deFigueiredo and Shaw (1987), involves use of a frequency domain transfer function evaluated at equidistant frequencies and obtaining a least squares estimate of an intermediate unknown parameter vector. This optimal estimate is then used to define the coefficients of a polynomial equation in a complex variable related to the amplitude scale factors. Solution of this polynomial equation yields the required delay time results. With the delays available, the amplitude scale factors are obtained from a set of linear equations. This method is referred to as the Prony method.

Note that the outcome of an algorithm may not provide satisfactory results that conform with other verification results, depending on the source of the given data. Figure 8.7 shows a successful enhancement of an underwater layers' image shown in the left side, while an expansion of a number of individual traces is shown on the right side to illustrate the sharpening effect of the filtering process, showing consistency between records. It is, therefore, important early in the analysis phase to determine which scheme to use. A knowledge-based system recognizing the selection criteria and limitations of each scheme is, therefore, of extreme importance. This part of the knowledge-based system can be enabled as production rules. Substantial information is available in the seismic waveform beyond that reflected in amplitude and delay time, which is currently unexploited yet available to describe the layers traversed by the interrogating medium. This part of the knowledge-based system can be encoded as pattern classifiers. Thus, an extremely novel artificial intelligence system can be designed as a hybrid system containing both expert system and pattern recognition technologies.

Time Domain Attribute Extraction Accounting for Losses

A discussion of attribute extraction in the time domain by identifying media properties accounting for attenuation effects, and a review of the basic problem and techniques available for processing the image follow. The inclusion of attenuation effects is discussed in terms of the added complexity. An algorithm to extract physically related information for amplitude, attenuation, and delay parameters contained in the reflected image records using an iterative inversion process based on Newton's method is described.

Attenuation causes the amplitude of a plane sine wave to decay exponentially with the distance traveled in the medium as shown graphically in Figure 8.8. There are at least three possible mechanisms that cause attenuation (Hamilton, 1971, 1972). These include anelastic losses in the frame of sediment particles, viscous losses due to relative movement of sediments and pore water, and losses due to volume scattering.

Assuming propagation in one layer at a given frequency ω, the amplitude of the sound pressure wave at distance d from the source is given by

$$M(d) = M_0 e^{-\alpha(\omega)d} \qquad (32)$$

The attenuation rate $\alpha(\omega)$ is a function of the frequency of the propagating pressure wave.

The product of the attenuation rate and distance traveled is assumed to be proportional to the radian frequency ω. The medium damping ratio D_i,

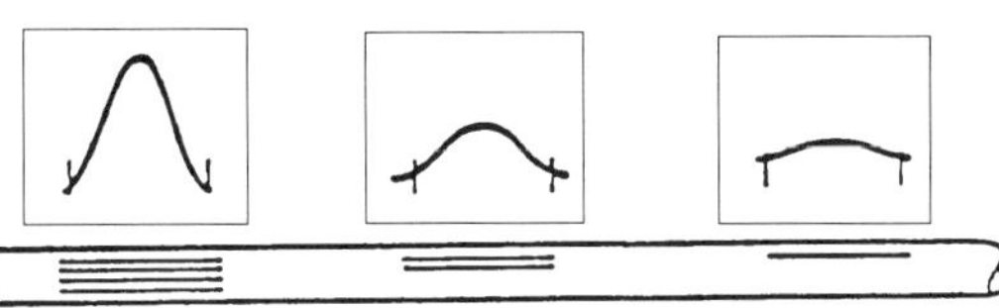

FIGURE 8.8. Typical illustration of signal decay due to attenuation.

which depends on the thickness of the layer and the attenuation rate α, is defined. Thus, an expression for the medium amplitude scale factor A_i in terms of frequency and damping ratio is

$$A_i = a_i e^{-\omega D_i} \qquad (33)$$

The new parameter D_i is assumed frequency independent. The attenuation model assumed here requires that a frequency domain approach in the image processing algorithm be used.

A Frequency Domain Model

The basic model adopted to represent the return pressure signal for each channel $y(t)$ assumes that $y(t)$ is the sum of delayed replicas of the source signal $x(t)$. The replicas are scaled and delayed versions of the original signal. The number of replicas is equal to the number of layers in the model N_L. The amplitude scale factors are denoted by A_i and vary with layer thickness and density, they are also frequency dependent as indicated in Eq. (33). The delay parameters are denoted by τ_i and are related to layer thickness and sound propagation velocity.

$$y(t) = \sum_{i=1}^{N_L} A_i x(t - \tau_i) \qquad (34)$$

Equation (34) must be written in the frequency domain by finding the Fourier transform of both sides. An approximation to the actual expression can be written as

$$Y(\omega) = \sum_{i=1}^{N_L} A_i e^{-j\omega \tau_i} X(\omega) \qquad (35)$$

The exact evaluation of the frequency response of Eq. (35) employs the modulation property of the Fourier transform. The approximation involved here assumes that the amplitudes are time invariant.

Now a frequency domain transfer function can be obtained and is defined by

$$Z(\omega) = \frac{Y(\omega)}{X(\omega)} \qquad (36)$$

As a result the complex valued transfer function is expressed as

$$Z(\omega) = \sum_{i=1}^{N_L} A_i e^{-j\omega\tau_i} \qquad (37)$$

The interpretation task involves finding the real parameters a_i, D_i, and τ_i for all layers considered $i = 1,2,\ldots,N_L$, given the transfer function $Z(\omega)$. As a preparatory step, we separate real and imaginary portions of the transfer function to obtain

$$Z_r(\omega) = \sum_{i=1}^{N_L} A_i \cos \omega \, \tau_i \qquad (38)$$

and

$$-Z_{Im}(\omega) = \sum_{i=1}^{N_L} A_i \sin \omega\tau_i \qquad (39)$$

Note that the Fourier transform of Z is obtained at the discrete frequencies ω_j over the pass band of the acoustic signal. Assuming that the frequency index j ranges from 1 to N_ω

$$Z_r(\omega = \sum_{i=1}^{N_L} A_i \cos \omega\tau_i + \eta_r \qquad (40)$$

and

$$-Z_{Im}(\omega) = \sum_{i=1}^{N_L} A_i \sin \omega\tau_i + \eta_{Im} \qquad (41)$$

Equations (40) and (41) are the basis for a non-linear parameter estimation algorithm. Note that noise components η_r and η_{Im} are present in the real and imaginary portions of the transfer function. Our search is for the parameters a_i, D_i, and τ_i for all layers considered $i = 1,2,\ldots,N_L$.

Layer Parameter Estimation

The aim of parameter estimation techniques is to evaluate the parameters to minimize the noise components in the sum of the squares sense

$$I = \sum_{j=1}^{N_\omega} \eta^2_r(\omega_j) + \eta^2_{Im}(\omega_j) \qquad (42)$$

The minimization is carried out with respect to the unknown layer parameters. The vector U of the unknown parameters is defined by

$$U = \begin{bmatrix} u_1 \\ u_2 \\ \cdot \\ \cdot \\ \cdot \\ u_{N_L} \end{bmatrix} \qquad (43)$$

Each individual subvector is given by

$$u_i = \begin{bmatrix} a_i \\ D_i \\ \tau_i \end{bmatrix} \qquad (44)$$

The condition for least-square parameter estimation is

$$f = \left[\frac{\partial I}{\partial U} \right] = 0 \qquad (45)$$

Equation (45) represents $3\,N_L$ individual equations, which are written as

$$f_{lk} = \sum_{j=1}^{N_\omega} \eta_r \left[\frac{\partial \eta_r}{\partial u_{lk}} \right] + \eta_{Im} \left[\frac{\partial \eta_r}{\partial u_{lk}} \right] = 0 \qquad (46)$$

Here Eqs. (40) and (41) are rewritten as

$$\eta_r(\omega_j) = Z_r(\omega_j) - \sum_{i=1}^{N_L} A_i \cos \omega_j\tau_i \qquad (47)$$

$$\eta_{Im}(\omega_j) = -Z_{Im}(\omega_j) - \sum_{i=1}^{N_L} A_i \sin \omega_j\tau_i \qquad (48)$$

Next obtain the partial derivatives required and expand Eq. (46) to the following form:

$$f_{l1} = \sum_{j=1}^{N_\omega} - e^{-\omega_j D_l} [\eta_r(\omega_j) \cos \omega_j\tau_l + \eta_{Im}(\omega_j)\sin \omega_j\tau_l] = 0 \qquad (49)$$

$$f_{l2} = \sum_{j=1}^{N_\omega} a_l\omega_j e^{-\omega_j D_l} [\eta_r(\omega_j) \cos \omega_j\tau_l + \eta_{Im}(\omega_j)\sin \omega_j\tau_l] = 0 \qquad (50)$$

and

$$f_{l3} = \sum_{j=1}^{N_\omega} a_l\omega_j e^{-\omega_j D_l} [\eta_r(\omega_j) \sin \omega_j\tau_l - \eta_{Im}(\omega_j) \cos \omega_j\tau_l] = 0 \qquad (51)$$

These equations completely specify the required estimates of the parameters.

An Iterative Inversion Process

In contrast to the problem that neglects attenuation, the parameter estimation Eqs. (49), (50), and (51) are nonlinear and require an iterative solution technique. The Newton–Raphson iterative technique may be employed to perform the inversion process. Starting with an initial guess of the unknown parameters, denoted by $U^{(0)}$, we obtain successively (hopefully) improved estimates of the unknowns according to Newton's iterative formula

$$U^{k+1} = U^k + \Delta^k \tag{52}$$

The vector of the incremental improvements Δ^k on the kth iteration is obtained as the solution of the following system of linear equations that involves the Jacobian matrix J evaluated at the current values of the unknowns. Thus

$$f(u^i) + J(u^i)\,\Delta^i = 0 \tag{53}$$

Newton–Raphson method is a powerful solution technique that requires the evaluation of the Jacobian matrix formed by taking the partial derivative of each equation with respect to the unknown variables of interest. For the given structure of the layered media, constructing the Jacobian with reference to partitioned submatrices that correspond to each layer and to the layers associated unknowns is more convenient. The required elements of each submatrix are detailed in El-Hawary (1988). Newton's method is known for its quadratic convergence provided that an initial guess, which is reasonably close to the actual solution is available.

The complexity of this iterative inversion process is due to the requirement of evaluating the Jacobian matrix J whose dimension is $3\,N_L \times 3\,N_L$ at each iteration. The elements involve multiple summations and trigonometric function evaluations as indicated earlier. A second complicating factor arises since the evaluation of the incremental improvements requires solving a set of linear equations that is not sparse. Finally, a good initial guess of the solution must be available to guarantee convergence of the iterations.

In order to avoid these complexities, a number of improvements can be introduced. Use of an initial guess of the unknown amplitudes and delays based on a prior solution for the same data record neglecting attenuation is possible. A second possibility involves the evaluation of the Jacobian only on the first few iterations and subsequently fix it in a quasi-Newton implementation. Third, for a given area start by assuming a minimum number of layers involved and increase the number further if warranted.

Physical Properties as Attributes

Obviously from the preceding discussion, time domain representations allow extracting features such as amplitude scale factors, travel delay times, and the damping ratio. An additional feature is offered by the normalized peak-to-trough time, which can sometimes be traced to the thickness of the underwater beds. Further attributes that are physical properties of each of the layers traversed can be inferred.

In essence, more attribute information about the media is extracted on the basis of peak and delay parameter detection techniques. The goal of the process is to extract physically related properties from amplitude and delay parameters contained in the reflected image records that have been compensated for heave effects. The amplitude scale factors are related to the layers' reflection coefficients and the delay times are related to layer depth and the velocity of sound propagation in the given medium.

The reflection coefficient at a boundary is related to the acoustic impedances of the layers by the fundamental relation

$$r = \frac{Z_2 - Z_1}{Z_2 + Z_1} \tag{54}$$

The acoustic impedance Z is the product of the layer density ρ and the compressional wave velocity C. An approximation that can be used is

$$r = 0.5\Delta\ln Z \tag{55}$$

As a result, the amplitude scale factors are related to the acoustic impedance by

$$a_i = 0.5 \ln \frac{Z_{i+1}}{Z_i} \tag{56}$$

Alternatively, the acoustic impedances can be found from the amplitude scale factors using the inverse relation

$$Z_{i+1} = Z_i \exp 2a_i \tag{57}$$

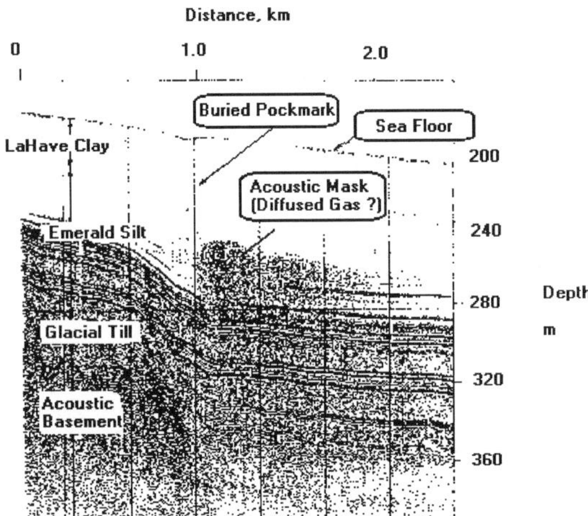

FIGURE 8.9. Typical underwater layers' image following interpretation.

Usually Z_1 corresponds to propagation through the water with known ρ and C. Clearly, knowledge of the acoustic properties of the media can be used to find geotechnical properties.

Additional Attributes

There is a wealth of trace features that can potentially serve in the classification and interpretation of the marine seismic images, in addition to the ones discussed in detail in this chapter. One such attribute is obtained by using a time series approach to model the trace in terms of an autoregressive (AR) model whose coefficients serve as the desired attributes

$$y(t) = \sum_{i=1}^{m} A_{im} y(t - i) + v(t) \tag{58}$$

The AR model is known as the one step ahead prediction error deconvolution model.

If the power spectrum of the trace is considered, then a number of attributes emerge such as frequency of maximum power, average power weighted frequency, and specific power points on the frequency scale. The autocorrelation function of the trace offers more attributes such as the ratio of ith lag value to the jth lag value of the autocorrelation, the ratio of first minimum value to the zero lag value, times of zero crossings, and time lag of first minimum value.

It is safe to assume that there are possibly over 100 trace attributes to consider. The challenge, however, is to reduce that number by identifying a small but yet effective set of features that aid in the interpretation task.

Approach for Marine Seismic Interpretation Expert System

An expert system approach is ideally suited for the task at hand. A wealth of recent literature documents progress of expert system implementations in diverse application areas. Most expert systems are rule-based (as an example: IF . . . THEN . . . : SITUATION . . . ACTION). This is valuable in marine seismic interpretation to allow reasoning under dynamic uncertainties and to enable inference of information on the structure of the underwater layered media. Possible causes of the uncertainties include errors due to source and sensor heave not accounted for by the preprocessing stage, errors due to multipath propagation, and in general errors in the modeling process. Figure 8.9 shows the desired outcome of an interactive interpretation expert system.

The ingredients of an expert system are a knowledge base and a control strategy. The knowledge base contains essentially the fundamental assumptions made, the types of output and their desired qualities, as well as all available a priori knowledge of data, signal, and layer classifications. A number of production rules are also required. These rules include available heuristic information for the decision process, algorithm selection procedures, and relevant consistency checks.

The expert system integrates three phases known as the analysis phase, pattern recognition phase, and artificial intelligence phase. The analysis phase incorporates algorithms for estimation and detection techniques. The pattern recognition phase includes a number of functions such as feature extraction, cluster analysis, and template matching. Inputs to the expert system include a priori knowledge for processing the results. The output of pattern recognition is a classification of uncertainties that the analysis phase is reducing. The artificial intelligence phase performs supervisory function and resolve issues such as determining the most appropriate filter to apply for a particular data set. In Figure 8.10, a functional block diagram of the elements of an expert system are given to illustrate the interactions involved.

Deep-knowledge rules can be developed to begin the screening process. This includes first estimating peak and delay times using the cross-correlation between source and received signals. In this step a threshold must be established on the amplitude of the extrema considered. Practical implementations rely on past experience in a trial and error procedure to set a threshold that is appropriate for the area surveyed. A data base of prior knowledge is an important component of an automated implementation of this procedure. In a second step, refine the estimates using minimum variance information.

Further rules are established through the linearized delay model detection. The choice of the esti mate of the delays plays a central role in determining the success of this procedure. Aids in choosing these values include consistency between successive records requirements and in some instances (especially in the initial processing phase) using results of a sequential correlation detector may be necessary.

Beyond such interpretation rules, it is possible to examine the signals in depth using pattern recognition methods. Often data representation through transformation of the transduced time series to other domains such as power, phase, cepstral, and correlation domains permit access to additional information in the waveform and represents information optimally for machine interpretation. In the first stage knowledge representation is studied for seismic signals. Knowledge representation is carried out at two levels: (1) representation of the waveform information to fully extract all useful information from the signal and (2) representation of the rules used by the analyst in interpretation of signals.

In a second stage, rules used by the analyst must be interpreted. At present, rules are reflected in two steps in the interpretation process. Signals are transformed into the domains of interest and features extracted in each domain. Using cluster analysis and discrimination techniques, supplemented by an experienced analyst, an optimal feature set is identified. This optimal feature set is then used with linear and parametric pattern

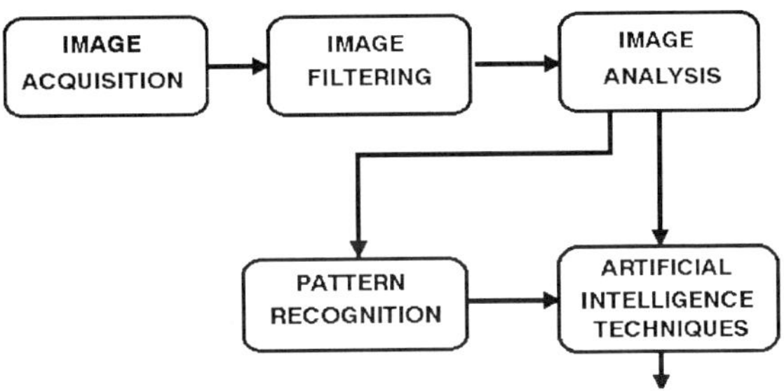

FIGURE 8.10. Functional block diagram of an expert system.

classifiers to design the marine seismic signal interpretation system.

Conclusions

Issues in underwater seismic signal analysis and interpretation on the basis of reflections and multiple reflections from the seabed and the underlying media have been discussed. An expert system approach to marine seismic interpretation is an extremely valuable tool in the search for hydrocarbons. Some basic issues involving acquisition and processing of the seismic signals were reviewed and elements of a knowledge-based interpretation system were discussed. The main software modules incorporated in the system including compensating for the dynamic heave component present in the signals were emphasized. Using conventional and parallel Kalman filtering proved to be useful tools. Considerations for modeling the heave process were highlighted. A number of information extraction algorithms that are central to the interpretation process were discussed. In particular, delay and amplitude parameter estimation using a combined cross-correlation-minimum variance filter, linearized recursive estimation, and an event enhancement filter were treated.

Also discussed was the problem involving identifying media properties while accounting for attenuation effects, and the basic problem and techniques available for processing the image were reviewed. The inclusion of attenuation effects was discussed in terms of the added complexity due to the use of an iterative inversion scheme based on Newton's method. A systematic algorithm designed to extract physically related information from amplitude, attenuation, and delay parameters contained in the reflected image records was described. Additional trace features were discussed as the basis for applying pattern recognition techniques as an assist in the process. Ingredients of a knowledge-based interpretation system for marine seismic image analysis and interpretation were discussed.

References

Aminzadeh, F., and Chatterjee, S., 1987, Application of clustering in exploration seismology: in Pattern Recognition and Image Processing, F. Aminzadeh, (Ed.): Geophysical Press, pp. 372–387.

Bhattacharya, R., 1978, Dynamics of Marine Vehicles: Wiley-Interscience, New York.

Bois, P., 1980, Autoregressive pattern recognition applied to the delimitation of oil and gas reservoirs: Geophys. Prosp. **28**, 572–591.

Bois, P., 1981, Determination of nature of reservoirs by use of pattern recognition algorithm with prior learning: Geophys. Prosp. **29**, 687–701.

Bonamini, R., De Mori, R., Lettera, A., and Rogerro, R., 1981, An electro-cardiographic signal understanding system: in Pattern Recognition, Theory and Applications, J. Kittler, K.S. Fu, and L.F. Pau, (Eds.): D. Reidel, Dordrecht, pp. 443–464.

Chen, C.H., 1985, Recognition of underwater transient patterns: Pattern Recognition, **18**(6), 485–490.

DeFigueiredo, R.J.P., and Shaw, S., 1987, Spectral and artificial intelligence methods for seismic stratigraphic analysis: in Pattern Recognition and Image Processing, F. Aminzadeh, (Ed.): Geophysical Press, London, pp. 426–445.

Dodds, D.J., 1980, Bottom-Interacting Ocean Acoustics, Vol. IV:5: Plenum Press, New York.

El-Hawary, F., 1982, Compensation for source heave by use of Kalman filter: IEEE J. Oceanic Eng. **OE-7**, 89–96.

El-Hawary, F., 1983, Accounting for attenuation in subsurface layered media parameter estimation: A frequency domain approach: presented at IEEE Oceans '83 Conference, 45–49.

El-Hawary, F., 1985, An approach to seismic information extraction: in Time Series Analysis: Theory and Practice, Vol. **6**, O.D. Anderson, J.K. Ord, and E. A. Robinson (Eds.): Elsevier, Amsterdam, pp. 223–238.

El-Hawary, F., 1987a, An approach to extract the parameters of source heave dynamics: Can. Elec. Eng. J. **12**(1), 19–23.

El-Hawary, F., 1987b, Image processing of underwater acoustic arrays geophysical exploration including attenuation effects: presented at International Conference on Digital Signal Processing, Florence, Italy, 623–627.

El-Hawary, F., 1987c, Parallel Kalman filtering for heave compensation using higher order models: presented at IEEE Pacific Rim Conference on Communications, Computer, and Signal Processing, Victoria, B.C., 355–359.

El-Hawary, F., 1987d, Image analysis methods from seabed reflections and multiple reflections: Int. J. Pattern Recognition Artificial Intelligence **1**(2), 261–272.

El-Hawary, F., 1988, Role of parameter estimation and correlation techniques in remote imaging with emphasis on signal attenuation in varied media: presented at SPIE Wave Propagation and Scattering in Varied Media Conference, **927**, 112–121.

El-Hawary, F., and Ravindranath, K.M., 1986, Application of array processing for parallel linear recursive kalman filtering in underwater acoustic exploration: presented at IEEE Oceans '86, **1**, 336–340.

El-Hawary, F., and Richards, T., 1987, Heave response modeling using higher order models: presented at Int. Symp. Simulation and Modeling, Santa Barbara, CA, 60–64.

El-Hawary, F., and Vetter, W.J., 1980, Spatial parameter estimation for ocean subsurface layered media: Can. Elec. Eng. J. **5**(1), 28–31.

El-Hawary, F., and Vetter, W.J., 1981, Heave compensation of shallow marine seismic reflection records by Kalman filtering: presented at IEEE Oceans '81.

El-Hawary, F., and Vetter, W.J., 1982, Event enhancement on reflections from subsurface layered media: IEEE J. Oceanic Eng. **OE-7**(1), 51–58.

Fu, K.S., 1982, Syntactic Pattern Recognition and Applications: Prentice-Hall, Englewood Cliffs, NJ.

Gelb, A., 1974, Applied Optimal Estimation: MIT Press, Cambridge, MA.

Hagen, D.C., 1981, The application of principal component analysis to seismic data sets: presented at Second Int. Symp. Computer Aided Seismic Analysis and Discrimination, 98–109.

Hamilton, E.L., 1971, Elastic properties of marine sediments: J. Geophys. Res. **76**, 2.

Hamilton, E.L., 1972, Compressional wave attenuation in marine sediments: J. Geophys. **37**, 4.

Hassab, J.C., and Chen, C.H., 1985, On constructing an expert system for contact localization and tracking: Pattern Recognition **18**(6), 465–474.

Huang, K.Y., and Fu, K.S., 1984, Detection of bright spots in seismic signal using tree classifiers: Geoexploration **23**, 121–145.

Hutchins, R.W., 1978, Removal of tow fish motion noise from high resolution seismic profiles: presented at

SEG-US Navy Symposium on Acoustic Imaging Technology and on Broad Data Recording and Processing Equipment, National Space Technology Lab., Bay St. Louis, MI.

Hutchins, R.W., Dodds, D.J., Parrott, R., and King, L.H., 1982, Characterization of sea floor sediments by geo-acoustic scattering models using high resolution seismic data: presented at Oceanology Int. Conference, Brighton.

Justice, J.H., Hawkins, D.J., and Wong, G., 1985, Multidimensional attribute analysis and pattern recognition for seismic interpretation: Pattern Recognition, **18**(6), 391–408.

Khattri, K., and Gir, R., 1976, A study of seismic signatures of sedimentation models using synthetic seismograms: Geophys. Prosp. **24**, 454–477.

Khattri, K., Sinvhal, A., and Awashti, K., 1979, Seismic discriminants of stratigraphy derived from monte carlo simulation of sedimentary formations: Geophys. Prosp. **27**, 168–195.

Kung, S.Y, 1985, VLSI signal processing: From transversal filtering to concurrent array processing: in VLSI and Modern Signal Processing, S.Y. Kung, H.J. Whitehouse, and T. Kailath (Eds.): Prentice-Hall, Englewood Cliffs, NJ, pp. 127–152.

Love, P.L., and Simaan, M., 1985, Segmentation of seismic section via image processing and AI techniques: Pattern Recognition **18**(6), 409–419.

Mathieu, P.G., and Rice, G.W., 1969, Multivariate analysis used in the detection of stratigraphic anomalies from seismic data: Geophysics **34**, 507–515.

McCormick, M.E., 1973, Ocean Engineering Wave Mechanics: Wiley-Interscience, New York.

Meditch, J.S., 1969, Stochastic optimal linear estimation and control: McGraw-Hill, New York.

Mendel, J.M., 1983, Optimal seismic deconvolution: An estimation based approach: Academic Press, New York.

Mendel, J.M., 1986, Some modeling problems in reflection seismology: IEEE ASSP **3**(2), 4–17.

Robinson, E.A., 1967, Multichannel Time Series Analysis with Digital Computer Programs: Holden Day, San Francisco.

Robinson, E.A. and Durrani, T.S., 1986, Geophysical Signal Processing: Prentice-Hall, Englewood Cliffs, NJ.

Severance, R.W., 1975, Optimum filtering and smoothing of buoy wave data: J. Hydronaut. **9**, 69–74.

Sinvhal, A., and Khattri, K., 1983, Application of seismic reflection data to discriminate subsurface lithostratigraphy: Geophysics **48**, 1498–1513.

Sinvhal, A., Khattri, K., Sinvhal, H., and Awashti, A.K., 1984, Seismic indicators of stratigraphy: Geophysics **49**, 1196–1212.

Price, W.G. and Bishops, R.E.D., 1974, Probabilistic Theory of Ship Dynamics, Wiley-Interscience, New York.

Travassos, R.H., 1985, Application of systolic array technology to recursive filtering: in VLSI and Modern Signal Processing, S.Y. Kung, H.J. Whitehouse, and T. Kailath, (Eds.): Prentice-Hall, Englewood Cliffs, NJ, pp. 375–388.

Trorey, A.W., 1962, Theoretical seismograms with frequency and depth dependent absorption: Geophysics **27**, 766–785.

White, J.E., 1965, Seismic Waves: Radiation, Transmission, and Attenuation: McGraw-Hill, New York.

9
Clustering of Attributes by Projection Pursuit for Reservoir Characterization

A.T. Walden

From the values of each trace occupying a reservoir interval (or other time segment) on a (flattened) seismic section a set of p attributes can be computed and assigned to, say, the CDP location of the trace. Attributes could, for example, include energy and bandwidth. For a set of N traces, N points in a p-dimensional space are obtained. The ability to determine the clustering of such points in the high-dimensional space then corresponds to recognizing the locations of traces that have similar properties (governed by the attribute values) over the reservoir interval. This ability provides a useful first step in characterizing the reservoir interval.

A technique for investigating high-dimensional point swarms that is exploratory in nature is "projection pursuit," now becoming established in statistics. This method and its computational implementation are the subject of this work.

A transformation that shifts, stretches, and rotates — known as "centering and sphering" — is applied to the raw points to standardize their structure. A procedure is described to make this step statistically robust by computing a robust estimate of the covariance matrix concerned.

The high-dimensional data are projected onto a line, the position of which changes until the projection is maximally interesting (at least locally, in an optimization sense). Interestingness is equated with a non-Gaussian projected distribution, which usually equates to strong bimodality. Non-Gaussianity is measured by an entropy index that is rendered location and scale free. The points are then split apart according to the mode into which their projection falls. The same procedure is applied to each of these groups and clusters begin to emerge that are not further separable.

The optimization stage, which involves repeated calculation of the entropy index and its derivatives, makes extensive use of the fast Fourier transform (FFT), and so suffers little from increasing the size of the data set.

Projection pursuit is similar in some structural respects to minimum entropy deconvolution (MED), where the latter has kurtosis as its "projection" index.

The technique is applied to some synthetic seismic data over a reservoir to demonstrate the various ideas. This is not the only possible geoscience application, since any area where multivariate data need to be clustered could benefit from the approach (e.g., sedimentology).

Introduction

One aspect of seismic interpretation is the ability to characterize a reservoir interval, i.e., to determine the locations of groups of traces that have similar properties. The human eye is often quite adroit at doing this but not all characteristics are equally apparent to the eye, and the process is quite time consuming. The method discussed forms the basis of a more computer-based approach to initial characterization of a reservoir interval. The method is *not* automatic, since user input and decisions are required; however, it is largely automated in the sense that tedious operations are computerized.

In attempting to cluster attributes extracted from seismic data in order to recognize structural patterns, one basic problem is the lack of any a priori detailed statistical model. For example, the data will not likely be multivariate Gaussian. An investigatory technique that makes only very loose

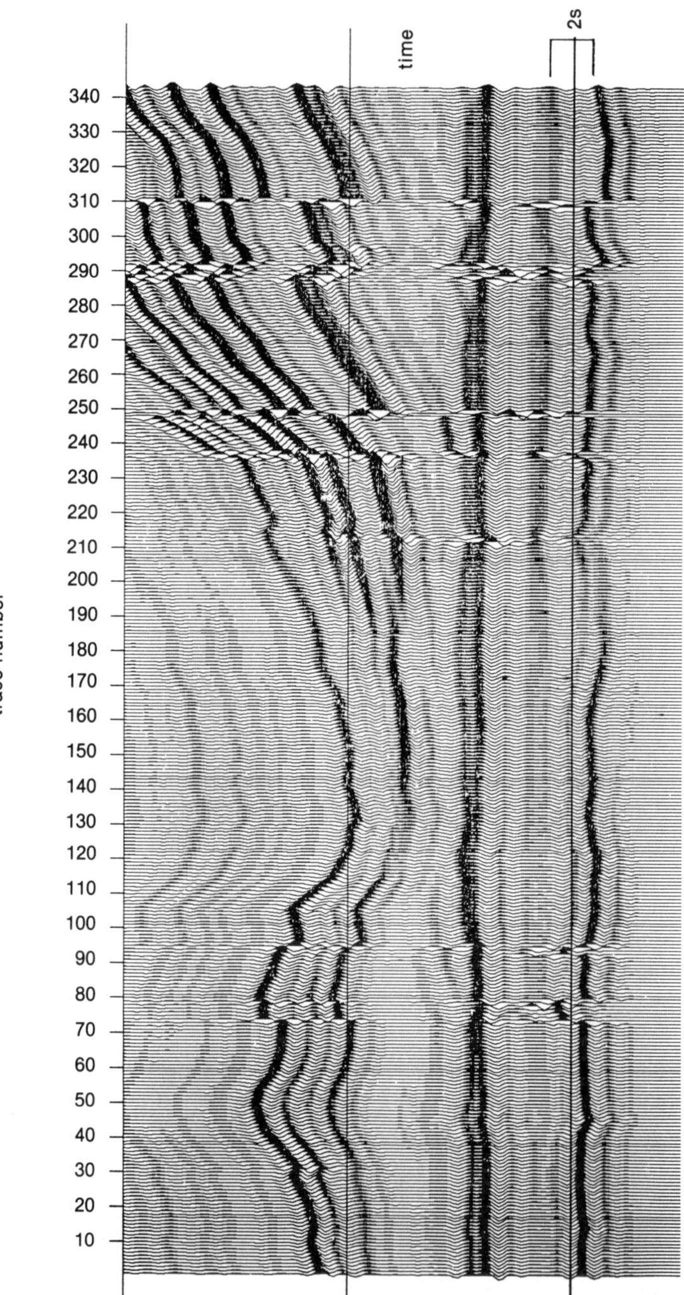

FIGURE 9.1. Flattened synthetic seismic section with reservoir interval marked by bracket on right side.

demands in the way of assumptions is that of projection pursuit, and the discussion focuses on this method of clustering.

Ideas are introduced that make up the method of projection pursuit. The ideas are illustrated using seismic data from a synthetic seismic model. In this example case study only two attributes or variables have been used. This simplification is to facilitate easy appreciation of the steps of projection pursuit through two-dimensional (2-D) plots; of course when used for other than illustration purposes higher dimensional data are much more likely to be used.

Figure 9.1 shows the synthetic seismic section after horizon flattening. The 40 ms interval of interest is delineated by the bracket on the right

FIGURE 9.2. (a) Scatterplot of the two variables in the original coordinates, (b) scatterplot of transformed variables following centering and sphering, and (c) as for (b) but using robust estimates in the centering and sphering with trimming at 5%.

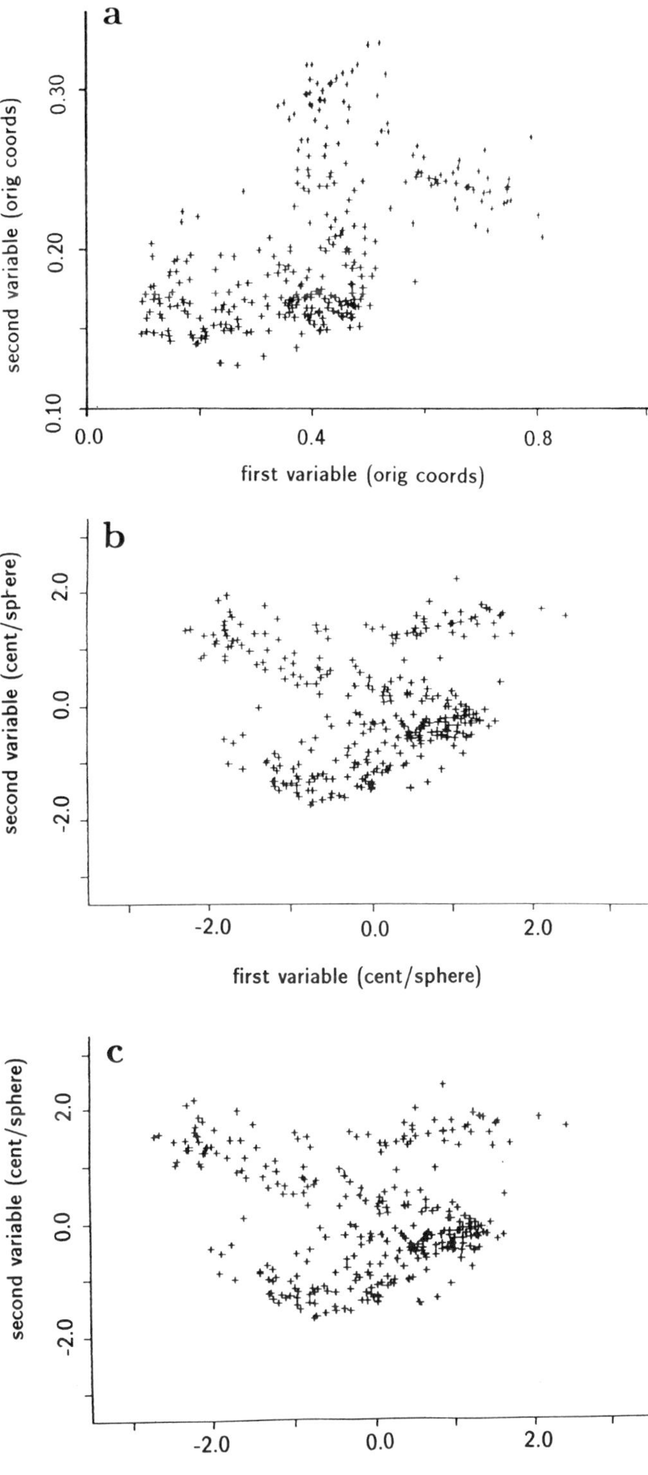

side. From each trace that had been successfully shifted two variables were produced. The first, energy, is defined as the sum of squared values. The second, the varimax norm, is the sum of the values to the fourth power, divided by the energy squared. The resultant scatter plot of some 309 pairs of variable values is given in Figure 9.2a. Human perception favors three separate clusters of points. The question then is whether projection pursuit can also perceive these clusters; if it can, it should also be fairly successful in the more complex higher dimensional scatters.

Basics of Projection Pursuit

The mechanics of clustering by projection pursuit as implemented here are as follows. The multivariate set of points is first centered and sphered so that the mean of the transformed variables is zero, and their covariance matrix is the identity matrix. This is the "standardization" step and the transformed (standardized) data set is then examined. The method of centering and sphering is given in Appendix A, and the idea is simply to put all variables onto an equal footing so that their influence is not, for example, due simply to a larger variance due to the particular units of measurement used. Tukey and Tukey (1981) point out that sphering the data "forces us to examine only aspects of our collection of data vectors other than its general ellipsoidal nature." The centered and sphered version of the scatter plot of Figure 9.2a is given in Figure 9.2b. The same three basic clusters are still apparent, but the point swarm has undergone rotational and scale changes. The points are now centered on (0,0), and the axes units are in multiples of a standard deviation. Note that the original variables are mixed by the transformation, to produce the two new centered and sphered variables (see Appendix A). Centering and sphering can be "robustified" by using robust estimates of the correlation matrix, variances, and means. The method for doing this is described later. Figure 9.2c incorporates this robustification (with a trimming level of 5%). The result is a slightly better separation of the clusters. Robust centering and sphering with 5% trimming has been used in all the examples presented.

Next, in the simplest approach (1-D projection) a projection axis is found such that the projection of the multivariate data onto this axis produces a maximally non-Gaussian smoothed distribution (i.e., smoothed histogram), as judged by the size of a projection index that assesses non-Gaussianity. This is the "optimization" step, and begins with an initial projection axis, and iterates to the "optimal" axis, which is a rotation of the initial vector about the origin. The form of the non-Gaussianity should ideally be that of bimodality i.e., two peaks (or perhaps even multi-modality). Figure 9.3a shows the orientation of the initial (guess) axis, and Figure 9.3b shows the projection of the points onto that line in the form of a smoothed histogram; Figure 9.3c shows the orientation of the final (optimized) axis, and the projection of points onto that line is given in Figure 9.3d. The projection index—which measures non-Gaussianity—is maximized, and is consequently larger for Figure 9.3d than Figure 9.3b. The ability to clearly distinguish two populations of points is much easier in Figure 9.3d.

The data are then split into two groups that (approximately) correspond to the two populations represented by the two peaks in the smoothed distribution. This is the "isolation" step. Either or both of these subgroups of the original data can now be independently pursued—i.e., the whole process, including centering and sphering is repeated on either group or on both groups—and subgroups are formed from these. This is illustrated in Figure 9.4. Figure 9.4a repeats Figure 9.3d, and shows that 101 points have projection values to the left of the split, and 208 points have projection values to the right of the split. The 101 points that have been isolated may now be subjected to the same procedure as the 309 initial points, and a bimodal distribution may be found (using an optimized projection axis) as shown in Figure 9.4b, which splits the 101 points into groups of sizes 58 and 43. The 208 points may also be decomposed in the same way. The process can be continued until further splitting does not occur, or is not helpful to the investigation in progress.

The Development of Projection Pursuit

Several aspects of the above method (in particular the isolation step) were described in Mattson and Dammann (1965). However, they always projected onto the eigenvector corresponding to the

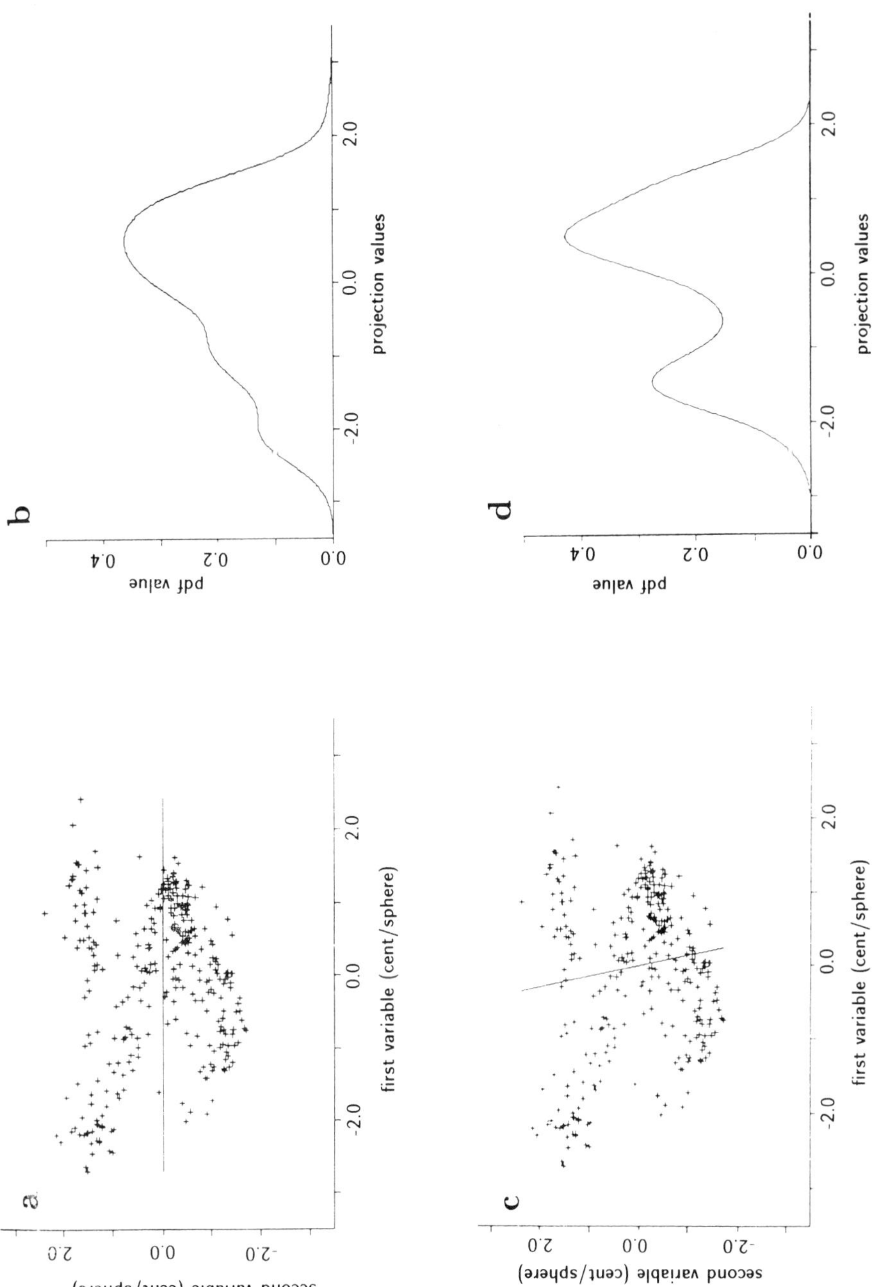

FIGURE 9.3. The improvement to be had in the optimization step. Robustly centered and sphered data with (a) an initial projection vector, (b) the corresponding smoothed histogram, (c) the optimized projection vector, and (d) its corresponding smoothed histogram from which two clusters are obvious.

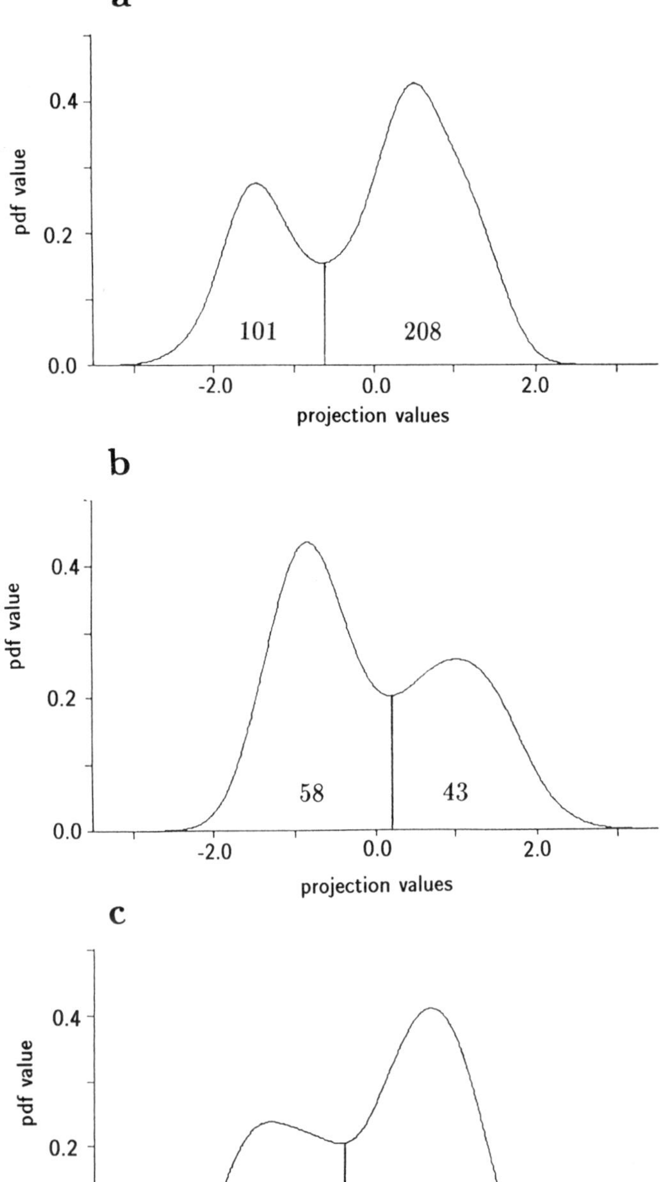

FIGURE 9.4. Illustration of the splitting apart of the points. (a) The smoothed histogram of Figure 9.3d showing bimodality and the number of projected values on either side of the trough, (b) a similar display for the 101 points in (a) projected onto their appropriate optimum vector, and (c) likewise, for the 208 points in (a).

largest eigenvalue (of the data covariance matrix) whereas, as will be demonstrated, the optimization of the projection axis really is essential. Major contributions were also made in Kruskal (1969), but the real impact of the method dates from Friedman and Tukey (1974), who demonstrated the power of the method by using it to isolate 15 spherical Gaussian clusters of 65 points, each centered at the vertices of a 14-dimensional (14-D) simplex. Since the projections used were *linear*, it was thus demonstrated that the same results could be obtained using linear projections as were obtained in Sammon (1969) using *nonlinear* methods. However, the clusters were very well separated, making the task relatively easy. The procedure used in Friedman and Tukey (1974) differs considerably in detail and emphasis from the scheme discussed here.

Donoho (1981) pointed out that projection pursuit has much in common with MED-type deconvolution methods. A projection index is defined (equivalent to an objective function in MED-type deconvolution) and a projection vector **a** is found that maximizes the projection index (hence, **a** plays the same role as the deconvolution filter in MED-type deconvolution). Details are given in Appendix D. A very interesting survey of projection pursuit (in its manifest forms) has been written by Huber (1985) who points out that projection pursuit avoids the "curse of dimensionality," i.e., that high dimensional space is mostly empty. However, because the method uses low-dimensional linear projections, it is poorly suited to deal with nonconvex clusters (e.g., curved and twisted sausage or banana shapes).

Probably the most practically useful look at projection pursuit is the doctoral thesis of Jones (1983), who gives advice on computational procedures in addition to discussion of theory. (Some of Jones' results have recently been published in Jones and Sibson, 1987.) Jones (1983) gives several ways to implement projection pursuit—different projection indices, different dimensions for the projection (e.g., projecting of p-D data onto 2-D rather than just one). In this discussion I shall look at perhaps the simplest approach of Jones, a projection index corresponding to (-1) times Shannon's entropy, and a 1-D projection.

The most recent work is that of Friedman (1987b), which appeared while the analyses discussed here were being carried out, and has some features in common with the work described here.

Hereafter projection pursuit will often be referred to by the abbreviation PP.

Projection Pursuit (PP) Using (Minus) Shannon's Entropy as the Index

As is well known, Shannon's expression for the entropy of a probability density function $f(z)$ is

$$- \int_{-\infty}^{\infty} f(z) \ln f(z)\, dz$$

Minimizing entropy corresponds to maximizing $\int_{-\infty}^{\infty} f(z) dz$, a quantity often called negentropy. Let us define

$$e = \int_{-\infty}^{\infty} f(z) \ln f(z) dz \qquad (1)$$

This expression is not scale invariant, so that changing the variance of $f(z)$ will change the value of e. It is a standard result that for distributions on the real line with *fixed variance*, the Gaussian distribution maximizes entropy and minimizes e. Thus in order to use (1) to assess how Gaussian is a random sample from some distribution, it is necessary to introduce scale invariance. This can be done in one of two ways. First the form of e can be modified to the scale invariant form e':

$$e' = \int_{-\infty}^{\infty} f(z) \ln f(z)\, dz + \ln s[f(z)]$$

(Hastie and Tibshirani, 1987) where $s[f(z)]$ is some measure of scale. This could be the usual standard deviation, or a robust measure such as the median absolute deviation. However, if e' is to be differentiable with respect to z, $s[f(z)]$ must be too. Alternatively, e can be used with the proviso that a fixed scale be used for z for all comparisons (say unity variance). For computational reasons (see later) this latter approach is adopted. Hence if seeking to "maximize e" a distribution very different from the Gaussian will be obtained. In the present context, what does "maximize e" mean?

Suppose N observations on each of p variables are available, $X_1, X_2, \ldots, X_p$. This generates data x_{ij}, $i = 1, \ldots, N$ for variable $j, j = 1, \ldots, p$. The data can be represented as the $N \times p$ matrix:

$$\mathbf{X} = \begin{bmatrix} x_{11} & x_{12} & \cdots & x_{1p} \\ x_{21} & x_{22} & \cdots & x_{2p} \\ \vdots & \vdots & \ddots & \vdots \\ x_{N1} & x_{N2} & \cdots & x_{Np} \end{bmatrix} \qquad (2)$$

with (i,j)th element x_{ij}. Note each row corresponds to the p variable values for a single observation. The idea of centering and sphering is to standardize the point cloud in p-dimensional space by a transformation so that it is centered at zero and has an estimated covariance matrix equal to the identity matrix (see Appendix A). Each variable $S_1, \ldots, S_p$ resulting from the transformation has estimated mean zero, variance unity, and is uncorrelated with the other variables. Given a projection vector $\mathbf{a}$, the linear projection of the standardized data, sample by sample, onto $\mathbf{a}$ is given by

$$z_i(\mathbf{a}) = \mathbf{a}^T \mathbf{s}_i = \sum_{j=1}^{p} a_j s_{ij} \qquad i = 1, \ldots, N \quad (3)$$

where $\mathbf{s}_i^T = (s_{i1}, s_{i2}, \ldots, s_{ip})$ and s_{ij} is the ith observation on the jth variable *after centering and sphering* (see Appendix A). The variance of the projected data is

$$\mathrm{Var}\{\sum_{j=1}^{p} a_j S_j\} = \sum_{j=1}^{p} a_j^2 \mathrm{Var}\{S_j\} \qquad (4)$$

since the S_j are uncorrelated. But since $\mathrm{Var}\{S_j\} = 1$ for all j due to the standardization carried out in the centering and sphering, the variance is simply $\sum_{j=1}^{p} a_j^2$. Thus if the projection vector is always normalized to unit length, $\sum_{j=1}^{p} a_j^2 = 1$, the variance will remain at unity, whatever the projection orientation. I shall denote a unit length projection vector by $\mathbf{a}_s$; the subscript emphasizes that it lies on the unit spheroid.

Now estimate the probability density function of the $z_i(\mathbf{a}_s)$ using a Gaussian kernel density estimator:

$$f_N(z; \mathbf{a}_s) = \frac{1}{Nh} \sum_{i=1}^{N} K\left(\frac{z - z_i(\mathbf{a}_s)}{h}\right) \qquad (5)$$

where h is the window width or smoothing parameter (see later) and K is the Gaussian kernel function given by

$$K(y) = [1/\sqrt{(2\pi)}] \, e^{-y^2/2}$$

Define the equivalent of Eq. (1) based on the kernel density function estimator by

$$e_N(\mathbf{a}_s) = \int_{-\infty}^{\infty} f_N(z; \mathbf{a}_s) \ln f_N(z; \mathbf{a}_s) \, dz$$

Thus e_N will measure departures from Gaussianity of the projected data, since the variance will be fixed at unity. With this proviso, e_N is scale invariant. It is also location invariant since changing the location of the multivariate sample merely moves the distribution of the projected data along the projection line; since the integral is from $-\infty$ to ∞ this is irrelevant to the value of e_N.

f_N in Eq. (5) may be computed using the fast Fourier transform method due to B.W. Silverman; from this an estimate $\hat{e}_N(\mathbf{a}_s)$ of $e_N(\mathbf{a}_s)$ may be found using Simpson's rule (see Appendix B).

Initial Projection Vectors for Projection Pursuit (PP)

The kth principal component of the sample of p-variate observations is given by $P_k = \hat{\Theta}_{1k} X_1 + \cdots + \hat{\Theta}_{pk} X_p$ where $\hat{\Theta}_{lk}$ is the lth element of the kth eigenvector $\hat{\Theta}_k$ of the sample *covariance* matrix of the original variables with corresponding eigenvalues $\hat{\theta}_k$ (see Appendix A). The kth principal component explains $\hat{\theta}_k / \sum_j \hat{\theta}_j$ of the total variance of the system.

The kth estimated eigenvector of the *correlation* matrix of the original variables with corresponding eigenvalue $\hat{\gamma}_k$ explains $\hat{\gamma}_k / p$ of the total "variance" in the scatter of dimensionless standard scores $y_{ij} = (x_{ij} - \bar{x}_j)/\hat{\sigma}_j$ (where $\bar{x}_j$ and $\hat{\sigma}_j$ are the estimated mean and standard deviation of the jth variable). For example, for $p = 2$ and $\hat{C}_{12}$ a positive correlation between the variables, the larger of the two eigenvalues takes the value $1 + \hat{C}_{12}$ with corresponding eigenvector $(1/\sqrt{2})(1,1)$, while the smaller takes the value $1 - \hat{C}_{12}$ with an eigenvector $(1/\sqrt{2})(-1,1)$. The sum of the two eigenvalues is thus 2 as required.

Following centering and sphering the correlation matrix is the identity matrix. The eigenvectors are now the unit vectors $(1, 0 \ldots, 0), (0, 1, 0, \ldots, 0), \ldots, (0, 0, \ldots, 1)$, and the corresponding eigenvalues are all unity. The unit vectors are, of course, equal to the eigenvectors of the original correlation matrix transformed using the rotation part of the centering and sphering transformation (see Appendix A). Unlike the ranking imposed on the eigenvectors of the original covariance or correlation matrix by their ability to account for different amounts of "variance," there is no hierarchical structure to the unit eigenvectors following centering and sphering.

These p unit vectors could provide starting projection vectors, but it is wise to try additional starting configurations such as systematically or randomly chosen vectors (standardized to unit length).

Figure 9.5 provides a graphic example of the need to consider a range of initial projection vectors. Figure 9.5a shows three initial vectors (dashed) and the optimum projection vector (solid line) obtained from all three. The corresponding smoothed histogram of the projection onto the optimum projection vector is shown in Figure 9.5b. However, a fourth

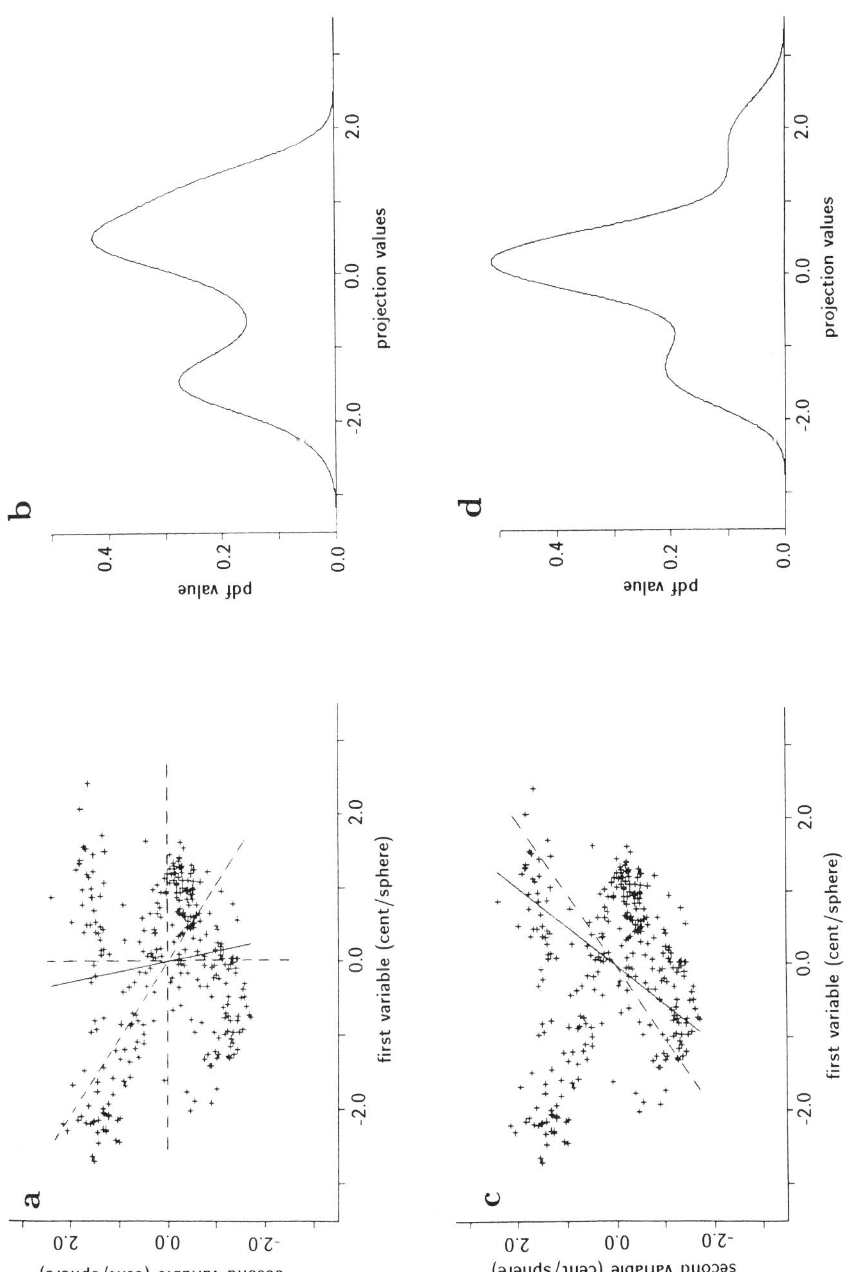

FIGURE 9.5. The effect of local maxima of the index. (a) All three (dashed) initial vectors lead to the (solid) optimum vector and its smoothed histogram for the projection in (b); however, a fourth initial vector (dashed) in (c) gives a different (solid) optimum vector, and different smoothed histogram (d).

initial vector, shown as the dashed line in Figure 9.5c, gives rise to the optimum projection line (solid line) in that figure. The corresponding smoothed histogram (Figure 9.5d) is quite different in this case (and, indeed, leads to poorer clustering). Of course, as pointed out earlier, the optimization usually finds only a "local" maximum to the projection index. In this example, the estimated value for e for the optimum projection vector of Figure 9.5a was -1.402, while for the optimum vector of Figure 9.5c it was -1.412. Hence this latter maximum is smaller than that for the preferred projection of Figure 9.5a, but not by much! Computational accuracy is thus clearly important if (what appears to be) the global maximum is to be located.

The Optimization Step

It is of course necessary in the optimization step to find a projection direction that maximizes, at least locally, the projection index. In two dimensions this could be carried out by simply rotating the projection vector in small increments, and finding the direction that maximizes the index. In higher dimensions this approach is clearly computationally unrealistic. Fortunately a very efficient computational scheme, again using the FFT, can be used. Denote an initial p-D unit-length projection vector by $\mathbf{a}^{(0)}_s$, where the subscript s emphasizes that it lies on the spheroid, since it is of unit length, and (0) denotes that it is the initial guess. Next compute $z_i(\mathbf{a}_s^{(0)})i = 1,\ldots,N$ from Eq. (3), and then $\hat{e}_N(\mathbf{a}_s^{(0)})$. To iteratively improve $\mathbf{a}_s^{(n)}$ the derivatives

$$g_m = d\hat{e}_N/da_{ms}, \qquad m = 1,\ldots,p$$

must be computed. This can be done using the FFT (see Appendix C). I denote the p-D vector of derivatives by $\mathbf{g}(\mathbf{a}_s^{(0)})$. Jones (1983) suggests that the updated projection vector be found using the method of Kruskal (1964):

$$\mathbf{a}^{(1)} = \mathbf{a}_s^{(0)} + \lambda^{(0)} \mathbf{g}_s(\mathbf{a}_s^{(0)})$$

(steepest ascent) where $\mathbf{g}_s(\mathbf{a}_s^{(0)}) = [1/\sqrt{(\sum_{m=1}^{p}g_m^2)}]$ $(g_1,g_2,\ldots,g_p)|_{\mathbf{a}_s^{(0)}} = (g_{1s},\ldots,g_{ps})$ say, i.e., a standardized version of the gradient vector which also lies on a p-D unit spheroid. The new projection vector is now restricted to the p-D unit spheroid by rescaling it so that $\sum_{j=1}^{p}(a_j^{(1)})^2 = 1$ in exactly the same way as for $\mathbf{g}$; we thus obtain $\mathbf{a}_s^{(1)}$.

In fact $\mathbf{a}^{(n)}$ is obtained as

$$\mathbf{a}^{(n)} = \mathbf{a}_s^{(n-1)} + \lambda^{(n-1)}\mathbf{g}_s(\mathbf{a}_s^{(n-1)})$$

where

$$\lambda^{(n-1)} = 0.65\lambda^{(n-2)}4^{\cos^3\theta}$$

and $\cos\theta$ is the angle between $\mathbf{g}_s(\mathbf{a}_s^{(n-1)})$ and $\mathbf{g}_s(\mathbf{a}_s^{(n-2)})$, i.e., $\cos\theta = \mathbf{g}_s^T(\mathbf{a}_s^{(n-1)})\mathbf{g}_s(\mathbf{a}_s^{(n-2)})$, where superscript T indicates the transpose. This particular formulation is due to Kruskal (1964), who also suggests taking $\lambda^{(0)} = 0.2$.

Sometimes instead of continuing to converge to a (local) maximum, the method begins to oscillate about in the region of the maximum. Hence if it is noted that $e(\mathbf{a}^{(n)}) < e(\mathbf{a}^{(n-1)})$ some action is required. Jones (1983) suggested modifying $\lambda^{(n-1)}$ and trying again:

$$\lambda_{\mathrm{mod}}^{(n-1)} = \lambda^{(n-1)} \frac{\mathbf{g}_s^T(\mathbf{a}_s^{n-1)})\mathbf{g}_s(\mathbf{a}_s^{(n-1)})}{\mathbf{g}_s^T(\mathbf{a}_s^{(n-1)})\mathbf{g}_s(\mathbf{a}_s^{(n-1)}) - \mathbf{g}_s^T(\mathbf{a}_s^{(n)})\mathbf{g}_s(\mathbf{a}_s^{(n-1)})}$$

i.e.,

$$\lambda_{\mathrm{mod}}^{(n-1)} = \lambda^{(n-1)} \frac{1}{1 - \mathbf{g}_s^T(\mathbf{a}_s^{(n)})\mathbf{g}_s(\mathbf{a}_s^{(n-1)})}$$

since $\mathbf{g}_s^T(\mathbf{a}_s^{(n-1)})\mathbf{g}_s(\mathbf{a}_s^{(n-1)}) = \sum_{m=1}^{p}[g_{ms}(\mathbf{a}_s^{(n-1)})]^2 = 1$. This method will work well when the surface about the maximum is close to quadratic. Given $\mathbf{a}_s^{(n-1)}$, a choice of $\lambda^{(n-1)}$ that is too large will cause $\mathbf{a}_s^{(n)}$ to overshoot the maximum, and give rise to a lower value of e. The gradient vector at $\mathbf{a}_s^{(n)}$ will be at $180°$ to that at $\mathbf{a}_s^{(n-1)}$, hence the cosine of the angle between them, $\mathbf{g}_s^T(\mathbf{a}_s^{(n)})\mathbf{g}_s(\mathbf{a}_s^{(n-1)}) = -1$, and $\lambda_{\mathrm{mod}}^{(n-1)} = \frac{1}{2}\lambda^{(n-1)}$. Thus, starting again at $\mathbf{a}_s^{(n-1)}$, the new choice of $\mathbf{a}_s^{(n)}$ using $\lambda_{\mathrm{mod}}^{(n-1)}$ should be much closer to the maximum. In practice this optimization technique has proven very satisfactory: convergence usually occurs after about six iterations, sometimes including a refinement step as just discussed.

There are almost certain to be several local maxima for e. It is thus very important to try several initial vectors $\mathbf{a}_s^{(0)}$ to attempt to locate the different local maxima.

Outliers and Robustification

The most contentious part of PP is probably the decision on how best to treat outliers. The distribution of projected data is a marginal distribution from a multivariate distribution, and not simply a univariate distribution in the usual sense. Friedman (1987a) points out that for a 5-D standard Gaussian distribution, 5% of the observations are distant 3.3 from the origin, and for 10 and 15 dimensions the corresponding 5% distances increase to 4.3 and 5.0. Thus vectors from the origin to each of these points define projection orientations for which the corresponding points will appear to be outliers, and are called "pseudooutliers." It is thus important that a projection index should not react too much to such pseudooutliers. Hence indices based on moments (such as kurtosis) could perform particularly unpredictably.

There is nothing in the approach of Jones and Sibson (1987), which is designed to accommodate outliers in the same sense of the inherently robustified approach of Friedman and Tukey (1974). How can, and how should, their approach be made robust to real outliers?

Again, there is no general agreement on either the necessity or implementation of robustification. However, if desired it should be carried out in a strictly multivariate sense, and thus one obvious approach is to robustify the centering and sphering. I implemented the technique of (ellipsoidal) multivariate trimming or MVT (e.g., Devlin et al., 1981). It also appears to be this technique that is recommended in Tukey and Tukey (1981, p. 197). The method is straightforward. It starts with the usual estimates of the mean and covariance matrix of the data, denoted here $\bar{\mathbf{x}}^{(0)}$ and $\hat{\mathbf{\Xi}}^{(0)}$ [see Eq. (A-1)]. The sample Mahalanobis distance d_i^2 of each point in p-D space from $\bar{\mathbf{x}}^{(0)}$ is then calculated using the matrix $\hat{\mathbf{\Xi}}^{(0)}$ to provide variances and covariances:

$$\{d_i^{(0)}\}^2 = (\mathbf{x}_i - \bar{\mathbf{x}}^{(0)})^\mathrm{T}\hat{\mathbf{\Xi}}^{(0)-1}(\mathbf{x}_i - \bar{\mathbf{x}}^{(0)}), \qquad i = 1,\ldots,N$$

where $\mathbf{x}_i$ is the ith row of $\mathbf{X}$, with $\mathbf{X}$ defined in Eq. (2). The points corresponding to the most extreme α percent of distances are then put aside, and new estimates of the mean and covariance matrix $\bar{\mathbf{x}}^{(1)}$ and (0) to provide variances and covariances: $\hat{\mathbf{\Xi}}^{(0)}$

$$\{d_i^{(1)}\}^2 = (\mathbf{x}_i - \bar{\mathbf{x}}^{(1)})^\mathrm{T}\hat{\mathbf{\Xi}}^{(1)-1}(\mathbf{x}_i - \bar{\mathbf{x}}^{(1)}), \qquad i = 1,\ldots,N$$

and the process iteratively continued. Each iteration the covariance matrix can be rescaled to a correlation matrix, $\hat{\mathbf{C}}$ and the off-diagonal elements of the correlation matrix for the current iteration compared with those from the previous iteration using

$$F_\mathrm{s} = \frac{1}{n_\mathrm{up}} \sum_{i<j} |F(\hat{C}_{ij}^{(k)}) - F(\hat{C}_{ij}^{(k+1)})|$$

where n_up is the number of terms in the upper (or lower) triangle of the correlation matrix, excluding the diagonal, and $F(\cdot)$ is the Fisher "z-transform" of the correlations, i.e.,

$$F(\hat{C}_{ij}^{(k)}) = \frac{1}{2} \ln \{(1 + \hat{C}_{ij}^{(k)})/(1 - \hat{C}_{ij}^{(k)})\}$$

Devlin et al. (1981) suggested that convergence of the process be assumed when F_s is less than 10^{-3}. Typically only about four iterations are required for this convergence criterion to be satisfied. The resultant estimates of the covariance matrix and the mean vector are then used for the calculations of Appendix A.

There is a hidden danger with this robustification step. It relates to the necessity of scale invariance for the projection index e. The variance of the projected data given in Eq. (4) will be unity only if $\mathrm{Var}\{S_j\} = 1, j = 1,\ldots,p$, but following robustification this, in general, will not be true since the robustification will scale the nonextreme values to have variance of unity, but the extreme points will cause the overall variance to exceed unity. While it would be easy to change the $\sum a_j^2$ to give Eq. (4) a value of unity, this would have to change continuously according to direction and would greatly complicate the derivatives (Appendix C). As can be seen from the examples reported here (which use the robustified centering and sphering) this need not be too troublesome – it has not caused any obvious problem here – but when the robustification has a greater effect than here, we can imagine that problems might arise. Friedman (1987b, section 6.1) simply deletes the points considered extreme by the robustification, thus preserving the unit variance. However, this seems rather dangerous in reservoir characterization, since these points might be highly significant. An alternative approach is initially to delete the extreme points, find the splitting line (with these points still deleted) and then restore the deleted points, which will consequently be assigned to one cluster or the other according to the splitting line.

Examples of Clusters Obtained

So far examples have been given of projection lines and the smoothed histogram of the projected data. Here, the resultant clustering will be emphasized. Figure 9.6a shows an initial vector (dashed) and

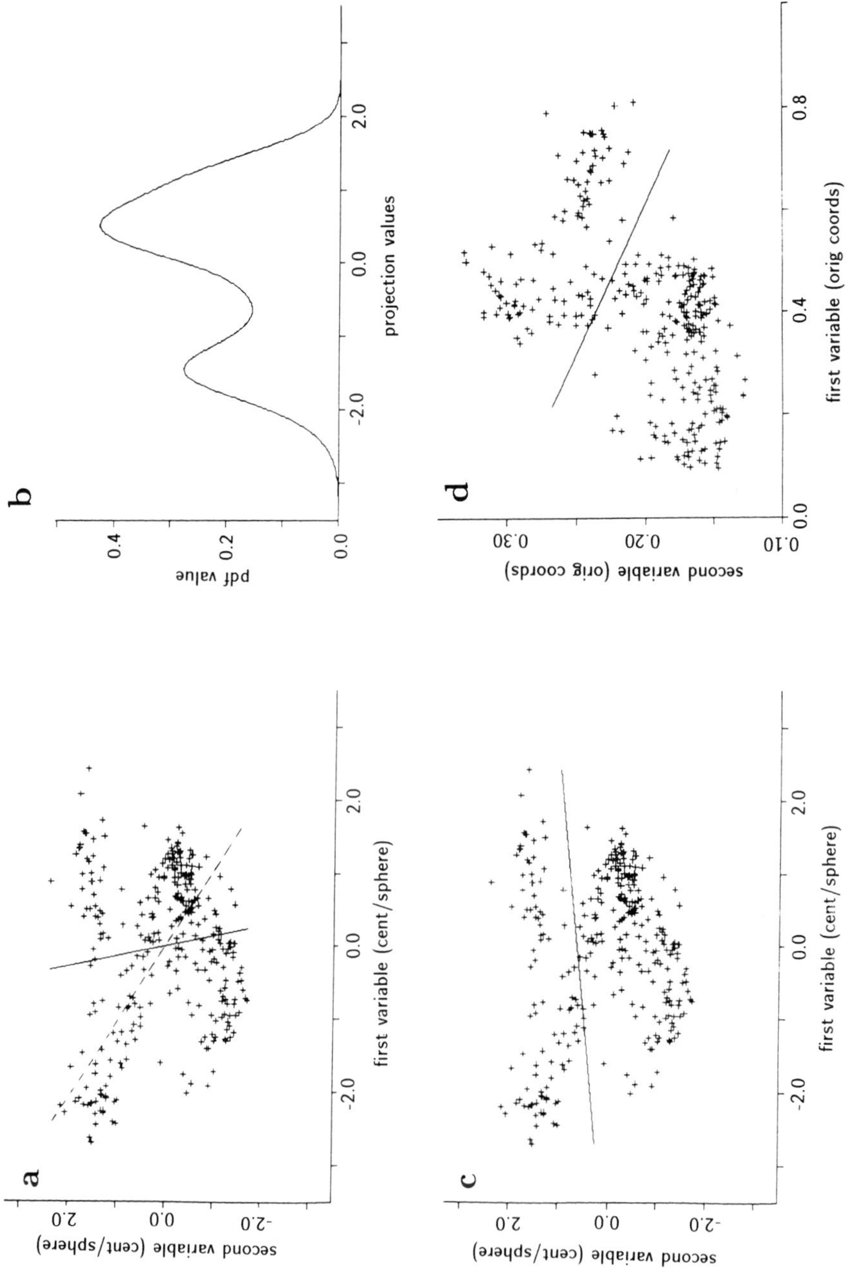

FIGURE 9.6. Illustration of splitting into clusters. (a) The initial projection vector (dashed) and optimum (solid), and (b) the smoothed histogram for the projection onto the optimum vector, (c) the splitting line defined by the trough of (b) in centered and sphered coordinates, and (d) the splitting similarly defined in the original coordinates.

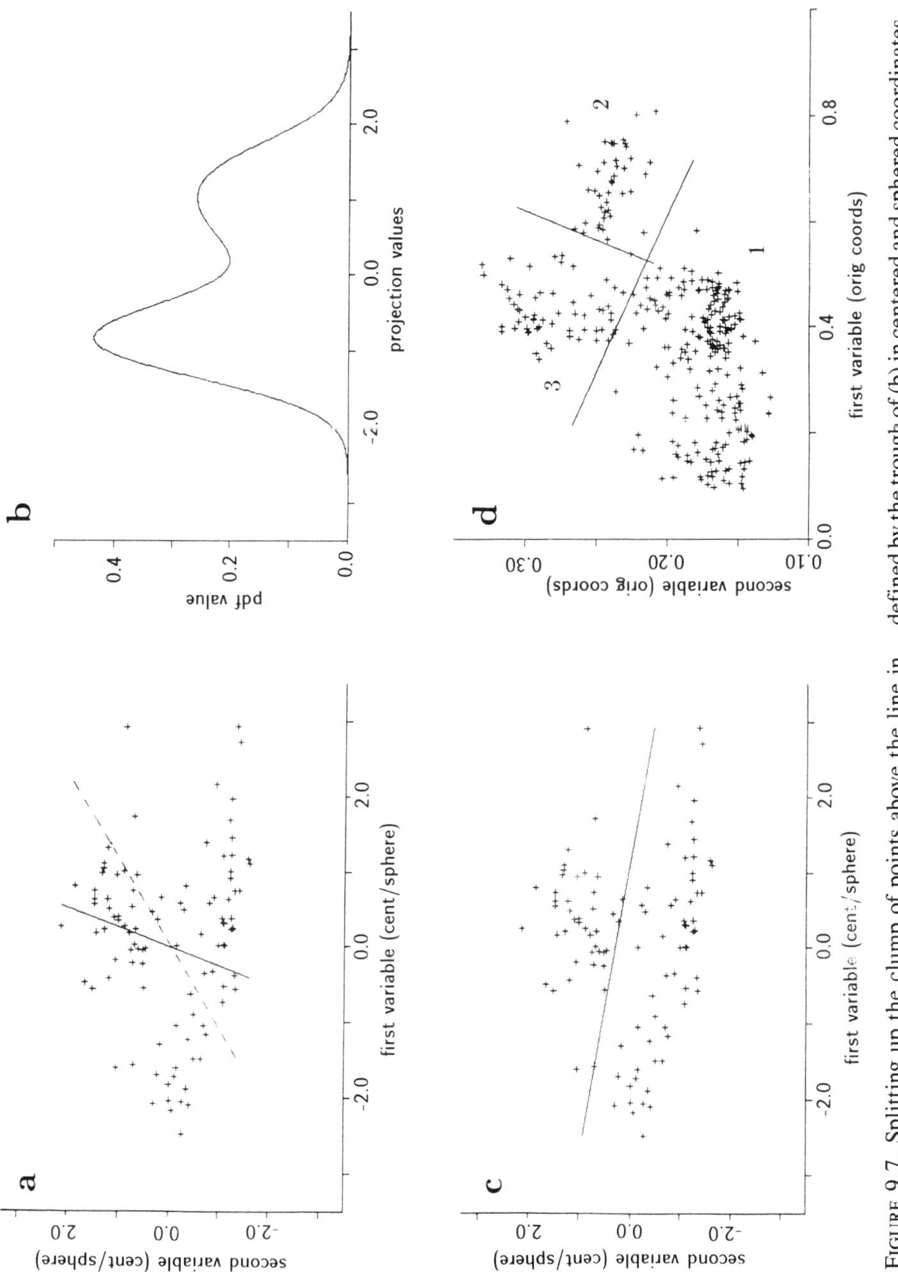

FIGURE 9.7. Splitting up the clump of points above the line in Figure 9.6d. (a) The 101 points and initial (dashed) and optimum (solid) projection vectors, (b) the smoothed histogram corresponding to the optimum vector in (a), (c) the splitting line defined by the trough of (b) in centered and sphered coordinates, and (d) the splitting of the whole set of points defined by this line and that of Figure 9.6d in original coordinates.

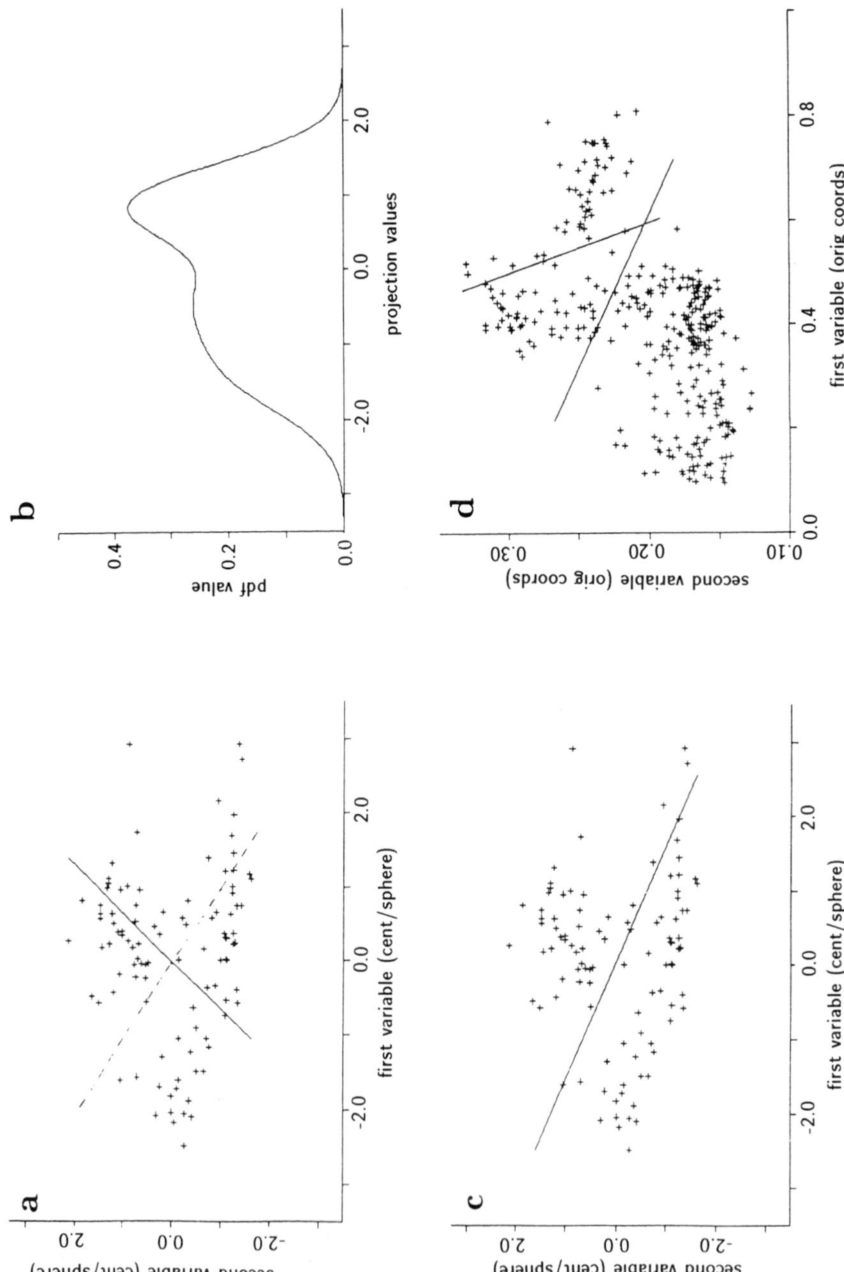

FIGURE 9.8. As for Figure 9.7 except a different initial vector used in (a) and resultant different optimum vector. The split in (c) is not as satisfactory as in Figure 9.7c, and in fact would not be selected as it corresponds to a smaller value of the projection index.

the resultant optimum projection line, while Figure 9.6b shows the resultant smoothed histogram. The projection onto the initial vector resulted in an e value of -1.436 and a kurtosis of 2.95; the projection onto the optimum increased the estimate of e to -1.402, and decreased the kurtosis to 1.84 (much more non-Gaussian). Splitting the data by those points with projection values less or greater than the projection value corresponding to the trough minimum (-0.64), gives the split of the data in Figure 9.6c, in centered and sphered coordinates, and Figure 9.6d in the original coordinates. This, then is the first separation of the data by projection pursuit.

Now examine the 101 points with projections less than -0.64. These are plotted in Figure 9.7a after (robust) centering and sphering. The initial vector shown (dashed) gives an e value of -1.498 and kurtosis of 2.11, while the resultant optimum vector gives an e value of -1.390 and kurtosis of 1.74. The corresponding smoothed histogram is given in Figure 9.7b. The trough occurs at a projection value of 0.2. Splitting the data accordingly gives Figure 9.7c (centered and sphered coordinates). In Figure 9.7d the whole data set has been plotted in original coordinates with the two splitting lines marked. This clustering *corresponds almost perfectly with that chosen by the human eye* and is very encouraging. However, results from other initial vectors that might have been chosen should be examined. Figure 9.8a gives another choice of initial vector and the corresponding "optimum" projection line. The initial vector gives an e value of -1.539 and kurtosis of 2.60, while the resultant optimum vector gives an e value of -1.445 and kurtosis of 1.91. Hence, this is definitely not a global optimum, and would not be chosen instead of the result in Figure 9.7a. The split in Figure 9.8c is clearly inferior to that in Figure 9.7c, as is the clustering in Figure 9.8d, compared to that in Figure 9.7d.

Only one side of the projection of Figure 9.7b has been pursued here, i.e., points to the left of the trough. Examining the 208 points with projections greater than -0.64 gives Figure 9.9 in the now familiar form. The result is that the large cluster is split (Figure 9.9d), which is perhaps visually justifiable, but not strongly so. This raises an important practical point. Clearly any data set consisting of N observations can be split up into N trivial clusters

each of which consists of a single observation. Faced with a plot such as in Figure 9.2a, the human eye would tend to stop clustering after only a few clusters have been discerned, because it would automatically allow for scattering within clusters, and hence would not continue clustering until N trivial clusters are produced. Automatic clustering such as carried out by projection pursuit determines clustering from bimodality of projected data, which in turn depends on the smoothing, h, used. (The human eye would likewise impose some smoothness value in determining whether the scatter of points within a cluster represents acceptable scatter within a cluster rather than the presence of true subclusters.) The following results are given in Jones (1983, pp. 167–168). The value of h that minimizes the mean integrated squared error (MISE) in estimating the univariate Gaussian distribution from a sample of size N taken from it, namely $h = 1.06\sigma N^{-1/5}$ [see Appendix E, Eq. (E-1)], will give multimodal estimates about 15% of the time. Formula (E-2), which was used in the examples given here, will give multimodal estimates about 19% of the time. The rule (E-3) will give multimodal estimates about 33% of the time (Silverman, 1986, p. 140). (All these percentages vary very little with N.) Hence even when the underlying distribution is a homogeneous Gaussian, it can appear not to be so for certain sample configurations and smoothing choices. The user of projection pursuit must thus often make a decision to "stop splitting," just as would the human eye. Generally a good strategy seems to be to let projection pursuit go "too far," and then start recombining clusters until the physical interpretation makes most sense.

In Figure 9.10a is shown in greater detail the reservoir interval corresponding to that marked in Figure 9.1. The variables or attributes were computed from these traces. As explained above, the preferred clustering by projection pursuit is that given in Figure 9.7d. If we attach flags 1, 2, and 3 to the three groups thus distinguished, and plot the flag values against the corresponding CDP numbers we obtain Figure 9.10b. (Here flag values of 0 correspond to traces not used in the clustering because of poor flattening.) Accordingly, the response in the reservoir interval may be considered to be of one type from CDP 1 to about 42, of a second type to CDP 102, then a mixture with

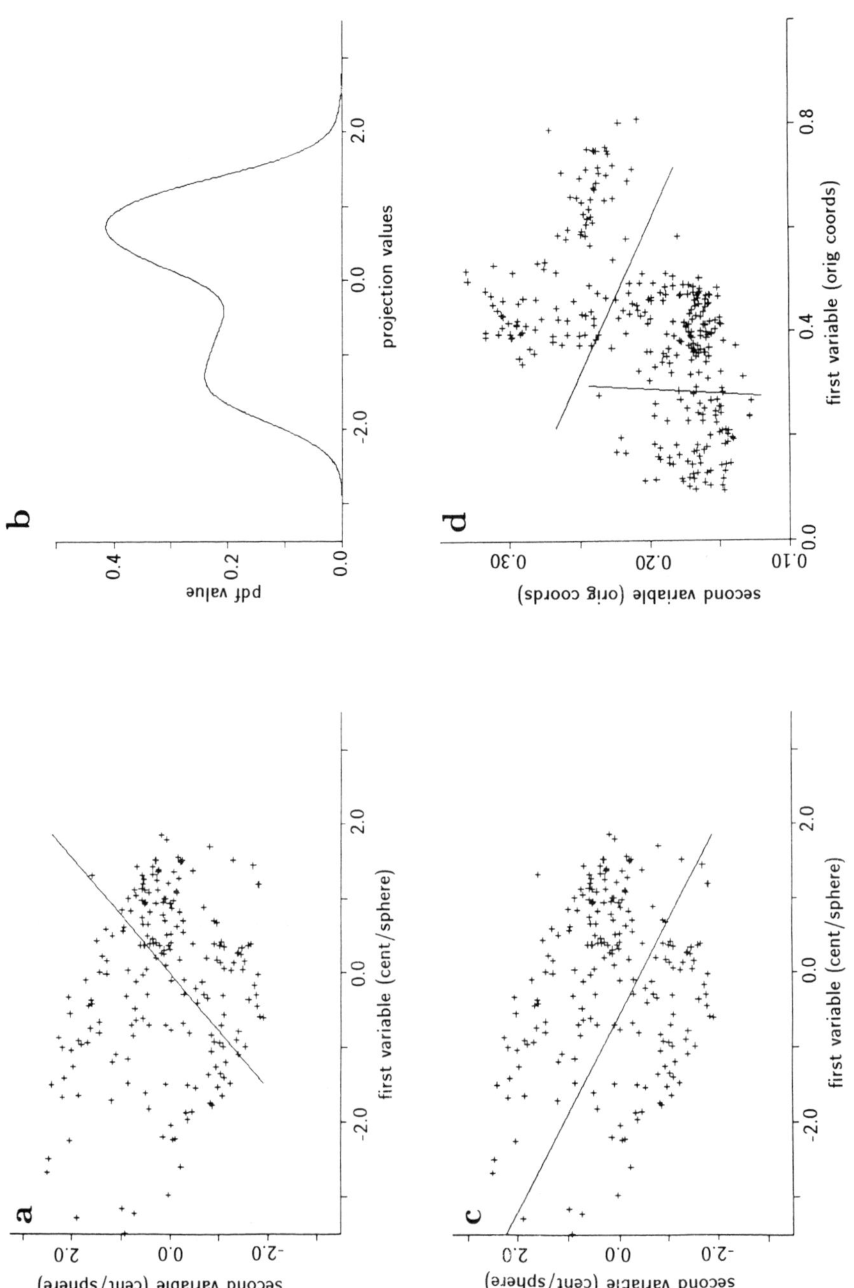

FIGURE 9.9. Splitting up the clump of points below the line in Figure 9.6d. (a) The 208 points and optimum (solid) projection vector, (b) the smoothed histogram corresponding to the optimum vector in (a), (c) the splitting line defined by the trough of (b) in centered and sphered coordinates, and (d) the splitting of the whole set of points defined by this line and that of Figure 9.6d in original coordinates.

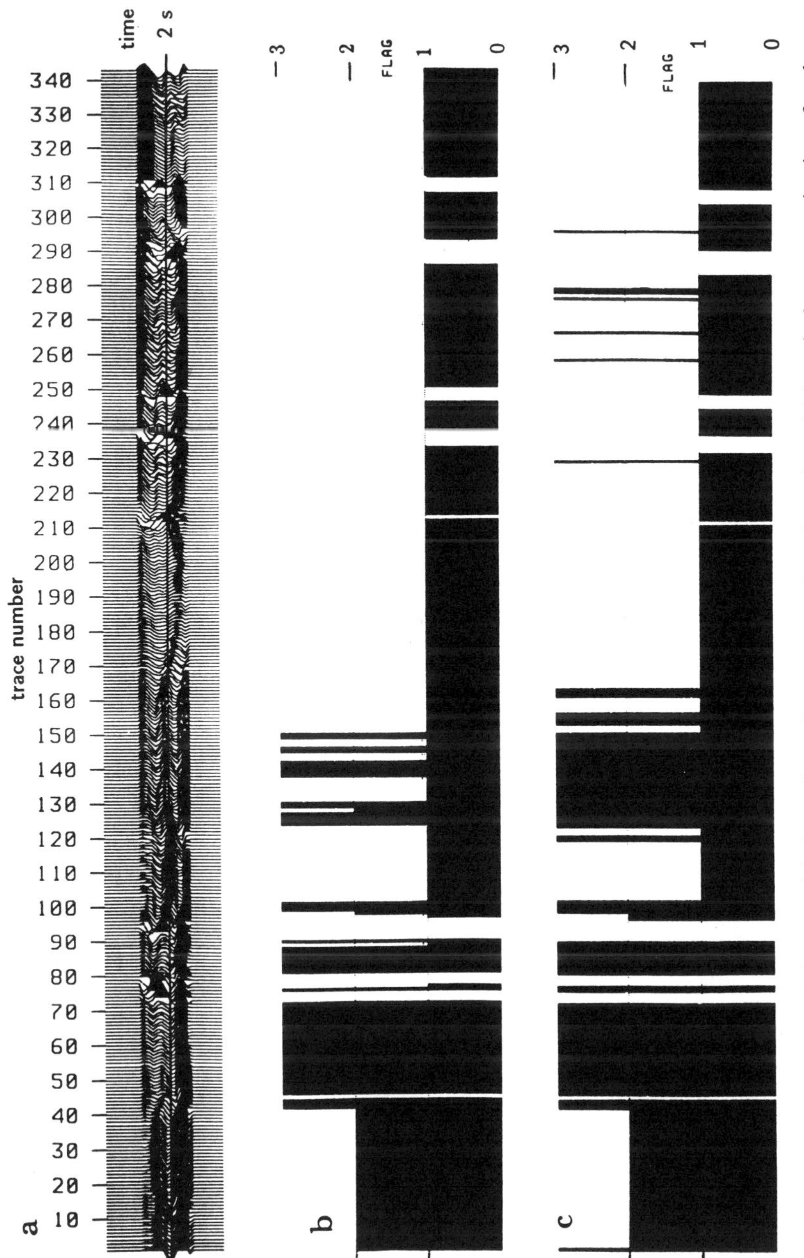

FIGURE 9.10. Comparison of clusters with those chosen by eye. (a) The reservoir interval from which the variables were computed, (b) categorization of each trace by a flag number 1, 2, or 3 for the three clusters found by PP (flag value of zero indicates a rejected trace), and (c) equivalent categorization of each trace by flag number for the three clusters discerned by eye by an independent observer.

mainly a third type to about CDP 150, and then the third type right to the end of the segment. Clear transition zones are thus indicated. Figure 9.10c shows a similar categorization carried out by an independent observer, by eye, using the scatterplot of variable values (Figure 9.2a). The two are very similar, with the PP result giving fewer isolated spikes, and a less homogeneous categorization around CDPs 120–160.

It is known that at CDP 30 the sand present in the interval is water wet (contains no gas) while at CDPs 105, 160, and 220, gas is present. There is thus a gas–water contact somewhere between CDPs 30 and 105, presumably in the section denoted by flag 3. Another successful mapping of a gas–water contact is given in Lendzionowski et al, 1990.

Conclusions

Projection pursuit has been shown to be an appealing way to search through data to find clusters. This has been demonstrated on 2-D data, but the methodology is the same for higher dimensions. Computations are extremely rapid, with all the hard work being done by efficient FFT algorithms. Large numbers of points can be handled easily. In a sense, the method is automated, but it is not automatic, since, for example, the user needs to consider (1) provision of several initial vectors, (2) whether convergence to a (local) optimum has occurred to within numerical accuracy, and (3) whether to continue splitting. The technique may find application in several areas of the geosciences.

Acknowledgments. The author thanks the Chairman and Board of Directors of The British Petroleum Company for permission to publish this paper.

References

Banks, D.L., and Young, G.A., 1987, Contribution to the discussion of Jones and Sibson: J.R. Stat. Soc., Ser. **A 150**, 23.

Darlington, R.B., 1970, Is kurtosis really "peakedness?": Am. Stat. **24**, 19–22.

Devlin, S.J., Gnanadesikan, R., and Kettenring, J.R., 1981, Robust estimation of dispersion matrices and principal components: J. Am. Stat. Assoc. **76**, 354–362.

Donoho, D.L., 1981, On minimum entropy deconvolution: in Applied Time Series Analysis II, D.F. Findley (Ed.): Academic Press, New York.

Friedman, J.H., 1987a, Contribution to the discussion of Jones and Sibson: J.R. Stat. Soc., Ser. **A 150**, 26–27.

Friedman, J.H., 1987b, Exploratory projection pursuit: J. Am. Stat. Assoc. **82**, 249–266.

Friedman, J.H., and Tukey, J.W., 1974, A projection pursuit algorithm for exploratory data analysis: IEEE Trans. Comput. **23**, 881–890.

Fukunaga, K., and Koontz, W.L., 1970, Application of the Karhunen-Loève transform to feature selection and ordering: IEEE Trans. Comput. **19**, 311–318.

Gentleman, J.F., 1978, Moving statistics for enhanced scatter plots: Appl. Stat. **27**, 334–358.

Hastie, T., and Tibshirani, R., 1987, Contribution to the discussion of Jones and Sibson: J.R. Stat. Soc., Ser. **A 150**, 27–28.

Hoaglin, D.C., 1983, Letter values: A set of selected order statistics: in Understanding Robust and Exploratory Data Analysis, D.C. Hoaglin, F. Mosteller, and J.W. Tukey (Eds.): John Wiley, New York, pp. 33–57.

Huber, P.J., 1985, Projection pursuit: Ann. Stat. **13**, 435–475.

Joe, H., 1987, Contribution to the discussion of Jones and Sibson (1987): J.R. Stat. Soc., Ser. **A 150**, 29.

Jones, M.C., 1983, The projection pursuit algorithm for exploratory data analysis: Ph.D. thesis, University of Bath (U.K.).

Jones, M.C., and Lotwick, H.W., 1984, A remark on algorithm AS176 kernel density estimation using the fast Fourier transform: Appl. Stat. **33**, 120–122.

Jones, M.C., and Sibson, R., 1987, What is projection pursuit?: J.R. Stat. Soc., **A 150**, 1–36.

Kruskal, J.B., 1964, Non-metric multidimensional scaling; a numerical method: Psychometrika **29**, 115–129.

Kruskal, J.B., 1969, Toward a practical method which helps uncover the structure of a set of multivariate observations by finding the linear transformation which optimizes a new 'index of condensation': in Statistical Computation, R.C. Milton and J.A. Nelder (Eds.): Academic Press, New York.

Lendzionowski, V., Walden, A.T., and White, R.E., 1990, Seismic character mapping over reservoir intervals: Geophys. Prosp. **38**, 951–969.

Levy, S., and Oldenburg, D.W., 1987, Automatic phase correction of common-midpoint stacked data: Geophysics **52**, 51–59.

Longbottom, J., Walden, A.T., and White, R.E., 1988, Principles and application of maximum kurtosis phase estimation: Geophys. Prosp. **36**, 115–138.

Longstaff, I.D., 1987, On extensions to Fisher's linear discriminant function: IEEE Trans. Pattern Anal. Mach. Intell. **9**, 321–325.

Mattson, R.L., and Dammann, J.E., 1965, A technique for determining and coding subclasses in pattern recognition problems: IBM J. Res. Dev. **9**, 294–302.

Nickerson, W.A., Matsuoka, T., and Ulrych, T.J., 1986, Optimum-lag minimum-entropy deconvolution: 51st Annu. Int. Mtg., Soc. Expl. Geophys., Expanded Abstracts, 519–522.

Sammon, J.W., 1969, A nonlinear mapping for data structure analysis: IEEE Trans. Comput. **18**, 401–409.

Silverman, B.W., 1982, Kernel density estimation using the fast Fourier transform: Appl. Stat. **31**, 93–97.

Silverman, B.W., 1986, Density estimation for statistics and data analysis: Chapman and Hall, London.

Tukey, P.A., and Tukey, J.W., 1981, Preparation; pre-chosen sequences of views: in Interpreting Multivariate Data, V. Barnett (Ed): John Wiley, New York, pp. 189–213.

Walden, A.T., 1985, NonGaussian reflectivity, entropy and deconvolution: Geophysics **50**, 2862–2888.

Wiggins, R.A., 1978, Minimum entropy deconvolution: Geoexploration **16**, 21–35.

Appendix A: Centering and Sphering Multivariate Data

Centering and sphering takes some ellipsoidal shaped scatter of points, with varying dispersion (variance) in orthogonal directions, and varying correlation between variables, and changes it to a spheroidal-shaped scatter, with fixed dispersion in orthogonal directions and zero correlation between variables. Hence the three operations incorporated in centering and sphering are shifting, stretching/shrinking, and rotation. The major eigenvector of the covariance matrix in the original coordinates represents the direction of maximum dispersion, the next-ranked eigenvector gives the direction of maximum dispersion, which is orthogonal to the first, etc. However, after centering and sphering all dispersion has been equalized, and the rotation involved makes the new eigenvectors (covariance now equals the correlation matrix) simply the unit vectors in centered-and-sphered coordinates.

The following method may be used to achieve centered and sphered data.

Step 1: Computation of Eigenvalues and Eigenvectors of Covariance Matrix

From $\mathbf{X}$ in Eq. (2) estimate the $p \times p$ covariance matrix $\hat{\mathbf{\Xi}}$ with j,lth element

$$\hat{\mathbf{\Xi}}_{jl} = \frac{1}{N-1} \sum_{i=1}^{N} (x_{ij} - \bar{x}_j)(x_{il} - \bar{x}_l) \tag{A-1}$$

where $\bar{x}_j = \frac{1}{N} \sum_{i=1}^{N} x_{ij}$ is the mean of the jth column, and hence the estimated mean of the jth variable. Next, assuming a full rank covariance matrix, the p eigenvalues $\{\hat{\theta}_j\}$ and $(p \times 1)$ eigenvectors $\{\hat{\mathbf{\Theta}}_j\}$ corresponding to the estimated covariance matrix $\hat{\mathbf{\Xi}}$ are calculated. These of course satisfy

$$\hat{\mathbf{\Xi}}\hat{\mathbf{\Theta}}_j = \hat{\theta}_j\hat{\mathbf{\Theta}}_j, \quad j = 1,\ldots,p \tag{A-2}$$

The scaling on the eigenvectors $\{\hat{\mathbf{\Theta}}_j\}$ is such that $\hat{\mathbf{\Theta}}_j^{\mathrm{T}}\hat{\mathbf{\Theta}}_j = 1, j = 1,\ldots,p$. The eigenvectors are thus orthonormal. A second set of eigenvectors is computed, by rescaling the $\{\hat{\mathbf{\Theta}}_j\}$ to give $\hat{\mathbf{\Phi}}_j = \hat{\mathbf{\Theta}}_j/\sqrt{(\hat{\theta}_j)}, j = 1,\ldots,p$, i.e., such that $\hat{\mathbf{\Phi}}_j^{\mathrm{T}}\hat{\mathbf{\Phi}}_j = 1/\hat{\theta}_j$.

Step 2: Location Shift (Centering)

Now subtract the mean from each of the columns of $\mathbf{X}$:

$$\begin{bmatrix} x_{11} - \bar{x}_1 & x_{12} - \bar{x}_2 & \cdots & x_{1p} - \bar{x}_p \\ \vdots & \vdots & \ddots & \vdots \\ x_{N1} - \bar{x}_1 & x_{N2} - \bar{x}_2 & \cdots & x_{Np} - \bar{x}_p \end{bmatrix} =$$

$$\begin{bmatrix} y_{11} & y_{12} & \cdots & y_{1p} \\ \vdots & \vdots & \ddots & \vdots \\ y_{N1} & y_{N2} & \cdots & y_{Np} \end{bmatrix} = \begin{bmatrix} \mathbf{y}_1^{\mathrm{T}} \\ \vdots \\ \mathbf{y}_N^{\mathrm{T}} \end{bmatrix}$$

say, where $\mathbf{y}_i^{\mathrm{T}} = (y_{i1},\ldots,y_{ip})$. This operation centers the data.

Step 3: Rescale and Rotate to Produce an Identity Covariance Matrix

Next, the data must be adjusted to a form that corresponds to a new set of variables with an estimated covariance matrix equal to the identity matrix. This may be done as follows. Form new *variables* $S_1,\ldots,S_p$ from $Y_1,\ldots,Y_p$ using

$$S_j = \sum_{l=1}^{p} \hat{\Phi}_{lj} Y_l = \hat{\Phi}_j^T \mathbf{Y}, \qquad j = 1,\ldots,p$$

where $\hat{\Phi}_{lj}$ is the lth element of the jth $1 \times p$ eigenvector $\hat{\Phi}_j$ and $\mathbf{Y}$ is the $p \times 1$ vector

$$\begin{bmatrix} Y_1 \\ \vdots \\ Y_N \end{bmatrix}$$

This transformation rescales and rotates the axes. In terms of the data *samples*, form:

$$s_{ij} = \hat{\Phi}_j^T \mathbf{y}_i, \qquad i = 1,\ldots,N; \qquad j = 1,\ldots,p$$

i.e.,

$$(y_{i1}, y_{i2}, \ldots, y_{ip}) \to (\mathbf{y}_i^T \hat{\Phi}_1, \mathbf{y}_i^T \hat{\Phi}_2, \ldots, \mathbf{y}_i^T \hat{\Phi}_p)$$
$$= (s_{i1}, s_{i2}, \ldots, s_{ip})$$

To show that the centered and sphered data have unity estimated covariance matrix, first denote the estimated covariance by $\widehat{Cov}$. Then:

$$\begin{aligned}
\widehat{Cov}\{S_j, S_l\} &= \widehat{Cov}\{\hat{\Phi}_j^T \mathbf{Y}, \hat{\Phi}_l^T \mathbf{Y}\} \\
&= \hat{\Phi}_j^T \hat{\Xi} \hat{\Phi}_l \\
&= \frac{\hat{\Theta}_j^T}{\sqrt{(\hat{\theta}_j)}} \hat{\Xi} \frac{\hat{\Theta}_l}{\sqrt{(\hat{\theta}_l)}} \\
&= \frac{\hat{\Theta}_j^T}{\sqrt{(\hat{\theta}_j)}} \hat{\theta}_l \frac{\hat{\Theta}_l}{\sqrt{(\hat{\theta}_l)}} \\
&= \begin{cases} 1 & \text{if} \quad j = l \\ 0 & \text{if} \quad j \neq l \end{cases}
\end{aligned}$$

Effect on Original Eigenvectors

What is the effect of the rescaling and rotation on the eigenvectors in the original coordinate system? If an eigenvector $\hat{\Theta}_j$ is transformed in the same way as the $\mathbf{y}_i$ then

$$\begin{aligned}
q_{ij} &= \hat{\Phi}_j^T \hat{\Theta}_i, \qquad i,j = 1,\ldots,p \\
&= \frac{\hat{\Theta}_j^T}{\sqrt{(\theta_j)}} \hat{\Theta}_i \\
&= \begin{cases} \dfrac{1}{\sqrt{(\hat{\theta}_j)}} & \text{if } j = i \\ 0 & \text{if } j \neq i \end{cases}
\end{aligned}$$

i.e.,

$$(\hat{\Theta}_{i1}, \hat{\Theta}_{i2}, \ldots, \hat{\Theta}_{ip}) \to (0, 0, \ldots, 0, \frac{1}{\sqrt{(\hat{\theta}_i)}}, 0, \ldots, 0, 0)$$

i.e., the transformation results in a scaled unit vector in the new coordinates.

Comments

The transformation that reduces a covariance matrix to diagonal form with the eigenvalues of the matrix on the diagonal is the Karhunen–Loève transform. Reduction of a matrix to an *identity* matrix is now becoming popular in classification techniques (e.g., Longstaff, 1987); when the transformation to diagonal form is applied to a matrix that is the sum of two covariance matrices it is sometimes called the Fukunaga–Koontz transform (Fukunaga and Koontz, 1970).

Appendix B: Calculation of the Kernel Density Estimator Using FFT Methods

The kernel density estimator is given by

$$f_N(z) = \frac{1}{Nh} \sum_{j=1}^{N} K\left(\frac{z - z_j}{h}\right)$$

where N is the sample size, h is the window width (smoothing parameter), and K is the kernel function, assumed here to be Gaussian in shape:

$$K(y) = [1/\sqrt{(2\pi)}] e^{-y^2/2}$$

The data $\{z_j\}$ may be standardized to unity *estimated* standard deviation for practical convenience. Direct evaluation of this expression is very slow if N is large, and FFT methods can be used.

The Fourier Transform of $f_N(z)$

Define the Fourier transform of $K(y)$ as

$$\tilde{K}(v) = \int_{-\infty}^{\infty} K(y) e^{-iyv} dy$$

and the Fourier synthesis of $K(y)$ by

$$K(y) = \frac{1}{2\pi} \int_{-\infty}^{\infty} \tilde{K}(v) e^{iyv} dv$$

Then,

$$K\left(\frac{z - z_j}{h}\right) = \frac{1}{2\pi} \int_{-\infty}^{\infty} e^{ivz/h} e^{-ivz_j/h} \tilde{K}(v) dv \tag{B-1}$$

so that

$$f_N(z) = \frac{1}{2\pi} \frac{1}{Nh} \sum_{j=1}^{N} \int_{-\infty}^{\infty} e^{ivz/h} e^{-ivz_j/h} \tilde{K}(v) dv$$

Now make the substitution $\omega = v/h$:

$$f_N(z) = \frac{1}{2\pi} \frac{1}{N} \sum_{j=1}^{N} \int_{-\infty}^{\infty} e^{iwz} e^{-i\omega z_j} \tilde{K}(\omega h) d\omega$$

$$= \frac{1}{2\pi} \int_{-\infty}^{\infty} \left(\frac{1}{N} \sum_{j=1}^{N} e^{-i\omega z_j} \right) \tilde{K}(\omega h) e^{i\omega z} d\omega$$

But

$$f_N(z) = \frac{1}{2\pi} \int_{-\infty}^{\infty} \tilde{f}_N(\omega) e^{i\omega z} d\omega$$

so that

$$\tilde{f}_N(\omega) = U_N(\omega) \tilde{K}(\omega h) \qquad \text{(B-2)}$$

where

$$U_N(\omega) = \frac{1}{N} \sum_{j=1}^{N} e^{-i\omega z_j}$$

the sample characteristic function, and

$$\tilde{K}(\omega h) = e^{-\frac{1}{2} h^2 \omega^2}$$

the Fourier transform of the Gaussian kernel (with argument ωh). Thus the kernel density estimator can be computed by multiplying the sample characteristic function by the Fourier transform of the Gaussian kernel, and then inverse Fourier transforming to obtain $f_N(z)$.

Note that $\tilde{K}(\omega h)$ is the Fourier transform of $\frac{1}{h} K \left(\frac{y}{h} \right) = K_h(y)$, say, and $f_N(z)$ can be more simply written as

$$f_N(z) = \sum_{j=1}^{N} \frac{1}{N} K_h(z - z_j)$$

Discretization onto a Grid

Computation of the transform domain components of Eq. (B-2), and the inversion of the product to obtain $f_N(z)$ is most efficiently carried out using the FFT (Silverman, 1982, 1986; Jones and Lotwick, 1984). To use the FFT the data must be discretized by mapping them onto a grid, with sufficient padding at each end to avoid wrap-around effects. The left end of the grid is denoted a and is set equal to $\min(z_j) - 3h$, while the right end is denoted b and set equal to $\max(z_j) + 3h$. Next define $M = 2^r$ equispaced points on the interval $[a,b]$. This has been implemented with $r = 9$ giving 512 points. The points are defined as

$$t_k = a + \left(k + \frac{1}{2} \right) \delta, \qquad k = 0, \ldots, M - 1$$

where $\delta = (b - a)/M$.

The data are discretized onto the grid using the rule that if z_j lies in $[t_k, t_{k+1}]$ then a weight $[1/(N\delta)][(t_{k+1} - z_j)/\delta]$ is applied at t_k, and $[1/(N\delta)][(z_j - t_k)/\delta]$ at t_{k+1}. Note that if $z_j = t_k$, then the weights are, respectively, $1/(N\delta)$ and 0. By applying the same rule for all points z_j, a set of total weights $\{w_k, k = 0, 1, \ldots, M - 1\}$ is obtained.

Estimation of the Sample Characteristic Function

Fourier transforming the w_k using the FFT gives

$$W_l = \frac{1}{M} \sum_{k=0}^{M-1} w_k e^{-i2\pi lk/M}, \qquad l = 0, \ldots, M - 1$$

Defining $\omega_l = 2\pi l/(b - a) = 2\pi l/(M\delta)$ results in

$$t_k \omega_l = [a + \left(k + \frac{1}{2} \right) \delta] 2\pi l/(M\delta)$$

Thus

$$2\pi kl/M = t_k \omega_l - \pi l(2a + \delta)/(M\delta)$$

Hence,

$$W_l = \frac{1}{M} \sum_{k=0}^{M-1} w_k e^{-i t_k \omega_l} [e^{i\pi l(2a + \delta)/(M\delta)}]$$

As already seen, when an z_j coincides exactly with a grid point t_k, a weight $1/(N\delta)$ is obtained at $t_k = z_j$. Hence,

$$W_l \approx \frac{1}{M} \frac{1}{N\delta} \sum_{j=1}^{N} e^{-i\omega_l z_j} [e^{i\pi l(2a + \delta)/(M\delta)}]$$

since the grid spacing is very fine. Thus

$$W_l \approx \frac{1}{M\delta} U_N(\omega_l) [e^{i\pi l(2a + \delta)/(M\delta)}]$$

i.e., W approximates U_N apart from a scaling and phase shift.

Multiplication and Inverse Transformation

Now multiply W_l by the Fourier transform of the kernel, i.e., by $e^{-\frac{1}{2} h^2 \omega_l^2}$ to obtain

$$W_l e^{-\frac{1}{2} h^2 \omega_l^2}$$

the discrete equivalent of Eq. (B-2) and take the inverse discrete Fourier transform of this product using the FFT:

$$y_k = \frac{1}{M\delta} \sum_{l=0}^{M-1} U_N(\omega_l) e^{-\frac{1}{2}h^2\omega_l^2} e^{i2\pi lk/M} \left[e^{i\pi l(2a+\delta)/(M\delta)} \right]$$

Now $2\pi kl/M = t_k\omega_l - \pi l(2a + \delta)/(M\delta)$; hence

$$
\begin{aligned}
y_k &= \frac{1}{M\delta} \sum_{l=0}^{M-1} U_N(\omega_l) e^{-\frac{1}{2}h^2\omega_l^2} e^{it_k\omega_l} \\
&= \frac{1}{2\pi} \frac{2\pi}{M\delta} \sum_{l=0}^{M-1} U_N(\omega_l) e^{-\frac{1}{2}h^2\omega_l^2} e^{it_k\omega_l} \qquad \text{(B-3)} \\
&\approx \frac{1}{2\pi} \int_0^{2\pi/\delta} U_N(v) e^{-\frac{1}{2}h^2v^2} e^{it_k v} \, dv \\
&\approx \hat{f}_N(t_k)
\end{aligned}
$$

i.e., an approximation to the kernel density estimator at $z = t_k$.

Estimation of e_N

Now $\hat{f}_N$ is available at a grid of equispaced points $t_0(\mathbf{a}_s),\ldots,t_{M-1}(\mathbf{a}_s)$ where M is a power of 2. Thus $e_N(\mathbf{a}_s)$ can be estimated by using Simpson's rule:

$$
\begin{aligned}
\hat{e}_N(\mathbf{a}_s) &= \frac{1}{2} \sum_{j=0}^{M-2} \{\hat{f}_N[t_j(\mathbf{a}_s)] \ln \hat{f}_N[t_j(\mathbf{a}_s)] + \hat{f}_N[t_{j+1}(\mathbf{a}_s)] \\
&\qquad\qquad \ln \hat{f}_N[t_{j+1}(\mathbf{a}_s)]\}\delta \\
&\approx \frac{1}{2} \sum_{j=0}^{M-2} (y_j \ln y_j + y_{j+1} \ln y_{j+1}) \, \delta
\end{aligned}
$$

where δ is the distance between adjacent $t_j(\mathbf{a}_s)$ values, and the ys are given in Eq. (B-3).

Appendix C: Calculation of the Derivatives of $\hat{e}_N$ Using FFT Methods

The derivatives

$$g_m = d\hat{e}_N/da_{ms}, \qquad m = 1,\ldots,p$$

are to be computed using the FFT. Now

$$e_N(\mathbf{a}_s) = \int_{-\infty}^{\infty} f_N(z;\mathbf{a}_s)\ln f_N(z;\mathbf{a}_s)\,dz$$

and hence

$$
\begin{aligned}
\frac{de_N(\mathbf{a}_s)}{da_{ms}} &= \int_{-\infty}^{\infty} \frac{d}{da_{ms}} \{f_N(z;\mathbf{a}_s)\ln f_N(z;\mathbf{a}_s)\}\,dz \\
&= \int_{-\infty}^{\infty} [\ln f_N(z;\mathbf{a}_s) + 1] \frac{df_N(z;\mathbf{a}_s)}{da_{ms}}\,dz \quad \text{(C-1)}
\end{aligned}
$$

(In what follows I shall sometimes drop explicit dependence on $\mathbf{a}_s$ to reduce notational complexity.) Now,

$$f_N(z) = \frac{1}{Nh} \sum_{j=1}^{N} K\left(\frac{z-z_j}{h}\right) = \frac{1}{N} \sum_{j=1}^{N} K_h(z - z_j)$$

where $K_h(y) = \frac{1}{h} K(\frac{y}{h})$ as in Appendix B. Hence,

$$\frac{df_N(z)}{da_{ms}} = \sum_{j=1}^{N} \left(\frac{1}{N}\frac{dz_j}{da_{ms}}\right)\left[\frac{dK_h(z - z_j)}{dz_j}\right] \quad \text{(C-2)}$$

Calculation of the First Term of the Summation in Eq. (C-2)

Now, the projection of the sphered data vector $\mathbf{s}_j$ onto the unit length projection vector $\mathbf{a}_s$ is given by

$$
\begin{aligned}
z_j(\mathbf{a}_s) &= \sum_{k=1}^{p} a_{ks}s_{jk} \\
&= \sum_{k=1}^{p} \frac{a_k}{\sqrt{(\sum_{l=1}^{p} a_l^2)}} s_{jk}
\end{aligned}
$$

where a_k are the raw coordinates and a_{ks} the sphered coordinates. The two are identical when $\sum a_l^2 = 1$. Differentiating with respect to a_m (*not* a_{ms}) gives

$$\frac{dz_j(\mathbf{a}_s)}{da_m} = \frac{1}{\sqrt{(\sum_{l=1}^{p} a_l^2)}} \left[s_{jm} - \left(\frac{a_m}{\sqrt{(\sum_{l=1}^{p} a_l^2)}}\right) z_j(\mathbf{a}_s) \right]$$

$\sum_{l=1}^{p} a_l^2 = 1$ gives $a_k \equiv a_{ks}$ and

$$\frac{dz_j(\mathbf{a}_s)}{da_{ms}} = [s_{jm} - a_{ms}z_j(\mathbf{a}_s)] \quad \text{(C-3)}$$

Thus Eq. (C-2) can be written as

$$\frac{df_N(z;\mathbf{a}_s)}{da_{ms}} = \sum_{j=1}^{N} \frac{dK_h[z - z_j(\mathbf{a}_s)]}{dz_j} \left\{ \frac{1}{N}[s_{jm} - a_{ms}z_j(\mathbf{a}_s)] \right\}$$

This result agrees with that quoted in Jones (1983, p. 87).

Calculation of the Second Term of the Summation in Eq. (C-2)

Now

$$
\begin{aligned}
\frac{dK_h(z - z_j)}{dz_j} &= \frac{1}{h}\frac{d}{dz_j} K\left(\frac{z-z_j}{h}\right) \\
&= \frac{1}{2\pi h} \int_{-\infty}^{\infty} e^{ivz/h} \frac{de^{-ivz_j/h}}{dz_j} \tilde{K}(v)\,dv
\end{aligned}
$$

$$= -\frac{1}{2\pi h} \int_{-\infty}^{\infty} e^{ivz/h} \left(\frac{iv}{h}\right) e^{-ivz_j/h} \tilde{K}(v)\,dv \tag{C-4}$$

using Eq. (B-1).

Calculation of the Summation of Eq. (C-2)

Inserting Eqs. (C-3) and (C-4) into Eq. (C-2) gives

$$\frac{df_N(z;\mathbf{a}_s)}{da_{ms}} = \frac{1}{2\pi} \frac{1}{Nh} \sum_{j=1}^{N} [a_{ms}z_j - s_{jm}] \int_{-\infty}^{\infty} e^{ivz/h} \left(\frac{iv}{h}\right) e^{-ivz_j/h} \tilde{K}(v)\,dv$$

Making the substitution $\omega = v/h$:

$$\frac{df_N(z)}{da_{ms}} = \frac{1}{2\pi} \frac{1}{N} \sum_{j=1}^{N} [a_{ms}z_j - s_{jm}]$$

$$\int_{-\infty}^{\infty} e^{i\omega z} i\omega\, e^{-i\omega z_j} \tilde{K}(\omega h)\,d\omega$$

$$= \frac{1}{2\pi} \int_{-\infty}^{\infty}$$

$$\left[\frac{1}{N} \sum_{j=1}^{N} (a_{ms}z_j - s_{jm})e^{-iwz_j}\right] i\omega\tilde{K}(\omega h)e^{iwz}\,d\omega$$

But, in terms of Fourier synthesis,

$$\frac{df_N(z)}{da_{ms}} = \frac{1}{2\pi} \int_{-\infty}^{\infty} \left[\frac{\widetilde{df_N}(\omega)}{da_{ms}}\right] e^{iwz}\,d\omega$$

so that

$$\left[\frac{\widetilde{df_N}(\omega)}{da_{ms}}\right] = \left[\frac{1}{N} \sum_{j=1}^{N} (a_{ms}z_j - s_{jm})e^{-i\omega z_j}\right] i\omega\tilde{K}(\omega h)$$

Comparison with Eq. (B-2) shows that instead of the characteristic function $U_N(\omega)$ the weighted sample characteristic function has been obtained:

$$R_N(\omega) = \frac{1}{N} \sum_{j=1}^{N} (a_{ms}z_j - s_{jm})e^{-i\omega z_j}$$

and instead of $\tilde{K}(\omega h)$, $i\omega\tilde{K}(\omega h)$ has been obtained, i.e.,

$$\left[\frac{\widetilde{df_N}(\omega)}{da_{ms}}\right] = R_N(\omega)i\omega\tilde{K}(\omega h) \tag{C-5}$$

Discretization onto a Grid

The data are discretized onto the same grid as in Appendix B using the rule that if z_j lies in $[t_k,t_{k+1}]$ then a weight

$$[(a_{ms}z_j - s_{jm})/(N\delta)][(t_{k+1} - z_j)/\delta]$$

is applied at t_k, and

$$[(a_{ms}z_j - s_{jm})/(N\delta)][(z_j - t_k)/\delta]$$

at t_{k+1}.

Estimation of the Weighted Sample Characteristic Function

Fourier transforming the new total weights w_k^* using the FFT:

$$W_l^* = \frac{1}{M} \sum_{k=0}^{M-1} w_k^* e^{-i2\pi lk/M}, \qquad l = 0,\ldots,M-1$$

and, as in Appendix B,

$$W_l^* = \frac{1}{M} \sum_{k=0}^{M-1} w_k^* e^{-it_k\omega_l} [e^{i\pi l(2a+\delta)/(M\delta)}]$$

When a z_j coincides exactly with a grid point t_k, a weight $(a_{ms}z_j - s_{jm})/(N\delta)$ is used at $t_k = z_j$. Hence,

$$W_l^* \approx \frac{1}{M} \frac{1}{N\delta} \sum_{j=1}^{N} (a_{ms}z_j - s_{jm})e^{-i\omega_l z_j}[e^{i\pi l(2a+\delta)/(M\delta)}]$$

$$= \frac{1}{M\delta} R_N(\omega_l) [e^{i\pi l(2a+\delta)/(M\delta)}]$$

Thus W^* approximates the weighted sample characteristic function, apart from a scale factor and a phase shift.

Multiplication and Inverse Transformation

Now multiply W_l^* by $i\omega_l e^{-\frac{1}{2}h^2\omega_l^2} = i\omega_l\tilde{K}(\omega_l h)$ to obtain the discrete equivalent of Eq. (C-5), i.e.,

$$W_l^* i\omega_l e^{-\frac{1}{2}h^2\omega_l^2}$$

and take the inverse discrete Fourier transform of the product using the FFT:

$$y_k^* = \frac{1}{M\delta} \sum_{l=0}^{M-1} \sum_{j=1}^{N} \frac{1}{N} (a_{ms}z_j - s_{jm})e^{-\omega_l z_j} i\omega_l e^{-\frac{1}{2}h^2\omega_l^2}$$

$$e^{i2\pi lk/M}[e^{i\pi l(2a+\delta)/(M\delta)}]$$

$$= \frac{1}{M\delta} \sum_{l=0}^{M-1} r_N(\omega_l)i\omega_l e^{-\frac{1}{2}h^2\omega_l^2} e^{it_k\omega_l}$$

$$\approx \frac{1}{2\pi} \int_0^{2\pi/\delta} r_N(v)iv\tilde{K}(vh)e^{it_k v}\,dv$$

$$\approx \frac{\widehat{df_N}(t_k)}{da_{ms}} \tag{C-6}$$

i.e., an approximation to the derivative of the kernel density function at $z = t_k$.

Estimation of the Derivative

Since $\widehat{df_N/da_{ms}}$ is available at a grid of equispaced points $t_0(\mathbf{a}_s),\ldots,t_{M-1}(\mathbf{a}_s)$ Simpson's rule can again be applied to estimate $de_N(\mathbf{a}_s)/da_{ms}$ as $\widehat{de_N}(\mathbf{a}_s)/da_{ms}$ given by [see Eq. (C-1)],

$$\frac{1}{2}\sum_{j=0}^{M-2}\left(\{\ln\hat{f}_N[t_j(\mathbf{a}_s)] + 1\}\{\frac{\widehat{df_N[t_j(\mathbf{a}_s)]}}{da_{ms}}\} + \right.$$
$$\left. \{\ln\hat{f}_N[t_{j+1}(\mathbf{a}_s)] + 1\}\{\frac{\widehat{df_N[t_{j+1}(\mathbf{a}_s)]}}{da_{ms}}\}\right)\delta$$

$$= \frac{1}{2}\sum_{j=0}^{m-2}[(\ln y_j + 1)\,y_j^* + (\ln y_{j+1} + 1)\,y_{j+1}^*]\,\delta$$

with the ys given in Eq. (B-3) and the y^*s given in Eq. (C-6).

Appendix D: Relationship of Projection Pursuit (PP) to Minimum Entropy Deconvolution (MED) and Maximum Kurtosis Phase Estimation (MKPE)

Suppose our centered and sphered data are represented by the $N \times p$ matrix:

$$\begin{bmatrix} \vdots & \vdots & \ddots & \vdots \\ s_{n1} & s_{n2} & \cdots & s_{np} \\ s_{n+1,1} & s_{n+1,2} & \cdots & s_{n+1,p} \\ \vdots & \vdots & \ddots & \vdots \end{bmatrix}$$

with (i,j)th element s_{ij}. Each row corresponds to the p centered and sphered variable values for a single observation. Given a unit length projection vector $\mathbf{a}_s$, the projection of the nth row of the matrix onto $\mathbf{a}_s$ is

$$z_n(\mathbf{a}_s) = \mathbf{a}_s^T\mathbf{s}_n = \sum_{j=1}^{p} a_{js}s_{nj}$$

$\hat{e}_N(\mathbf{a}_s)$ is maximized over $\mathbf{a}_s$ where

$$\hat{e}_N(\mathbf{a}_s) \approx \int f_N[z(\mathbf{a}_s)]\ln f_N[z(\mathbf{a}_s)]dz(\mathbf{a}_s) \quad (D-1)$$

To carry out the optimization, an initial projection vector is found, $\mathbf{a}_s^{(0)}$ say, and then an iterative scheme used to find a (local) maximum. Typically, the initial projection vector would be one of the eigenvectors. Compare this approach with that used in Minimum Entropy Deconvolution (MED) due to Wiggins (1978), or Maximum Kurtosis Phase Estimation (MKPE) (Levy and Oldenburg, 1987; Longbottom et al., 1988).

Taking the p-D data points to be p successive values of a seismic data trace, each point corresponding to incrementing the start time from which the p trace values are selected by one sample interval, the matrix becomes

$$\begin{bmatrix} \vdots & \vdots & \ddots & \vdots \\ s_n & s_{n-1} & \cdots & s_{n-p+1} \\ s_{n+1} & s_n & \cdots & s_{n-p+2} \\ \vdots & \vdots & \ddots & \vdots \end{bmatrix}$$

where $s_n,s_{n-1},\ldots,s_{n-p+1}$ are consecutive values of the seismic trace. Projecting nth line of this matrix onto a vector $\mathbf{d} = (d_0,d_1,\ldots,d_{p-1})^T$ results in

$$z_n(\mathbf{d}) = \mathbf{d}^T\mathbf{s}_n = \sum_{j=0}^{p-1} d_j s_{n-j}$$

$$= (d * s)_n$$

where "$*$" denotes convolution. Next, maximize sample kurtosis

$$\frac{\sum z_n^4(\mathbf{d})}{[\sum z_n^2(\mathbf{d})]^2} \quad (D-2)$$

over $\mathbf{d}$. This index is also location and scale invariant in this application since the mean of seismic data is always approximately zero, and Eq. (D-2) is scale invariant. Note that $\mathbf{d}$ need not be of unit length here. In the standard approach to MED (i.e., Wiggins') a first approximation to the maximizing filter is found, $\mathbf{d}^{(0)}$ say, and then a recursive scheme used to refine the estimate of the deconvolution filter. Wiggins' original suggestion for $\mathbf{d}^{(0)}$ was a filter consisting of all zeros with a unit spike in the middle location. However, Nickerson et al. (1986) showed that such an arbitrary choice can result in the sample kurtosis being maximized at a local extremum giving rise to a deconvolution filter with inappropriate phase.

In the seemingly more promising recent approach (MKPE) a two-stage procedure is used. Firstly standard whitening deconvolution (incorporating a correction for nonwhiteness if desired) is used to whiten the amplitude spectrum. Then, estimated kurtosis is maximized over a single parameter — a phase shift — the assumption being that the phase spectrum is approximately linear. The separation of the amplitude and phase estimation steps gives rise to more stable and clearly understood deconvolution.

In Figure 9.D-1, PP and MED/MKPE are schematically contrasted. On the left side of the figure (PP) a near-Gaussian distribution is seen being changed into a bimodal distribution through PP. The near-Gaussian distribution represents the projection of the data onto the initial vector $\mathbf{a}_s^{(0)}$. The bimodal distribution represents the projection of the data onto the optimum vector, i.e., the one maximizing Eq. (D-1), at least locally. How near the distribution is to Gaussian for the initial vector depends on the $\mathbf{a}_s^{(0)}$ chosen. If the chosen $\mathbf{a}_s^{(0)}$ was close to the optimum, then the distribution might look more bimodal than Gaussian. In general though, I would expect to have to refine $\mathbf{a}_s^{(0)}$ significantly and hence expect the initial distribution to look quite Gaussian. (These ideas are illustrated later.) The entropy is seen to decrease with PP; this is of course because the index in Eq. (D-1) being maximized is the negative of Shannon's entropy. We have also indicated a decrease in kurtosis with bimodality induced by PP. This follows from Darlington (1970), who shows that symmetrically adding weight to a Gaussian distribution in the region of ± 0.74 to ± 2.33 standard deviations about the mean will *decrease* the value of the kurtosis from 3 toward 1. (The point of Darlington's article is his claim that kurtosis is best described not as a measure of peakedness versus flatness, but rather as a measure of unimodality versus bimodality.) It should be noted that bimodality well outside the region of ± 1 standard deviation (e.g., ± 5 standard deviations about the mean) could give an *increase* in kurtosis, because at such extremes extra probability increases tailweight. However, at ranges likely to be encountered in PP, a decrease in kurtosis is expected, and has been observed in all cases we have examined.

On the right side of Figure 9.D-1 (MED/MKPE) the near-Gaussian distribution is seen being changed into a cusp-like heavy-tailed distribution. The near-Gaussian distribution represents the projection of the data onto the initial projection vector or deconvolution filter $\mathbf{d}^{(0)}$. The cusp-like distribution represents the projection of the data onto the optimum deconvolution filter, i.e., the one

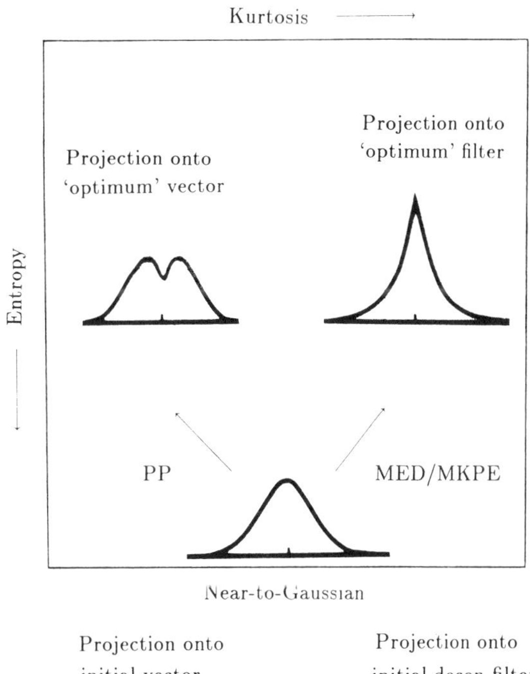

FIGURE 9.D-1. Diagram showing the relationship of PP to MED/MKPE.

maximizing Eq. (D-2), at least locally. Again, the nearness of the distribution to Gaussian depends on the $\mathbf{d}^{(0)}$ chosen. In the case of MED, the initial filter $\mathbf{d}^{(0)}$ might be of the type suggested by Nickerson et al. (1986), while for MKPE it would be the minimum-delay whitening operator. Again entropy decreases (as indicated by the name MED; but formally as discussed in Donoho, 1981, or Walden, 1985) but kurtosis increases due to the increasing tailweight.

An interesting and important question is "why does PP typically produce a bimodal distribution, while MED/MKPE produces a cusp-like heavy-tailed distribution?" Why doesn't PP produce a cusp-like distribution, or MED/MKPE produce a bimodal distribution? The most likely explanation is as follows (but requires further investigation). Starting from a near-to-Gaussian distribution it seems natural, *in the absence of smoothing*, for MED/MKPE to produce a cusp-like distribution because this will maximize its projection index [sample kurtosis in Eq. (D-2)] at least locally, rather than a bimodal distribution with weight well out in the tails. PP uses as projection index minus

Shannon's entropy, and hence seemingly could go to the left or right, i.e., toward a bimodal or cusp-like distribution. However, the smoothing h imposed in the kernel density estimator means that cusp-like distributions cannot occur, and the technique is pushed naturally toward bimodal distributions.

Appendix E: Choosing the Smoothing h in Kernel Density Estimation

Every time the multivariate data are projected onto a projection line, be it the initial projection vector or one produced in the optimization procedure, a kernel density estimate is produced from the projected sample. This requires specification of the smoothing constant h in Eq. (5). The problem is thus not the same as smoothing a random sample from a single distribution—here a marginal distribution of a multivariate distribution is being estimated, and of course a different marginal distribution for each different projection. There is no clear-cut best approach to doing this (e.g., Banks and Young, 1987). For the case when we have a random sample from a univariate distribution h is usually chosen to minimize the mean integrated square error (MISE):

$$\text{MISE} = E \int [f_N(z) - f(z)]^2 \, dz$$
$$= \int \{E[f_N(z)] - f(z)\}^2 dz$$
$$+ \int \text{Var}[f_N(z)]dz$$

with f_N the kernel density function estimator as in (5). As shown in Silverman (1986) this sum of a bias and variance component is given by

$$\text{MISE} \approx (\{h^4 [\int t^2 K(t) dt]^2\}/4)$$
$$\int [f''(z)]^2 dz + N^{-1}h^{-1}$$
$$\int [K(t)]^2 dt$$

Since changing h acts in opposite ways on the bias and variance components, it follows that the minimization of MISE involves the usual tradeoff between variance and bias, or random error and systematic error so familiar in spectrum estimation.

The value of h that minimizes this formula is given by

$$h = [\int t^2 K(t) dt]^{-2/5} \{ \int [K(t)]^2 dt \}^{1/5}$$
$$\{ \int [f''(z)]^2 dz \}^{-1/5} N^{-1/5}$$

Given a sample from a univariate Gaussian distribution and using a Gaussian kernel K the forms of K and f'' are known and the optimal h becomes

$$h = 1.06 \sigma N^{-1/5} \qquad \text{(E-1)}$$

where σ is the standard deviation of the Gaussian data. In their approach Jones and Sibson (1987) took

$$h = N^{-1/5} \qquad \text{(E-2)}$$

throughout. Since the data had been sphered $\sigma \approx 1$ and 1.06 was replaced by 1. Jones and Sibson (1987) justified this most simple choice by finding the entropy index to be insensitive to window width h. Joe (1987) claims this need not always be the case, especially for long-tailed distributions. Note that after the isolation step when the data have been split, the smoothing applied to projections of either subset will depend on its size through N. However, within the optimization stage h is fixed no matter what the projection orientation. A fixed value of h appears linked to the desire to *jointly* estimate the marginal distributions.

Banks and Young (1987) suggest that if the distribution of projected data is to be thought of in a univariate sense, then rather than use an h optimal for the Gaussian, one ought to optimize for a density intermediate between dull (Gaussian) and interesting (say bimodal). Unfortunately, this requires finding the minimum of the chosen MISE by numerical integration, and must be done for each different N (i.e., at each isolation step). In general the small effect of the new optimal h would seem unlikely to justify the amount of computation involved. Also the distribution that is chosen for the minimization would still be somewhat arbitrary.

The simple formula $h = N^{-1/5}$ that is designed for the smoothing of a Gaussian distribution is not ideally suited for bimodal distributions. Silverman (1986) recommends the (adaptive) optimum h:

$$h = 0.9 N^{-1/5} \min(\text{standard deviation},$$
$$\text{interquartile range}/1.349) \qquad \text{(E-3)}$$

(The interquartile range, IR, for a Gaussian distribution is 1.349 times the standard deviation, i.e., IR = 1.349σ.) When estimating this, the standard deviation would be estimated by the usual formula

$$\hat{\sigma} = \sqrt{\frac{1}{(N-1)} \sum (z_i - \bar{z})^2}$$

while the interquartile range could be estimated as $Q(0.75) - Q(0.25)$ where $Q(p)$ is the sample quantile defined by (e.g., Gentleman, 1978):

$$
\begin{aligned}
z_{(u_t)} + u_f [z_{(u_{t+1})} - z_{(u_t)}] & \quad (2N)^{-1} && \leq p \leq 1 - (2N)^{-1} \\
z_{(1)} & \quad 0 && \leq p \leq (2N)^{-1} \\
z_{(N)} & \quad 1 - (2N)^{-1} && \leq p \leq 1
\end{aligned}
$$

The $z_{(i)}$ are the data values ranked in increasing size order, u_t is the integer part, and u_f the fractional part of $u = Np + 0.5$.

Silverman's formula works well for a wide range of distributional shapes, including long-tailed and bimodal distributions. When the data are Gaussian, the *estimate* of σ obtained from IR/1.349 will usually be close to $\hat{\sigma}$. However, the estimate IR/1.349 will be more resistant in the case of long-tailed distributions than will $\hat{\sigma}$. In general then, if the distributioin is long-tailed the minimum will occur for IR/1.349, while if it is bimodal, the minimum will occur for $\hat{\sigma}$. IR/1.349 is virtually the same as the so-called "F-pseudosigma" of Hoaglin (1983). Note that when this formulation for h is used in the optimization stage, for given N, the smoothing can change depending on the orientation of the projection vector because of the adaptive nature of the formula. Jones and Sibson (1987, p. 6) prefer to fix h, and trials by this author gives tentative support for this. The formula (E-1) was used for the examples given herein. However, it seems wise to have the facility for adaptively choosing h within the optimization stage if desired.

10
Exploring the Fractal Mountains

Brian Klinkenberg and Keith C. Clarke

Introduction

The form of the earth's surface, reflected by the variation in elevation over space, is very often used as one of the best examples of a naturally fractal phenomenon. The simulated landscapes that are included in Mandelbrot's now classic texts (Mandelbrot, 1977, 1982), and in the work of many others conducting fractal simulation, are very realistic (e.g., Peitgen and Saupe, 1988). Are the models, however, correct? Can we explore the synthetic mountains of the fractal models, just as we can explore the real mountains on earth, and look for natural structure, form, and process? In this chapter we examine the applicability of fractal methods to natural topography from a geoscience, rather than a mathematical, viewpoint. We consider the question "is the land surface fractal?"

What, then, are fractals? Theoretically, fractal shapes have no characteristic size, and are said to be self-similar (statistically so, for land surfaces), and independent of scale—or scaling. Statistical self-similarity means that the results of processes that act on the land surface (at different scales) cannot be statistically distinguished, regardless of the scale at which we view the surface. Within the geosciences, fractal concepts have already been applied to the land surface in a wide variety of ways. A summary of the literature is given that illustrates the impact that fractals have had in the field, and highlights their potential.

As there are many different types of landscape, so there are many different definitions of fractal. Thus, it is important that we first clearly define the type of fractal that we will be working with—the

fractional Brownian motion (fBm) class of fractals. This is the most common group of landscape simulation models in use today, although different models (such as the midpoint displacement method) have been developed. Acceptance of the fBm model allows us to formulate certain statistical tests that can be used to investigate the goodness of fit of the model to selected land surfaces. A brief review of the many different tests that have been used to measure the goodness of fit summarizes the current literature on the subject, and notes some discrepancies between techniques.

It is also important to consider a typology of landscapes, based on the processes that formed them, and the scale at which the processes operate, for the fractal property of self-similarity does not apply to all land surface types or across all possible scales of investigation. Results of some empirical fractal tests using a number of United States Geological Survey Digital Elevation Models show that, for certain land surfaces, and given some scale assumptions, the fBm model does provide a reasonable fit.

Fractals in the Geosciences Today

The simplest use of the fractal dimension in geomorphology is as a comparative parameter: "If two objects are the same they must at least have the same fractal dimension" (Kadanoff, 1986, p. 7). This use of the fractal dimension D as a simple descriptive statistic has received some attention from geomorphologists, and may eventually receive much wider attention if more reliable methods for its determination can be developed. As Goodchild and Mark (1987, p. 267) state:

The numerical value of D may be the most important single parameter of an irregular cartographic feature, just as the arithmetic mean and other measures of central tendency are often used as the most characteristic parameters of a sample.

Fox and Hayes (1985) considered the use of D and an associated amplitude parameter as descriptors for the seafloor profiles they were investigating. In soil-covered landscapes, Culling and Datko (1987, p. 384) noted that D can be used to differentiate between differing evolutionary surfaces. Determination of D will provide a means of comparing two landscapes; however, the fractal dimension does not provide a unique identification as two dissimilar landscapes can have identical values of D.

Burrough (1981, 1983a) reviewed a number of studies in fields such as hydrology, climatology, soils, and geology, and considered them in light of the fractal model. He observed that coastlines, topography, river discharges, climatic variation, island and lake size distributions, and soil variation all share one characteristic: that, on first sight, no single length, area, or time scale can be applied to them. The fractal model did not provide a universal fit for all of the data Burrough analyzed, but its application revealed aspects of the data that were not obvious previously (e.g., Burrough 1983b,c).

Theoretically, fractals should be most applicable to phenomena that are the result of processes that are scalable—such as turbulence and wind speed. Conversely, if the processes are structured or nonscalable, greater deviations from the fractal model would be expected (Burrough, 1983a), or, possibly, nonrandom fractal models could be used. As real landscapes are the result of interactions of many (possibly nested) processes, each of which may be dominant over a different range, the land surface may be best described as a multifractal phenomenon, with the different dimensions corresponding to the dominance ranges of the different processes—if each process is scaling (i.e., fractal) within its own limits.

Thus, the fractal dimension of a landscape is not expected to be constant—it is expected to apply to a range through time or space or to vary continuously according to some function (Scholz and Aviles, 1986; Lovejoy and Schertzer, 1986; Man-

delbrot, 1977, 1984; Kennedy and Lin, 1986). Mark and Aronson (1984) noted that the scales at which the dimensions change could be of use in defining homogeneous geomorphological provinces; Fox and Hayes (1985) used a similar concept when applying their "province picker" to a spectral analysis of seafloor terrain. Kent and Wong (1982), in a study on the fractal nature of the shorelines of lakes on the Canadian shield, felt that the change in dimension occurred where the dominant process switched from a large-scale glacial corrosive process to a smaller scale erosional process.

A fractal-based style of analysis, which eventually may have a significant impact in geomorphology, was first used in an analysis of Martian ejecta deposits (Woronow, 1981). In quantitative geomorphological studies the different resolution capabilities of measuring devices, coupled with the different resolutions associated with the physical size of the landform itself, can lead to measurements that are dependent on the measurement scale. For example, if the perimeters of a number of areal features that are identical in plan outline, but of varying size, are measured, the relationship of their perimeters to their (area)$^{0.5}$ will not be linear. This nonlinearity would be the result of including details in the perimeter measurements of the larger features that are too small to be included in the perimeter measurements of the smaller features. This dependency of quantitative measurements on the measuring resolution "can present considerable obstacles to meaningful quantitative description, classification, and modeling of landforms because the data may appear internally inconsistent" (Woronow, 1981, p. 202).

However, the concept of the fractal dimension allows for the interplay of quantitative measurements, size differences, and the resolution of the measuring device. For example, Woronow (1981) used a fractal measure to determine if functional relationships exist among impact craters on Mars. He concluded that the three classes of craters observed on Mars are the result of different physical processes, and are not different simply because of the effects that the variation in their sizes has on our measurement capabilities.

Woronow's analysis was based on a measure derived from a simple perimeter–area relationship. For simple Euclidean shapes the relationship between the area and perimeter is independent of changes in the resolution of the measuring instrument, and can be expressed as

$$\frac{P}{A^{\frac{1}{2}}} = K \tag{1}$$

where P is the perimeter, A is the area, and K is a constant. Woronow developed a similar equation for fractal curves:

$$\frac{P^{\frac{1}{D}}}{A^{\frac{1}{2}}} = K \tag{2}$$

Using nonlinear regression the fractal dimensions D for the three classes of craters were determined (the dimensions were 1.03, 1.08, and 1.27). Woronow (1981) concluded that the fractal dimension captured the crenulation effects, while the value of K was an indication of the overall geometric shape (e.g., circular, elliptical).

There have been relatively few studies of the fractal characteristics of the Earth's land surface. One of the first was by Goodchild (1982), who studied Random Island, off the eastern coast of Newfoundland, and determined the fractal dimensions of the shoreline, selected contours, and lake outlines using three different methods. The three methods produced different D values for the same features, but all methods exhibited the same trend of an increasing D value with increased elevation. Goodchild (1982) stated that the systematic trend in the fractal dimension was due in part to the different geomorphic processes dominating at different elevations. Other studies have since reported similar findings.

For example, Mark and Aronson (1984) studied the fractal geometry of 17 USGS digital elevational models. They found that within any given physiographic province there were usually consistent distances at which the fractal dimension changed, a reflection of the characteristic slope length and structural control of the province. They concluded that both conventional geomorphic wisdom (landscapes have characteristic scales) and the fractal model (geomorphic surfaces are statistically self-similar) are applicable.

In another study, Roy et al. (1987) investigated the fractal nature of a single USGS digital elevational model (DEM) in some detail. They found that although the entire DEM exhibited properties of self-similarity, sections and horizontal and vertical slices — or isosets — of the DEM exhibited differing characteristics. That is, the fractal dimension varied spatially within their study area.

As the above studies illustrate, fractals are most commonly used by geomorphologist in the study of landscape form. The fractal dimension quantifies one aspect of form, but differs from traditional measures in that it provides a description that is independent of the sample. Most other measures of form are sample dependent (Brown and Scholz, 1985; Scholz and Aviles, 1986; Mark, 1975). Scale-invariant measures "are expected to have a fundamental physical significance, since the ensemble average of their spatial means do not depend on the scale (or dimension of space) over which they are averaged" (Lovejoy and Schertzer, 1988, p. 182). Scale-invariant or fractal measures allow extrapolation from the observed scale to unobserved scales (Gilbert, 1987).

However, these statements are somewhat at odds with conventional geomorphological philosophy, in which scale is one of the fundamental concerns. It is through the links between size and form, and function and scale, that geomorphic systems are traditionally characterized (Mark and Aronson, 1984). However, consider Lovejoy's (1982) work on the fractal dimensions of clouds. This work produced results that were counter to the then current theories of mesoscale atmospheric modeling. But Lovejoy's findings were subsequently confirmed and greatly extended (e.g., Lovejoy and Schertzer, 1985; Lovejoy and Mandelbrot, 1985). Thus, it would seem too early yet to reject the use of fractal concepts in the geosciences, even though they presently appear counterintuitive.

In many fields, such as geomorphometrics, quantitative methods are used to investigate the links between form and process. Physical models of the process must contain the scaling and self-similarity properties inherent in the geometry if they are to be truly functional (Scholz and Aviles, 1986; Lovejoy and Mandelbrot, 1985). Knowing that a phenomenon appears fractal (between certain limits) allows us to make certain assumptions about the physical processes that produced the phenomenon. Knowledge of the fractal dimension will help define the types of questions that might produce interesting results when looking for the links between form and process (Kadanoff, 1986). For example, within geomorphology, what is needed is a geomorphologically derived mathematical link between landform and the mechanics of geomorphic processes (Mark and Aronson, 1984). In the absence of such a link, the use of the fractal model as a method of simulation that produces reasonable results becomes a viable alternative.

As an interesting aside, Pentland (1984) asked a number of people unfamiliar with the concept of fractals to subjectively rank a number of images on a roughness scale from 1 (smooth) to 10 (rough). He reported a near perfect correlation between the subjective roughness ranking and the curve's fractal dimension. The fractal dimension of an object thus appears to represent perceptual roughness extremely well. This provides another reason for the intuitive appeal of D as a summary statistic, and also provides support for Mandelbrot's (1982, p. 581) statement that "the basic proof of a stochastic model of nature is in the seeing."

Measuring the Fractal Dimension

The key parameter for fractal analysis is the fractal dimension [which is not necessarily equivalent to the Hausdorff–Besicovitch dimension (Mandelbrot, 1984)]. This value is a real number, which differs from the more familiar Cartesian dimension. The latter is an integer, with a value of one for a line and two for a surface. The fractal dimension varies from one for a straight line to two for a surface to three for an infinitely contorted surface, or any real value in between. A good analogy is to a sheet of paper. A line drawn on a sheet of paper, with Cartesian dimension one, actually is an area of ink covering the surface. As the "wiggliness" of the line increases, so the amount of area that the line occupies increases. When the wiggliness increases enough, the paper appears to be entirely covered with ink, giving the line a dimension better represented by the area of the sheet of paper. Thus real lines lie somewhere between the idealized Cartesian line, with infinitely thin width, and the extreme of complete surface cover. Similarly, for surfaces, the sheet of paper when flat could represent the Cartesian plane with dimension 2. As we bend and crinkle the paper, however, the surface resembles more and more a volume, with dimension 3. Eventually, we can produce a "ball" of paper, best described as a volume.

The fractal dimension, therefore, is a real number, and lies between one and two for a line, and between two and three for a surface. Lines were the first features in the geosciences to have their fractal dimensions computed. Typically the lines were coastlines, contour lines, and cross-sections of terrain. The interest in the latter two types of line resulted from an assertion of Mandelbrot's that any fractal object can be cut to produce a fractal object with a fractal dimension of the original object, minus one. Thus a cross-section should have the fractal dimension of the terrain minus one. Coastlines were the first to have their fractal dimensions measured (e.g., Richardson, 1961; Mandelbrot, 1967; Goodchild, 1982). The dimensions were found to be real values in the range 1.1 to 1.4 (implying that the land surface had a fractal dimension between 2.1 and 2.4).

Two fundamentally different methods have been used to compute the fractal dimension of a line. Of these, the most commonly used is the walking dividers method. This method involves taking a cartographic line and repeatedly measuring its length using "instruments" of different resolution. A good analog is walking a pair of dividers with a given spacing along the line while counting the number of whole and part steps necessary to reach the end. [It should be noted that there is considerable controversy over how to handle the partial steps that remain at the end (see Aviles et al., 1987 and Gilbert, 1987, for example).]

A computer implementation of the algorithm has been presented in Shelberg et al. (1982) and in Kennedy and Lin (1986). For every walk along the line, the distance spacing of the dividers is recorded, as is also the new measured length of the line. A linear regression between the log transforms of these two values is then performed. Since the divider spacing is log transformed, it is best to use values that increase as a power of two so that the observations on the independent variable for the regression are equally spaced. The fractal dimension is then estimated as $1 - b$, where b is the slope of the regression line. Note that b is negative, since as the distance between the dividers increases, more and more of the detail in the line is lost, and the total line length diminishes. The log–log plot of line length versus resolution has become known as the Richardson plot.

The alternative technique to the walking dividers is to examine the distribution of intersegment angles along a line (Eastman, 1985). Intersegment angles are the angles formed between successive triplets of points, that is, pairs of connected line segments. This is one of the few methods that yields the fractal dimension without statistical estimation by regression. The fractal dimension is given by

$$D = \frac{log\ (2)}{log\ (2) + log\left[\dfrac{\bar{c}}{(\bar{a}+\bar{b})^2}\right]^{0.5}} \tag{3}$$

where $\bar{a}$ and $\bar{b}$ are the mean segment lengths to the central point and $\bar{c}$ is the mean distance between the outside two points. Eastman (1985) found the results of this method for cartographic lines to be consistent with the walking dividers method.

Another technique used to determine the fractal dimension uses the area enclosed by a line that closes on itself. This approach is not satisfactory for coastlines or profiles, but has been used extensively for contour loops and the outline shapes of particles and other objects. The most common method using this approach is box counting. In this method the area is gridded and the number of edge grid cells is tabulated. The logarithm of the average number of boundary cells is regressed against the logarithm of the cell size, with the absolute value of the slope of the line yielding the fractal dimension. Cell counting has been used in the analysis of forest fires in Albinet et al. (1986), for contours and shorelines in Goodchild (1980, 1982), for lakes in Hakanson (1978), for radar images of rain storms in Lovejoy et al. (1987), for vegetation patterns in Morse et al. (1985), and for digital elevation models in Shelberg et al. (1983).

Covering the area with circles of different sizes is a variation on the box counting method. In this case, the phenomena are point locations, and the average number of points within a certain distance is tabulated for different distances. The same log–log regression, with a slope giving the fractal dimension, is used. Lovejoy et al. (1986) used the method for meteorological data, and Kagan and Knopoff (1980) used it for earthquake foci.

Finally, at least for fractal dimensions between one and two, several scientists have used the area–perimeter relationship to compute the fractal dimension. For this, values of shape area and shape perimeter are tabulated at different resolutions. The log of the area is regressed against the log of the perimeter of the shape, yielding a gradient that is two over the fractal dimension. These values are simple to calculate, and the method is far less computationally complex than the walking dividers method. Kent and Wong (1982) and Goodchild (1982) used the area–perimeter relationship to determine the dimensions of shorelines and contours, Lovejoy (1982) used it for cloud outlines, Skoda (1987) used it for radar echoes of rain, and Woronow (1981) used it to analyze impact crater distributions on the planet Mars.

However, in spite of this, the walking dividers method remains the most widely used technique. It has been applied to contours, fault line traces (Aviles et al., 1987), the outlines of urban areas (Batty and Longley, 1986), and to a number of analyses of the cartographic line generalization process (e.g., Muller, 1986).

Three techniques have been used to compute the fractal dimension of surfaces. Mark and Aronson (1984) were the first to use the surface's variogram to measure the fractal dimension in their empirical testing of the validity of the fractional Brownian model for terrain. The fractal dimension was estimated from the slope of a log–log plot of the mean-squared elevation difference between points and the distance between the points.

The variogram method has been used for a large number of geoscience variables, including soils and rock surfaces (Burrough, 1981, 1983a), soil acidity (Culling, 1986), and, particularly, for topography in Culling and Datko (1987), Roy et al. (1987), and Klinkenberg (1988). In the case of the variogram, the fractal dimension is $3 - (b/2)$, where b is the slope of the line from the log–log regression. Generally, the fractal dimensions computed from variograms have been high, and occasionally have been summarized as consisting of marked segments with different slopes, implying that different dimensions apply to different scale ranges.

The second method for computing the fractal dimension for surfaces involves the use of the power spectrum. Pentland (1984) and Burrough (1981) used the Fourier power spectrum of the surface and surface cross sections to estimate the fractal dimension. The log of the observed power spectrum is regressed against the log of the spacing [which can be computed from the length of the two-dimensional (2-D) series divided by the harmonic number] as a function of distance in one dimension. Pentland performed this computation for blocks of pixels of decreasing size and used the resulting value as a spatial regionalization technique. Clarke (1988) published a 2-D discrete Fourier method that used the 2-D power spectrum for estimating the fractal dimension, for use in the simulation of topographic texture. The two latter methods share in common a complex measurement process involving enormous numbers of calculations and significant processing time on a large computer.

The final surface based method is the Triangular Prism Surface Area method (Clarke, 1986). This method computes the area of four surface triangles for blocks of four grid cell elevations. After each surface area measurement, the size of the cells is halved, and the surface area calculation is repeated, until the grid cell is the size of the grid resolution. The fractal dimension is then calculated as $2 - b$, where b is the slope of the line from the log–log regression of surface area versus the area of the square. This part of the computation is identical to the 2-D case, however, one is added to the value to reflect the higher geometric dimension. This method has computed very low dimensions for topography, especially where the topography is smooth, yet yields good results for images and large-scale objects such as particles and molecules. The program printed in Clarke's (1986) paper was applied to a small area of topography, although the grid spacing was not accounted for in the computations. This method provides different values for the fractal dimension than for the variogram and the power spectrum.

Clearly, further comparative study of the methods is required, a process that has been hampered by the lack of a good, truly fractal, data set. Discrepancies between the methods may be attributable to the way in which the regression line reacts to the effect of peaks at significant spatial scales. In some cases, a highly significant scale could give the impression of two straight line segments in the regression line, while in others the effect may be to reduce the r^2 value, and therefore the utility, of the regression line.

New methods to determine the fractal dimension, and new applications of the methods, continue to appear, and the active literature has expanded beyond the journals traditionally associated with the geosciences. In general, methods that use cross-sections or individual contour lines to compute the fractal dimension of terrain leave much to be desired. It is virtually impossible to select a representative profile or contour. Directionality and the elevation effect of contours prevent this method from generating valuable results. On maps, mountain peaks are almost always shown as small, circular contour loops, which obviously would not be well represented as fractals. In addition, as indicated above, a comparative study of the various methods, using identical data sets with known fractal parameters, is still lacking.

The generation of fractal data sets is less simple than it seems. Klinkenberg (1988), for example, was unable to achieve satisfactory matches between the surfaces that were generated using cascading mechanisms with specific fractal dimensions (Mandelbrot, 1975) and the fractal dimensions that were calculated using some of the methods previously discussed. In addition to the difficulties of finding suitable data sets, possibilities for misapplication of the techniques exist, and few scientists have published or released as shareware either their algorithms or their computer programs. For example, the r^2 value is intended only as a rough indication of the fit and not as a definitive measure, and fractal dimensions given to four significant digits should be treated as suspect.

Some Empirical Results

A number of empirical studies have been made on the fractal dimensions of natural landscapes. In most studies, the landscapes are placed into groups on the basis of their geographic location (i.e., into physiographic provinces). The purpose of the grouping is to see if insight into the relationship between landscape form and fractal dimension can be obtained if the results are grouped into natural regions.

The division of the United States into natural physiographic regions has been intensively researched (e.g., Hunt, 1967; Thornbury, 1965), and the landforms of each physiographic province reflect their distinctive structural framework. It is expected that traditional morphometric characteristics of 1:24,000 USGS Digital Elevation Models

TABLE 10.1. Summary of the analyses of 58 USGS DEMs, grouped into their respective physiographic provinces.

Physiographic province (number of DEMs analyzed)	Entire-surface variogram D (SD)	Sectional varioigram D (SD)	Angle variogram D (SD)	Alternative dividers method D (SD)	Traditional dividers method D (SD)
Coastal Plain	2.65	2.66	2.62	1.63	1.20
(3)	(0.075)	(0.070)	(0.119)	(0.102)	(0.017)
Blue Ridge	2.32	2.42	2.31	1.23	1.18
(10)	(0.068)	(0.093)	(0.059)	(0.070)	(0.060)
Valley and Ridge	2.36	2.52	2.34	1.32	1.16
(12)	(0.137)	(0.122)	(0.083)	(0.087)	(0.031)
Appalachian Plateaus	2.30	2.38	2.31	1.23	1.18
(14)	(0.080)	(0.125)	(0.074)	(0.088)	(0.024)
Great Plains	2.35	2.40	2.34	1.40	1.18
(2)	(0.092)	(0.107)	(0.048)	(0.102)	(0.014)
Interior Low Plateau	2.41	2.45	2.42	1.28	1.19
(3)	(0.208)	(0.136)	(0.083)	(0.130)	(0.013)
Rocky Mountains	2.28	2.35	2.28	1.24	1.12
(1)	(n.a.)	(0.194)	(0.059)	(0.037)	(0.021)
Colorado Plateau	2.34	2.32	2.31	1.21	1.18
(4)	(0.024)	(0.088)	(0.010)	(0.085)	(0.031)
Basin and Range	2.29	2.33	2.25	1.21	1.13
(10)	(0.087)	(0.128)	(0.078)	(0.043)	(0.035)

Source: Klinkenberg (1988).

(DEMs) (Elassal and Caruso, 1983) within each province would be relatively consistent; it remains to be seen whether their fractal characteristics are consistent also. However, as the fractal dimension has been found to be closely tied to visual qualities of form (Pentland, 1984), some agreement is expected. Klinkenberg (1988) analyzed 58 DEMs, which together represent 9 of the 24 physiographic provinces that make up the coterminous United States, and a summary of these results follow.

The methods used in Klinkenberg (1988) to determine the fractal dimensions are as follows. The variogram method was applied to a single 256 by 256 cell subarea of each DEM in three different ways: (1) the entire 256 by 256 cell subarea was analyzed, (2) the four quarters of the subarea were individually analyzed (referred to as the sectional variogram results), and (3) the entire subarea samples were placed into one of six angular classes, and the variograms for each angular class analyzed. (The angular swaths were each 30° wide — the symmetry allows for the 180° to 360° values to be mapped into the 0° to 180° values.) The walking dividers method was applied to the individual contours, and the average dimension determined for each contour level (i.e., for the five horizontal cross sections analyzed, the average dimension was determined).

A variation of the dividers method — in which the relationship between the step size and the contour line is reversed — was also used. In the traditional implementation, the step size changes while the scale of the data remains constant. In this alternative implementation, the scale of the data varied while the step size remained constant, and the total length of all contour lines for a given elevation was the dependent variable, not the length of the each individual contour. The "step size" for a given scale was equated to the average chord length (total length of all contour lines at that elevation divided by the total number of coordinate pairs). This alternative implementation of the dividers method will be referred to as the surficial contours method.

Apparently Klinkenberg's analyses produced results that differ according to the physiographic region (Table 10.1). The results also show considerable method-produced variation (e.g., the sectional variograms generally produce dimensions consistently different from the other variogram methods). The tabulated dimensions for the variogram methods are those associated with the first straight-line segment present in the variogram plot

TABLE 10.2. Average dimensions obtained from the first and second linear segments from the different variogram methods applied to 58 DEMs.

	Mean / standard deviation of D	
Variogram method	First linear segment	Second linear segment
Entire surface	2.34 / 0.123	2.66 / 0.220
Sections	2.42 / 0.133[a]	2.68 / 0.167
Angles	2.33 / 0.101	2.70 / 0.080

Source: Klinkenberg (1988).
[a]Significantly different from the other two first linear segment values, using a standard t test (0.05 level).

only. The average distance to which this dimension applies is 5.2 km for the entire surface results, 2.9 km for the sectional results, and 4.0 km for the angle variogram results. (A minimum distance of 30 m was set by the spacing of the DEM.) The averages of the dimensions produced by the second linear segment (if present) are not significantly different (Table 10.2) (also see Roy et al., 1987). Note, however, that only 34% of the 58 entire surface variograms exhibited more than one definite linear segment in the variogram plot, 32% of the 348 angle variogram plots exhibited more than one definite linear segment, and only 15% of the 232 sectional variogram plots exhibited more than one definite linear segment.

The patterns in the variances of the dimensions produced by the various methods used in Klinkenberg (1988) were distinct. As the average dimension of the DEM subarea increased, the sectional variograms were more likely to produce more similar dimensions for the four sections, whereas the contour-based methods were more likely to produce dissimilar dimensions for each contour level. Thus, the rougher the landscape the more similar individual sections of it appeared, whereas horizontal cross sections appeared more dissimilar. It was also found that surfaces with higher fractal dimensions appeared slightly more anisotropic than surfaces with lower fractal dimensions. This result is contrary to expectations, because regional (isotropic) trends in the land surface should be more apparent on lower dimensional surfaces.

The walking dividers method produced dimensions very different from those produced by the other methods (Table 10.1). The constancy and consistent low values produced by the dividers method mimic the results reported in other studies (e.g., Muller, 1986).

Klinkenberg (1988) also took the analyses one step further. To see how homogeneous the grouped DEM subareas were—from a traditional morphometric point of view—a discriminant analysis was performed using four traditional morphometric parameters [mean elevation, the standard deviations and kurtosis of the elevations, and a coefficient of dissection (Strahler, 1952)]. Fifty (86%) of the DEM subsamples were assigned to their correct provinces; four of the misassigned DEM subsamples occurred on the borders between two physiographic provinces. Then, using the fractal dimension obtained from the entire surface variogram as the dependent variable, and the physiographic region as the categorical variable, a one-way analysis of variance was used to test whether the groups were distinguishable solely on the basis of their fractal dimensions. The results indicate that it is possible to distinguish the physiographic provinces on the basis of their fractal dimensions ($p < 0.01$) (Klinkenberg, 1988).

Mark and Aronson (1984) analyzed 17 DEMs using the variogram method. Klinkenberg analyzed the same DEMs using a slightly different implementation of the variogram method. A comparison of these two sets of results reveals that about half were in close agreement, with differences in the dimensions generally less than 0.10. Some of the other results, however, were substantially different, with differences in the dimensions of up to 0.36. The methodology used by Mark and Aronson in their study might explain some of the differences in the results.

Mark and Aronson determined the breaks in the slopes by visual inspection, the "best fitting" lines were hand drawn, and the slopes determined by careful measurement. In the study by Klinkenberg (1988), the breaks in the slopes were determined visually using an interactive, computer-based procedure, and least-squares regression was used to determine the slopes of the lines. Therefore, it is not surprising that some differences in the values of the slopes, and thus in the dimensions, were reported.

Some of the conclusions that Mark and Aronson (1984) made can be tempered somewhat when considered in light of the results presented in Klinkenberg (1988). Because most of the DEMs they investigated had poor fits to the fractal model, their conclusions were mixed. When considered in light of the larger number of DEMs analyzed by Klinkenberg, it can be seen that the fractal model does provide a reasonable fit for a majority of the cases. Furthermore, consideration of the nature of the physiographic province in which the DEM resided provides a reasonable explanation for the lack of fit in those cases where the fractal model was found to be deficient.

An analysis of the sensitivity of the variogram method was performed by one of the authors (Clarke) using a DEM from Copake, New York. The number of distance classes, and the number of cases within each class, was allowed to range from a low of 5 to a high of 30. The r^2 value associated with the variogram regression line ranged from a low of 0.76 to a high of 0.96, while the value of D concurrently ranged from a low of 2.36 to a high of 2.74. This illustrates the sensitivity of the variogram method to the distribution of the data. However, Klinkenberg found that when using a constant number of distance classes (and cases within each class) the variogram method produced consistent results.

In another study, Roy et al. (1987) analyzed one DEM in the White Mountains of New Hampshire. Using the variogram method they obtained an overall surface dimension of 2.15, a dimension that applied up to a distance lag of 2.0 km. At greater distances the dimension was much higher, $D = 2.82$. The average dimensions of the derived contours, obtained by the dividers technique, was found to be 1.09, with a range from 1.01 to 1.28. A similar range of results was also present in the analyses by Klinkenberg (Tables 10.1 and 10.2).

In their study Roy et al. (1987) also found that the fractal dimension generally decreased with elevation. They reasoned that "the crenulations associated with fluvial erosion" decrease with increasing elevation (Roy et al., 1987, p. 75). Klinkenberg

(1988) also found a general decrease in dimension with elevation in 58% of the cases, although the relationship was found to be fairly weak. However, the opposite behavior was sometimes observed—the fractal dimension sometimes increased with increased elevation. This was also found to be the case on Random Island by Goodchild (1982). Interestingly, 5 of the 10 DEM subareas that exhibited the opposite behavior are spatially adjacent to each other, although in different physiographic provinces. This raises the possibility that there might be some physical reason for the different relationship between the dimension and elevation.

One of the DEMs analyzed by Klinkenberg was previously analyzed by Steyn and Ayotte (1985) using a standard 2-D discrete Fourier transform software package—although their work was not explicitly concerned with fractals. In their study Steyn and Ayotte were particularly concerned with the degree of directionality in the terrain. Figure 10.1b in Steyn and Ayotte's paper illustrates that the amplitude spectrum shows marked asymmetry in the northeast to southwest direction—a finding that corresponds to the direction of anisotropy identified by the directional variograms applied to the same DEM by Klinkenberg.

It is possible to transform the values from Steyn and Ayotte's (1985) study into dimensional values. Although the overall average fractal dimensions obtained from the two studies agree fairly well, the individual directional dimensions do not (Table 10.3). However, if the averages of the values obtained from the three directional variograms that exhibited two definite linear sections in their variograms are computed (2.47, 2.54, 2.54, respectively), then the average dimensions agree very closely.

Steyn and Ayotte (1985, p. 2887) noted that variograms "seem more capable of detecting the scale breaks in topographies" than Fourier-based techniques; the limited results presented in Table 10.3 support that claim. Based on these limited results, it would appear that spectral analysis, even when considered on a directional basis, produces values that are spatial averages. It should be noted that Steyn and Ayotte explicitly mention that they looked for evidence of multiple spectral domains in their analyses. Furthermore, Clarke (1988) found that Fourier analysis responded more to the finer

TABLE 10.3. Comparison of the angle variogram dimensions with those obtained from a Fourier-based technique, for the Blairs Mills, PA, DEM.[a]

Direction (in degrees, from north going east)	Angle variogram D		Fourier transform D
	First linear segment	Second linear segment	
0–30	2.35	–	2.47
30–60	2.47	–	2.27
60–90	2.33	–	2.65
90–120	2.19	2.76	2.47
120–150	2.21	2.87	2.52
150–180	2.22	2.87	2.55
Average (entire surface)	2.39		2.45

Sources: Klinkenberg (1988) and Steyn and Ayotte (1985).
[a]Only 180° are reported, due to the symmetric nature of the relationship.

texture in the landscape than to the coarser structure, and consistently returned a much higher dimension than did a variogram analysis on the same land surface (also see Culling and Datko, 1987).

The application of the fractal model to a number of DEMs from a range of physiographic provinces produced a range of results—from instances where the model fit the landscape well, to instances where the model fails to fit the landscape at all. Nonetheless, the use of the fractal dimension as a morphometric parameter was found to provide some discriminating power. However, the various techniques used to determine the fractal dimension produced a wide range of values, with the variogram methods providing the most reasonably consistent results. The dividers method applied to individual contours did not provide meaningful results. The dimensions determined by this method did not vary even though a wide variety of landscape types were looked at. The surficial contours methods produced results that lie between the individual contours results and the variogram results.

Conclusion

The application of fractal geometry to natural landscapes has achieved mixed results. Nonetheless, the fractal model appears to be an acceptable model, within limitations, and as a morphometric parameter, the fractal dimension should prove to be very useful. As Hobson (1972) notes, a morphometric parameter should be conceptually descrip-

tive, easily measurable and suitable at a variety of scales—the fractal dimension satisfies all three of Hobson's concerns.

To determine if fractals and land surfaces are truly compatible, more varied analyses should be performed, including examining a broader range of scales and grouping adjacent DEMs to allow for analyses at much larger scales. In addition, because of the sensitivity of the measurement methods, the algorithms used to implement the various techniques, and the exact details of any analyses, should be presented in the literature. Furthermore, a stricter geomorphological approach should be used in analyzing the fractal nature of the land surface.

Conventional geomorphology considers the drainage basin as the fundamental geomorphic unit, and many of the analyses of the land surface appear to indicate a relationship between fluvial processes and the fractal dimension (Roy et al., 1987; Klinkenberg, 1988). Fluvial processes act on the fractal dimension of the land surface in two conflicting ways. On landscapes with high fractal dimensions, erosion will tend to decrease the dimension. Conversely, on landscapes with an initially low fractal dimension, erosional processes will tend to increase the dimension. Thus, a landscape may be the result of a number of processes, with each process producing a specific fractal domain. Poor results may be caused by working in different dimensional regimes associated with drainage basins of different evolutionary ages. Removing the confounding effects of working across drainage basins may produce very different results.

Methods currently used to generate simulated fractal landscapes produce reasonable visual approximations of natural landscapes. However, natural land surfaces contain scale-dependent features that most simulation methods fail to model (e.g., Mandelbrot, 1975). The combined approach to landscape simulation used by Clarke (1988) would appear to be an appropriate one to take. Clarke suggested using spectral analysis to model the scale-dependent aspects of the landscape (as also suggested in Fox and Hayes, 1985), and a fractal model, such as the one suggested in Mandelbrot (1975), to simulate the scale-independent aspects. Realistic simulation models — which capture both the scale-free and scale-dependent aspects of natural landscapes — would prove useful in a wide variety of fields, such as mesoscale climatic modeling (Skoda, 1987), drainage simulation (Goodchild et al., 1985), and the building of error models in cartography. With models such as these, we will truly be able to explore the peaks and valleys of our synthetic mountains, and never have to leave our office.

References

Albinet, G., Searby, G., and Stauffer, D., 1986, Fire propagation in a 2-D random medium: J. Phys. 47, 1–7.

Aviles, C.A., Scholz, C.H., and Boatwright, J., 1987, Fractal analysis applied to characteristic segments of the San Andreas fault: J. Geophys. Res. 92(B1), 331–344.

Batty, M., and Longley, P., 1986, The fractal simulation of urban structure: Environ. Planning A 18, 1143–1179.

Brown, S.R., and Scholz, C.H., 1985, Broad bandwidth study of the topography of natural rock surfaces: J. Geophys. Res. 90, 12,575–12,582.

Burrough, P.A., 1981, Fractal dimensions of landscapes and other environmental data: Nature (London) 294, 240–242.

Burrough, P.A., 1983a, The application of fractal ideas to geophysical phenomena: paper presented at a conference organized by the Institute of Mathematics and Its Applications, 8 June 1983.

Burrough, P.A., 1983b, Multiscale sources of spatial variation in soil. I. The application of fractal concepts to nested levels of soil variation: J. Soil Sci. 34, 577–597.

Burrough, P.A., 1983c, Multiscale sources of spatial variation in soil. II. A non-Brownian fractal model and its application in soil survey: J. Soil Sci. 34, 599–620.

Clarke, K.C., 1986, Computation of the fractal dimension of topographic surfaces using the triangular prism surface area method: Comput. Geosci. 12, 713–722.

Clarke, K.C., 1988, Scale-based simulation of topographic relief: Am. Cartog. 15, 173–181.

Culling, W.E.H., 1986, Highly erratic spatial variability of soil-pH on Iping Common, West Sussex: Catena 13, 81–89.

Culling, W.E.H., and Datko, M., 1987, The fractal geometry of the soil-covered landscape: Earth Surface Processes Landforms 12, 369–385.

Eastman, J.R., 1985, Single-pass measurement of the fractal dimension of digitized cartographic lines: paper presented at the annual conference of the Canadian Cartographic Association, 1985.

Elassal, A.A., and Caruso, V.M., 1983, USGS digital cartographic standards: Digital Elevation Models, U.S. Geological Survey Circular 895-B.

Fox, C.G., and Hayes, D.E., 1985, Quantitative methods for analyzing the roughness of the seafloor: Rev. Geophys. 23(1), 1–48.

Gilbert, L.E., 1987, Are topographic data sets fractal?: paper presented at MGUS-87, Redwood City, CA.

Goodchild, M.F., 1980, Fractals and the accuracy of geographical measures: Math. Geol. 12, 85–98.

Goodchild, M.F., 1982, The fractional Brownian process as a terrain simulation model: Model. Simulation 13, 1133–1137.

Goodchild, M.F., and Mark, D.M., 1987, The fractal nature of geographic phenomena: Ann. AAG 77, 265–278.

Goodchild, M.F., Klinkenberg, B., Glieca, M., and Hasan, M., 1985, Statistics of hydrologic networks on fractional Brownian surfaces: Model. Simulation 16, 317–323.

Hakanson, L., 1978, The length of closed geomorphic lines: Math. Geol. 10, 141–167.

Hobson, R.D., 1972, Surface roughness in topography: A quantitative approach: in Spatial Analysis in Geomorphology, R.J. Chorley, (Ed.): Methuen, London, pp. 221–245.

Hunt, C.B., 1967, Physiography of the United States: Freeman, San Francisco.

Kagan, Y.Y., and Knopoff, L., 1980, Spatial distribution of earthquakes: The two-point correlation function: Geophys. J. R. Astron. Soc. 62, 303–320.

Kadanoff, L.P., 1986, Fractals, where's the physics?: Phys. Today, February, 1986.

Kennedy, S.K., and Lin, W.-H., 1986, Fract—A fortran subroutine to calculate the variables necessary to determine the fractal dimension of closed forms: Comput. Geosci. 12, 705–712.

Kent, C., and Wong, J., 1982, An index of littoral zone complexity and its measurement: Can. J. Fish Aquatic Sci. 39, 847–853.

Klinkenberg, B., 1988, Tests of a fractal model of topography: Ph.D. thesis, University of Western Ontario.

Lovejoy, S., 1982, Area-perimeter relation for rain and cloud areas: Science 216, 185–187.

Lovejoy, S., and Mandelbrot, B.B., 1985, Fractal properties of rain, and a fractal model: Tellus 37(A), 209–232.

Lovejoy, S., and Schertzer, D., 1985, Generalized scale invariance in the atmosphere and fractal models of rain: Water Resources Res. 21(8), 1233–1250.

Lovejoy, S., and Schertzer, D., 1986, Scale invariance, symmetries, fractals, and stochastic simulations of atmospheric phenomena: Bull. Am. Meteorol. Soc. 67, 21–32.

Lovejoy, S., and Schertzer, D., 1988, Extreme variability, scaling and fractals in remote sensing: analysis and simulation: in Digital Image Processing in Remote Sensing, J.P. Muler, (Ed.): Taylor & Francis, pp. 177–212.

Lovejoy, S., Schertzer, D., and Ladoy, P., 1986, Fractal characterisation of inhomogeneous geophysical measuring networks: Nature (London) 319, 43–44.

Lovejoy, S., Schertzer, D. and Tsonis, A.A., 1987, Functional box-counting and multiple elliptical dimensions in rain: Science 235, 1036–1038.

Mandelbrot, B.B., 1967, How long is the coast of Britain? Statistical self-similarity and fractional dimension: Science 156, 636–638.

Mandelbrot, B.B., 1975, Stochastic models for the Earth's relief, the shape and the fractal dimension of coastlines, and the number-area rule for islands: Proc. Nal. Acad. Sci. U.S.A. 72, 3825–3828.

Mandelbrot, B.B., 1977, Fractals: Form, Chance and Dimension: Freeman, San Francisco.

Mandelbrot, B.B., 1982. The Fractal Geometry of Nature: Freeman, San Francisco.

Mandelbrot, B.B., 1984, Fractals in physics: Squig clusters, diffusions, fractal measures, and the unicity of fractal dimensionality: J. Stat. Phys. 34, 895–929.

Mark, D.M., 1975, Geomorphometric parameters: A review and evaluation: Geograf. Ann. Ser. A 3, 165–177.

Mark, D.M., and Aronson, P.B., 1984, Scale-dependent fractal dimensions of topographic surfaces: An empirical investigation, with applications in geomorphology: Math. Geol. 16, 671–683.

Morse, D.R., Lawton, J.H., Dodson, M.M., and Williamson., M.H., 1985, Fractal dimension of vegetation and the distribution of arthropod body lengths: Nature (London) 314, 731–733.

Muller, J.-C., 1986, Fractal dimension and inconsistencies in cartographic line representations: Cartog. J. 23, 123–130.

Pentland, A.P., 1984, Fractal-based description of natural scenes: IFFE PAMI 6, 661–674.

Peitgen, H.-O., and Saupe, D., Eds., 1988, The Science of Fractal Images: Springer-Verlag, Berlin.

Richardson, L.F., 1961, The problem of contiguity: An appendix to 'Statistics of deadly quarrels': General Sys. Yearbook 6, 139–187.

Roy, A., Gravel, G., and Gauthier, C., 1987, Measuring the dimension of surfaces: A review and appraisal of different methods: Proc. Auto-Carto Eight, 68–77.

Scholz, C.H., and Aviles, C.A., 1986, The fractal geometry of faults and faulting: Earthquake Source Mechanics (Geophysical Monograph 37), M. Ewing Ser. 6.

Shelberg, M.C., Moellering, H., and Lam, N.S., 1982, Measuring the fractal dimensions of empirical cartographic curves: Proc. Auto-Carto Five, 481–490.

Shelberg, M.C., Moellering, H., and Lam, N.S., 1983, Measuring the fractal dimensions of surfaces: Proc. Auto-Carto Six, 319–328.

Skoda, G., 1987, Fractal dimension of rainbands over hilly terrain: Meteorol. Atmos. Phys. 36, 74–82.

Steyn, D.G., and Ayotte, K.W., 1985, Application of two-dimension terrain height spectra to mesoscale modeling: J. Atmos. Sci. 42, 2884–2887.

Strahler, A.N., 1952, Hypsometric (area-altitude) analysis of erosional topography: Bull. Geol. Soc. Am. 63, 1117–1142.

Thornbury, W.D., 1965, Regional Geomorphology of the United States: John Wiley, New York.

Woronow, A., 1981, Morphometric consistency with the Hausdorff-Besicovitch dimension: Math. Geol. 13, 201–216.

11
Image Analysis of Particle Shape

John Starkey and Sandra Rutherford

Image analysis can provide data on particle shape, size, orientation, and distribution from a digitized, two-dimensional image. Data from several sections through a specimen may be combined for a three-dimensional analysis.

Digitization can be accomplished using a variety of devices, which fall into two general categories, scanners and tablets. Scanners, such as microdensitometers, video cameras, and, more recently, image scanners, divide the image into a matrix of pixels and allocate a gray level to each. Tablets come in a variety of forms, which depend on the method of detection used to locate the cursor or stylus; the image is usually represented by a one-dimensional array of x–y coordinates.

The algorithms required to identify the particle boundaries depend on which digitizing device is selected. Where a scanner is used the perimeter of the particle must be extracted from the pixel matrix by comparing adjacent pixels. In contrast, because digitizing tablets are interactive, the x–y coordinates of the perimeter of the particle are obtained sequentially as the operator follows the boundary with the cursor.

Geological applications of particle shape analysis often require that the shapes be described by fitting geometric forms to the particles. The best-fit ellipse provides a meaningful description and is amenable to further quantitative analysis.

Introduction

Image analysis techniques may be applied to geological studies of particle shapes to collect data automatically. The parameters that are commonly determined include size, center of gravity, long dimension and width at right angles to it, perimeter length, and orientation of the long dimension. The three basic steps of image analysis are digitization, the conversion of images from a continuous form to a discrete form; enhancement and restoration, the improvement of degraded or noisy images; and segmentation and description, the conversion of images into simplified "maps" to enable measurements of properties of image segments (Rosenfeld and Kak, 1976).

The following discussion relates primarily to the first and third of these steps and is illustrated using examples from the study of the shapes of mineral grains in rocks. However, the techniques can be extended to other applications where similar problems arise.

An image analysis system was described previously (Starkey and Simigian, 1987) in which a microdensitometer was used to digitize the particles and a mainframe computer was used to analyze the digital data. This system was subsequently modified for use with a microcomputer and a tablet-type digitizer (Simigian and Starkey, 1989). The image analysis techniques described below, and their application, are illustrated using these earlier techniques, which have been further modified for use with a video camera.

Digitization

Digitization is the necessary first step in all image analysis applications. The digitizing devices can be categorized as either scanners or tablets. Scanners determine the light intensity at discrete intervals on

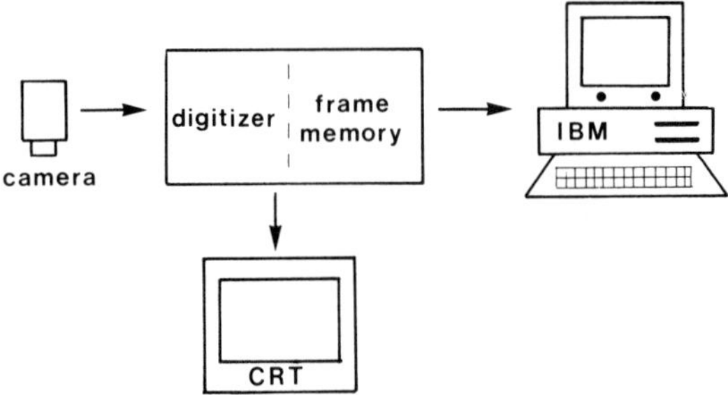

FIGURE 11.1. A schematic diagram of the video camera, frame grabber, CRT, and microcomputer.

a raster, which is usually square, and the data are sent to the computer as values indicating levels of gray. Each value, or picture element, is termed a "pixel." In contrast, digitizing tablets record the x–y coordinates of points selected by the operator either continuously or at selected intervals.

Scanners include such devices as microdensitometers, flying spot scanners, video cameras, flatbed scanners, and, more recently, image scanners (Fabbri, 1984). The optical microdensitometer used in the system described by Starkey and Simigian (1987) allows the pixel size and the grid spacing of the raster to be set at 25, 50, 100, or 200 μm. The number of gray levels that are detectable is 256. Therefore, the microdensitometer has high resolution. The digitized information from the microdensitometer is written onto a magnetic tape. There is no provision for any operator control and storing the data on magnetic tape to be transferred to the main frame computer is time consuming and inconvenient.

These shortcomings are alleviated by using a video camera in conjunction with a frame grabber board installed in a microcomputer. The frame grabber stores the gray levels from the video camera as digital data in frame memory, which is located on the board (Figure 11.1). The operator is able to modify the image on the video monitor prior to capture by the grabber board by manipulating the light levels. The captured image is either processed in the microcomputer memory, using appropriate software, or the digitized image is written to disc.

The resolution of the system is determined by the specifications of the video camera and the frame grabber board. The system used here consists of a Sony AVC-D5 CCD camera and an Imaging Technology Incorporated PCVISION *plus* frame grabber board. The image raster is 512 by 480 pixels. The correspondence between a pixel and the image depends on the magnification of the camera. The video camera is capable of discriminating 256 gray levels, however, only 128 gray levels were used in this application, with white equal to 128 and black equal to 0. The image is represented in the computer as a two-dimensional matrix of numbers representing the individual gray levels, or light intensity values, associated with a discrete area or pixel (Figure 11.2). The location of data within the raster corresponds to the x–y coordinates of the pixel in the image. The different phases present in the image are characterized by specific gray levels or discrete ranges of gray levels. Thus, one can identify black objects, or white objects, or a variety of grey objects by sets of pixels with the corresponding gray level.

Digitizing tablets are fundamentally quite different from scanners, they record x–y coordinates directly and do not identify the phases present. They use several techniques to locate the x–y positions of a stylus or hand-held cursor on the tablet. The most common type of detector is the antenna-transmitter; others use sonic-pulse transmission or a potentiometer (Stanton, 1987). The digitizer

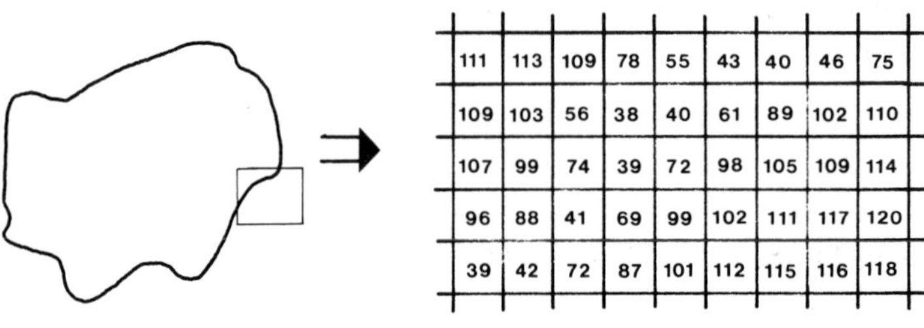

FIGURE 11.2. A schematic diagram of a square matrix of gray level values illustrating a portion of a particle boundary. White is equal to 128 and black is 0.

used by Simigian and Starkey (1989) uses an optical encoder that does not require that the stylus be used in conjunction with a special tablet surface. Therefore, a rock slab or thick photograph can be digitized. The interactive nature of digitizing tablets requires the operator to select what is to be digitized therefore producing a one-dimensional array of x–y coordinates associated with an individual particle boundary. The array of x–y coordinates can be collected either in stream mode, by following the perimeter of the particle, or in point mode, by selecting discrete points on the perimeter. The phase being digitized has to be recorded by the operator.

Image Manipulation

The method of identification of particle boundaries differs depending on whether the data are obtained from a tablet or a scanner. Where a tablet is used the boundaries of the particles are followed by the cursor and the consecutive stream of x–y coordinates between the initial and final positions on the particle boundary delineate the boundary without further manipulation.

Where the data are obtained from a scanner the particle boundaries must be identified within the matrix of gray levels. This requires that there be detectable differences in the gray levels between adjacent particles, which may not always be the case. Figure 11.3a illustrates a photomicrograph of a thin section of quartz grains viewed between crossed polarized light in a petrographic microscope. A few grains can be identified by differing gray levels. However, on rotation of the thin section (Figure 11.3b) many more grains appear. In such situations some form of image enhancement is required. Even where the difference in gray level between adjacent grains may appear imperceptible, as in Figure 11.3a, the difference may still be sufficient for the boundary to be enhanced by edge finding techniques (Niblack, 1986). However, particularly when dealing with thin sections under the petrographic microscope, there is often insufficient contrast across at least some boundaries for edges to be detected. Therefore, it has proved necessary to trace the particle boundaries of the specimen to produce a line drawing.

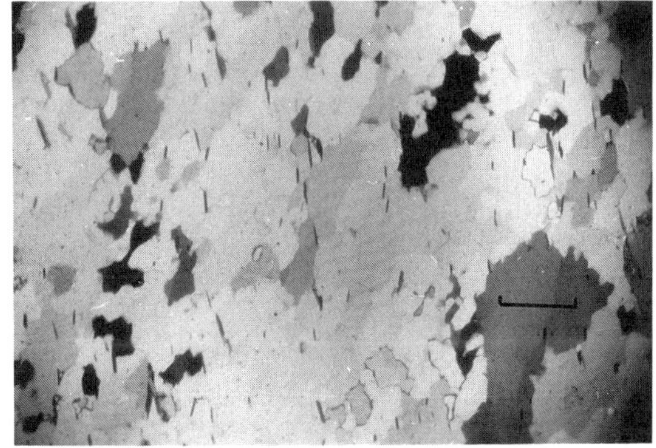

a

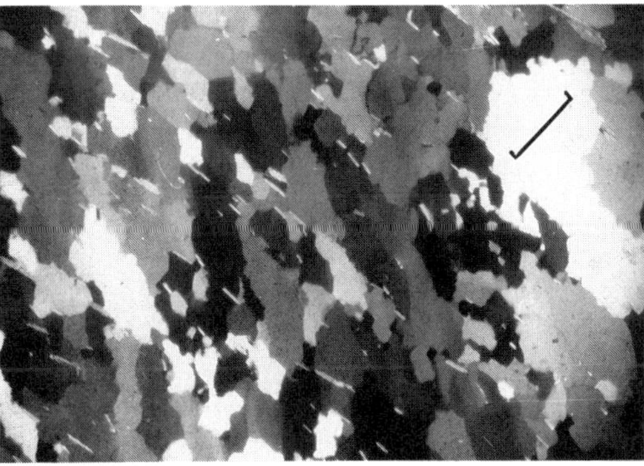

b

FIGURE 11.3. (a) A photomicrograph of deformed quartz grains. The scale bar is equal to 0.5 mm on the specimen. (b) A photomicrograph of the deformed quartz grains rotated 45°. Note the increased number of grains observed. The scale bar is equal to 0.5 mm on the specimen.

Fabbri (1984) delineated the desired particle boundaries by projecting an image onto a sheet of paper and tracing the boundaries by hand; he subsequently photographed the traced image and used the photographic negative for analysis. Starkey and Simigian (1987) prepared photographs of the specimen and traced the particle boundaries onto a transparent acetate film, or tracing paper, and analyzed the tracing directly. Alternatively, the image may be traced on the video monitor by the operator directly from a camera mounted on the microscope (Allard and Benn, 1989; Lapique, et al., 1988).

Analyzing an image that has been digitized using a scanner requires that the particle boundaries be located within the matrix of gray values. Where the gray values of individual particles are distinct, the change from one gray level to another is sufficient to locate the boundary. In practice, this is not a

FIGURE 11.4. A line tracing of Figure 11.3b enhancing the grain boundaries. The scale bar is equal to 0.5 mm on the specimen.

single step function, since the square areas of the image represented by the pixels are occupied by varying proportions of the two particles on either side of the boundary. Therefore, there is usually a change in gray level from one particle to the next over one or more pixels. This requires that a threshold gray level be selected that separates the two particles. The value of the threshold may change along a particle boundary as particles that are characterized by different gray levels occur in juxtaposition with the particle whose boundary is being determined.

The problem of identifying particle boundaries is somewhat different where the boundaries of particles are enhanced by tracing and the resulting line drawing is digitized. Here, although the line drawing consists only of black lines on a white background (Figure 11.4), the digitized image still contains a range of gray levels because the square area of a pixel may contain any proportion of black line

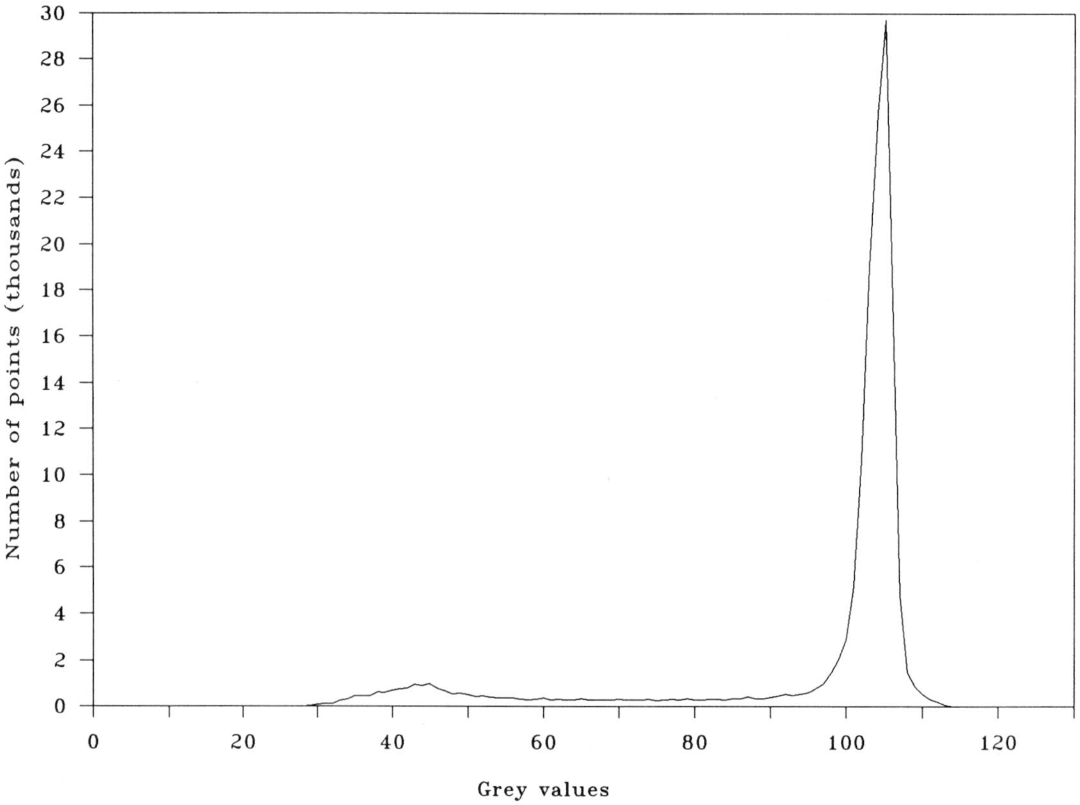

FIGURE 11.5. A frequency distribution of gray level values for the entire image shown in Figure 11.4. The threshold was found to be 95.

FIGURE 11.6. A computer plot of the line traced grains from Figure 11.4.

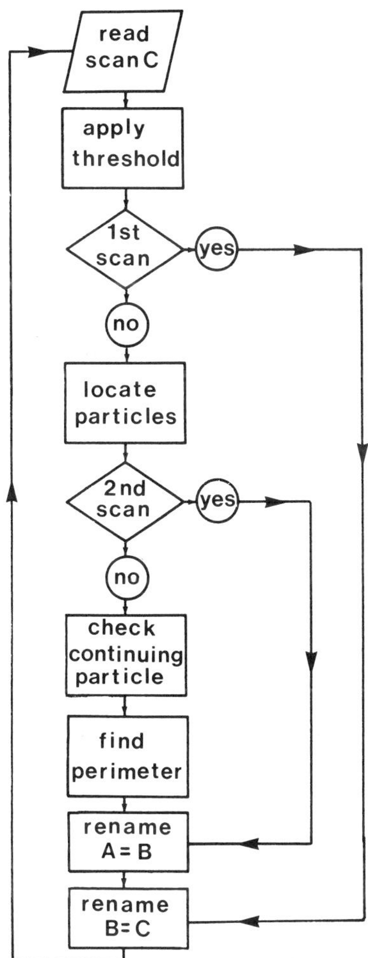

FIGURE 11.7. A schematic flowchart illustrating the reading, storing, and processing of matrix scans.

and white background. A typical bimodal frequency distribution of gray levels obtained from such a line drawing is shown in Figure 11.5. The gray levels corresponding to the black lines are responsible for the small peak and those corresponding to the white background produce the larger peak. The analysis of such an image requires that a threshold value be chosen that separates the two modes; the values above and below the threshold are set to one and zero, respectively, which reduces the matrix data to binary form.

In practice, the difference between the maximum gray level value and the mode of all the values of the image is subtracted from the mode to obtain a suitable threshold value, in this case 95 (Figure 11.5). The mode of all the pixel values is close to the mode of the larger white peak, since most of the values are due to the background. The difference between the mode of all pixel values and the maximum value slightly overestimates the half width of the peak due to background values. Setting the threshold at the overall mode minus this difference truncates the data below the background peak.

The particle boundaries are located within the binary matrix using the method previously published in Starkey and Simigian (1987), which is designed to conserve computer memory. The boundaries determined from Figure 11.3b are shown in Figure 11.6. At any one time, only three rows of the matrix, corresponding to three scans of data, are processed. The matrix rows residing in memory are labeled A, B, and C. Figure 11.7 is a flowchart of the process by which the scans are processed. The data of the first scan are read into row C and the threshold value is applied, the values above the threshold, which are those which approach white, are set to 1 and those below are set to 0. Row C is then stored as B and the data of the second scan are read into row C. Since B is the first row of the matrix all particles are considered to be new; therefore B is analyzed pixel by pixel and each new particle, which is represented by a string of 1's separated by 0's, is located and a unique sequential label is assigned. Row B is then stored as A and row C is stored as B. The third scan is read into row C and the threshold applied, particles are

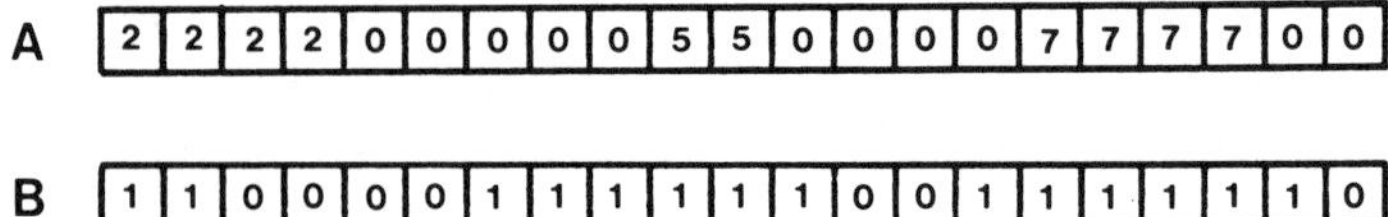

FIGURE 11.8. A schematic diagram of a portion of row A and row B. Note row B is binary with the particles shown as 1's separated by 0's. The identification numbers corresponding to those particles are shown in row A.

located in B and the continuation of particles from the row A is recognized by comparing B with A (Figure 11.8). The identification label for each particle found in A is used to label the continuation of the particle in B. Row B is then stored as A and C is stored as B and a new scan is read into row C. The effect of this procedure is to replace the 1's that fill each particle with an identification number.

Once the process of storing and analyzing three rows of the matrix is initiated the perimeter pixels of a particle are identified by considering the three scans together. A perimeter pixel is one that is located in row B and has one or more of the four immediately adjacent pixels occupied by a zero (Figure 11.9). The y coordinates of the perimeter pixels are determined from the number of the row of the matrix in which they occur; the x coordinates are indicated by the position of the pixels within the row. The identification number assigned to the particle is used to label the x and y coordinates as they are stored. The process of locating particles and storing scans continues until all the rows have been read.

Geometry

Algorithms other than those required to obtain particle boundaries may also differ depending on which input device is selected. For instance, the algorithms to calculate the center of gravity of a particle and the area can be identical if the x–y

coordinates of the particle boundaries are used; however, if a scanner is used simpler algorithms can be applied that do not require extra manipulation of the data.

As the data in the scanned image are analyzed to locate particle boundaries, by the method described above, the area and center of gravity of the particle are determined simultaneously. The number of pixels of a particular particle in successive B rows are summed to derive the area. The x and y coordinates of each of these pixels are also summed. Once the entire perimeter of the particle has been determined the summed x and y coordinates are divided by the area to arrive at the x coordinate and y coordinate of the center of gravity.

As previously noted, the tablet-type digitizer obtains the x–y coordinates of individual particle boundaries directly. Therefore, there is no information about the interior of the particle and the area and center of gravity must be calculated by integration over the area of the particle (Tough and Miles, 1984). Figure 11.10 is a schematic diagram of a particle with perimeter points numbered 1 through N, which is overlain with cartesian coordinate axes. Note that the perimeter points are unevenly spaced and that the particle is divided into $N-1$ polygons by constructing lines through each perimeter point perpendicular to the x axis. The area of each polygon is derived from the following formula:

$$\text{Area} = \text{base} \times \text{height}$$

where base is the distance between two consecutive points on the perimeter in the x direction and height is the distance between the x axis and the midpoint between the two perimeter points. Therefore, if the coordinates of point 1 are X_k, Y_k, and of point 2 are X_{k+1}, Y_{k+1}, the area for the polygon defined by those points is

$$\text{Area} = \tfrac{1}{2}(X_{k+1} - X_k) \times (Y_{k+1} + Y_k)$$

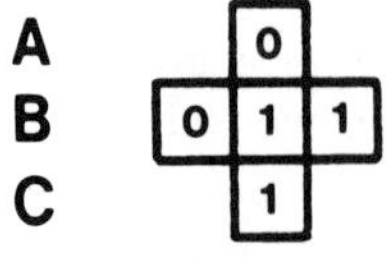

FIGURE 11.9. A schematic diagram illustrating a perimeter pixel.

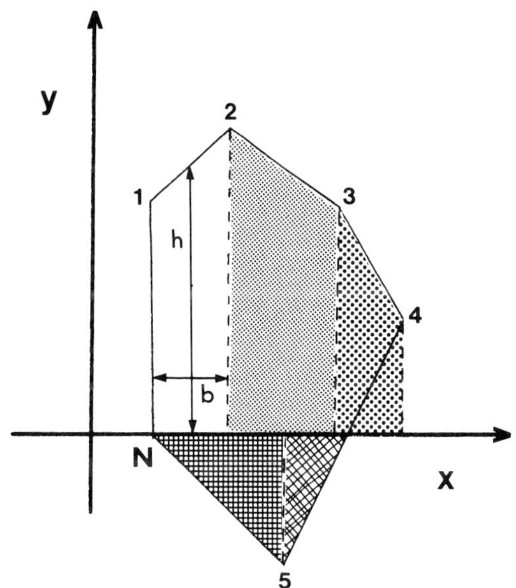

FIGURE 11.10. A schematic diagram of a polygon with cartesian coordinate axes overlain. The height of polygon 1 and 2 is shown with h and the base of the polygon with b.

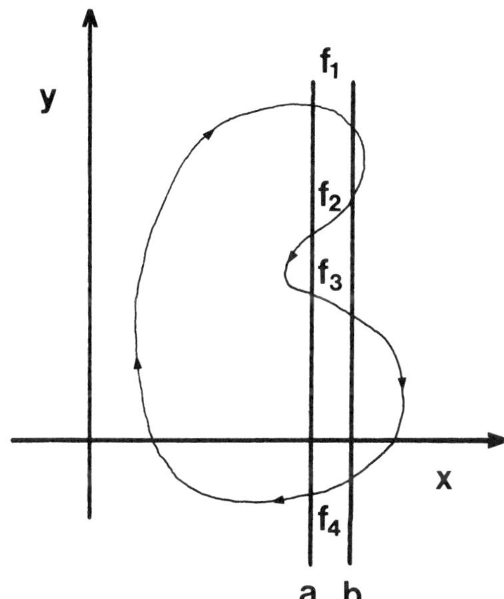

FIGURE 11.11. A schematic diagram of a particle with cartesian coordinate axes overlain. The functions f_1, f_2, f_3 and f_4 are the portions of the strip ab.

Note that the surplus area of the polygon between points 3 and 4 is canceled out from the total area by the computation of the polygon between points 4 and 5, thus the correct area is calculated. The area of the particle is the sum of the areas of the polygons.

Figure 11.11 illustrates the computation of the center of gravity for an irregularly shaped particle. The basic procedure is to divide the shape into "strips," calculate the center of gravity for each strip separately, and then sum these values in order to determine the center of gravity for the aggregate shape. In Figure 11.11 a particle is overlain by a cartesian coordinate system to illustrate the contribution of a single strip, which extends from $x=a$ to $x=b$. In Figure 11.11, functions $f_1(x), f_2(x), f_3(x)$, and $f_4(x)$ represent the y values on the particle boundary that cuts through this strip. Given explicit forms for these functions, the center of gravity contribution for this strip (Δx_c) is given by

$$\Delta x_c = {}_a\!\int^b x[f_1(x) - f_2(x)]\,dx + x[f_3(x) - f_4(x)]\,dx \tag{1a}$$

and

$$\Delta x_c = {}_a\!\int^b x\,f_1(x)\,dx - {}_a\!\int^b x\,f_2(x)\,dx$$
$$+ {}_a\!\int^b x\,f_3(x)\,dx - {}_a\!\int^b x\,f_4(x)\,dx \tag{1b}$$

In practice, the integration in Eq. (1a) is performed in separate pieces, as suggested by the four separate terms shown in Eq. (1b). Different signs apply to the terms in Eq. (1b). This can be explained by imagining the calculation as being conducted along the perimeter of the particle (as suggested by the arrows in Figure 11.11). In traversing the strip along the arcs represented by f_1 and f_3 the movement is from left to right, while along the arcs represented by f_2 and f_4 movement is from right to left.

Functions representing the particle boundary can be approximated using unevenly spaced x–y coordinate data. In a single strip, the boundary locus can then be approximated by an "affine" function, $f(x) = \alpha x + \beta$ in which α is the slope between two adjacent perimeter points and β is the y-intercept of the line through those points. (A higher order approximation is typically unnecessary given the resolution of tablet digitizers.)

Ordered perimeter data (x_k, y_k) determine parameters of the approximating functions as follows:

$$\alpha_k = \frac{y_{k+1} - y_k}{x_{k+1} - x_k} \tag{2}$$

FIGURE 11.12. A computer plot of the best-fit ellipses corresponding to the grains in Figure 11.6.

$$\beta_k = y_k - x_k \frac{y_{k+1} - y_k}{x_{k+1} - x_k} \qquad (3)$$

At this point, substituting the approximating function for $f(x)$ in Eq. (1b) leads to the following contribution from a strip from x_k to x_{k+1}:

$$\Delta x_c(k) = \int_{x_k}^{x_{k+1}} x(\alpha_k x + \beta_k)\, dx \qquad (4)$$

Equation (4) can be integrated and simplified to

$$\Delta x_c(k) = \frac{\alpha_k}{3}\,[(x_{k+1} - x_k)(x_{k+1}^2 + x_k^2 + x_{k+1}\, x_k)]$$
$$+ \frac{\beta_k}{2}\,[(x_{k+1} - x_k)(x_{k+1} + x_k)] \qquad (5)$$

Finally, Eqs. (2) and (3) can be substituted for α_k and β_k to arrive at the final formula for the x coordinate of the center of gravity.

The y coordinate of the center of gravity is computed in an entirely analogous fashion. The formula for $\Delta y_c(k)$ is equivalent to the formula for $\Delta x_c(k)$, apart from a sign change on the leading term:

$$\Delta y_c(k) = -\frac{\alpha_k}{3}\,[(y_{k+1} - y_k)(y_{k+1}^2 + y_k^2 + y_{k+1}\, y_k)]$$
$$+ \frac{\beta_k}{2}\,[(y_{k+1} - y_k)(y_{k+1} + y_k)] \qquad (6)$$

The center of gravity for the complete particle (x_c, Y_c) is given by

$$x_c = \sum_{k=1}^{N} \Delta x_c(k) \quad \text{and} \quad y_c = \sum_{k=1}^{N} \Delta y_c(k) \qquad (7)$$

The algorithms for these computations have been published elsewhere (Simigian and Starkey, 1989).

Other parameters may be calculated from the area, center of gravity, and the perimeter coordinates. For instance, the shape of a particle can be described by deriving geometric forms. An enveloping rectangle is obtained from the maximum dimension of a particle and the width at right angles to it. Such a rectangle does not maintain the area of a particle but it has been used in some studies (Ostwald and Lusk, 1978). A variety of ellipses can be calculated that do maintain the area of the particle and Simigian and Starkey (1986) have shown that the best-fit ellipse is a good, objective geometric form to use in shape analysis.

The best-fit ellipse corresponds to the particle boundary in such a way that the sum of the squares of the distance between each point on the boundary and the ellipse is minimized. The perimeter x–y coordinates and center of gravity are used to calculate the second central moment, the eigenvalues of the second central moment, and the eigenvector. The calculation of the long and short axis using eigenvalues yields only the aspect ratio. Therefore, the area of the ellipse is normalized to the area of the particle by applying the following scale factor, $a/\pi\ \sqrt{R}$, where a is the area and R is the ratio of the eigenvalues (Simigian and Starkey, 1986). This allows the best-fit ellipse corresponding to each particle to be drawn (Figure 11.12). The aspect ratio (long axis versus short axis) and the orientation of the long axis of the ellipse provide useful parameters with which to describe the particle. The data for individual particles can be used to compute average geometric forms that describe the total aggregate of particles.

Conclusions

The statistical analysis of particle shapes and aggregates requires that sufficient, quantitative data be obtained in an objective manner. Affordable hardware is now available to accomplish this, based on either scanning or tablet-type digitizers. Such devices digitize either the samples themselves or images of the samples. Scanners are faster and the desired x–y coordinates are intrinsically more accurate and reproducible. Scanners also record the gray levels from the image that may be useful for subsequent identification of the particles. However, there are occasions when the operator interaction required with tablet digitizers can prove useful by permitting data to be selectively recorded.

Particles are identified as they are digitized using a tablet. Scanners, on the other hand, record the image as a matrix of pixel values corresponding to gray levels and the particles have to be identified within the matrix by computation. To assist in this identification enhancing the particle boundaries either by digital edge finding techniques or by physical enhancement prior to digitization is often advantageous.

Once the particle boundaries are located and the x–y coordinates of the data points along their periphery determined, a variety of parameters can be calculated to describe individual particles and these can be combined to characterize the aggregate of particles. Among the most analytically useful of such descriptive parameters is the ellipse, which can be computed for a particle such that the mean squared difference between the ellipse and the particle boundary is minimized. Such best-fit ellipses reflect the aspect ratio of a particle shape, the length to width ratio, while an analysis of the ellipses for all the particles provides information on any preferred shape orientation of the aggregate.

The techniques described and the algorithms developed have broad application in the area of particle shape analysis. Further general discussion of image analysis techniques can be found in the work of Inoué (1986), Pratt (1978), Rosenfeld and Kak (1976), Serra (1982), and Underwood (1970).

Acknowledgments. This research was supported by operating Grant A3555 from the National Science and Engineering Research Council of Canada, which is gratefully acknowledged. Thanks are also extended to Thomas Fox Rutherford for his encouragement and help in preparing some of the figures.

References

Allard, B. and Benn, K., 1989, Shape-preferred-orientation analysis using digitized images on a microcomputer: Comput. Geosci. 15, 441–448.

Fabbri, A.G., 1984, Image Processing of Geological Data: Van Nostrand Reinhold, New York.

Inoué, S., 1986, Video Microscopy: Plenum Press, New York.

Lapique, F., Champenois, M., and Cheilletz, A., 1988, Un analyseur vidéographique interactif: description et applications: Bull. Minéral. 111, 679–687.

Niblack, W., 1986, An Introduction to Digital Image Processing: Prentice-Hall, Englewood Cliffs, NJ.

Ostwald, J., and Lusk, J., 1978, Sulfide fabrics in some nickel sulfide ores from Kambalda, Western Australia: Can. J. Earth Sci. 15, 501–515.

Pratt, W.K., 1978, Digital Image Processing: John Wiley, New York.

Rosenfeld, A., and Kak, A.C., 1976, Digital Picture Processing: Academic Press, New York.

Serra, J., 1982, Image Analysis and Mathematical Morphology: Academic Press, New York.

Simigian, S., and Starkey, J., 1986, Automated grain shape analysis: J. Struct. Geol., 8, 589–592.

Simigian, S., and Starkey, J., 1989, Image: Modified for use on a microcomputer based system: Comput. Geosci. 15, 237–254.

Stanton, T., 1987, Tablets for precision graphics: PC Mag, 6, 159–181.

Starkey, J., and Simigian, S., 1987, Image: A fortran V program for image analysis of particles: Comput. Geosci. 13, 37–59.

Tough, J.G., and Miles, R.G., 1984, A method for characterizing polygons in terms of the principal axes: Comput. Geosci. 10, 347–350.

Underwood, E.E., 1970, Quantitative Stereology: Addison-Wesley, Reading, MA.

12
Interactive Image Analysis of Borehole Televiewer Data

Colleen A. Barton, Lawrence G. Tesler, and Mark D. Zoback

This chapter describes an interactive graphics system designed for borehole televiewer (BHTV) image analysis. The software provides one of the first comprehensive tools for borehole image data analysis available to exploration and research scientists on a low cost and easy-to-use personal computer, the Apple Macintosh II.

The program, called BHTVImage, provides an integrated environment for analyzing borehole shape and features. Images of BHTV data are displayed in false color on a graphics screen and are manipulated with a mouse pointing device. A variety of two- and three-dimensional displays of borehole radius and acoustic reflectivity are used to display the data. Tens of meters of borehole wall can be rapidly viewed by scrolling through the data within a graphics window. Interactive measuring tools are provided to quickly measure and record wellbore features. The values of the scales, units, and grid intervals can be modified interactively during program execution. Gross scale features can be easily extracted from the images through interactive thresholding to produce a scaled comparison with complementary log data such as resistivity or sonic recordings. Alternatively, full resolution images can be analyzed to investigate the fine details of fractures or cross-bedding intersecting the well.

One of the reasons Macintosh applications are easy to use is that there is remarkable human interface consistency among applications. To help meet that standard of consistency, the BHTV analysis software was implemented using an Apple product called MacApp. MacApp is an "object-oriented framework," i.e., a "generic" application that defines standard "objects" like windows and views. Any desired Macintosh application can be derived from MacApp by describing in code only the differences between that application and the generic one. The resulting program achieves consistency with other Macintosh applications without special effort on the part of the programmer. MacApp, as well as the bulk of the BHTV analysis program, is written in the object-oriented language Object Pascal. Certain BHTV analysis subroutines are written in the language C.

The software discussed has provided the primary analysis tool for the BHTV image data recorded in the Cajon Pass and KTB research wells. The Cajon Pass Well, located in Southern California near the San Andreas fault, was drilled to investigate the paradox of low stress, low heat flow measurements along an active plate boundary. The KTB deep drilling project located in the Oberpfalz zone of West Germany was undertaken to evaluate an ancient continental suture. The accurate determination of borehole shape and the characterization of wellbore breakouts and fractures with depth have been essential to understanding fault mechanics and tectonic stresses and to the interpretation of geophysical and core data from these and many other wells.

Introduction

The borehole televiewer is an ultrasonic well-logging tool useful for imaging lithostratigraphic features and for measuring the orientation and distribution of fractures as well as the orientation and

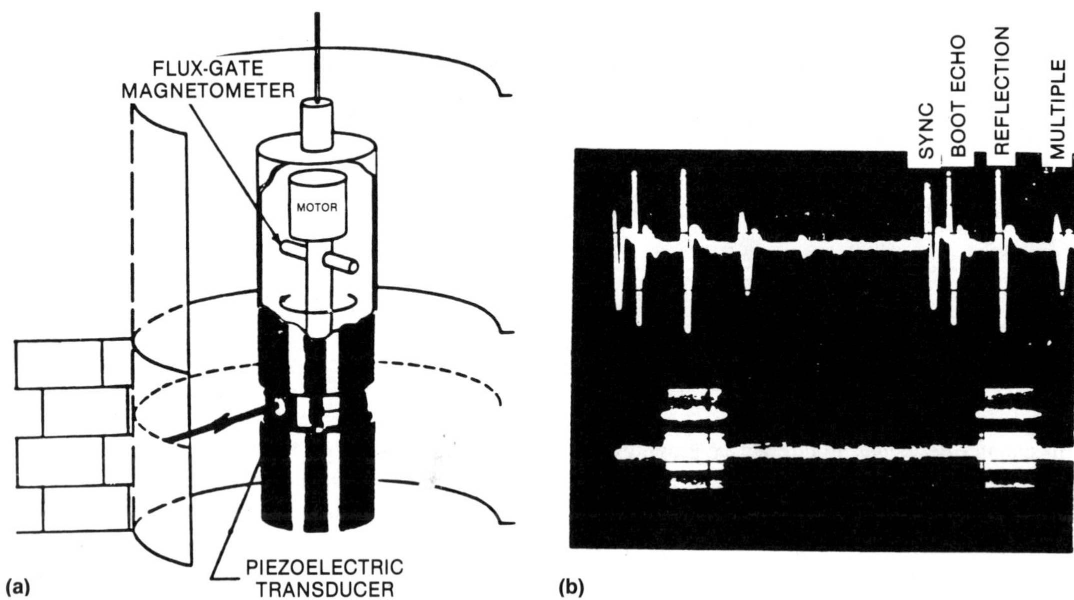

FIGURE 12.1. (a) Schematic diagram of the acoustic mechanism of the borehole televiewer (from Zemanek et al., 1970). (b) Analog wave train of the televiewer signal.

width of stress-induced wellbore breakouts. The analog televiewer, originally designed by Mobil Oil, Inc. (Zemanek et al., 1970), contains a rotating transducer that emits an acoustic pulse at the rate of 1800 times a second. Figure 12.1 is a schematic diagram of the acoustic mechanism of the televiewer tool. The 1.4-MHz transducer rotates at three revolutions per second and moves vertically up the borehole at a speed of 2.5 cm/s. The transducer diameter is about 1.27 cm (0.5 in.), however the emitted sound is focused to a narrow beam of about 3° due to the high frequency. A fluxgate magnetometer within the tool fires at each crossing of magnetic north, making it possible to orient the data.

Prior to the 1980s and the emergence of downhole digital imaging tools, borehole wall images were limited to analog photographs. Image analysis was a static, tedious process. With the inception of the microprocessor, digital images of the borehole wall have become the standard data presentation format.

Image photographs are still routinely collected in standard field operation of the analog BHTV tools. The analog televiewer tool transmits ultrasonic seismograms through a standard wireline logging cable which are recorded on videotape. The analog data are channeled into a three-axis oscilloscope where the horizontal sweep represents the scan of the rotating transducer around the borehole, the vertical sweep represents the rise of the tool in the borehole, and the intensity is modulated by the reflected energy from the borehole wall. The oscilloscope then produces an unwrapped 360° gray scale image of about 1.52 m (5 ft) of the borehole wall. The BHTV "log" is constructed from Polaroid photographs taken of successive oscilloscope screens as the tool progresses uphole during field logging. Each 3×5 photograph is then assembled into a composite log. The analog "north pulse" is recorded on one of the two audio channels of the video recorder. Depth is also recorded analog on the remaining audio track of the video recorder.

The analog televiewer tool has not undergone any major design changes since it was first introduced to geophysical logging in 1970. Interpretation of the photographic log has been limited to estimations of gross structure from the reflected image. The primary use of televiewer data had been to determine the orientation of fractures in

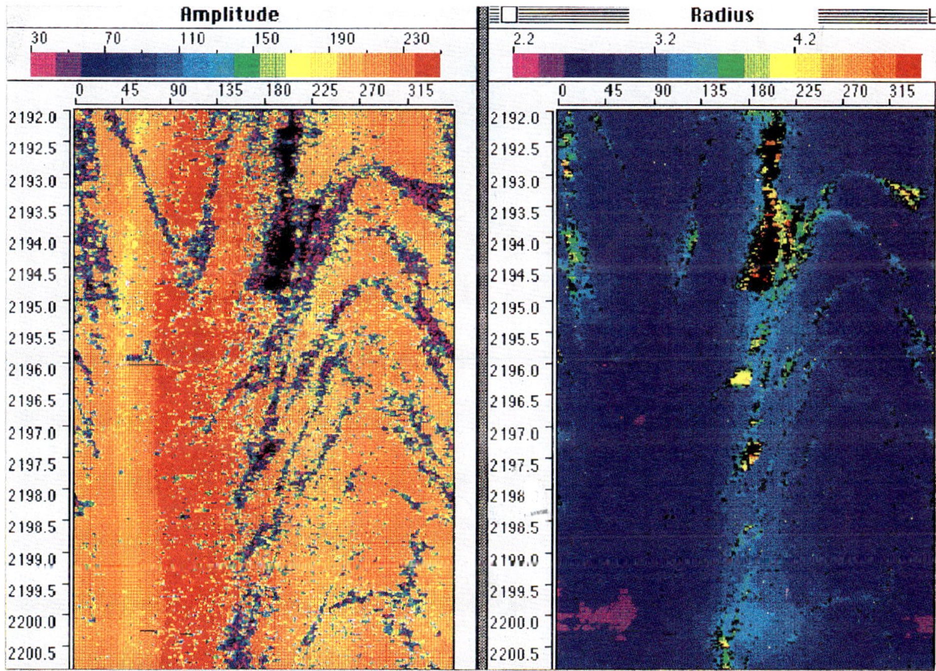

FIGURE 12.3. Standard unwrapped 360° images of the borehole wall. Vertical axis is depth, horizontal axis is azimuth, and color represents the amplitude of the reflected pulse (left window) or the borehole radius (right window).

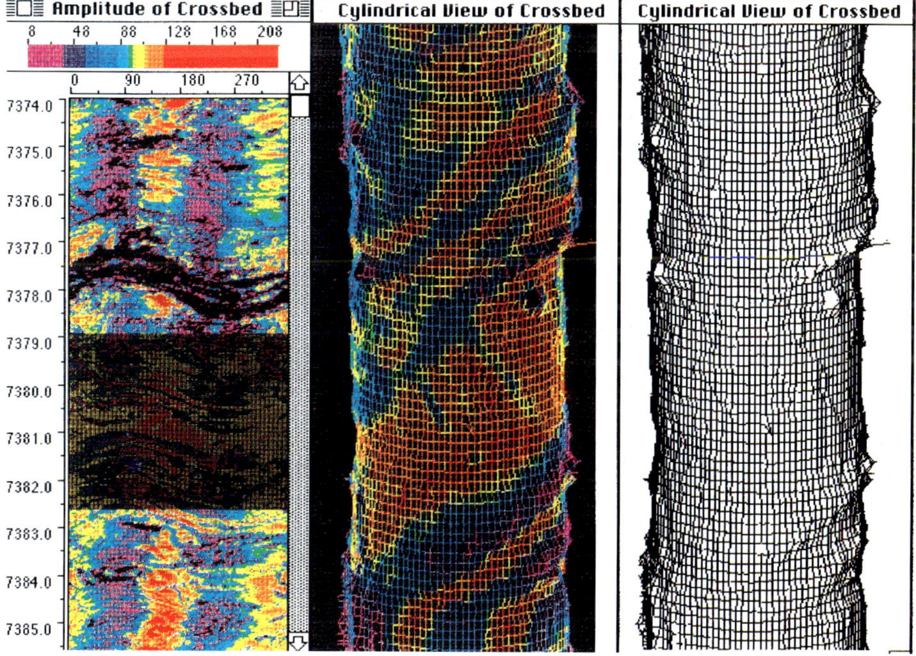

FIGURE 12.4. Highlighting in 2-D view (left window) over a cross bedded interval of BHTV data recorded in a Gulf Coast well indicates user selection of the image data for rendering 3-D projections. 3-D projection where surface color represents the amplitude value (middle window) and 3-D projection in black and white (right window) where differences in weathering of the beds is apparent.

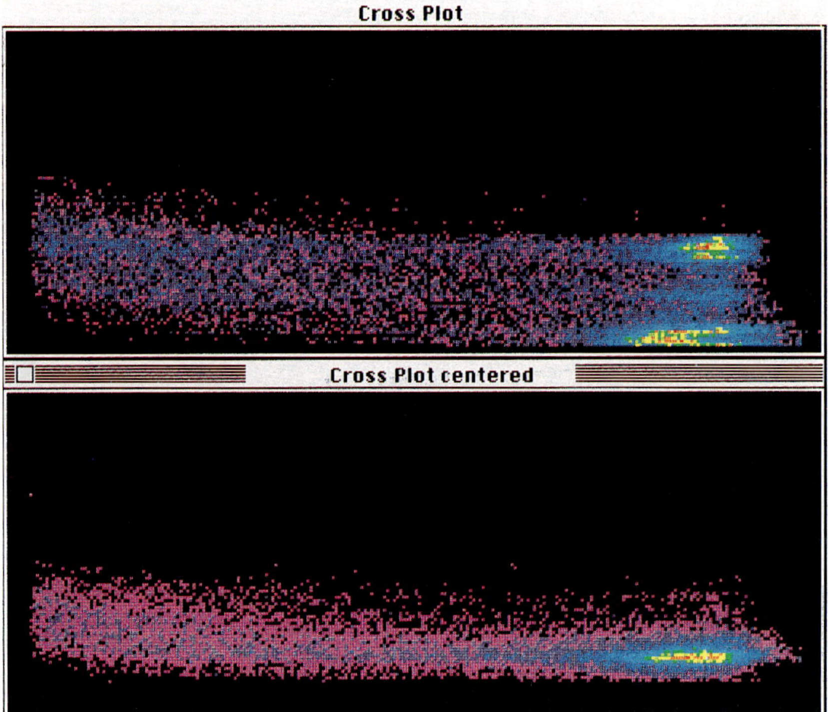

FIGURE 12.6. Top window is a cross plot of the reflectivity values (x-axis) versus traveltime values (y-axis) over an interval in the Cajon Pass well where the BHTV tool was off-center in the borehole. The bottom window is the cross plot over the same interval after the data have been corrected for off-center effects.

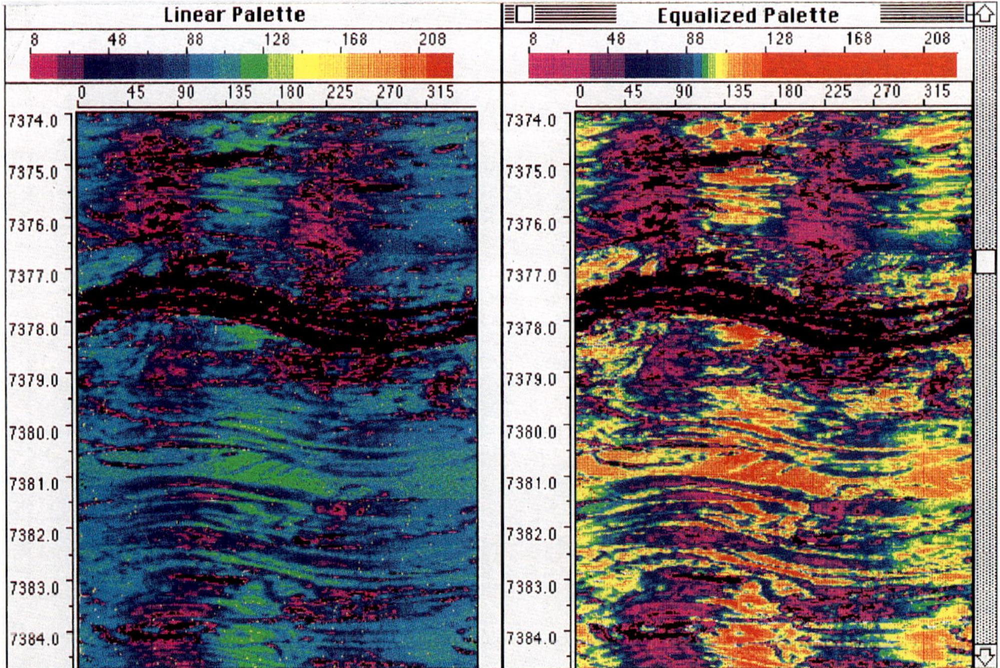

FIGURE 12.7. An example of the effects of histogram equalization. The left display shows the standard linear color distribution and the right panel the equalized color distribution. Note that the details of the cross bedding are enhanced at depth 7380.5 ft using the equalized color scale.

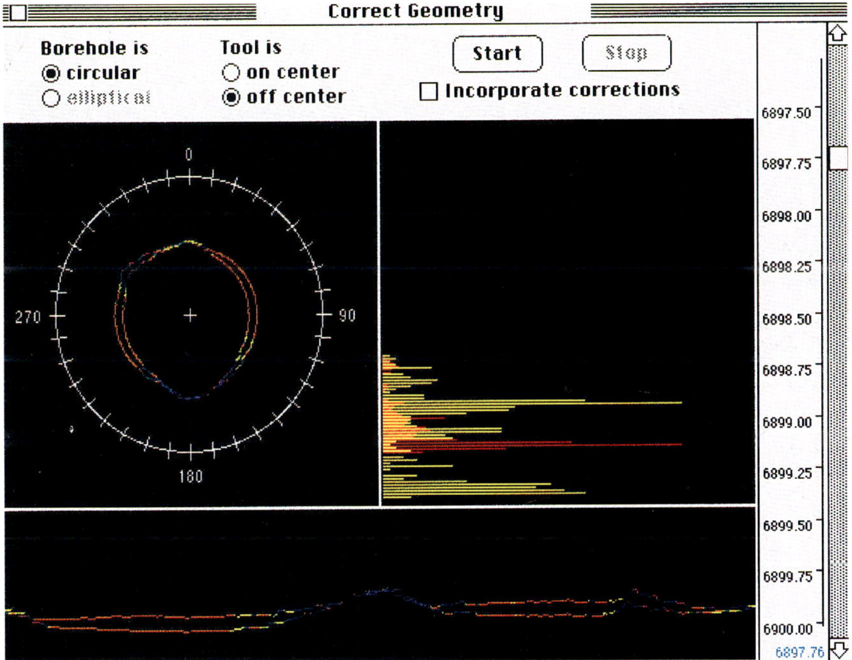

FIGURE 12.11. Workstation screen showing the interactive data correction of borehole geometric effects. Upper left polar plot shows original data scan and corrected scan of data. Histograms of the original travel time data (yellow) and the corrected data are shown in the upper right plot. The lower plot represents the original data and the corrected data in cartesian coordinates where the x axis is azimuth and the y axis is radius.

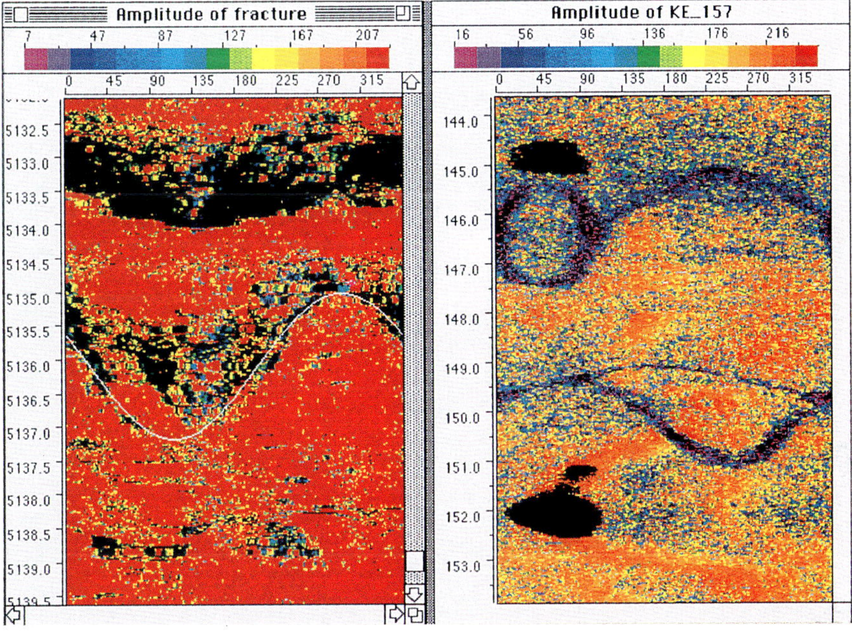

FIGURE 12.16. Left window shows intersecting fractures at depth 5135.5 ft in the Auburn Geothermal well, New York. White curve represents the interactive fit to a steeply dipping fracture. Right window is an example of data recorded over a complex fracture zone where fractures are not perfectly planar in a well located near Anza, California.

granitic rocks, analyses that involved tedious and inaccurate measurements. No capability existed in the analog data to correct the measured orientations of planar features for the effects of off-center tools or elliptical boreholes, conditions that can lead to substantial measurement error. With the rising use of wellbore breakouts to determine *in situ* stress, televiewer data have become increasingly important.

Digital televiewers have recently been designed that contain downhole processors to reduce the received signal to a peak amplitude and associated travel time before digital transmission (Hinz and Schepers, 1985; Schlumberger, Inc.). In digital tools, the horizontal resolution of the data is generally reduced by about one half of that of the analog tools. However, precision is gained by eliminating noise problems such as analog signal degradation in the cable. The marker pulse at magnetic north and depth readings are encoded automatically in the data recorded by the digital tools. The tilt of the tool is monitored by two inclinometers and an accelerometer provides continuous data on tool speed. Although the resolution is completely dependent on hole size and logging speed, the televiewer typically will provide horizontal resolution of a few millimeters and a 1 cm vertical resolution in a 30.5-cm (12-in.) borehole.

Uphole data processing schemes for analog borehole image data have been operational since the early 1980s. Early postlogging processing of the images was accomplished by fiber optics recording hardware (Wiley, 1980; Broding, 1982). At the outset of the development of uphole digitization systems specialized equipment was built; today inexpensive off the shelf data acquisition boards are available for workstations and most personal computers. In postlogging digitization the recorded BHTV signal is examined within a specified window to determine the peak amplitude and its associated travel time. The determination can be performed by efficient software algorithms (Barton, 1988) or electronically measured by edge detection hardware (Pasternack and Goodwill, 1983; Taylor, 1983; Rambow, 1984; Wong et al., 1989; Faraguna et al., 1989). The time required to capture and record each pair of values is generally within 0.5 μs.

Digitization of standard analog data results in dual measurement of acoustic reflectivity of the borehole wall and the ultrasonic travel time of the imaging pulse at a spatial resolution well above conventional logging tools. The software and techniques described herein to analyze and interpret BHTV data are applicable to any digital televiewer data set whether obtained directly from a digital tool or derived by analog-to-digital conversion from an analog wave train (Figure 12.1b).

Advantages of Digital Interactive BHTV Data Analysis

Digitizing and digital processing of the analog televiewer data have opened a new dimension in the interpretation of BHTV data, most significantly, the precise measurement and display of the travel time of the reflected pulse. Assuming a known mud velocity along the raypath for the source and reflected pulses, the two-way travel time can be converted to distance to give the detailed topography of the borehole wall. Another advantage of digital data is that it can be enhanced and corrected for geometric effects. Borehole features can be displayed in a way that best suits the type of feature under investigation. For example, wellbore breakouts are best represented in polar cross section. Finally, it is possible to make systematic quantitative measurements quickly and accurately.

BHTVImage utilizes various two- and three-dimensional displays of the borehole radius and acoustic reflectivity to facilitate data analysis. It allows the user to make quantitative measurements of borehole features such as breakouts, fracture orientation and apparent aperture, and lithologic features. The analysis software has been implemented on a Macintosh II personal computer. The program provides an integrated environment for analyzing borehole shape and features where images of BHTV data are displayed in false color on a graphics screen and are manipulated by graphics mouse and keyboard commands.

The primary advantages to interactive data analysis are that it gives the geophysicist the ability to (1) interactively manipulate the data to obtain an optimal view of a particular feature, (2) look at the same data interval simultaneously from a variety of perspectives, and (3) make decisions as the analysis proceeds.

TABLE 12.1. Header specifications.

Data attribute	Definition
Well name	Well identification used as title for graphics output
Well location	Well location used as title for graphics output
Log date	Date BHTV data were recorded
Mag dec	Magnetic declination at site in degrees east of north
Marker	Azimuth correction for data collected in marker mode
Deviation	Borehole deviation in degrees
Max depth	Bottom depth of digitized interval
Min depth	Top depth of digitized interval
Tool spec	Calibration adjustment
Tool units	Units of the calibration adjustment
Fluid vel	Velocity of the borehole fluid during logging
Vel units	Units of the fluid velocity
Ddelay	Time delay for initiation of digitization in μs
Tool offset	Transit time from the transducer to its housing in μs
Samprt	Sample rate in samples/μs
Window	Sampling window length in μs
Samples	Number of samples digitized for each pulse
Max TT	Maximum recorded travel time
Max AMP	Maximum recorded amplitude of the reflected pulse
Min TT	Minimum recorded travel time
Min AMP	Minimum recorded amplitude of the reflected pulse
Cutoff TT	Array of histogram equalization bins for travel time
Cutoff AMP	Array of histogram equalization bins for amplitude
Width	Number of pulses per scan
Height	Total number of scans (full revolutions) in the data file

Menu-driven commands and analysis operations provide an ease of use factor to the software. Tens of meters of borehole wall image can be rapidly viewed by scrolling through the data within a graphics window. Images can be enlarged to the full resolution of the screen and scrolled to search for fine features. Alternatively, larger sections of the well can be displayed at the scale of standard logs. Interactive measurement tools are provided to quickly measure wellbore features. The values of the scales, units, and grid intervals can be modified interactively during program execution. Menu-operated window management provides easy access to the various views the user is interpreting.

BHTV File Format

A common aspect to all borehole image data is the physical size of the data set and the associated difficulties in managing the volume of data generated. The voluminous size of image data controls the configuration of the computer hardware and impacts software design along every step of the acquisition, display, enhancement, and analysis process. Computer speed and memory capacity are rapidly increasing, however, these constraints remain a significant consideration in the design of an image processing system. Data compression is a widely used and widely researched topic in computer science. Compression and decompression algorithms exist for computer systems common to oil industry, research, and government facilities. They are at present too costly in computational time to be useful in interactive image analysis.

The data input consists of header information followed by arrays of amplitude and travel time values. Careful thought is required in header format design given the extreme variability in image data and the fact that other scientists may use the data. For example, image enhancement involves filtering and other processing that results in permanent transformation of the data. It is important to keep a history of the transformations that a particular image data set has undergone to preserve the ability to interpret the data. In header format design it is beneficial to consider extensibility, that is, the flexibility to extend the header at a later time it to include more, fewer, or different header parameters. Table 12.1

summarizes header specifications used in this system. The header parameters can be inspected by choosing a menu command that presents a dialog box, a display of the current settings of the header parameter. These parameters can be edited as necessary to adjust the data to the proper depths, compass azimuth, etc.

In the uphole digitization scheme developed at Stanford the returned reflection is windowed and discretely sampled (Figure 12.1b). The peak amplitude and associated travel time are measured for each pulse firing. The conversion of travel time to true borehole radius is completely dependent on the BHTV tool and transducer frequency used to record the data. The mud fluid velocity in the borehole, V_f, and the correct radius of the televiewer tool, T_r, must also be known. The transit time between the transducer and the outside of the window housing the transducer, T_{off}, must be known for the particular tool and transducer frequency used to record the data. For example, T_{off} is 16 μs for the large diameter high frequency M & W Instruments, Inc. televiewer and 9.81 μs for the Schlumberger Inc. high frequency large diameter tool. Conversion of the travel time data to borehole radius is then

$$R = \frac{(S \times T_i \times W/N) + D - T_{off}}{2} \times V_f + T_r \qquad (1)$$

where T_i is the two-way travel time, S the sample rate in samples/μs, D the digitizing delay time in μs, W the digitization window length in μs, and N the number of samples digitized per pulse. To assure accuracy in the conversion of travel time to radius for each logging run a calibration adjustment can be added to Eq. (1). The calibration constant is obtained from a calibration test of the tool in a specially designed tank before each logging run. A calibration tank requires at least two different radius values, for example, a smaller radius from 0° to 180° circumference and a larger radius from 180° to 360°. The travel time values measured at known radius values can be used to compute the calibration constant and the fluid velocity. Data recorded in this tank may also be used to calibrate the compass azimuth of the transducer with respect to magnetic north.

Conversion to the format for BHTVImage includes resampling of variable scan length data to a fixed length, shifting the data to correct for the

magnetic declination at the logging site or any other azimuthal data shift, and scaling of the amplitude and travel time values to the range 0 to 256 expected by the program. An essential consideration during data acquisition is the threshold detection level imposed during data acquisition. When no reflection is detected above the noise threshold that particular pulse must be flagged within the data set to signify that no energy returned to the transducer. These values of missing data are important elements of the BHTV data format.

Human Interface Design

Experience with a predecessor to BHTVImage on a different computer had shown that the utility and appeal of the program would be quite limited unless the user could interact freely with an image, i.e., display it alone or alongside other images, scroll through it to study its features, decide whether to correct for geometric distortion and do so if desired, decide whether to enhance the image in one or more ways and do so if desired, choose views that seem at the time to be useful, and make measurements of any desired features in any order at all. Achievement of these goals required a highly interactive graphical user interface.

Another goal we chose was to make the BHTVImage software accessible to other researchers. Two factors were important to achieve this goal: affordability and ease of use. The Macintosh II computer met all the above criteria as well as technical requirements such as sufficient memory capacity to hold multimegabyte images and a high-fidelity color display.

One of the reasons Macintosh applications are easy to use is that there is remarkable human interface consistency among applications. After a user has learned one or two applications, the knowledge he or she has gained about how to interact with them can be generalized to other applications. Studies have shown this to result in reduced learning time. To accrue the advantages of consistency, BHTVImage was designed to follow Apple's user interface guidelines wherever feasible.

The Macintosh user interface is based on a number of principles. One is modelessness. A mode is a state of an interactive system that affects the meaning of user actions. If the user takes an action in one mode thinking the system is in another mode, undesired effects can result. As a result, the Macintosh guidelines discourage modes, and insist that when modes are unavoidable, the current mode be inescapably obvious to the user. If BHTVImage, for example, had been designed so that the user specified a filter command and then the borehole section to filter, then between the first and second specification, the system would have been in a mode expecting the section to be specified. Instead, in analogy with other Macintosh applications, BHTVImage was designed so that the user first specifies ("selects") a borehole section and then any operation on it—or several sequential operations, or none. Between the selection and the operation, there is no mode. The user may scroll or move the window, run other Macintosh applications, peruse the menus in search of a desired command, select a different section of the borehole for the operation, or even change one's mind and never specify the operation.

Another principle of the Macintosh user interface is that everything the user must think about should be visible. When we added the ability to measure a fracture or breakout, our first thought was to write the measurement data directly to a text file. However, if we had done so, the user would not have been able to see the measurements until after the file was saved and another program was run to examine it. Instead, we decided to display the measurements in a text window immediately after they were made. In addition to providing the desired visibility, we discovered the additional benefit that the measurements could be edited using a standard text window editor to delete a mistaken measurement or for other useful purposes.

One other principle of the Macintosh user interface is that applications should be tested on typical users, revised in response to difficulties they encounter, retested, revised again, etc., until the ease of use level meets expectations. In the case of BHTVImage, one target user (an author of this paper) was involved in the detailed design and implementation of the program. Thus, it was not surprising that all of the 20 or so geophysicists who have tried it thus far—some with no help but a 10-page preliminary manual—have reported extreme satisfaction with the user interface.

Key components of the Macintosh user interface, and their expression in BHTVImage (discussed below), include the following:

Macintosh	BHTVImage
Windows	Amplitude window, borehole radius window, correction window, etc.
Views	Unwrapped view, polar cross-sectional view, cylindrical projection, etc.
Selections	Selection of text, of one fracture, or of all data between two depths
Menus	File, Edit, Windows, Views, Settings, Palette, Analyze, and Profile
Dialogs	Dialogs to provide control over the display of scales, filter options, etc.
Keyboard shortcuts	The escape key to record a fracture, command-X to remove a section
Palettes	Palette of 16 hues, gray scale or black and white

Overview of Program Usage A Scenario of the Analysis of BHTV Data

Open a Data File

The standard representation of borehole image data has been a 2-D unwrapped 360° view of borehole azimuth versus depth. The interpretation of borehole image data involves the display of different borehole features in an optimal geometry so that they can be viewed and most accurately measured. Planar features that intersect the borehole appear as sinusoids on the 2-D view so they are best analyzed in the 2-D split images (Figure 12.2, after Zemanek, et al., 1970). In Figure 12.3 (see Color Plate I), nearly vertical fractures that strike about 180° are associated with low reflectivity values and large values of borehole radius. Other borehole parameters that can be extracted from image data, for example, caliper information from the BHTV, may be represented as geophysical log profiles.

Intervals of the data may be selected from the 2-D unwrapped image, using the mouse much like selecting text in a Macintosh word processor program. The selected interval can then be viewed in a variety of projections. The cylindrical geometry of the data is best represented by polar cross sections of the data scans or isometric 3-D cylindrical projections (Figure 12.4, see Color Plate I). In the cylindrical projection, the radius of the cylinder is modulated by the borehole radius value at each pulse providing a true scale reconstructed image of the borehole wall. Three-dimensional wire frame cylindrical

FIGURE 12.2. Intersection of a plane with a cylinder and the effective sinusoidal curve on the 2-D unwrapped cylindrical surface (after Zemanek et al., 1970).

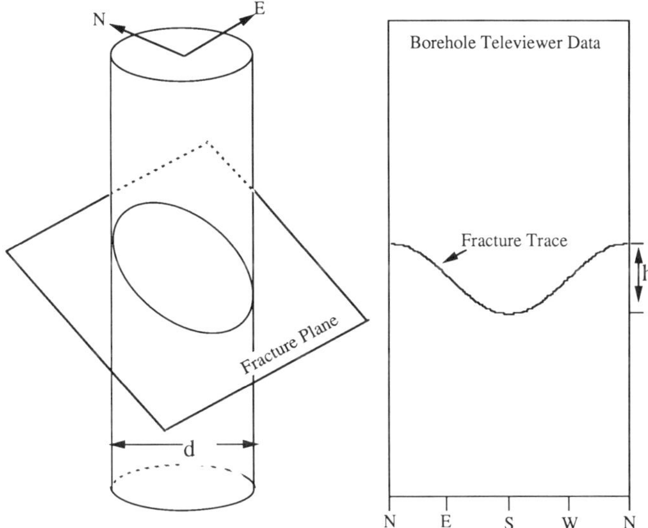

projections of the data show fine detail of the borehole wall not visible in conventional 2-D displays. The cylindrical projection may be scrolled vertically to view the reconstructed "core like" image with depth or scrolled horizontally to rotate the image 360° about the vertical axis. The amplitude values can be used to color modulate the surface of the cylindrical projection for a composite image of the data (Figure 12.4, middle window). These views can also be interactively displayed as black on white to investigate only the topography (Figure 12.4, right window). The data displayed in Figure 4 were recorded over a cross-bedded interval in a well located in the Gulf Coast. A 3-D wire frame image with reflectivity values superimposed in color has become an important interpretative tool. As shown in Figure 12.4 there is more variability in the reflectivity image than in the corresponding travel time image indicating that the impedance contrast over this interval exceeds the small variations in the travel time due to the differential erosion of the interbedded sand and shale. There is an option in this view to exaggerate the surface of the cylinder to amplify small variation in the topography of the surface. The differential erosion between the sand and shale layers is evident in the black and white view of Figure 12.4 (right window).

Polar cross sections of the data scans can give an accurate view of circumferential wellbore conditions. The polar projection also uses the travel time values to modulate the radius of the polar cross section. Scrolling the data in polar cross section allows a rapid view of borehole shape changes with depth (Figure 12.5, middle window). A compass plot around the polar scan is used to reference the data to geographic north. A graticule can be used in the polar plot for a more accurate assessment at the borehole dimensions. These alternative views of the data can be simultaneously displayed in other windows.

Additional views of any selected interval of data include a plot of the scans in Cartesian cross section showing azimuth versus borehole radius and a histogram view of the travel time data. These views are useful for evaluating borehole shape and noise contamination. Each view can be scrolled to inspect the data for changes with depth.

Data intervals can also be plotted as a cross plot of amplitude (x axis) versus distance (y axis) to look for anomalous relationships in the two variables that could indicate the presence of a particular borehole feature or the need to make a geometric correction to the data. There are expected relationships between reflectivity and travel time in the borehole, for example, low amplitude values and high travel time values within fracture zones, anomalous relationships such as high travel time and high amplitude isolated in a cross plot such as the upper plot in Figure 12.6 (see Color Plate II) are the result of anomalies in the data. These plots can be used to interactively determine intervals of problematic data such as the off-center effects characteristic of

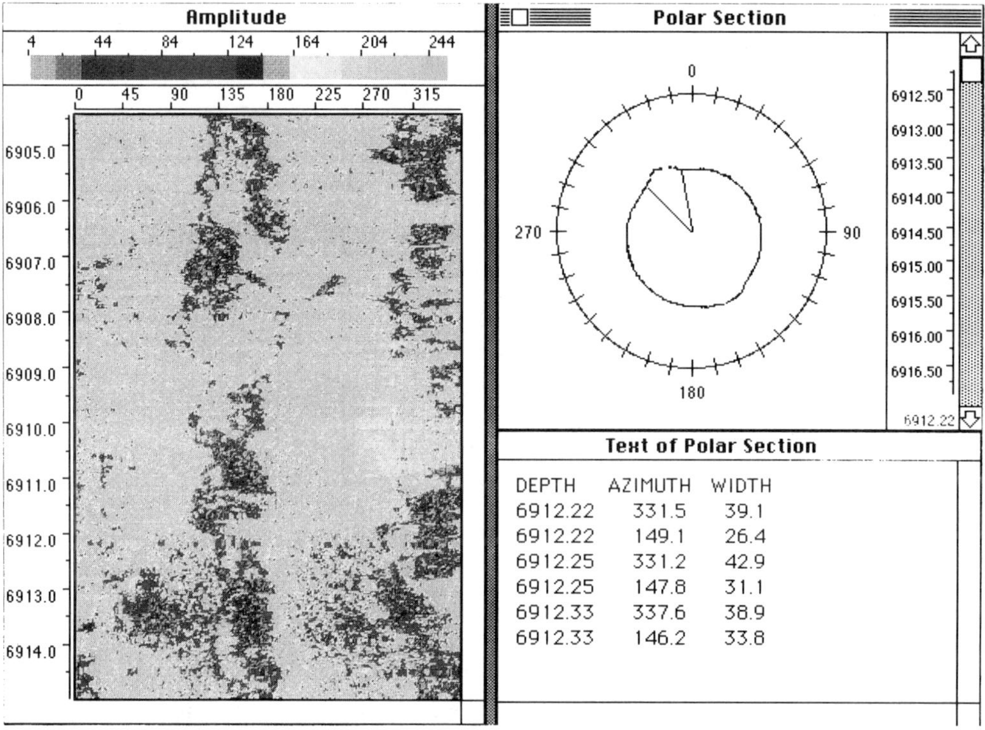

FIGURE 12.5. The right window is an unwrapped view of borehole radius over a 14 ft interval of breakouts in the Cajon Pass well. The left window shows data scans in polar cross section and the interactive analysis of breakout azimuth and width for this data.

the low travel time high amplitude cluster in Figure 12.6. After correction for off-center effects the cross plot shows a normal distribution of data (Figure 12.6, lower plot).

Color or gray scale palettes are used to false color map the data to an image. Gray scale palettes can sometimes be a more intuitive tool using simple light to dark gray shades to indicate increasing values in the data. Data are color or gray scale mapped to either a linear palette where the data values are mapped to 16 different hues or to an enhanced palette that utilizes histogram equalization. Histogram equalization generates an image that assigns the available brightness levels or colors to equal numbers of samples and produces a display of higher color contrast that can accentuate features not visible in a linear scale. The same image over an 11 ft (36.9 m) section of a Gulf Coast well in Figure 12.7 (see Color Plate II) shows the effects of histogram equalization. The left image has a linear color distribution and the right image has an equalized color

map. Histogram equalization has been an extremely useful method to enhance borehole wall details.

Due to the variability in data, value ranges, and data quality, the linear or equalized color maps may not always provide the optimum display for interpretation in all cases. For this reason interactive palette manipulation was implemented that allows the palette hues to be interactively adjusted using a graphics mouse. This technique allows for expansion or reduction of the color contrast as needed for optimum enhancement.

Data thresholding is an additional palette manipulation. This technique was developed to examine gross scale features of the borehole reduced to black on white images. With thresholding the color palette is replaced by a sliding scale of the range of values in the data. A cutoff threshold value is interactively selected from this scale. Figure 12.8 shows an interval of the Cajon Pass well where breakouts and fine scale fractures are evident in both the reflectivity and borehole radius views.

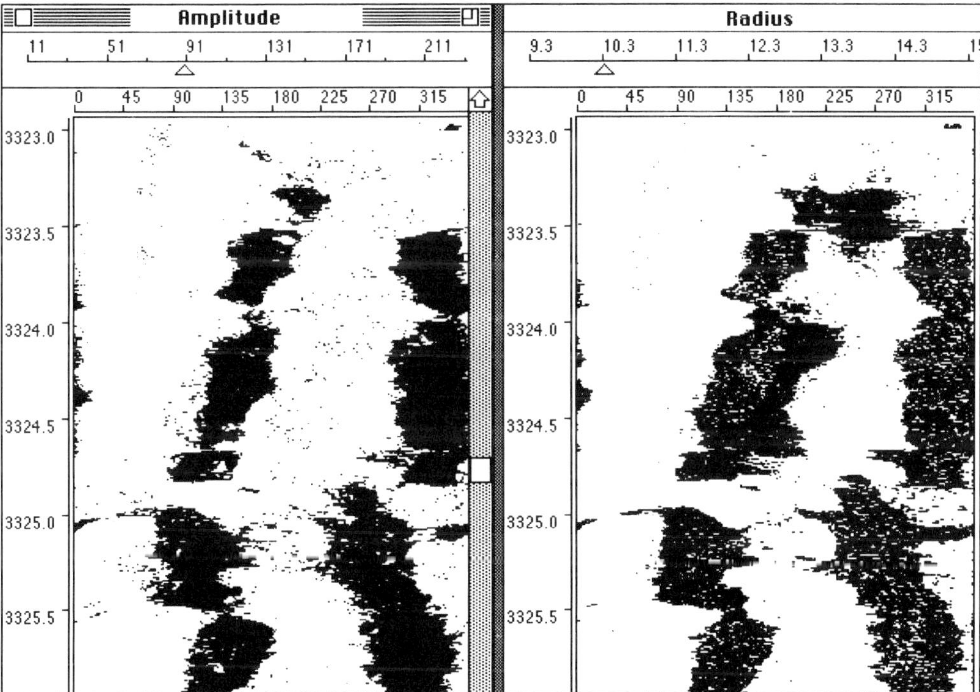

FIGURE 12.8. Right window is the amplitude of the reflected energy and left window is the borehole radius where cutoff threshold levels have been used to suppress all data except borehole features.

For reflectivity, the cutoff threshold marks the value below which all reflectivity pixels will be black and above which reflectivity pixels will be white. Similarly, a radius threshold can be used above which pixels will be black and below which they will be white. This view is useful for the correlation of BHTV logs with companion log data described below.

Low-pass filtering and subtractive smoothing (high-pass filtering) routines have been implemented as an aid in the detection of fine scale features. These filters are horizontal and 1-D. In subtractive smoothing the data are low-pass filtered to attenuate all of the high frequency features, such as edges and lines, then the smoothed image is subtracted from its original resulting in a difference image that has only edges and lines substantially remaining.

Standard Processing

The introduction of digital image data brought the possibility of correcting data for distortions caused by less than optimal recording geometry. Unfavor-

able logging conditions are a pervasive aspect of borehole image data acquisition. Instability of the borehole, washouts, fault zones, and the high pressures and temperatures at depth can lead to poor quality data. One objective for processing borehole image data is to identify these environmental or tool problems that distort the data and to provide the information required to make corrections to the data.

Analog data acquisition onto video tape is sensitive to human error as well as to poor borehole conditions. The missynchronization of recorder operation and winch operation can lead to missing or repeat log sections. When logging at sea ship's heave is seldom entirely mechanically compensated and data recorded at ocean drilling sites often show repetition of logged features. In addition, the tool may "stick" in the borehole for several rotations of the acoustic transducer then rapidly slip uphole creating a characteristic blocky image (Figure 12.9, left window).

For these reasons data need to be preprocessed and edited prior to interpretation to avoid erroneous

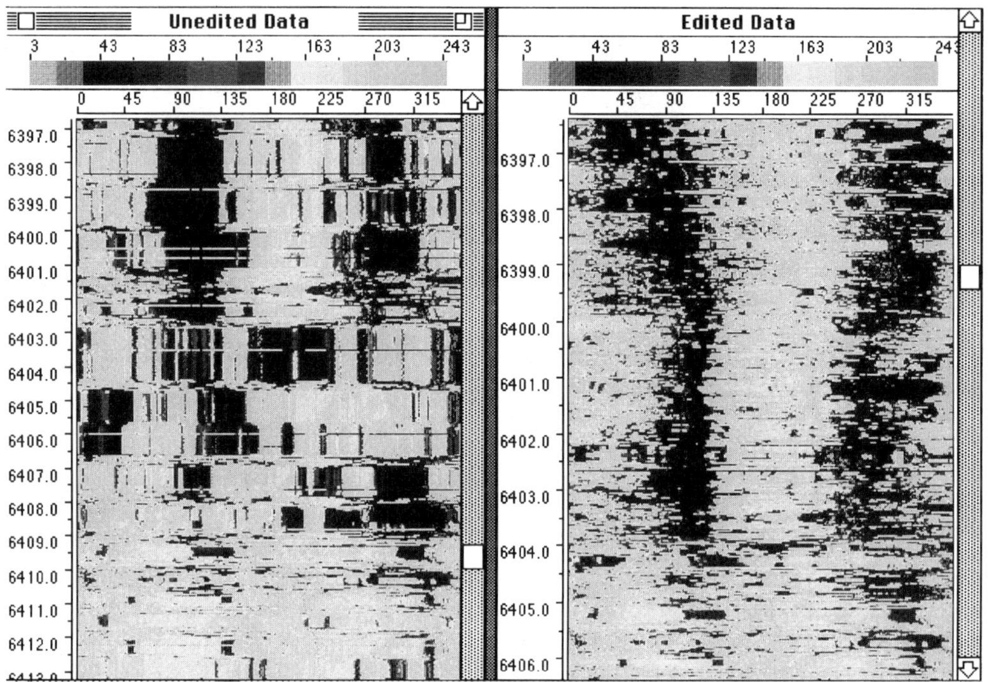

FIGURE 12.9. Right window shows unedited BHTV data recorded in HOLE 504B located in the East Pacific Rise; left window shows the results of visual editing where breakouts are now visible in the data centered at azimuths 100° and 280° between depths 6406 and 6401 ft.

conclusions and the masking of subtle features. An interactive visual editing of the digital data has been implemented in the BHTV system to allow the user to delete data that should not be included in an interpretation. By selecting the unwanted interval with the mouse and using the standard Macintosh Edit Menu command "Cut" the selected scans are removed from the data file. Figure 12.9 shows the application of this interactive editing on data collected at Hole 504B in the East Pacific Rise. The blocky intervals in the left image of unedited data correspond to repeat scanlines due to ships heave; the right image of edited data reveals an interval of stress-induced wellbore breakouts that cannot be distinguished on the unedited image.

A fundamental effect of logging on the quality of BHTV image data is random noise. Whether it is electrical noise from the motor driving the rotation of the transducer or the interference from spurious voltage surges inscribed onto the videotape the result is anomalous spikes in the data. Scattering of energy due to the sampling geometry is another major source of noise. The optimum raypath geom-

etry consists of a centralized tool in a circular borehole where each incident pulse has normal reflection to the tool. Nonnormal incidence of a stray pulse from a rough surface can create complex raypaths and multiple reflections. These translate to anomalously high travel times and spurious reflectivity values. Noise spikes can interfere with data interpretation and they must be eliminated where automatic fitting or feature analysis routines are implemented. Alternatively, if the spikes actually represent valid data, they may contain important information.

A median filter is used to mitigate this type of noise in the data. A median filter was selected among the myriad of smoothing operations in image processing because it is a smoothing in which the edges in an image are maintained. In median filtering a template is slid along the scans of data and the center value of the template is given the median value of all the values covered by the template. This type of filter is particularly good for removing impulse-like data because the pixels corresponding to noise spikes in their neighborhood are replaced by the most typical pixel in that neighborhood. A horizontal median template of any number of pixels can be used to preprocess BHTV data. A three point to median was determined to be the most appropriate because it provides sufficient filtering while having the advantage of being the fastest type of computational sort for large image files (Richards, 1986).

To diagnose the degree of noise in a given data set the data scans may be examined in either polar cross section or Cartesian cross section to look for spikes. Symptomatic speckles in the unwrapped views also indicate noise contamination.

Geometric Corrections

A significant source of error in the interpretation of BHTV data can arise from the apparent location of borehole features from a noncentered tool that has the effect of "moving" features to incorrect positions on the borehole wall (Georgi, 1985). These effects can lead to errors in calculating orientations of planar features as well as incorrect azimuth measurements for features such as well-bore breakouts. Corrections to the travel time images for the effects of off center tools in circular boreholes are made through geometric correction of the travel time data (Barton, 1988).

Off-center tools in elliptical boreholes present another, more complicated, geometric problem and the potential loss of a larger percentage of the returned reflection. This error is dependent on the orientation of the ellipse and dip direction. If the long axis of the ellipse is coincident with the dip direction the dip is overestimated; if the long axis is perpendicular to the dip direction it is underestimated (Georgi, 1985). A solution to this geometric problem using inversion was presented by Lysne (1986).

The borehole televiewer is designed to acquire data with the tool vertically positioned in a vertical borehole. Bowspring centralizers are positioned above and below the tool to stabilize it in the center of the borehole. Ideally the tool should be centrally located in a circular borehole. In reality, this optimum geometry is often not achieved. The possible raypaths for the BHTV pulse launched from the transducer are shown in Figure 12.10 (after Georgi, 1985). When the tool is significantly off-center in a circular drillhole vertical bands of missing data

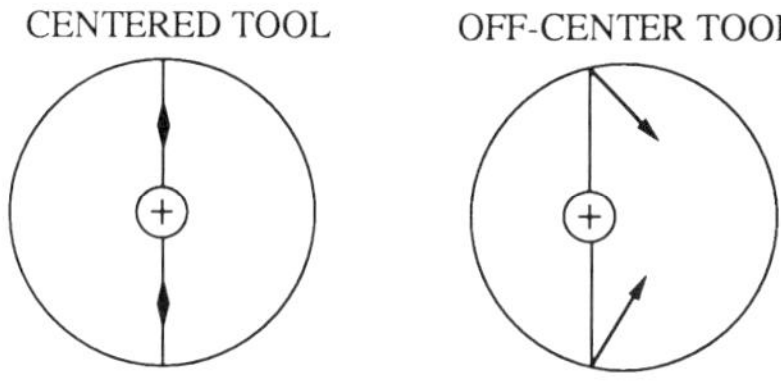

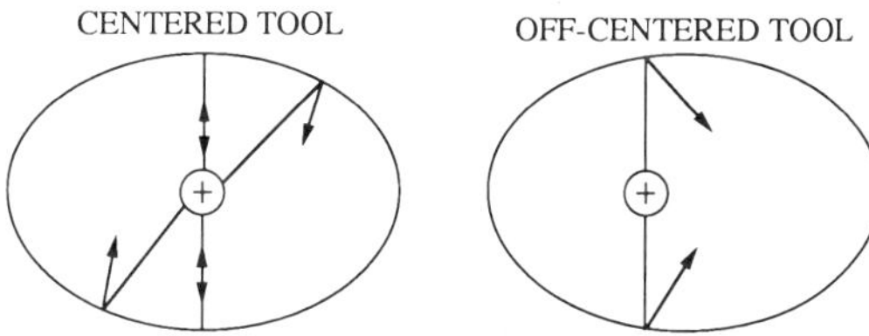

FIGURE 12.10. The possible raypaths for the BHTV pulse for circular and elliptical boreholes with centralized and off-centered tools (after Georgi, 1985), with permission of SPWLA.

result from nonnormal incidence of the pulse at the borehole wall and subsequent deflection of the returned signal away from the transducer.

To diagnose off-center effects the user can look for uneven vertical color banding in the unwrapped views or an off-center position in the polar cross section. A histogram of the travel time values will show a bimodal distribution if the tool is off-center and the Cartesian cross section will have a warped appearance instead of maintaining a horizontal line. The cross plot can also be used to diagnose off-center effects where anomalous data will occur as clusters off the main diagonal.

The polar, histogram and Cartesian views have been combined in an interactive correction window to facilitate the correction process. The upper left plot of Figure 12.11 (see Color Plate III) displays the borehole radius in polar cross section, the lower plot in Cartesian cross section, and the upper right plot as a histogram in a single window for the user to evaluate potential off-center problems. The bimodal distribution of borehole radius, in yellow, and the curvature of the uncorrected Cartesian cross section indicate these data are from an off-centered tool in a circular borehole.

Geometric corrections for decentralized tools in circular boreholes are made using an algorithm

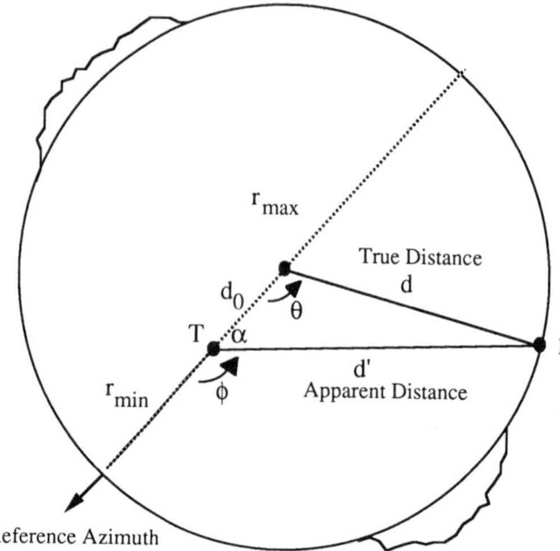

FIGURE 12.12. Geometry of the correction of BHTV data for the effects of an off-centered tool in a circular borehole.

The initial estimate for the tool position is accurate only for featureless circular boreholes. Where features are present the minimum radius may have no relationship to the true borehole center. To solve this problem the center of the borehole is iteratively determined until the routine converges to the true center. When the difference between the current iteration and the previous iteration, δd_0, is roughly zero the routine has converged. For most data sets convergence to the true center of the borehole occurs within 3 or 4 iterations. For extremely variable data, where adjacent travel time values vary more than 35%, stability in finding the true center is achieved by evaluating successive d_0 values. The smallest value of d_0 within a given number of iterations was found to provide the best estimate of the true borehole center. Once the borehole center and radial distances are calculated for each pulse the routine interpolates the amplitude values to their correct spatial position.

In the more complicated case of an off-center tool in an elliptical borehole a Marquardt inversion is used to determine the tool distance and azimuth from the borehole center (after Lysne, 1986). It was noted by Lysne (1986) that this inversion technique is unstable where the tool was off-center by less than 10% of the nominal borehole radius, in other words, as the tool approaches the center of an elliptical borehole the off-center direction becomes

that finds the true center of the borehole then calculates the corrected azimuth and radius for each reflected pulse. In the case of an off-center tool in a circular borehole the correction can be made by a relatively simple forward model. In an ideal circular borehole the point on the circumference with the minimum radial distance to the tool defines the tool position. Borehole features complicate this ideal scheme and, instead, an iterative search for the center is used to determine the true center. The minimum radius, r_{min}, initially defines the center of the borehole (Figure 12.12). Here d' is the recorded distance from the tool to the point P, d is the true distance from the center of the borehole to the point P, ϕ is the angle from the reference azimuth to the point P. The maximum radius, r_{max}, is taken as the diameter at the azimuth ϕ minus r_{min}. The distance the tool is offset from the true center, d_0, is simply $(r_{max} - r_{min})/2.0$. With this initial guess of the borehole center the radial distance, d, and angle to true center, θ, are computed for each pulse through:

$$d^2 = (d' - d_0 \cos \alpha)^2 + d_0 \sin^2 \alpha \qquad (2)$$

and

$$\theta_0 = (\phi + \pi/2) + \cos^{-1}(d_0 \sin \alpha) \qquad (3)$$

where $\alpha = \pi - \phi$ (see geometry in Figure 12.12).

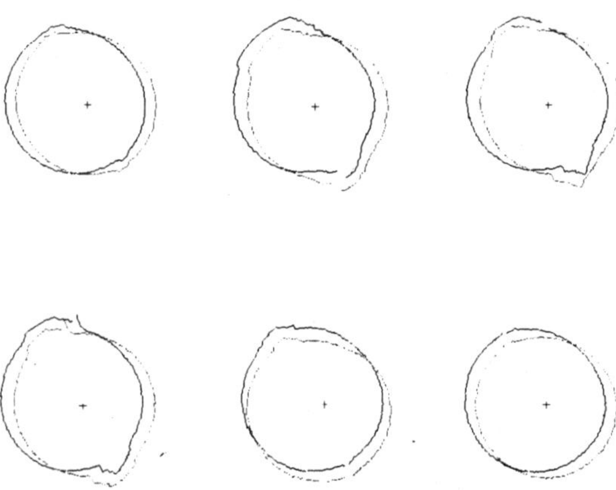

FIGURE 12.13. Polar cross sections are sample scans of the synthetic data (black) and data corrected for the effects of an off-center tool in an elliptical borehole (gray). The cross-hairs mark the center position for each scan.

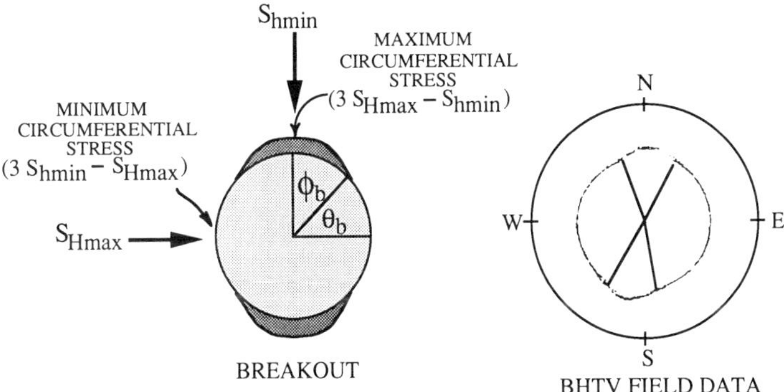

FIGURE 12.14. Schematic representation of the breakout process showing the angle of breakout initiation ϕ_b. Also shown is BHTV data over a breakout interval where polar cross sections delineate the breakout shape. The radial lines indicate the picks of breakout width. Breakout azimuth is the bisector of this angle.

less unique. This was found to be the case when this algorithm was applied to field data where the tool was not sufficiently off axis in an elliptical borehole. This technique was found to successfully correct the data (Barton, 1988), however, field data recorded in an elliptical borehole where the tool was well off-axis were not available and synthetic data had to be generated to test this algorithm. The success of this algorithm applied to the synthetic data is shown in polar cross section in Figure 12.13 where the polar projection plotted in black is the synthetic data and the projection plotted in red is the corrected data and the cross-hairs mark the center for both the original and corrected data. Note that this correction modifies both the azimuth and shape of the breakouts, features that are important to stress measurements.

Applications of BHTV Image Analysis

Perhaps the most significant potential benefit of digital BHTV data is to make accurate quantitative measurement of physical phenomena that are imaged by this logging tool. The measurements of planar features, bed boundaries, foliation, and fractures are some of the primary targets for borehole image data analysis. The use of wellbore breakouts to measure *in situ* stress has made data recorded with the borehole televiewer an extremely valuable component in many investigations of crustal stresses.

Analysis of Fine Scale Features

Although wellbore breakouts, washouts, and key seating can be seen on most 2-D images they are best distinguished in polar section of BHTV data (Broding, 1981; Barton; 1988; Menger and Schepers, 1988). The detailed borehole shape available through polar cross sections of BHTV data and interactive measurement tools permit the rapid analysis of stress induced breakout orientation or the orientation of hydraulic fractures (Barton, 1988). Figure 12.14 is a conceptual drawing of the development of a breakout (see theoretical shapes in Zoback et al., 1985) along with a polar cross section of data from the Cajon Pass well, which demonstrates the picking technique. The greater travel time within the breakout defines the characteristic breakout shape. θ_b is the angle of breakout initiation with respect to S_{Hmax}. Several superimposed scans of travel time data (representing a vertical distance of several centimeters in the well) are plotted in polar cross section to allow measurement of breakout azimuth and minimum breakout width.

The minimum width of a breakout is a function of the initial failure of the borehole wall on breakout formation and erosional effects on the initial shape of the breakout, such as fluid circulation or tool trips. Superimposing data scans before

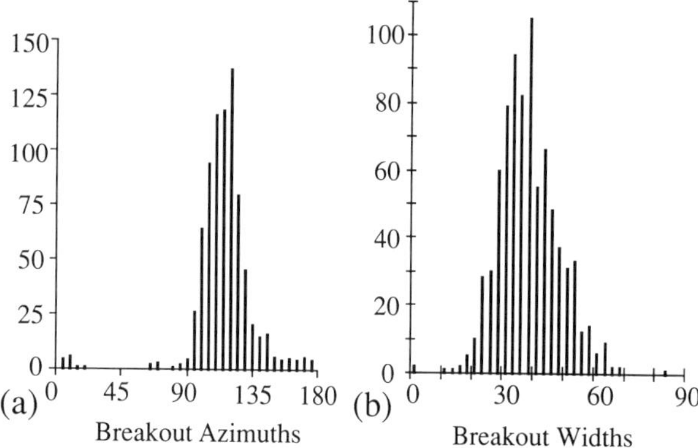

FIGURE 12.15. (a) Histogram of breakout azimuths over a 262 m interval in the EE-3 well of the Fenton Geothermal Field. (b) Histogram of breakout widths over the same 262 m interval.

plotting was found to be preferable to averaging scans to preserve the minimum width of the breakout over the vertical interval. The data must be corrected for tool position and magnetic declination before plotting. The objectives of this analysis technique are to obtain precise breakout azimuths and the minimum breakout width and to utilize all of the recorded data. Figure 12.5 shows the systematic process of the data analysis where a 360° unwrapped view of the data (left window) has been color tuned to enhance the breakouts in this interval. Polar cross sections of the borehole are plotted in the left window. The user visually picks the breakout widths using the graphics mouse to drag through the breakout angle. Where there are continuous reflections from the broken out sections, as in the example in Figure 12.14, the breakout widths are easily determined. The breakout width for each interval is picked interactively and stored in a text window. The data window can be edited and miscellaneous notes entered regarding the data quality or other information as needed. The data window can be saved for further statistical analysis or for plotting of the depth distribution of breakouts. Several off-the-shelf Macintosh software packages are available for statistical reductions and graphics plotting.

In the data analysis, the two sides of a breakout are picked independently. The two radial lines in Figures 12.5 and 12.14 represent the picked angle of the breakout width, the breakout azimuth bisects this angle. Breakout azimuth should coincide with the direction of least horizontal principal stress; breakout width is shown in the following discussion to be important for estimation of stress magnitude. Figure 12.15a is a typical histogram of the distribution of breakout azimuths and widths measured from data recorded in well EE-3 at the Fenton Geothermal site, New Mexico (Barton et al., 1988). A total of 928 separate breakout azimuths and 644 breakout widths were measured in the analysis of the Fenton data where the mean direction is 119°, and the standard deviation 11° resulting in an S_{Hmax} direction of N30°E.

The occurrence of wellbore breakouts may be used to fully determine the *in situ* stress state. With knowledge of C_0, S_{hmin}, and breakout width (Figure 12.15b) the magnitude of maximum horizontal principal stress can be estimated by the breakout analysis. A detailed description of the analysis of breakout data can be found in Barton et al. (1988). Using independently determined values for C_0 and S_{hmin}, the analysis of breakout widths in the Fenton well constrains the value of maximum horizontal principal stress to be approximately S_v, as predicted from the occurrence of both strike-slip and extensional earthquakes in response to fluid injection at Fenton Hill (Fehler et al., 1986). The Fenton well has been the site of extensive research and available information on the minimum horizontal principal stress and on rock strength made the site an excellent test of the calculation of stress magnitude using the angle of breakout initiation.

Detailed analyses of borehole shape using televiewer data can provide well-resolved orientations of the horizontal principal stresses. The magnitude of the maximum horizontal compressive stress can be constrained by these data. The analysis of breakout width may be a promising technique to estimate stress magnitude in drillholes where other techniques are not useful.

Fracture Analysis

The primary use of BHTV data since it became operational in geophysical logging has been the measurement of the orientation and distribution of planar features in a drillhole. As mentioned, planar features that intersect the borehole appear on the unwrapped 360° view as sinusoids (see Figure 12.2). These sinusoids are often discontinuous for fine scale fractures and they can show very complex patterns at points where several fractures intersect or where fractures are not perfectly planar. At least three fractures intersect at depth 5135.5 ft (1565.3 m) in the left window of Figure 12.16 (see Color Plate III). The circular features in the right window of Figure 12.16 are interpreted as fracture planes that enter and exit the borehole as a chord. The steeply dipping fracture at depth 149.2 m (489.5 ft) appears to merge with a shallow fracture above. It is important in these cases to use the various enhancement techniques discussed above to best resolve the trace of the fracture before measuring its orientation. Because of the flexibility designed into the interactive analysis system different enhancement techniques can be explored until the optimum resolution of the fracture of interest is achieved before determining the fracture orientation.

The BHTV image of a fracture surface can be extremely irregular due in part to the true topography of the surface, the limits of the resolution of the transducer, and multiple reflections from within the fracture surfaces. Because the eye can often perform the best least-squares fit to a curve within a complex network of fractures an interactive routine has been implemented to measure the amplitude and phase of the sinusoid (Figure 12.16). The graphics mouse is used to drag an adjustable curve over the imaged fracture; up and down mouse movements control the amplitude of the sinusoid and left to right movements control its phase. These measurements give the strike of the plane as $90° - \phi$ where ϕ is the minimum of the sinusoid and the dip angle, α, as $\tan^{-1} h/d$ where h is the peak to trough amplitude and d the borehole diameter. Interpreting fracture orientations using this interactive method reduces the possibility of errors from improperly fit sinusoids that can occur with other automatic methods in highly fractured intervals. It also aids in the detection of fracture planes not readily visible in the data. For example,

in Figure 12.16 the trace of the steeply dipping fracture below the intersecting fractures at 5135.5 ft (1565.3 m) is not initially apparent; however placement of the flexible sinusoid reveals the fracture trace well below the intersection of the three fractures.

With the amplitude and phase of the sinusoid determined, a wire frame cylindrical projection of the BHTV data over the fracture interval is generated that is mathematically rotated into the plane of the fracture so that the fracture width can be measured perpendicular to the fracture plane (right side of Figure 12.17). This rotation is required to correct for apparent dip of the fracture. Measuring the width at a number of points around the wellbore where the aperture is smallest minimizes the increase of apparent aperture caused by spalling during drilling and localized erosion. Alteration zones that characteristically surround a fracture are susceptible to damage during drilling. The drill bit causes an initial destruction of the fracture zone at its intersection with the drillhole. This is followed by erosion from fluid flow, reamer tools, and logging tool trips. This means that fracture aperture as preserved in the BHTV log is an upper bound to the true aperture of the fracture at some distance away from the borehole. The term "apparent fracture aperture" is used in this text to describe what is, in effect, an average of a number of "minimum width" measurements.

Once the fracture orientation and, if possible, the apparent aperture have been measured the measurement is stored in a text window for further data reduction. The text window (Figure 12.17) can be edited to add field notes or additional information as to the quality of the log. The data recorded are depth, dip direction, and dip and aperture (if measured). The ASCII data can be saved using the standard Macintosh Save command in the File menu and later plotted in stereographic projection or as a profile to investigate trends in the fractures with depth or trends controlled by fracture width.

Measurement of the dip direction of fractures vertical fractures that intersect a portion of the borehole as a chord (Figure 12.16, right window) is accomplished by selecting and recording the azimuth of each fracture intersection. The dip direction of this fracture is the midpoint of these two azimuths.

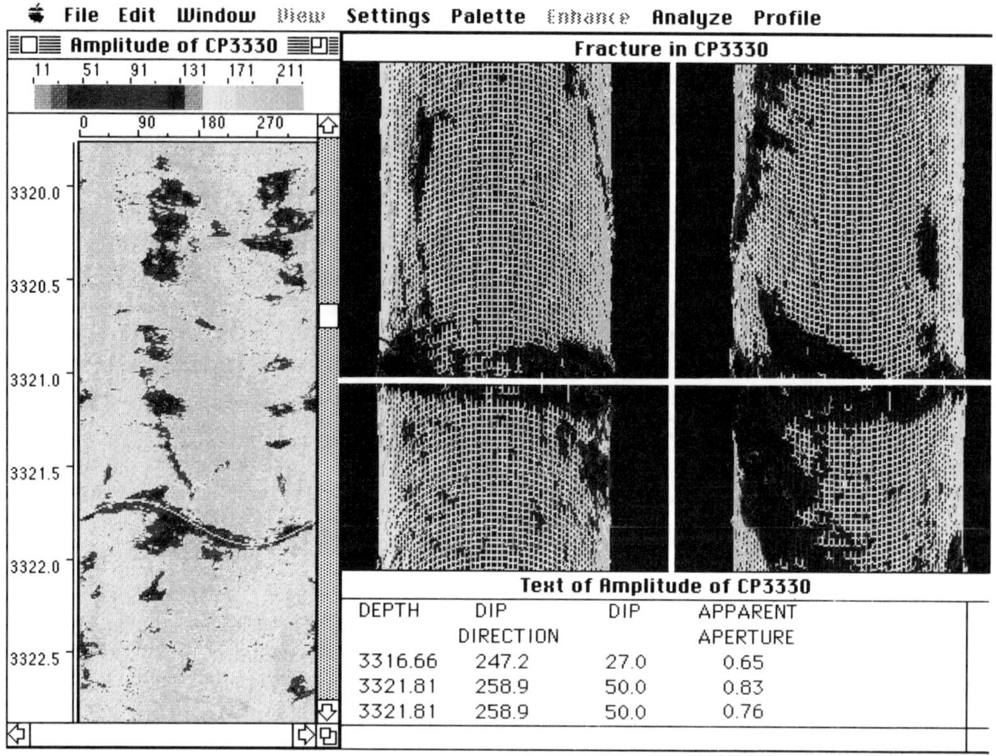

DEPTH	DIP DIRECTION	DIP	APPARENT APERTURE
3316.66	247.2	27.0	0.65
3321.81	258.9	50.0	0.83
3321.81	258.9	50.0	0.76

FIGURE 12.17. Interactive fit of a flexible sinusoid to a shallow dipping fracture at depth 3321 m in the Cajon Pass research well is shown in the left window. In the right window is the 3-D cylindrical image of the fracture plane.

A comprehensive analysis of the natural fractures intersecting the Cajon Pass research well in Southern California is a good example of the application of the fracture analysis software to field data. The Cajon Pass Well has been drilled to investigate the low stress low heat flow paradox along a major plate boundary. The fracture study was completed in order to characterize the macroscopic structure of the crust over the depth interval 1840 to 3450 m (6036.7 to 11,318 ft) (Barton and Zoback, 1991).

Orientations of all macroscopic fractures, corrected for fracture dip are shown in Figure 12.18 in a lower hemisphere stereographic projection. On the left side the data are presented as poles to fracture planes and on the right side they are presented in a contour diagram of pole densities using spherical Gaussian statistics (after Kamb, 1959). The Kamb method of contouring pole densities calculates the number of standard deviations from a uniform distribution of points on the projection. The contour shading shown in the legend in Figure 12.18 represents the standard deviation from a random distribution. Although there is clearly a random component of fracture orientations, a statistically significant concentration of fractures strikes north–south and dips to the west. The primary concentration of fractures strikes at an azimuth of 175° and dips 63° west. Also shown in Figure 12.18 is the N60°W local trend of the San Andreas Fault (SAF) and the average orientation of S_{Hmax} determined through the analysis of *in situ* wellbore breakouts and hydraulic fractures (N59°E, Shamir and Zoback, 1988). Although the San Andreas Fault is the dominant tectonic feature at this site the majority of the macroscopic fractures in the Cajon Pass well are neither aligned with the San Andreas nor parallel to the direction of maximum horizontal stress as is often assumed (e.g., Engelder, 1982).

The fracture distribution in the Cajon Pass well does not decrease with depth. Figure 12.19 shows a fracture frequency profile over a subset of the fracture data measured. Aside from the interval

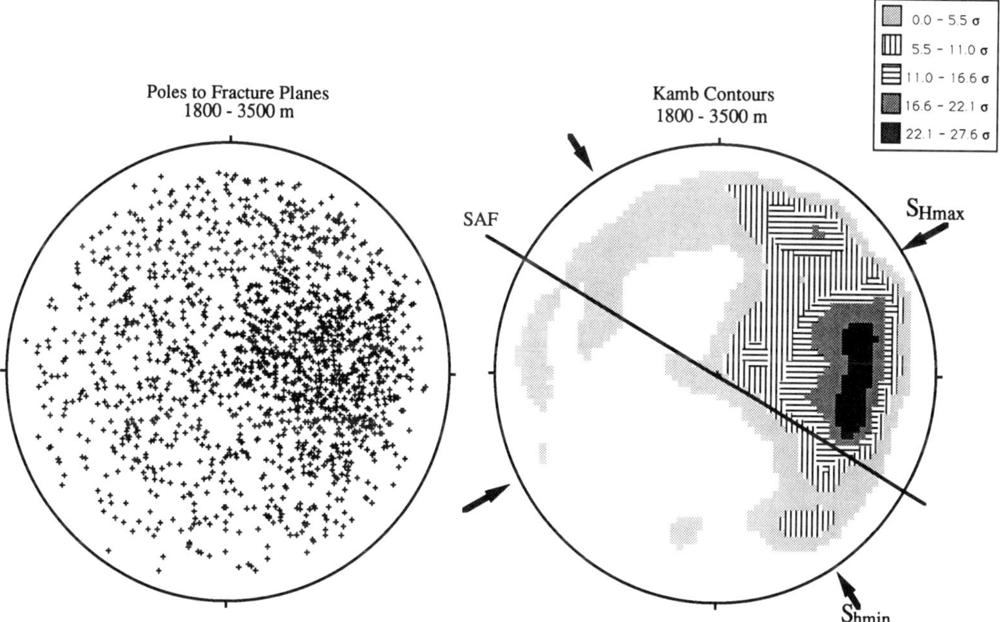

FIGURE 12.18. (a) Lower hemisphere equal area projection of poles to fracture planes over the interval 1829 to 3454 m in the Cajon Pass reserach well. (b) Kamb contours of the poles to fracture planes over the same interval. Also shown is the orientations of the San Andreas Fault (SAF) and the principal stress directions determined from wellbore breakout and hydraulic fracturing experiments in the well.

from 1940 to 1960m (6364.8 to 6430.4 ft) of stress-induced wellbore breakouts (Shamir et al., 1988), which contain only a few observable fractures, the total fracture population shows no trend in size or distribution with depth over the logged interval. These results are in contrast with a fracture analysis from an Illinois drillhole where fracture density and hydraulic conductivity decrease with depth (Haimson and Doe, 1983). Fracture studies of 10 wells located in the western Mojave Desert just west of the Cajon Pass well by Seeberger and Zoback (1982) found only a slight decrease of fracture density with dcpth for most wells studied. They found some wells to have a uniform fracture density distribution while others showed concentrations of fractures at various depths as is the case in the Cajon Pass well.

Where fractures are open along the extent of their intersection with the borehole, apparent fracture aperture could be measured. The double asterisks in the second profile of Figure 12.19 correspond to zones where the fracture width per meter is greater than 10 cm; the single asterisks correspond to zones with fracture width is between 5 and 10 cm. The remaining intervals have nominal apparent aperture per meter. In several intervals with a high frequency of fractures the cumulative width per meter is quite low. However, where the apparent aperture per meter is large there are usually a large number of fractures. Although mechanisms acting to close fractures would be expected to reduce the number of open fractures with increasing depth, the data do not show this trend.

Figure 12.19 also shows the fracture population correlated with lithology and sonic velocities. The Cajon Pass experiment has provided an ideal complementary data set to examine the acoustic waveform response to fluid-filled fractures. The complete BHTV log and the digital analysis of fracture geometry were used in conjunction with the full waveform sonic data to determine the response to fractures (Barton and Moos, 1988). Note that there is little apparent relationship between fracture density and sonic velocity. A much greater correlation exists between the velocities and the apparent apertures determined from the BHTV logs. The

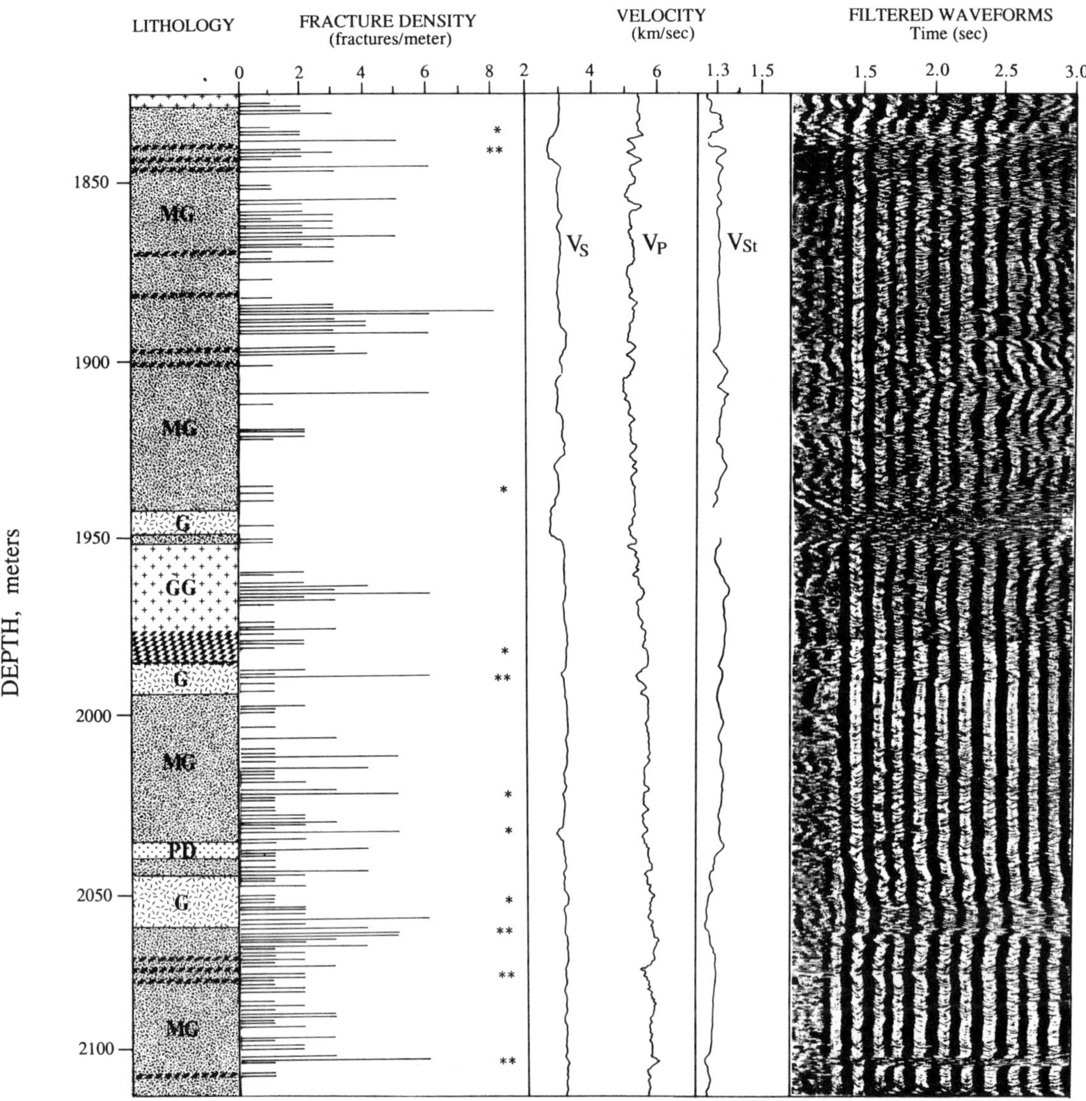

FIGURE 12.19. Lithology, fracture density, apparent aperture, sonic velocities, and filtered sonic waveforms over the interval 1829 to 2115 m in the Cajon Pass research well.

relationship between Stoneley wave amplitude and permeability has been demonstrated by a number of authors (Paillet, 1980; Burnes et al., 1988). The amplitude of the Stoneley wave is particularly sensitive to the presence of hydraulically conductive fractures intersecting the borehole wall (Paillet and White, 1982; Rosenbaum, 1974), but it is also affected by variations in the borehole diameter and the seismic properties of the rock. The Stoneley arrivals in the sonic waveforms recorded in this interval of the Cajon Pass well (Figure 12.19) are an excellent example of this response.

Measurements of the orientation, distribution, and apparent aperture of macroscopic fractures in the Cajon Pass well are extremely important to the interpretation of other geophysical and core data. Current research into the statistical distribution of fracture aperture, fracture orientation, and fracture spacings is providing insight into the mechanical character of the brittle crust at the Cajon Pass site (Barton and Zoback, 1991).

Correlation of BHTV "Log" Data

Log interpretation has long been the basis for discriminating changes in lithology, fluid content of pore space, structural horizons, and general physical properties of rock both for industry exploration and scientific research. Geophysical logs that are generally recorded in a freshly drilled well include the caliper, natural gamma, spectral natural gamma (from which K, U, and Th concentrations can be determined), induction resistivity, neutron porosity, gamma–gamma density, and sonic travel time. Correlation of these logs and extracted core provides ideal information for physical properties measurements with depth.

BHTV travel time data can be used to generate extremely fine scale "logs" of variations in downhole parameters such as pseudo 4-arm caliper emulation, cumulative cross-sectional area, and wellbore eccentricity. These profiles may be correlated with dipmeter, full waveform, resistivity, and other diagnostic logs to assess borehole stability and to define permeable zones or lithologic boundaries. The depth resolution of image data is generally much finer than the 0.5 ft depth resolution of geophysical logs, however, digital smoothing or decimation of image data to the scale of conventional logs provides excellent "ground truth" to the complex process of geophysical log analysis.

BHTV data provides information at about 1 cm (0.3 in) intervals whereas typical logging produces a data point for every 15 cm (6 in) of wellbore. For this reason, analog BHTV images are commonly photo-reduced to the scale of log data (5 in. = 100 ft) before correlations can be made. As mentioned previously interpretation of an analog BHTV image usually involves overlays and tedious drafting. Digital data provide several advantages over analog data for this type of coarse scale data interpretation and correlation with standard logs. The first is the simplicity of scaling with a minimal loss of information. The second advantage is the ease of constructing an interpreted BHTV log. In a coarse scale analysis only the large scale features such as breakout intervals or through-going fractures are required for interpretation. The location of these features can be plotted through interactive thresholding of amplitude and borehole radius where black represents a feature and all other variations in the data have been suppressed into a white background (Figure 12.20, right profiles). Fractures may also be interactively picked and the interpreted sinusoids plotted. Figure 12.20 shows standard geophysical logs over the interval 9600 to 9800 ft (2926 to 2987 m) recorded in the Cajon Pass Research well with the corresponding digitally interpreted section of BHTV data. Major features include the sinusoidal trace of fractures and wellbore breakouts, which are represented by discontinuous black zones at 180° azimuths and fractures.

The following preliminary correlation of the analog field data over this interval was made by F. Paillet (written communication, 1988). The continuous interval of wellbore breakouts below depth 9800 ft (2987 m) is evident on the caliper log (Figure 12.20, track 4, curve 1) and the elongation of the borehole is very pronounced in the x–y caliper log (Figure 12.20, track 1, curves 1 and 2). The fracture zone just below 9700 ft (2956.6 m) has caused a marked decrease in both the LLS and LLD resistivity curves (Figure 12.20, track 2, curves 1 and 2) however this zone has a minimal effect on the neutron porosity logs (Figure 12.20, track 4, curves 2 and 3). Severe breakouts at 9665 ft (2945.9 m) cause an anomaly in the caliper (Figure 12.20, track 4, curve 1) and density logs (Figure 12.20, track 3, curve 3). Between 9600 and 9650 ft (2926 and 2941 m) breakouts are interrupted by small fractures; breakouts often terminate at the intersection of a fracture.

It is interesting to compare the digital interpretation of BHTV logs with an "interpreted section" made from the black and white Polaroid field logs by a scientist who is very familiar with BHTV data and very experienced in log interpretation. In addition, this particular expert has artistic talents useful to this sort of data reduction. In Figure 12.20, the far right window is the manual interpretation of the interval 9400 to 9600 ft (2865 to 2926 m) in the Cajon Pass well, sketched by F. Paillet (written communication, 1988). Thick lines represent the interpreted trace of larger fractures visible in the analog records and thin lines the trace of smaller fractures. Breakouts are again represented by vertical black bands. Fractures located in the digital counterpart to the interpreted interval compare well with the manually picked fractures. Breakouts reduce with good precision and correlate very well with the manually interpreted section. The most logical use of the digital interpretation is to provide

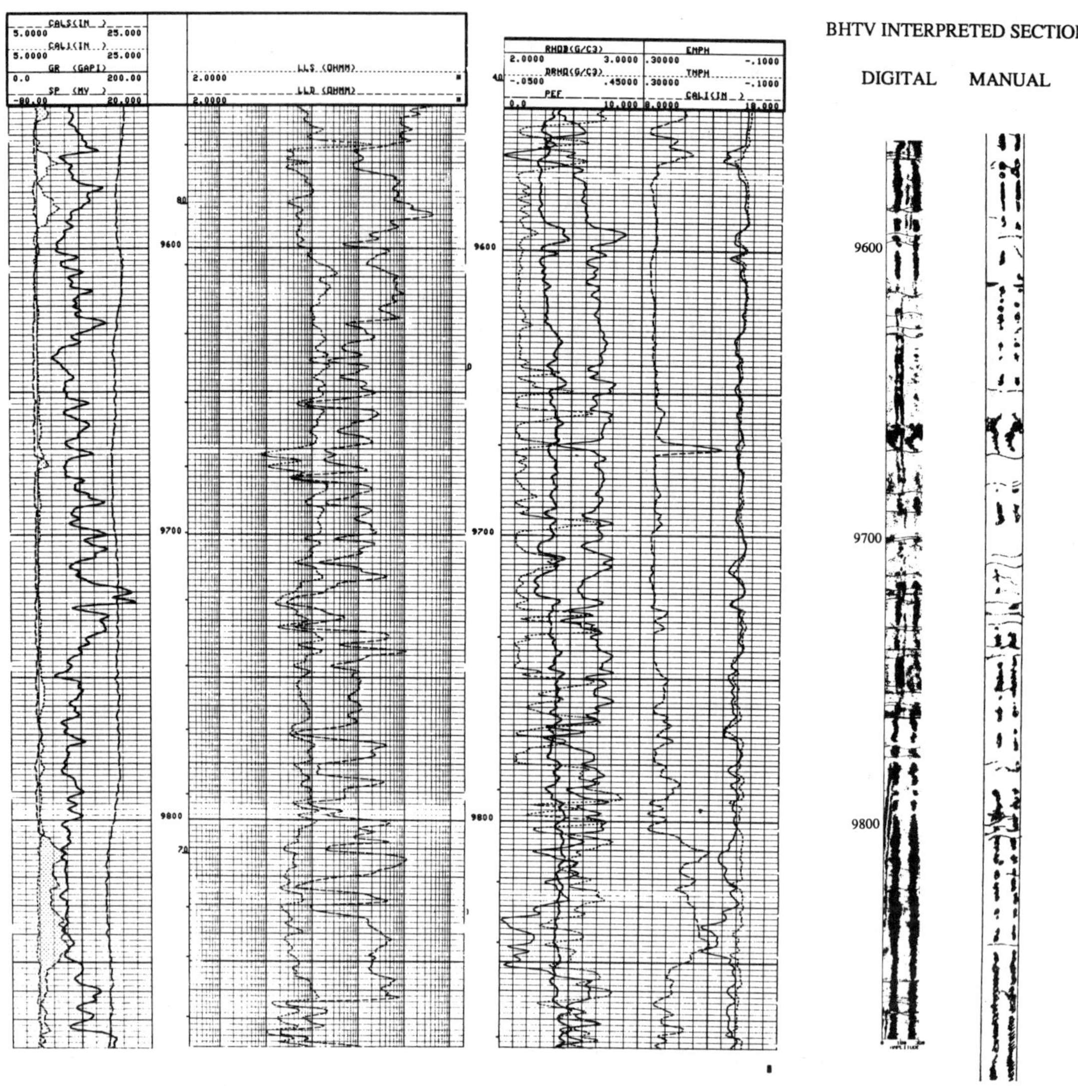

FIGURE 12.20. Standard Schlumberger logs over the interval 9600 to 9800 ft in the Cajon Pass research well correlated with BHTV data recorded over the same interval. The BHTV data have been both digitally and manually interpreted.

an easily generated, accurate base plot either for further computer-based interpretation or manual interpretation. Scientists that lack substantial experience with interpreting BHTV data or general log interpretation (or artistic capabilities) can produce more reliable interpretations with the help of the digital BHTV system.

BHTV Generated "Logs"

Caliper tools usually have two to four mechanical arms that measure the borehole diameter with depth. The BHTV log can be thought of as having several hundred "arms" that provides data at a scale that is 15 times finer than the standard caliper log. To emulate the HDT dipmeter tool, the horizontal and vertical resolution of the BHTV data must be made comparable to that of the HDT dipmeter pad. Performing a horizontal 15 point median filter is generally sufficient to mimic the HDT dipmeter pad width. The maximum value in each scan of the filtered data then provides one radius and azimuth of the long axis caliper pair. The orthogonal values provide the short axis caliper pair. Averaging successive caliper values with depth provides the correct vertical adjustment for comparison with the

HDT dipmeter tool. Figure 12.21a represents the fit of the BHTV generated "four arm" caliper to a typical scan of BHTV data and 12.21b the fit over an interval of wellbore breakouts measured from the same well. Black represents the data scan and gray the measured four arm radii. Caliper information in this form can be easily correlated with other log data and with dipmeter data. As is the case with dipmeter data the caliper data can be used to discriminate between breakout zones, washout zones, zones of key seating, and perfectly in-gauge zones (Plumb and Hickman, 1985). The long caliper arm measured in this way in many cases corresponds to the direction of minimum horizontal principal stress and in such a log provides a rapid method for breakout orientation analysis with depth. The left window in Figure 12.22 shows the image data for a 12 ft (3.7 m) interval of wellbore breakouts in the Auburn Geothermal Well, New York. The adjacent profile shows the emulation of the 4 arm dipmeter tool over this depth interval. The scatter plot indicates the orientation wellbore breakouts determined from this analysis.

There are a variety of causes for a noncircular borehole and wellbore breakouts may or may

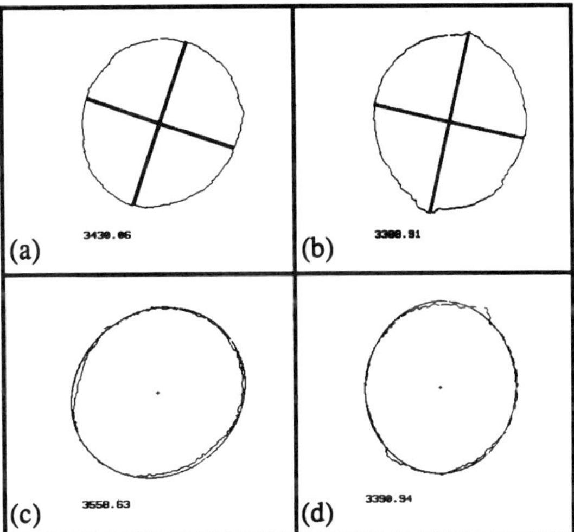

FIGURE 12.21. (a) Black represents the data scan and gray the "four arm" caliper measurement for a typical scan of data. The major axis of the ellipse is shown for reference. (b) Black represents the data scan and gray the "four arm" caliper measurement over an interval of wellbore breakouts. (c) Black represents the data scan and gray the fit of an ellipse to a typical scan of data. (d) Black represents the BHTV data scan and gray the fit of an ellipse over an interval where breakouts are evident.

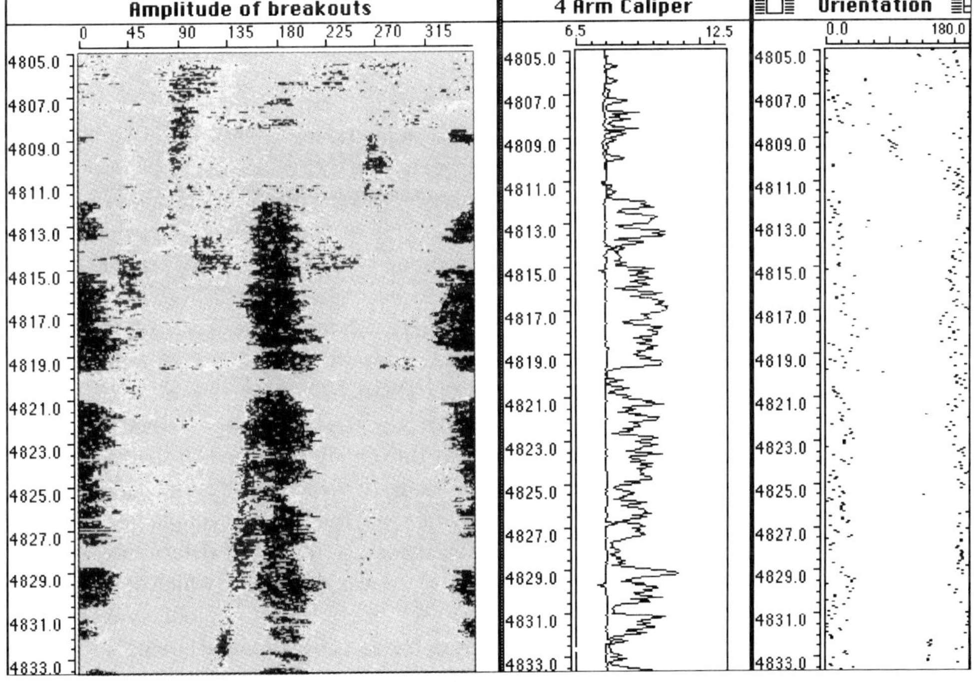

FIGURE 12.22. Caliper analysis over a 30 ft interval of wellbore breakouts in the Auburn Geothermal well, New York.

not coincide with the overall ellipticity of the borehole. For example, ellipticity can be drilling induced where the bit has tended to move up dip during drilling. The borehole wall may preferentially spall or washout in response to *in situ* stresses. At least-squares fit of an ellipse to the BHTV scans is used to measure the eccentricity of the borehole. The algorithm fits an ellipse to a field of points after each radius value is transformed to Cartesian x,y space from the polar r,θ pair by $x = r \cos \theta$, $y = r \sin \theta$. The angle of the ellipse is given by

$$\phi = \tfrac{1}{2} \tan{-1} \frac{\sum xy}{\sum x^2 - \sum y^2} \qquad (4)$$

A least-squares algorithm is used to determine the major, a, and minor, b, axes of the ellipse. From the equation of an ellipse

$$b^2 x^2 + a^2 y^2 = a^2 b^2 \qquad (5)$$

we have

$$y^2 = b^2 \left(1 - \frac{x^2}{a^2} \right) \qquad (6)$$

or

$$\hat{Y}_i = \left[b^2 \left(1 - \frac{X_i^2}{a^2} \right) \right]^{1/2} \qquad (7)$$

where $\hat{Y}_i$ is the estimated value of Y_i at specified values of X_i.

We want to have $\sum\limits_{i=1}^{n} Y_i^2 = (\hat{Y}_i - Y_i^2) = $ minimum:

$$\sum_{i=1}^{n} Y_i^2 = b^2 n - \frac{b^2}{a^2} \sum_{i=1}^{n} X_i^2 \qquad (8)$$

Setting $b = b_0$ and $b_1 = b/a$ we have

$$\sum_{i=1}^{n} Y_i^2 = b_0^2 n - b_1^2 \sum_{i=1}^{n} X_i^2 \qquad (9)$$

$$\sum_{i=1}^{n} X_i^2 Y_i^2 = b_0^2 n \sum_{i=1}^{n} X_i^2 - b_1^2 \sum_{i=1}^{n} X_i^4 \qquad (10)$$

Equations (9) and (10) are then inverted to determine the unknown values of b_1 and b_0. Generally only 20 equally spaced data points are required to accurately determine the shape of the ellipse. The percent eccentricity of the ellipse, defined as $(a-b)/(a+b)/2$, is calculated for each scan. Borehole televiewer data are capable of resolving borehole ellipticity in much finer detail than conventional caliper logs. Without some knowledge of borehole ellipticity the separation between caliper curves (C1

and C2 in standard HDT presentation) may be erroneously interpreted as wellbore breakouts.

Another profile computed with the BHTV travel time data is borehole cross-sectional area. The log of borehole cross-sectional area is calculated by integrating adjacent triangles defined by the travel time values and the center of the borehole. Borehole irregularities, washouts, or key seats, for example, are indicated by large fluctuations in this log. A profile of cumulative cross-sectional area with depth is also available. These logs are important for the determination of wellbore stability. A featureless borehole will have a smooth slope in its cumulative profile of cross-sectional area whereas breaks in slope may indicate borehole failure.

The amplitudes of reflectivity values are used to generate a reflectivity log as an aid in the detection of lithologic and structural boundaries. Acoustic reflectivity of the borehole wall as measured by the BHTV tool has incorporated the effects of the impedance contrast from various rock types and the borehole fluid along with the effects of energy scattering from irregularities in the borehole wall. The true reflection amplitude is proportional to the reflection coefficient at the borehole wall plus an unknown variation in amplitude due to scattering (the angle between the beam and the reflecting surface), thus no quantitative value of the reflectivity can be associated with a particular lithology. Because of these combined effects however, the BHTV image of the amplitude of the reflected energy often shows more detail than does the surface topography. This is particularly true for logged intervals of sedimentary rock where rock properties vary over small distances.

To determine the reflectivity value to represent a given depth interval a histogram of the amplitude values is computed for each scan. A single scan in the BHTV data has values that represent the reflectivity of unperturbed rock as well as reflectivity values associated with structural or erosional features (e.g., fractures, tool reamer marks). Interactive thresholding is used to determine the cutoff reflectivity value that best isolates the particular feature of interest. For example, to sample only the intact rock, a threshold determined cutoff reflectivity value is used above which histogram bins will be used to compute the mode value of the amplitude for each data scan as a single representative value. Figure 12.23 shows the results of this

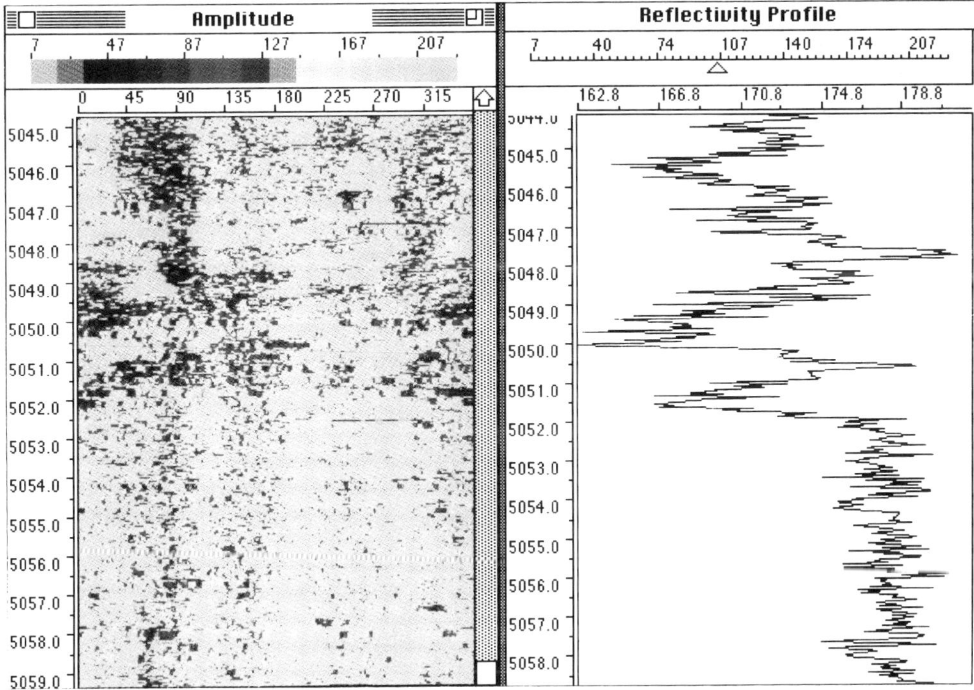

FIGURE 12.23. Reflectivity log over the contact between the basement marble formation and the Potsdam sandstone from data recorded in the Auburn Geothermal Well, New York.

statistical approach to an interval containing the contact between a basement marble formation and an overlying sandstone unit at 5051 ft (1539.5 m) in the Auburn Well, New York. The marble has a very different reflectivity character that the overlying sediment and the contact can be more easily picked in the "log" plot than in the image itself.

Reflectivity logs can be used for correlation with companion log data. Sharp decreases in reflectivity track breakout zones exactly and the decrease has been found to be proportional to the width of the breakout (Zoback and Moos, 1988). Reflectivity logs prove to be quite useful for correlation with other log data to isolate borehole features such as breakouts and fractures. Using reflectivity logs generated from the analysis of BHTV data Zoback and Moos (1988) found that low reflectivity values correlate with breakout zones in the Moodus Research Well located in New York state. Reflectivity profiles used in conjunction with the BHTV caliper log are useful for discriminating between natural through going fractures and wellbore breakouts. A sharp low reflectivity response associated with a strong kick in the eccentricity log character-

izes a natural fracture whereas low reflectivity values over a finite length of the borehole associated with a consistent azimuth of the major elliptical axis generally represents breakout zones.

Software Implementation

As discussed previously, user interface consistency among application programs is an important factor in the ease of use of Macintosh computers. To help achieve the desired consistency in BHTVImage, the program was implemented using an Apple product called MacApp, in whose original development one of us played a part. MacApp is a so-called "object-oriented framework," i.e., a "generic" application that defines standard "objects" like windows and views. Any desired Macintosh application can be derived from MacApp by describing in program code only the differences between that application and the generic one. The resulting program achieves consistency with other Macintosh applications without special effort on the part of the programmer.

MacApp, as well as the bulk of the BHTV analysis program, is written in the object-oriented language Object Pascal. Object Pascal is an extension of the Pascal language that provides a record-like structure type called an "object type." One type of object can be defined as being a subtype of another type. If type "dog" is defined as a subtype of "mammal," it means that the characteristics of dogs are generally the same as those of mammals. Only differences need be specified. One must specify that dogs wag their tails and bark, but one need not specify that dogs breathe—that characteristic is inherited from mammals.

MacApp provides a type of object called a "document." A document is a file containing text, graphic, and/or image data. The definition of type document specifies attributes that every document must have, e.g., a file name. The definition also defines operations that can apply to any document, such as "opening" it for examination and possible change, "printing" it onto paper, and "saving" changes to the disk.

BHTVImage, in turn, defines an object type "BHTVDocument" that is a subtype of the standard MacApp document. A BHTVDocument is like the generic document except it sports additional attributes like well name and fluid velocity, and it defines additional operations such as the ability to convert a travel time measurement to a radius.

MacApp provides another type of object called a "view," used to display part or all of a document in a form desired by the user. BHTVImage also defines three major types of views: the unwrapped view, which displays false-color image data; the profile view, which can graph a profile of one parameter sampled at each depth in a range; and the plot view, which can graph data sampled at numerous places at each depth. BHTVImage further defines several types of plot views, e.g., the travel time histogram, the polar cross section, the Cartesian cross section, and the 3-D cylindrical projection. A cross section view *inherits* the characteristics of a plot view and adds its own. A plot view *inherits* the characteristics of a generic view and adds its own.

The document and the view are the two most important object types defined by MacApp. The third most important object is called the "command." The generic command provides generic ways of dealing with doing the command, undoing it, and redoing it. If the command is issued by mouse action (like drawing a box in MacDraw), then the generic command also provides generic ways of tracking the movements of the mouse and giving the user an advance look at what will result when the mouse action terminates. In BHTV-Image, subtypes of the generic command object are defined to support the following user actions, among others: changing a threshold value through a sliding scale, changing the hue of a palette color, and cutting selected scans from the data file.

Certain BHTV analysis subroutines are written in the language C. The Macintosh Programmer's Workshop, a suite of software development tools, allows modules written in different languages (e.g., Pascal, C, Fortran, and Assembler) to be linked together into a single program and for procedure calls to be made between the modules.

Conclusions

The analysis of image data requires a high level of interactive manipulation by the analyst. Various features need to be observed in an optimal view to detect subtle features and the analyst must be able to make decisions as the analysis proceeds. To meet these goals for the image analysis of borehole televiewer data we have taken advantage of the interactive user interface of the Apple, Inc. Macintosh II computer. In addition, conform to the Macintosh user interface standards we used an Apple, Inc. product called MacApp, a developer's tool that provides easy programming access to the windows, menus, and dialog boxes of the Macintosh user interface. The program design defines a set of tools specific to borehole cylindrical geometry and to the particular features that are imaged by and can be quantified by BHTV data (e.g., fractures, wellbore breakouts). The program also provides a base for experimentation with new techniques to improve data quality, develop new measurement techniques, and explore unusual features encountered in a particular data set.

BHTVImage has been used as the primary analysis tool for BHTV data recorded in the Cajon Pass scientific research well in southern California and is also being used at the KTB ultradeep well site in Germany for the analysis of wellbore breakouts and the evaluation of wellbore stability.

Although the software is specific to BHTV data the analysis tools extend to other types of image data, for example, the Schlumberger, Inc., Formation Microscanner (FMS). The human interface approach, however, extends beyond image analysis programs. Geophysicists often encounter data that are voluminous and complex. They require the freedom to pursue various paths of reasoning and curiosity to make scientific discoveries without the burden of extensive computer programming at each turn in the path. One of the ways to achieve this freedom is to develop software tools that permit interaction with the data and minimize interaction with the computer. Our current work includes the development of similar analysis tools for the analysis of FMS image data and standard geophysical logs.

Acknowledgments. The authors would like to thank Apple Computer, Inc. for their cooperation in the development of BHTVImage. This work was supported in part by the Continental Lithosphere Program of the National Science Foundation through DOSECC, by the Kontinentales Tiefbohrprogramm of the Federal Republic of Germany, and by the contributors to the Stanford Rock Physics and Borehole Geophysics Project.

References

Barton, C.A., 1988, Development of in situ stress measurement techniques for deep drillholes, Ph.D. Thesis, Stanford University.

Barton, C.A., and Moos D., 1988, Analysis of macroscopic fractures in the Cajon Pass Scientific Drillhole over the interval 1829–2115 meters: Geophys. Res. Lett. 15(9), 1013–1016.

Barton, C.A., and Zoback, M.D., 1991, Self-similar distribution and properties of macroscopic fractures at depth in crystalline rock in the Cajon Pass scientific drillhole: J. Geophys. Res. (in press).

Barton, C.A., Zoback, M.D., and Burns, K.L., 1988, In situ stress orientation and magnitude at the Fenton Geothermal site, New Mexico, determined from wellbore breakouts: Geophys. Res. Lett. 15, 467–470.

Broding, R.A., 1981, Volumetric scanning well logging: Log Analyst 23(1), 14–19.

Burnes, D.R., Cheng, C.H., Schmitt, D.P., and Toksoz, M.N., 1988, Permeability estimation from full waveform acoustic logging data: Log Analyst 29, 112–122.

Engelder, T., 1982. Is there a genetic relationship between selected regional joints and contemporary stress within the lithosphere of North America?: Tectonophysics 1, 161–177.

Faraguna, J.K., Chace, D.M., and Schmidt, M.G., 1989, An improved borehole televiewer system–image acquisition analysis and integration: S.P.W.L.A. Annu. Log. Symp., 30th, Denver, CO., Trans, paper UU.

Fehler, M., House, L., and Kaieda, H., 1986, Seismic monitoring of hydraulic fracturing: 27the U.S. Symp. Rock Mech., Univ. of Alabama, 606–613.

Georgi, D.T., 1985, Geometrical aspects of borehole televiewer images: SPWLA Log. Symp. Transact., paper 1-C.

Haimson, B.C., and Doe, T.W., 1983, State of stress, permeability and fractures in the Precambrian granite of northern Illinois: J. Geophys. Res. 88, 7355–7371.

Hinz, K., and Schepers, R., 1985, SABIS–The digital version of the borehole televiewer: Eighth Eur. Formation Evaluation Symp., 1–20.

Kamb, W.B., 1959, Ice petrofabric observations from Blue Glacier, Washington, in relation to theory and experiment: J. Geophys. Res., 64, 1891–1910.

Lysne, P., 1986, Determination of borehole shape by inversion of televiewer data: Log Analyst 26, 64–71.

Menger, S., and Schepers, R., 1988, Method to derive high-resolution caliper logs from borehole traveltime data: Abst. 58th Annu. Int. SEG Meeting, Oct 30-Nov. 3, Annaheim, CA. 554–556.

Paillet, F.L., 1980, Acoustic propagation in the vicinity of fractures which intersect a fluid filled borehole: SPWLA, 21st Annu. Log. Symp., paper D.

Paillet, F.L. and White J.E., 1982, Acoustic modes of propagation in the borehole and their relationship to rock properties: Geophysics 47(8), 1215–1228.

Pasternack, E.S., and Goodwill, W.P., 1983, Applications of digitial borehole televiewer logging: S.P.W.L.A. Annu. Logging Symp., 25th, Calgary, Canada, Trans., paper C.

Plumb, R.A. and Hickman, S.H., 1985, Stress induced borehole elongation: A comparison between the four-arm dipmeter and the borehole televiewer in the Auburn Geothermal well: J. Geophs. Res. 90(B7), 5513–5521.

Rambow, H.K., 1984, The borehole televiewer–some field examples: S.P.W.L.A. Annu. Log. Symp, 25th, New Orleans, LA., Trans., paper C.

Richards, J.A., 1986, Remote Sensing Digital Image Analysis: Springer-Verlag, New York.

Rosenbaum, J.H., 1974, Synthetic microseismograms: Logging in porous formations: Geophysics 39, 14–32.

Seeburger, D.A., and Zoback, M.D., 1982, The distribution of natural fractures and joints at depth in crystalline rock: J. Geophs. Res. 87, 5517–5534, 1982.

Shamir, G., and Zoback, M.D., 1988, In situ stress orientation near the San Andreas fault: Preliminary results to 2.1 km depth from the Cajon Pass Scientific Drillhole: J. Geophys. Res., 989–992.

Taylor, T.J., 1983, Interpretation and application of borehole televiewer surveys: S.P.W.L.A. Annu. Log. Symp., 24th, Calgary, Canada, Trans., paper QQ.

Wiley, R., 1980, Borehole televiewer—revisited: S.P.W.L.A. Annu. Log. Symp, 21st, Lafayette, LA, Trans., paper HH.

Wong, S.A., Startzman, R.A., and Kuo, T-B, 1989, Enhancing borehole image data on a high-resolution PC: Soc. Petr. Eng. SPE 19124, 37–48.

Zemanek, J., Glenn, E.E., Norton, L.J., and Caldwell, R.L., 1970, Formation evaluation by inspection with the borehole televiewer: Geophysics 35, 254–269.

Zoback, M.D., Moos, D., Mastin, L., and Anderson, R.N., 1985, Wellbore breakouts and in situ stress: J. Geophys. Res. 90, 5523–5530.

Zoback, M.D., and Moos, D., 1988, In situ stress, natural fracture and sonic measurements in the Moodus, Connecticut scientific research well: report prepared for Woodword-Clyde Consultants, Wayne, N.J.

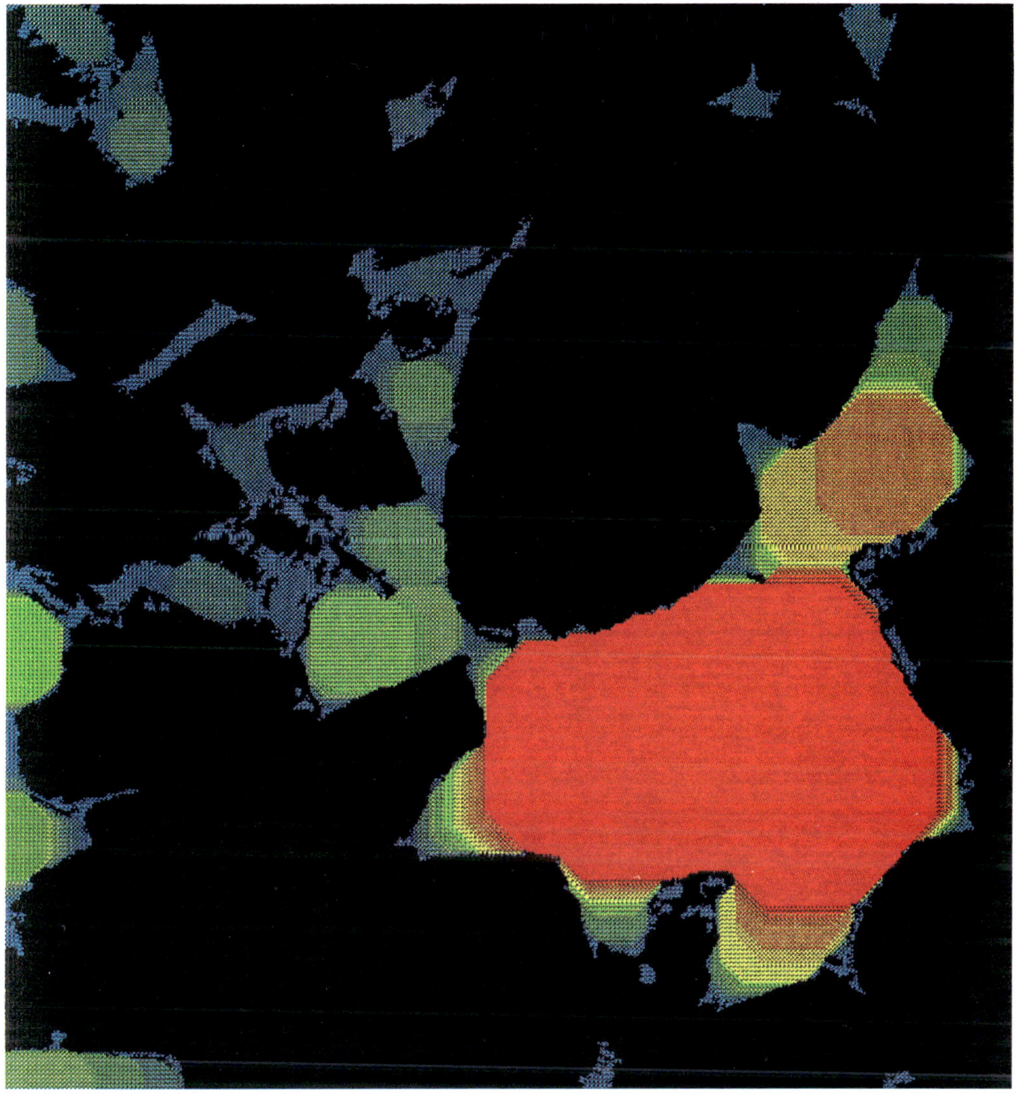

FIGURE 13.5. Composite image showing the influence of the ruler size on fractal dimension computation.

(a)

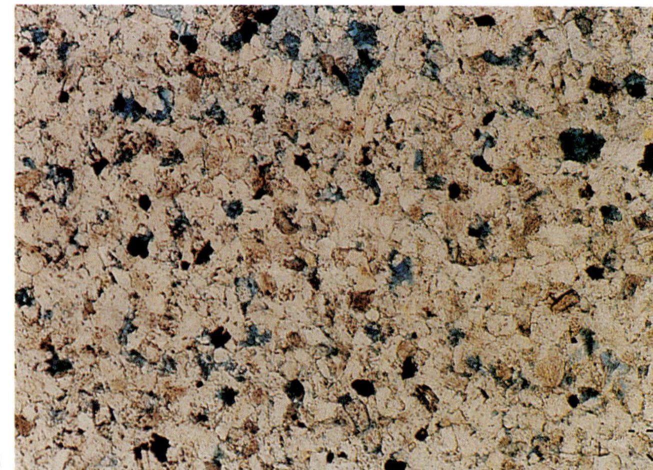

(b)

FIGURE 13.8. (a) Thin-section photomicrograph of a conventional core sample shows the very fine grain size and the poorly developed pore system. (b) Thin-section photomicrograph of a relatively unaltered drill cutting from the same interval reveals a coarser grained sandstone with a moderately well developed pore system.

FIGURE 13.10. Thin-section photomicrograph of a sidewall core sample reveals both altered and unaltered (the "button") portions typical of this sample type.

Petrographic Image Analysis: An Alternate Method for Determining Petrophysical Properties

R.E. Gerard, C.A. Philipson, F.M. Manni, and D.M. Marschall

Image analysis uses two-dimensional (2-D) geometric parameters of a pore network cross section to derive estimates of three-dimensional (3-D) petrophysical properties. A purpose of petrographic image analysis (PIA) is to obtain estimates of permeability (k) and porosity (ϕ) in cores that are difficult to analyze by conventional methods, e.g., sidewall cores, drill cuttings, and laminated and friable samples.

To achieve this purpose, two major objectives must be accomplished. First, an image must be acquired and segmented so that a single light intensity is assigned to all pores and a different light intensity is assigned to all rock material, thereby resulting in a binary image. Second, an equation must be determined to statistically relate 2-D geometric parameters measured on a thin section to 3-D petrophysical properties. These equations, which are presently limited to sandstones, are formation specific and magnification dependent.

Using conventional multiple linear or log-linear regression, a sizing technique has been developed to relate the 2-D geometric parameters for an image to the 3-D petrophysical properties. A glass bead model was used to determine which geometric parameters should be included in the multiple regression for the sizing technique. Relationships between the 2-D (visual) porosity or permeability and their 3-D counterparts are defined by integrals averaging the 2-D parameters over many random slices.

Fractal mathematics can be applied to measure 2-D parameters if a segmented image of a sandstone thin section is viewed as a stochastic Sierpinski carpet. For the 990 images analyzed with this method, the correlation coefficient for the best fit line has been 0.95 or better. Generalized least-squares factor analysis was performed to determine the relationship between porosity, permeability, and variables derived from fractal analysis. Validity of the clusters was checked by performing a discriminant analysis, which showed that 97.66% of the cases were correctly classified.

Sizing and fractal techniques allow PIA to be extended to other petrophysical parameters, namely formation factor and capillary pressure.

Data determined by PIA have been used successfully to help solve exploration and production problems. Improved water saturation values, identification of productive intervals, and reservoir characterization are some of the benefits of PIA. Thus, PIA offers an alternate method for deriving petrophysical data from difficult sample types.

Images from Petrographic Thin Sections

Introduction

In this application, PIA concerns the study of 2-D geometric properties of a pore network cross section to derive estimates of 3-D petrophysical properties, namely permeability and porosity. The purpose is to obtain k and ϕ estimates for samples such as sidewall cores, cuttings, and very small samples that are difficult to analyze by standard core measurement.

There are two major objectives that must be accomplished to successfully obtain k and ϕ

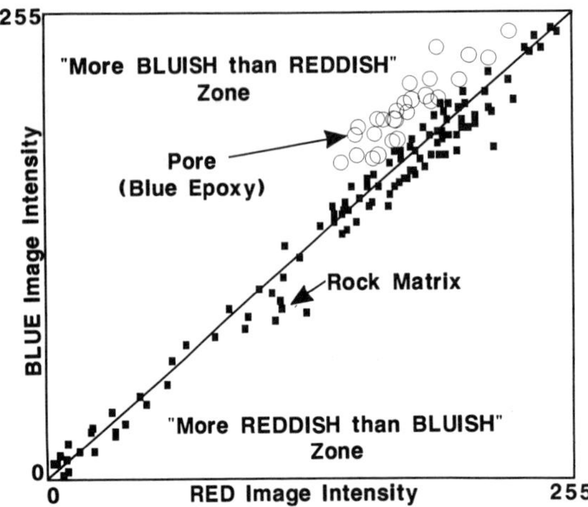

FIGURE 13.1. Schematic diagram showing scatter plot of blue intensities and red intensities for the transmitted light method.

estimates. First, pores and rock material must be separated by identifying a pore and defining its edges (a pore sectional can have inclusions and, therefore, more than one edge). This process, called "segmentation" of the image, leads to a binary image where all pore sectionals have the same light intensity (e.g., visually white) and the rock material has a different intensity (e.g., visually black).

The second objective is to establish a set of equations and/or algorithms relating geometric descriptors of observed pore sectionals to petrophysical properties of interest. Because the relationships in this image analysis application are statistical in nature, they yield empirical estimates of derived petrophysical properties.

Acquisition of an Image

Rock samples are analyzed by PIA in the form of a 30-μm-thick petrographic thin section. Each sample is examined through a microscope under a magnification selected to display the greatest range of pore sizes possible. In practice, 100 frames of the scene are captured by a black and white video camera and transmitted to an image analyzer where they are averaged. These averaged frames yield a single, digitized image composed of 512 × 512 pixels, each pixel having one of 256 levels of

gray. The process lasts only a few seconds and produces a gray level image that is stored in the analyzer video memory.

Two techniques have been used to acquire gray level images:

1. transmitting visible light through conventional petrographic thin sections where the rock is impregnated with blue dyed epoxy, and
2. reflecting incident UV fluorescence on specially prepared petrographic thin sections where the rock is impregnated with epoxy containing a fluorescent dye (Gies, 1987).

Image Segmentation

Incident UV Fluorescence

By carefully choosing the fluorescent dye, the excitor filter, and the barrier filter, mineral fluorescence can be eliminated. This method yields a quasibinary image: the rock material appears black and the pore sectionals appear gray to white. A rudimentary thresholding algorithm is sufficient to segment the image.

There are, however, two disadvantages to the method:

1. The thin sections have to be specially prepared, and
2. small pores may go undetected because their fluorescence can be quite faint. If the camera is not sensitive enough, these small pores will not be seen. For the same reason, edges formed on shallow wedges or shelves may be poorly defined.

Transmitted Light

The transmitted light method (Crabtree et al., 1984) takes advantage of a particular phenomenon that can be observed when a scatter plot is built of the gray levels of two images of the same scene, one taken through a blue filter and the other through a red filter (Figure 13.1). Most of the gray level values in a scatter plot are grouped along the first diagonal in a plume-shaped cluster. The cluster forms because blue intensity approximately equals red intensity for any pixel in this cluster. A small satellite cluster occurs off the first diagonal and contains most of the pixels associated with pore sectionals (i.e., they are more "bluish"). For a given spectral response of the filters, the position, shape, and density of the satellite cluster can be changed by adjusting the intensity of the light transmitted through the thin section, thereby achieving optimum separation between pixels belonging to pore sectionals and pixels belonging to rock material.

An unsigned subtraction of the "red" image from the "blue" image quasiextinguishes all the pixels grouped around the first diagonal leaving the pore sectional pixels fairly bright. A simple thresholding algorithm is then sufficient to segment the resulting image.

The segmented image can be false colored to resemble the thin section. The operator can compare the image with the thin section to ensure that wedges, overlain or overlying pores, and other artifacts have been resolved to satisfaction. Iterations of this process are permitted until a satisfactory segmentation is achieved. A trained operator needs only a few iterations (e.g., one to four or five) to achieve effective segmentation. The method, which is reasonably fast, requires between less than 1 min for most thin sections and 10 min for difficult thin sections.

There are several disadvantages of this method. First, the method involves a fair amount of subjectivity and, consequently, cannot be automated. Furthermore, the resolution of the satellite (pore) cluster from the main (rock) cluster becomes very poor, if not impossible, when either or both of the following conditions are realized:

1. Both the ratio of pore sectional area to perimeter and the area itself are large, and the rock matrix is mostly opaque, and
2. the rock material is composed of large, very clear crystals embedded in a dark matrix.

The main advantage of the method is that it uses conventional petrographic thin sections.

Applicability

It should be noted that conditions (1) and (2) are encountered most frequently in carbonate rocks. Consequently, incident UV fluorescence is used for carbonates. Transmitted light, however, is used for siliciclastic rocks because of ease of implementation.

Empirical Estimates of Petrophysical Parameters

Empirical estimates of petrophysical parameters have been limited to sandstones. The transmitted light method for capturing and segmenting a scene is used throughout. The idea is to establish an equation statistically relating 2-D geometric descriptors measured on a thin section to the 3-D petrophysical parameter of interest. This process is similar to a calibration, but is formation sensitive. Equations established for the Gulf Coast area, for example, generally cannot be used elsewhere. Moreover, the process supposes that the proper magnification has been established for the formation being calibrated. Magnification is an important factor that governs how well the range of pore sizes observed under the microscope represents the pore sizes of the entire sample.

Two techniques have been developed to derive these equations. One uses geometric descriptors such as pore sectional area and perimeter (i.e., essentially size parameters) and the other uses geometric descriptors derived from fractal mathematics (Jacquin and Adler, 1987). The first technique is called sizing and the second fractal. The fractal technique was employed after the sizing technique in hopes of finding a method that would not be formation sensitive. Although the fractal technique has proven to enhance the sizing method, it has not completely removed formation specificity from PIA.

Both sizing and fractal techniques use conventional multiple linear or log-linear regression to establish calibration equations. The method used is backward elimination with the following criteria:

1. maximize the correlation coefficient, and
2. minimize the standard deviation of the regression by rejecting any variable with a coefficient that has a too high significance level in the corresponding t-test.

Sizing Concepts

The first task in developing the sizing method was to determine which geometric descriptors should be included in the multiple regression. These descriptors were selected by studying the

(a)

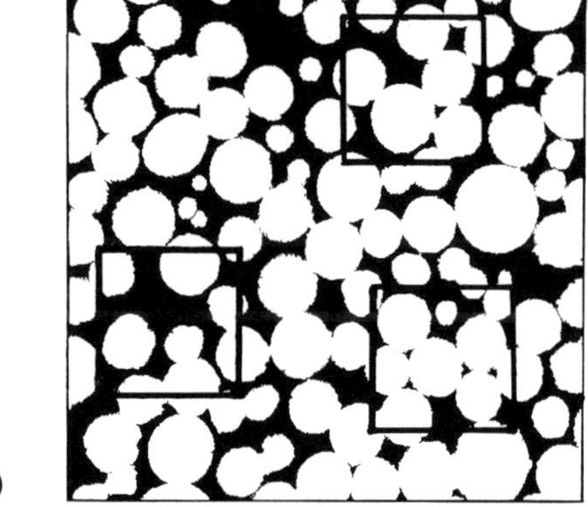

(b)

FIGURE 13.2. (a) 3-D model of cubic packing cut by random parallel planes. The rectilinear pore is a "preferred" linear path throughout a regular packing. (b) Random views selected from a thin section may be considered equivalent to one of the parallel planes in the 3-D model.

behavior of well-ordered deterministic models of clastic formations made of packed spheres of the same size: the glass bead models (Figure 13.2a). It is easily shown that such models are generated by replicating an original cell. Only an original cell needs to be considered for the purpose of this study.

If such an artificial formation is cut by a plane simulating a thin section (Figure 13.2b), one can define

1. a (visual) 2-D porosity,

$$\varphi = \frac{\text{cell area} - \text{total bead sectional area}}{\text{cell area}}$$

2. a (visual) 2-D permeability,

$$k_i = \frac{(\text{cell area} - \text{total bead sectional area})^2}{\text{cell area}}$$

The general expression for φ is

$$\varphi = 1 - \lambda^2 \left(1 + \alpha_1 \frac{h}{R} - \alpha_2 \frac{h^2}{R^2} \right) \quad (1)$$

where

λ^2 = a dimensionless coefficient defined by the type of packing (e.g., cubic, orthorombic),

h = an altitude coordinate defining the slicing plane,

R = the bead radius, and

α_1 and α_2 = dimensionless coefficients depending on the angle of the slicing plane relative to the frame of reference.

The 3-D porosity (ϕ) is obtained easily by integrating

$$<\varphi> = \int_0^1 \varphi \, d\left(\frac{h}{R} \right) = \phi \quad (2)$$

A general expression for k is much more difficult to establish. Berg (1970) suggests combining the Darcy equation with the Purday equation for flow in the privileged direction of the rectilinear pore. The combined equations result in the following expressions:

$$k_i = k \left(\frac{\gamma^2}{\lambda^2} \right)^2 \left[1 - \lambda^2 \left(1 + \alpha_1 \frac{h}{R} - \alpha_2 \frac{h^2}{R^2} \right) \right]^2 \quad (3)$$

and

$$k = K_0 \frac{d^2}{\delta^2} \quad (4)$$

where

k = the 3-D permeability (in the privileged direction),

λ^2 and

δ^2 = dimensionless coefficients depending on the type of packing,

K_0 = a coefficient depending on the units,

d = the diameter of the rectilinear pore,

while h, R, λ^2, α_1, and α_2 are defined in Eq. (1). It can be shown that

$$<k_i> = \int_0^1 k_i \, d\left(\frac{h}{R} \right) = \Gamma^2 k \quad (5)$$

*where Γ^2 is a combined coefficient depending on the type of packing. A physical test of the glass bead model justifies the selection of geometric descriptors for permeability (Figure 13.3). Both physical and theoretical data yield similar results.

FIGURE 13.3. Comparison of permeability calculated by the Berg model and the Krumbein model with permeability measured on a physical glass bead model. Berg's permeability is calculated for an orthorhombic packing.

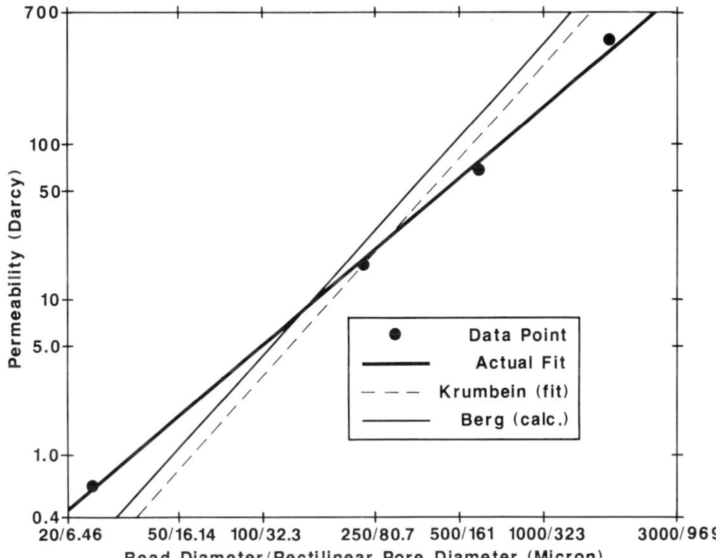

Several conclusions have been drawn from the above exercise. First, the relationships between 2-D (visual) porosity or permeability and their 3-D counterparts are defined by integrals averaging the 2-D parameter over many slices. Therefore, for the present technique, the estimates for k and ϕ cannot be determined from a single scene taken from a thin section. We assume that the arrangement of grains in a real rock is random enough so the average of the 2-D parameter measured on several views taken randomly on a single thin section will simulate the average taken over several slices. In other words, it is hoped that for sandstones $E(\varphi) = \langle \varphi \rangle = \phi$, at least over a small volume. In mathematical terms, we have assumed that φ and k_i describe stationary and ergodic random processes.

Second, the variables included in the regression for calculating the porosity estimate should be

1. the visual porosity φ (always present in the regression), and
2. a variable representing the arrangement of pores reflecting the packing and sorting of grains. For this purpose, we have chosen the pore area distribution represented by the following statistics: the median, the mean, the standard error of the mean, and the maximum.

Third, the variables included in the regression for calculating the permeability estimate should be

1. the mean pore area (always present in the regression),
2. the pore area distribution represented by the following statistics: the median, the mean, the standard error of the mean, and the maximum, and
3. a variable representing the pore shape configuration. For this purpose, we have chosen the distribution of the pore shape factor [defined as $(1/4\pi) \times (\text{Perimeter}/\text{Area})^2$] represented by the following statistics: the mean, the median, the standard deviation, and the standard error of the mean.

Finally for both estimates, we have assumed that the factors were separated, thereby adopting a log-linear model for the independent variables. To approximate a normal distribution for the dependent variables we have chosen the following transforms:

1. for porosity: $0.5 \log [(100 + \phi)/(100 - \phi)]$, and
2. for permeability: $\log (\sqrt{k} + \sqrt{k+1})$.

The visual porosity φ is calculated within an area enclosed by a guard region. Therefore, the pore sectionals whose center of mass is outside the guard region are excluded from the calculation. The number of scenes used in the averaging process is at least five.

FIGURE 13.4. Sierpinski carpet is the basic fractal model. The actual fractal model for sandstones has an added element of randomness as shown here with replication.

Fractal Concepts

A regular Sierpinski carpet (Figure 13.4), which is a fractal object, may be constructed in the following way: a black square of area W is divided into b^2 squares, and m of these subsquares are whitened (b and m are integers). These processes continue for the remaining squares as many times as desired.

At stage n the whitened area is given by the expression

$$S_n = W - W\left(\frac{b^2 - m}{b^2}\right)^n \tag{6}$$

This expression may be written as follows:

$$\frac{W - S_n}{W} = 1 - P_n = \left(\frac{b^2 - m}{b^2}\right)^n \tag{7}$$

where P_n is the visual 2-D porosity at stage n. By defining $a_n = 1/(b^2)^n$ as the Mandelbrot ruler (Mandelbrot, 1982), the above equation can be rewritten as follows:

$$\log(1 - P_n) = \left[1 - \frac{\log(b^2 - m)}{\log(b^2)}\right] \log(a_n)$$

$$= \left(1 - \frac{D}{2}\right) \log(a_n) \tag{8}$$

where D is the fractal dimension of the Sierpenski carpet. The fractal dimension is derived from a geometric object with a topological dimension of 2 (in this case a square).

A segmented image of a sandstone thin section can be viewed as a stochastic Sierpinski carpet if it is submitted to the following process.

1. Perform n erosions followed by n dilations at each stage (Figure 13.5, see Color Plate IV).
2. Calculate at each stage
   a. the visual porosity

$$P_n = \frac{\text{total pore sectional area}}{\text{scene area}}, \text{ and}$$

   b. a ruler size $a_n = (n/512)^2$, where the image is 512×512 pixels wide.
3. Fit a line to $\log(1 - P_n)$ versus $\log(a_n)$,

$$\log(1 - P_n) = \alpha \log(a_n) + \beta \tag{9}$$

which defines the fractal dimension as $D = 2(1 - \alpha)$. This is known as the Mandelbrot–Richardson plot (Figure 13.6).

In the program testing the applicability of the above process, 198 sandstone thin sections were examined, taking five random images per section. For these 990 images, the correlation coefficient for the best fit line is 0.95 or better. For many images, however, the data could be divided into subsets, each with its own best-fit line. These subsets indicate the possible mixture of several fractal objects with different fractal dimensions. To date, this property has not been used to evaluate samples. Only an average fractal dimension is calculated for each image. The best-fit line obtained by the above process is necessarily bounded by $n = 1$ (the smallest ruler possible is one pixel) and $P_n = 0$ (which occurs when n erosions have completely erased the image).

Primary variables defined in the proceeding discussion are

1. D, the fractal dimension,
2. f, the visual porosity that is calculated without a guard region before any erosion–dilation cycle is performed, and
3. c, an area corresponding to the x-intercept of the best-fit line when $P_n = 0$. This variable is calculated as follows:

$$c = (\text{pixel area}) (512)^2 \exp\left(\frac{-\beta}{\alpha}\right) \tag{10}$$

To determine whether a relationship existed between porosity, permeability, and the variables derived from the fractal analysis, a factor analysis

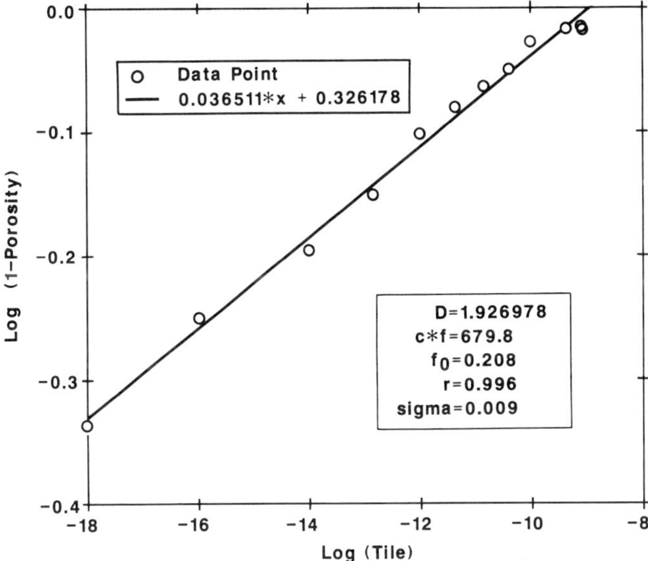

FIGURE 13.6. A Mandelbrot-Richardson plot for an actual sandstone sample. In this case, the ruler is actually a tile. Note that log is log base 2.

was performed using the data gathered on the 198 test samples, namely: D (the fractal dimension), C [defined as $\log(\sqrt{c \cdot f} + \sqrt{c \cdot f + 1})$], f (the visual porosity), J [defined as $f \cdot (m - D)/(2 - D)$, m is an experimentally defined coefficient close to 2], P [the permeability transformed by $\log(\sqrt{k} + \sqrt{k+1})$], and ϕ (the porosity) (Figure 13.7).

The factor extraction method was the generalized least-squares and the factor rotation method was Oblimin. Two factors were extracted: the first composed of P, C, and J and the second of ϕ, D, and f. This exercise shows that the fractal properties of a pore network are indeed correlated to permeability and porosity. The independent variables are D, C, and J. The dependent variables are the porosity and the transformed permeability, $\log(\sqrt{k} + \sqrt{k+1})$.

The regression for permeability is actually several regressions, one for each cluster or group of data points. The variables used for clustering were D and C. A least-squares method was used for clustering (Bezdek et al., 1984), the distance being the Mahalanobis distance. The validity of the clusters was checked by performing a discriminant analysis, which showed that 97.66% of the cases were correctly classified.

Extension to Other Petrophysical Parameters

Formation Factor

For clean sandstones, the dc conductivity is governed by the geometry of the pore space. This was described by Archie as formation factor and is a function of porosity according to the equation $F = 1/\phi^2$. As the pore morphology changes, however, the correlation between formation factor and porosity changes so that the cementation exponent (m) becomes a variable in the Archie equation. Thus, $F = 1/\phi^m$. Formation factor in sandstone reservoirs correlates well to porosity and, to a lesser extent, to permeability.

Consequently, formation factor is included among petrophysical parameters that can be estimated from image analysis. The following variables are used in the calibrating regression: visual porosity, a variable representing the arrangement of pores, and a variable representing the pore shape configuration. The visual porosity always will be present in the equation. As in the porosity equation, we have used the pore area distribution represented by the following statistics: the mean, the median, the standard deviation, the standard

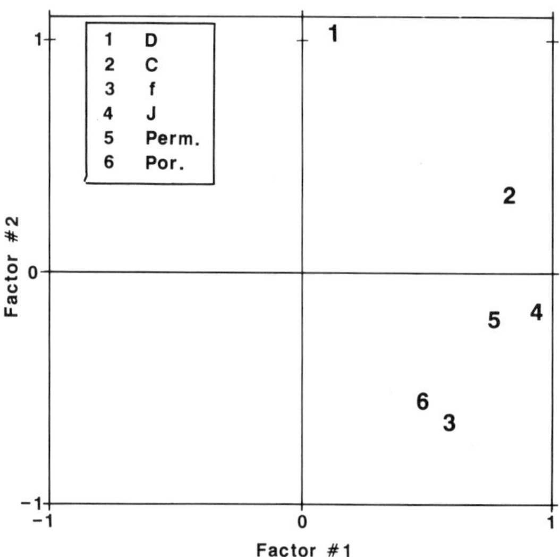

FIGURE 13.7. Factor analysis for 198 cases shows a definite correlation between porosity, permeability, and fractal parameters. Note that the Kaiser–Meyer–Olkin measure of sampling adequacy equals 0.6706. Oblimin convergence is 22 iterations.

error of the mean, and the maximum. As in the permeability equation, we have used the shape factor distribution represented by the following statistics: the mean, the median, the standard deviation, and the standard error of the mean. The model is log-linear for the independent variables and the dependent variable is $\log(F)$.

Capillary Pressure

Mercury capillary pressure curves can be viewed as cumulative distributions of the pore volume controlled by throats of diameter, d, which are transformed by the well known relation $P = 4\sigma\cos\theta/d$. To estimate the number and size of pore throats from a 2-D thin section is difficult because a small or narrow feature could be either a throat or a small pore. Since only a slice can be seen, whatever is above or below the plane is missing from the analysis. Thus, the incoming or outgoing throats occurring in planes outside the thin section cannot be counted and/ or measured.

There are several steps in the approach used to derive capillary pressure curves. Our approach has been as follows:

1. calculate a pseudodiameter from an estimate of the specific surface,
2. construct a cumulative histogram of the transform of this diameter (with the cumulative percentage on the x axis), and
3. take advantage of the correlation that exists between certain features of the capillary curve and permeability (Swanson, 1978; Thomeer, 1983) to adjust (calibrate) the above computed histogram. According to Ruzyla (1986), who quotes Underwood (1970) and Weibel (1979), an estimate of specific surface could be calculated as follows:

$$\text{specific surface} = \frac{4}{\pi} \cdot \frac{\sum \text{perimeter}}{\sum \text{area}} \quad (11)$$

The reciprocate of the estimate multiplied by the aspect ratio is a pseudodiameter suitable for this application. Moreover, the abscissas of the quasi-asymptote of the capillary pressure curve show a strong correlation with permeability (and a weaker correlation with porosity). This property has been used to derive a calibration equation. Independent variables in the regression are the estimate of porosity and log of the estimate of permeability (as derived earlier). The dependent variable is the saturation corresponding to the quasiasymptote of the capillary curve. The equation represents a mapping transform that is applied to the whole curve previously defined.

Benefits of Petrographic Image Analysis

PIA can be performed on most types of core samples. PIA, however, is ideally suited to provide petrophysical indices for samples such as sidewall cores, drill cuttings, and laminated samples that are difficult to analyze by conventional methods. Furthermore, reservoir quality can be quantified using PIA parameters. Data determined by PIA have been used successfully to help solve exploration and production problems in sandstone formations.

Drill Cuttings

The quality of drill cuttings is determined by the type of drill bit, the mud used, and the lithology of the formation. PIA on drill cuttings allows the

FIGURE 13.9. Depth plot of permeability and porosity. Locations of conventional cores are shown by boxes. Data points for drill cuttings samples are derived from PIA.

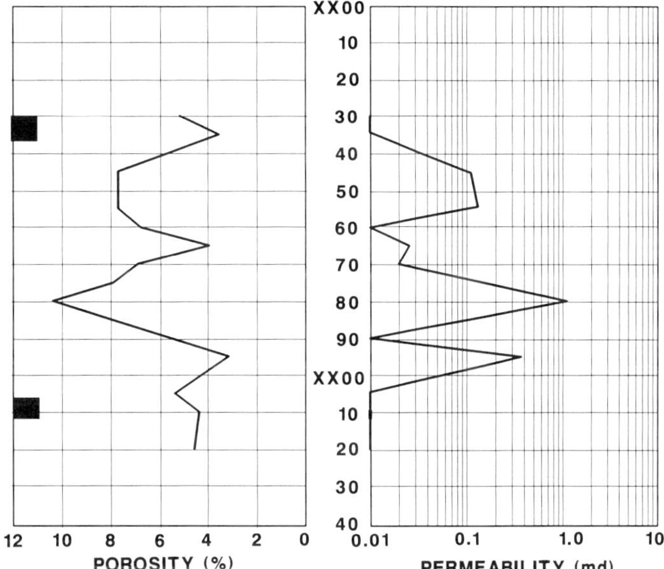

evaluation of reservoir quality on representative samples of a formation in areas where conventional cores have not or cannot be recovered.

Previous experience in a tight gas sand in the Gulf Coast has shown that 5% porosity will provide the 0.03 md minimum permeability required for fracture stimulation and production. Two conventional cores from a tight gas sand were measured for porosity and permeability and examined in thin section. Measured data gave porosity and permeability values below minimum limits for economic production. A thin-section photomicrograph of a conventional core sample (Figure 13.8a, see Color Plate V) shows the very fine grain size and poorly developed pore system (blue) that resulted in low permeability. Because neither conventional core had sampled possible productive zones, drill cuttings were selected to represent the interval between the two conventional cores. A thin section photomicrograph of a relatively unaltered drill cutting from this interval (Figure 13.8b) reveals a coarser grained sandstone with a moderately well-developed pore system (blue) compared to conventional core samples.

Porosity and permeability values for drill cuttings from the uncored intervals were determined by PIA and are illustrated with a depth plot (Figure 13.9). A total of 65 ft (19.18 m) of the section has porosities greater than 5%. Porosities greater than 5% and permeabilities greater than 0.03 md occur in 31 ft (9.45 m) of the section. Fourteen additional feet (4.27 m) of the section have permeabilities greater than the minimum requirement of 0.03 md, but porosities below 5%. PIA performed on drill cuttings from the uncored intervals identified three distinct lobes of sufficient permeability and porosity for fracture stimulation. Thus, PIA enhanced conventional coring and petrophysical techniques by evaluating and identifying three productive zones from drill cutting samples.

Percussion Sidewall Cores

Sample Damage

Percussion sidewall cores suffer from damage due to energy dispersed through the formation when the sampling chamber (bullet) impacts the wellbore wall. Two major changes occur in the formation being sampled:

1. The original formation pore morphology is altered from the shot impact and the mud solids being forced into pore spaces, and
2. mud filtrate is pushed more deeply into the sidewall core adding to the degree of filtrate flushing that occurred in the sample during drilling operations.

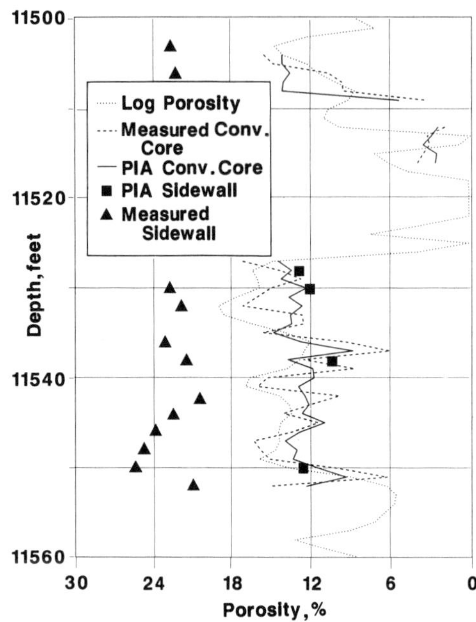

FIGURE 13.11. Depth plot for the Tuscaloosa Formation showing the relationship between log porosity, conventional core helium porosity, sidewall core fluid porosity, and PIA porosity index.

Alteration of the pore morphology is primarily a function of the intensity of the bullet charge, bullet configuration, and most importantly the elasticity of the rock type. Typically, poorly consolidated formations are compacted by the percussion sidewall, reducing porosity and permeability from the original formation value. Well-consolidated formations are shattered by the impact of the bullet, which increases porosity and permeability from the original formation value.

The magnitude of change due to percussion sampling is variable and formation specific. For example, heavy oil-producing unconsolidated sands in California suffer little compaction because viscous pore fluid has the ability to absorb the shot's energy. Well-consolidated sandstone of the Tuscaloosa Formation, however, shatters so extensively that a 6 to 10 porosity percent increase is common. Often, even in the most sensitive formation, a portion of the sidewall is undisturbed. This portion is typically toward the center and the deep end of the sidewall sample and is referred to as the "button." On a microscopic level the undisturbed portion of the sidewall shot can be identified and PIA can be used

to relate its pore features to rock properties of the undisturbed formation. The degree of damage to the altered portion can be determined as well.

The microscopic vantage point of PIA allows altered and unaltered portions of sidewall core to be analyzed individually. For example, a consolidated rock type that is shattered by shot impact will typically show higher porosity indices and lower permeability indices for the altered portion of the sidewall as compared to the unaltered portion. PIA of the altered sidewall will also show lower permeability indices than permeabilities measured by standard methods (excluding empirically derived permeabilities). This is due to differences in the microscopic view point of PIA compared to the macroscopic view point of standard core measurements. From a macroscopic view, permeability measured on altered sidewall core would be primarily controlled by the fracture system, thus giving higher permeabilities. Because PIA is performed on thin sections, fractures may have been excluded from the analyzed portions of the sample.

The power of PIA's ability to determine petrophysical indices on the unaltered portion of sidewalls is shown in Figure 13.10 (see Color Plate V). The photomicrograph of a sidewall core shows an extensively altered portion of the sample with shattered grains and mud solids invasion. In contrast, the unaltered button portion of the sidewall maintains the formation's integrity as indicated by preserved quartz overgrowths and clay morphology. The pore network remains intact so that data generated on this portion will be representative of the formation's pore morphology. Since PIA requires a sample area of 5–10 mm^2, this example shows that even a small button existing in the sidewall core allows a microscopic analysis of the unaltered pore morphology.

Comparisons and Applications of PIA on Sidewall Cores

Porosity and permeability were measured on conventional core samples taken in the Tuscaloosa Formation. Thin sections from end pieces of each plug were prepared for PIA. Sidewall core samples were also shot into the same cored interval and analyzed using standard techniques as well as PIA. Data generated by PIA include porosity, permeability, formation factor, and capillary pressure (pore size profile) indices.

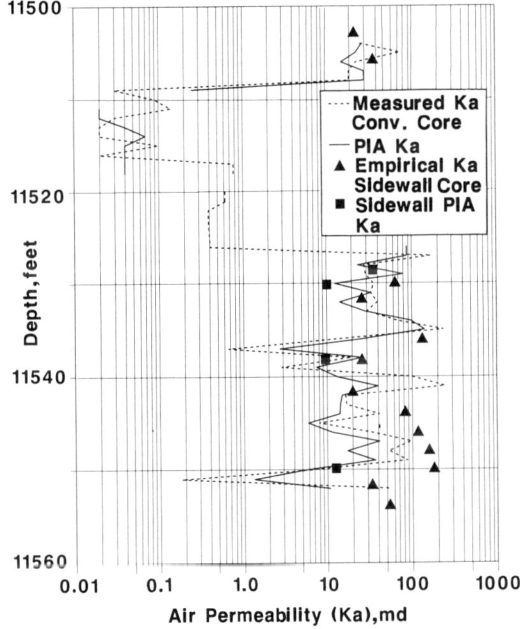

FIGURE 13.12. Depth plot for the Tuscaloosa Formation showing the relationship between conventional core air permeability, sidewall core empirical permeability, and PIA permeability index for conventional and sidewall core samples.

Figure 13.11 shows the comparison of porosities. Comparing neutron/density log porosities to conventional core porosities (ambient) yields a very good agreement. Sidewall core porosities measured via a standard core measurement method yield porosities that are 2 to 10 porosity percent too high due to the shot's shattering effect. This commonly occurs in a well-cemented sandstone, such as the Tuscaloosa Formation, which has porosities ranging from 8 to 16%. PIA performed on conventional core samples also shows very good agreement with core and log-derived porosities. PIA porosities derived from sidewall samples show a marked reduction from standard sidewall core measurement values, and compare very well to log and conventional core porosities.

Figure 13.12 shows the comparison of permeabilities. Conventional core measured permeabilities compare very closely to permeabilities derived from PIA on both conventional core and sidewall core samples. Empirical permeability values derived from standard sidewall core technique (particle size analysis) also compare well, but are higher

than PIA data and conventional core measurements. If sidewall core samples had been measured for permeability, the fractured nature of these shattered samples would have yielded permeabilities even higher than the empirical values.

Improved permeabilities and porosities from altered sidewalls, along with capillary pressure and formation factor indices generated by PIA, provide information that not only aids log interpretation, but also enhances reservoir exploration and exploitation. Figure 13.13 shows a plot of formation factor from PIA versus porosity from PIA. Selecting an intercept of one (at 100% S_w, $R_o = R_w$), the slope of a line drawn from the right-most point defines the highest cementation exponent m of the pore systems measured. Repeating this procedure for the left most point defines the lowest cementation exponent m. A series of lines drawn between the maximum and minimum values represents the variability of pore systems found in the data set. Appropriate m values then can be selected for each sample according to the relative position within this series of lines. These m values are plotted as a function of depth in the left track in Figure 13.14.

Water saturations were initially calculated using the Archie relationship of

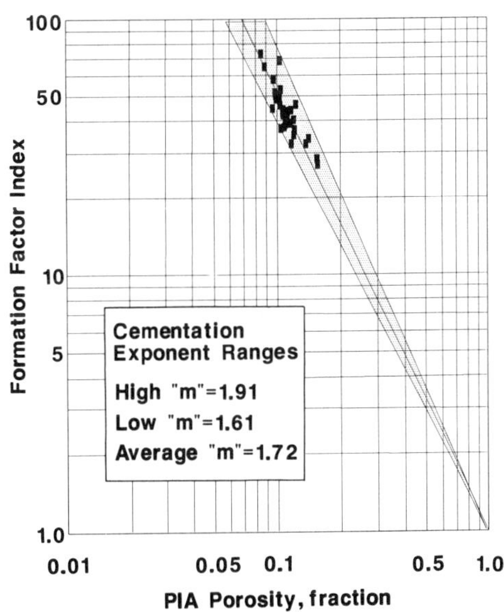

FIGURE 13.13. Plot of PIA porosity versus PIA formation factor illustrates the variability of pore systems found in the Tuscaloosa data set.

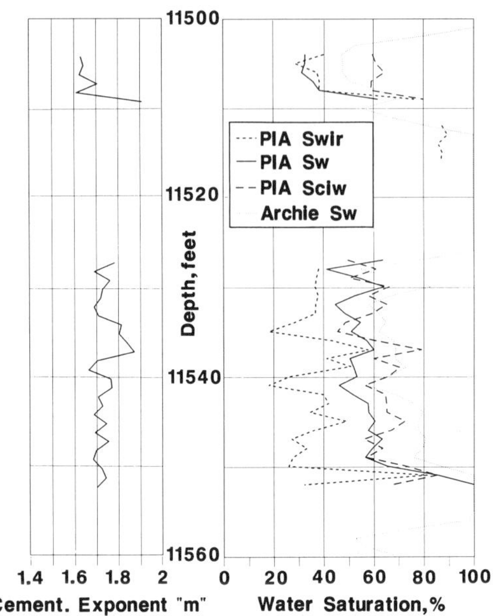

FIGURE 13.14. Depth plot showing the relationship between log (Archie) water saturation, PIA S_w, PIA S_{wir}, and PIA S_{ciw} for Tuscaloosa samples.

$$F = 1/\phi^2 \qquad (12)$$

where $m = 2$ is the assumed cementation exponent, and the water saturation equation,

$$S_w{}^n = (F \times R_w)/R_t \qquad (13)$$

where $n = 2$ is the assumed saturation exponent. Porosity and formation resistivity (R_t) values from log data (Figure 13.14, track two, the dotted line) were used to satisfy Eqs. (12) and (13).

Capillary pressure curves generated by PIA were used to correlate irreducible water saturation to PIA permeability and porosity values. Using permeability and porosity values for each sample, an irreducible water saturation (S_{wir} PIA) was selected from the correlation and plotted as a function of depth in the right track of Figure 13.14 as noted by the dot/dash line. Comparing irreducible water saturations (S_{wir}) to calculated log S_w shows that log S_w had higher values. This indicates that either pore water is mobil or the assumed value for m was incorrect.

Water saturations were then recalculated using m values that varied as a function of depth. Thus,

$$F = 1/\phi^m \qquad (14)$$

These water saturations (S_w PIA) are represented by the solid line on the right track in Figure 13.14. The S_w PIA compares closely in the upper section to S_{wir} PIA. Further down the log, however, S_w PIA values becomes lower than S_{wir} PIA, indicating that part of the lower section may have some water production.

Critical water saturations (S_{ciw} PIA), plotted in the right track of Figure 13.14, were selected from a well-cemented sandstone correlation by using PIA porosity and permeability indices. Critical water saturations, based on a 5% water cut for this formation, represent the maximum water that can be present in the formation so production can remain at a 5% water cut or less. Comparing critical water saturations to S_w PIA shows that the lower zone is productive down to 11,556 ft (3522.3 m).

High log S_w values were calculated for the lower zone of this example (Figure 13.14). Without PIA information to aid log interpretation, the lower zone may have looked nonproductive. Comparing log S_w (Archie) to the S_{ciw} PIA, this zone appears to be productive only at water cuts in excess of 5%. When m values from PIA data, however, are used to calculate S_w, both the upper and lower zones appear to be productive at a 5% water cut or less. The data also indicate an approaching free water level since irreducible water saturations and S_w PIA diverge with increasing depth. Fairly constant permeability and porosity values throughout this same interval identify a transition zone rather than a change in formation lithology occurring at the bottom of the lower zone. This information would not have been available using only conventional methods to determine petrophysical properties of percussion sidewalls. Thus, PIA provides additional information about the petrophysical properties of these altered sidewall cores and enhances the ability to correctly interpret log data.

Reservoir Characterization

PIA provides a definitive way to quantify and rank hydrocarbon reservoir quality. PIA data can be used to characterize pore complexes and to determine geologic controls on reservoir quality. A case study from the Gulf Coast Norphlet Formation shows the power of this PIA application.

The Upper Jurassic Norphlet Formation extends over much of the northern Gulf of Mexico Basin from east Texas to Florida. The formation occurs only in the subsurface and most commonly is found at depths greater than 12,000 ft (3657.6 m). The clastic rocks of the Norphlet Formation were deposited on the Middle to Upper Jurassic ,Louann Salt and lie beneath the Upper Jurassic Smackover Formation. Deposition of the Norphlet clastic rocks was controlled by differential subsidence and the presence of pre-Jurassic paleohighs (Mancini et al., 1985). The Norphlet Formation consists of localized basal marine shale overlain by eolian cross-bedded dune and massive interdune sandstones capped by a massive marine sandstone sequence. These eolian sediments tract updip into alluvial fan conglomerates, sandstones, and siltstones. Compaction and diagenesis are thought to be major factors controlling the development of porosity in the Norphlet Formation.

Petrographic analyses were performed on 257 Norphlet samples and results of these analyses were used to identify samples for PIA. The samples chosen for PIA represent rocks from the four depositional facies in the Norphlet Formation. Two samples each were selected from the eolian dune facies, the marine facies, and the alluvial fan facies; one sample was selected from the eolian interdune facies. These samples typify those rocks with the best reservoir quality within each facies.

Samples selected for PIA describe the variability of rocks within each depositional facies. Samples 11-1 and 16-18 are the best examples of the eolian dune facies; 16-18, the more immature of the two, contains more clay and rock fragments. Both samples are laminated. Samples 9-10 and 8-6 represent the marine facies. Both are massive; sample 9-10, however, is heterogeneous due to the presence of patchy cement. By contrast, sample 8-6 has a more homogeneous distribution of cement, as well as a greater range in grain size. Samples 14-1a and 23-2 typify the alluvial fan facies. Sample 14-1a represents the proximal alluvial fan conglomerates and sample 23-2 represents the more distal alluvial fan sandstones. Sample 3-8, from the eolian interdune facies, is a massive sandstone with a bimodal grain size distribution.

Pore Network Characterization

Pore size and pore shape measurements describe pore network geometry. PIA parameters were determined for each sample in this study, as well as for

TABLE 13.1. Cluster membership for Norphlet reservoir samples.

Sample No.	Cluster 1	Cluster 2	Cluster 3
3–8			X
8–6	X		
9–10	X		
11–1			X
14–1a		X	
16–18	X		
23–2	X		

individual laminations in the laminated samples.

Diameter distributions and pore size spectra characterize pore size. Feret and Heywood diameter distributions are useful in differentiating pore systems. For example, within laminated samples the coarser laminae tend to have a broader range of pore diameters than the finer laminae. Pore size spectra indicate that porosity in marine samples is associated with finer pores compared to eolian dune and alluvial fan samples. The eolian interdune sample has a bimodal pore size distribution.

Aspect ratio, shape factor, specific surface, and pore roughness spectra distributions characterize pore shape. These image analysis parameters all suggest that Norphlet pores tend to be rounded and fairly smooth.

Cluster Analysis

Cluster analysis was performed on a subset of PIA parameters to quantitatively classify rocks into reservoir populations. Parameters selected for cluster analysis include maximum, minimum, and mean grain size, mean pore shape factor, mean aspect ratio, three coefficients of the equation used to describe the uncalibrated capillary pressure curves, the frequency of each class interval in the pore size spectra, and the frequency of each class interval in the pore roughness spectra. Table 13.1 shows the cluster membership for the Norphlet reservoir samples. The partition into clusters is hard with no overlap of membership.

Cluster 1 is characterized by a unimodal pore size spectrum. Permeability indices for Cluster 1 range from 171 to 347 md; porosity indices range from 7.5 to 20.6%. The members represent marine, eolian, and alluvial fan facies. Cluster 2 is characterized by a skewed trimodal pore size spectrum. Only the Norphlet alluvial fan conglomerate belongs to Cluster 2. This sample has

TABLE 13.2. Image analysis and petrographic parameters for Norphlet reservoir clusters.

Parameter	Cluster 1	Cluster 2	Cluster 3
Mean k index (md)	212.0	184.0	1135.0
Mean ϕ index (%)	17.2	21.6	28.9
Mean shape factor	3.32	4.83	4.15
Mean aspect ratio	2.0	2.15	1.94
Mean total cement (%)	7.8	5.5	3.2
Mean silicate cement (%)	4.8	3.9	0.6
Mean carbonate cement (%)	1.3	0.8	1.6
Mean clay cement (%)	1.7	4.4	1.2
% total $\phi > 29$ µm	22.4	51.7	62.1

permeability and porosity indices of 184 md and 21.6%, respectively. Cluster 3 is characterized by a bimodal pore size spectrum. The members represent eolian dune and eolian interdune facies. The samples belonging to Cluster 3 have excellent permeability (979–1291 md) and porosity (23.1–34.6%) indices.

The fact that several different depositional facies are included in both Cluster 1 and Cluster 3 suggests that depositional environment alone does not control cluster membership. Differences in mean aspect ratio and, in particular, mean shape factor (Table 13.2) indicate that cluster membership is indeed a function of pore-complex geometry. These differences suggest reservoir quality ranking as Cluster 3 > Cluster 2 > Cluster 1.

Petrographic Control on Reservoir Quality

In addition to quantitatively classifying rocks into reservoir populations, PIA reveals information about diagenetic controls on reservoir quality. Pore shape factor, which is most sensitive to cement abundance, is the PIA parameter that best separates this data set into reservoir populations (Table 13.2). All Norphlet sandstones, except conglomerate sample 14-1a, exhibit an inverse correlation between shape factor and silicate cement (Figure 13.15). Samples with greater amounts of silicate cement tend to have smoother, more regular pores. This is due to the presence of smooth authigenic quartz crystal faces that have reduced pore surface irregularities.

This correlation between shape factor and silicate cement indicates that the pore-complex geometry is controlled by cement abundance and distribution. Petrographic data support this finding. Figure 13.16 shows the relative cement abundance for each reservoir population. Samples belonging to Cluster 3 have the lowest occurrence of both silicate cement and total cement (predominately silicate and carbonate cements). Cluster 1 samples, by contrast, with the greatest relative abundance of both silicate cement and total cement, are typical of poorer quality reservoir rocks. Low mean shape factor associated with Cluster 1 samples implies that silicate and carbonate cements produce smoother pore walls and reduce effective porosity. The fact that Cluster 1 samples have the smallest amount of total porosity greater than 29 µm (Table 13.2) further supports the finding that silicate and carbonate cements exert a diagenetic control on reservoir quality.

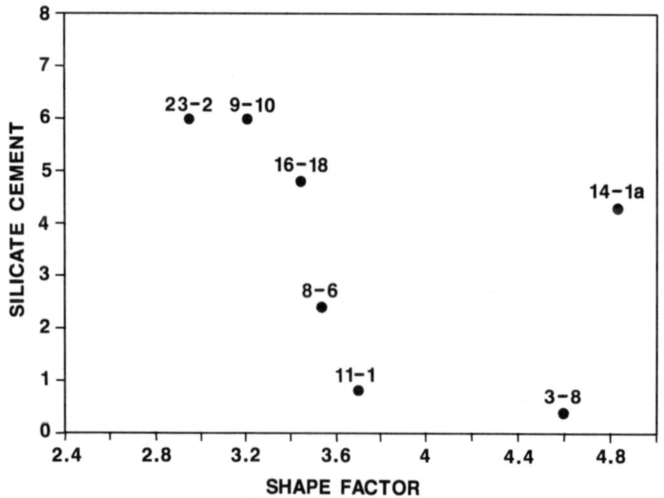

FIGURE 13.15. Plot of shape factor versus silicate cement for Norphlet reservoir samples selected for PIA.

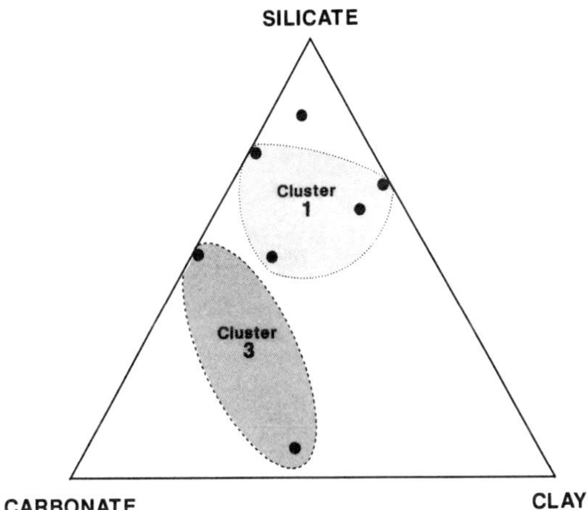

FIGURE 13.16. Ternary diagram of silicate, carbonate, and clay cement for selected Norphlet reservoir samples showing the location of cluster populations.

Applications to Conventional Core

There are several applications of PIA to finely laminated sequences in conventional cores. Laboratory-measured plug permeability and capillary pressure curves provide an average measurement of all laminations in the plug. PIA petrophysical properties can be determined from specific laminae. To use PIA successfully, however, plug samples must be taken parallel to strike. If a plug is not oriented parallel to strike, the measured plug permeability value can be pessimistic.

In Offshore Gulf of Mexico, many productive reservoirs exist in very shaly rock types. Frequently, interbedded sands/silts and shales are difficult to sample. When taking plug samples, the friable nature of this lithology often results in samples that are unsuitable for standard core measurements. PIA can accurately determine petrophysical data for sand/silt laminae only millimeters in thickness. In laminated sequences, production is dependent on the coarser laminae. Because laboratory analyses provide a bulk measurement, it is impossible to determine the petrophysical properties associated only with the productive laminae. PIA allows for this type of detailed analysis. It is also possible to digitize a representative whole core piece megascopically and determine a net sand percentage for the interval very quickly and easily.

Conventional core that loses structural integrity during coring operations, such as a rubble zone, is not suitable for standard core measurement methods. PIA offers an alternate method for deriving petrophysical data from these difficult sample types.

Acknowledgments. The authors wish to express their gratitude to the management of Core Laboratories, a Division of Western Atlas International, for permission to publish this manuscript. We thank Chris Strunk and Josi Perez for preparing the figures.

References

Berg, R.R., 1970, Method for determining permeability from reservoir rock properties: Transact. Gulf Coast Assoc. Geol. Soc. 20, 303–317.

Bezdek, J.C., Ehrlich, R., and Full, W., 1984, FCM: The fuzzy c-means clustering algorithm: Comput. Geosci. 1, 191–203.

Crabtree, S.J., Jr., Ehrlich, R., and Prince, C., 1984, Evaluation of strategies for segmentation of blue-dyed pores in thin sections of reservoir rocks: Comput. Vision, Graphic Image Process. 28, 1–18.

Gies, R.M., 1987, An improved method for viewing micropore systems in rocks with the polarizing microscope: Soc. Petr. Eng. Formation Eval. 2, 209–214.

Jacquin, C.G. and Adler, P.M., 1987, Fractal porous media II: Geometry of porous geological structures: Transport Porous Media 2, 571–596.

Mancini, E.A., Mink, R.M., Beardon, B.L., and Wilkerson, R.P., 1985, Norphlet Formation (Upper Jurassic) of southwestern and offshore Alabama: Environments of deposition and petroleum geology: Bull. Am. Assoc. Petr. Geol. 69, 881–898.

Mandelbrot, B.B., 1982, The Fractal Geometry of Nature: Freeman, San Francisco.

Ruzyla, K., 1986, Characterization of pore space by quantitative image analysis: Soc. Petr. Eng. Formation Eval. 1, 389–398.

Swanson, B.F., 1978, A simple correlation between air permeabilities and stressed brine permeabilities with mercury capillary pressures: Presented at the 53rd Annu. Fall Tech. Conf. Soc. Petr. Eng., Am. Inst. Min., Metall. Petr. Eng.

Thomeer, J.H., 1983, Air permeability as a function of pore-network parameters: J. Petr. Tech. 35, 809–814.

Underwood, E.E., 1970, Quantitative Stereology: Addison-Wesley, Reading, MA.

Weibel, E.R., 1979, Stereological Methods, Vol. 1—Practical Methods for Biological Morphometry: Academic Press, New York.

14

Image Processing of Magnetic Data and Application of Integrated Interpretation for Mineral Resources Detection in Yiesan Area, East China

Wu Chaojun

This paper deals with the principles and methods of image processing techniques that are used for integrated interpretation of magnetic data and other geodata. A practical procedure of image processing is suggested for observed magnetic data. The ground survey magnetic data in the Yiesan area of East China is used as an example of an application of integrated interpretation and mineral resources detection.

The classical potential data processing is the fundamental preprocessing and the results were transferred into the image file for further image processing.

Various filters were employed to enhance the boundary of the geological unit and of the linear feature that was interpreted as faults, axis of folds, and other linear geological structures.

Overlaying is a convenient and effective way to realize the integrated interpretation with other geodata or different data processing results.

Density slicing is used to obtain an interpreted sketch map automatically. Slicing makes the final result neater than the original image.

In the example, the boundary of intrusion from the results of magnetic data image processing is nearly the same as in the geological map, and some faults that were not interpreted before can be recognized clearly according to the linear feature on images. Some local areas were chosen for quantitative interpretation and for further exploration.

Therefore, image processing is a convenient and effective way for qualitative interpretation of magnetic data, and for integrated interpretation with other geodata.

Introduction

Different scales of gravity survey and aeromagnetic or ground magnetic survey have been carried out in most parts of China. A great amount of gravitational data and magnetic data were acquired and geological interpretation of these data is an important task. Therefore, in recent years integrated interpretation in combination with other geophysical data and geodata has been emphasized to make full use of these gravity and magnetic data.

Image processing techniques and different kinds of map presentation were introduced as an effective means to realize this aim. The principles and methods of image processing are discussed in this paper.

The new research project in the Yiesan area was designed to provide drilling positions and to detect new copper mines by use of geophysical data and other geodata. Therefore, the magnetic data and other geodata in the Yiesan area were used to apply the principles and methods of image processing.

Several steps may be followed in image processing of magnetic data. Proper potential transformation of observed magnetic data was considered as the basic preprocessing. The results were presented by various forms of presentation, such as gray scale map, color map, and artificial shadow map. These maps have more advantages than contour maps alone. The general geological information was shown directly and clearly on these maps.

The interface programs were designed to transfer the preprocessed results into standard image files that can be read by the Integrated Land and Watershed Information System (ILWIS system), designed by the International Institute of Aerospace Survey and Earth Sciences (ITC), The Netherlands. The ILWIS system integrates image processing capabilities, tabular databases, and conventional geographic information system characteristics (Meijerink et al., 1988). Further image processing was carried out under the ILWIS system.

Different image enhancement processings were applied to these data. Additional geological detail not interpreted before can be recognized clearly after image processing. Several new faults were interpreted.

Registering magnetic data to some selected geochemical data showed the correlations between these data. According to the pattern of known mineralization in the area, several areas were chosen for further investigation.

Significant local anomalies were selected for quantitative interpretation and integrated with electrical sounding data. A new ore body was interpreted to satisfy the observed magnetic anomaly and the VES anomaly.

Principles and Methods

One main objective of geophysical data image processing is displaying results on the screen or hard-copy image by using a color printer directly. Instead of a contour map the color map or gray scale map is more acceptable for human vision, because (1) the interpreter can recognize high or low anomaly according to color, and (2) can find anomaly patterns and linear features with quick comprehension. Therefore, image processing is very convenient for qualitative interpretation.

Also the spatial filtering or other enhancement of image processing is a real-time processing. The interpreter can monitor the result, compare the result with known geological information or his experiences, and promptly change different filter operators or other parameters to improve the results. Flexible designing of the filter will satisfy the different enhancement for different targets of interpretation. Hence, image processing is more effective.

Finally, it is advantageous to make an integrated interpretation by overlaying different data sources and known information by overlaying different reasonable data images together. Some correlations among these data can be found easily, and can provide some constraint conditions for the inverse problems of these data. After density slicing an interpreted sketch map may be obtained automatically.

The key to obtaining good results from image processing is to apply techniques properly. Therefore, it is necessary to discuss image processing techniques in detail.

Preprocessing of Geophysical Data

Unlike remote sensing data, observed geophysical data are not in square gridded size. For instance, aeromagnetic data have a much smaller sampling interval along the flight line than is found in the interval between flight lines. Therefore, gridding observed data into a proper gridded cell size is the first procedure of processing. The grid cell size should be compatible with the other geodata images or with the requirement of potential data processing. Different source data should be registered in the same geographical coordinates.

Potential transform processing is still the foundation of magnetic data processing. Potential transform processing transforms the original observed data into data in different observation levels, into data magnetized in different inclinations, into different components or gradients of potential fields, and so on. These data possess a certain physical meaning and differ from original observation conditions. The enhancement of image processing only improves the appearance of geophysical data for human vision (Jensen, 1986). Image processing cannot replace potential processing. Therefore, the classical data processing, either in frequency domain or in space domain, is the basic processing.

According to the practical situation of the data, some potential transform processing, such as upward continuation, downward continuation, reduction to the pole, vertical gradient derivative, apparent susceptibility, apparent density, various filtering, etc. may be chosen as preprocessing. The results of these processing steps can be used for further image processing, and some of them also can be treated as one kind of data image during overlaying (as discussed later).

If the cell size needed for potential transformation is different from other geodata image pixel size, the processed data should be resampled or desampled again to satisfy other geodata. In general, the spline interpolation is applied commonly for geophysical data instead of linear interpolation for remote sensing data.

After these preprocessings, the magnetic data should be transferred into standard image files. According to the distribution histogram, the magnetic data should be stretched into the integer values, which range from 0 to 255 to satisfy the demand of image processing.

Shadow of Artificial Illumination

Mapping is an important way to present the results. Besides color and gray scale maps, the shadow of anomalies under certain artificial illumination shows the anomaly reliefs clearly. This map presentation gives a stereo visual sensing of the anomalies and is useful for recognizing the pattern of anomalies.

The brightness of the illuminated surface representing the field of anomaly at every data point is proportional to the cosine of the angle α between the normal anomaly surface and the vector, which points to the specified illumination source (Kowalik and Glenn, 1987). The brightness can be calculated by the following expression:

$$S_{ij} = \frac{1+P_0 P_{i,j}+Q_0 Q_{i,j}}{(1+P_{i,j}^2+Q_{i,j}^2)^{1/2} \times (1+P_0^2+Q_0^2)^{1/2}},$$

$$i=2,3,\ldots,M\text{-}1$$
$$j=2,3,\ldots,N\text{-}1 \tag{1}$$

where S_{ij} is the brightness value of the calculating point,

M is the numbers of lines,

N is the numbers of points per line,

p_0 and Q_0 are the east–west and north–south components of the vector pointing to the illumination source,

P_0 is $-\tan(90°-\theta)\cos(\phi)$

Q_0 is $-\tan(90° -\theta)\sin(\phi)$

θ is the elevation angle of the illumination source, in degrees above the horizon,

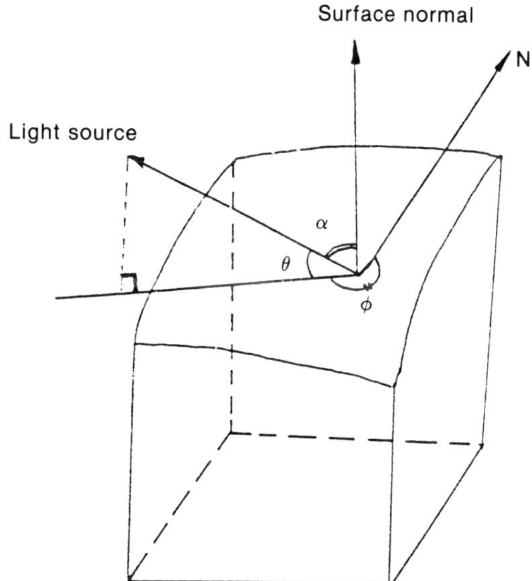

FIGURE 14.1. Illustration of illumination vector and the anomaly surface normal vector.

ϕ is the azimuth of the illumination source, in degrees clockwise from north.

The terms P_{ij} and Q_{ij} are the east–west and north–south components of the anomaly surface slope, respectively. The slope can be calculated according to the anomaly value. If a 3×3 window is applied,

$$P_{ij} =$$
$$\frac{A_{i-1,j+1}+2A_{i,j+1}+A_{i+1,j+1}-A_{i-1,j-1}-2A_{i,j-1}-A_{i+1,j-1}}{m}$$

and

$$Q_{ij} =$$
$$\frac{A_{i-1,j-1}+2A_{i-1,j}+A_{i-1,j+1}-A_{i+1,j-1}-2A_{i+1,j}-A_{i+1,j+1}}{m} \tag{2}$$

where $i = 2,3,\ldots, M-1$
$j = 2,3,\ldots, N-1$
$A_{ij} =$ the anomaly value, and
$m =$ a scalar slope factor that controls the effective slope of the anomaly surface.

The angles of θ, ϕ, and α are shown in Figure 14.1.

Overlaying a color map with the gray scale shadow map will show not only the relief of anomalies but also the amplitude of anomalies. Sometimes even more information is visible.

Spatial Filtering

Spatial filtering, a common way of enhancement in image processing, is performed by convolution between the original image and a certain filter in space domain. Therefore, different aims of image processing need different kinds of filters.

In general, these filters can be divided into two types, low-pass and high-pass filters. Low-pass filters of image processing are similar to the simple moving average filter in classical data processing of magnetic data. The typical filter operator, assuming the filter is in a 3×3 matrix, is

$$\begin{matrix} 1/9 & 1/9 & 1/9 \\ 1/9 & 1/9 & 1/9 \\ 1/9 & 1/9 & 1/9 \end{matrix}$$

The operators have the same value, 1/9, and the summation of these operators is equal to 1. This type of filter operator is always used to eliminate surface disturbances and to enhance the information of a source in a larger dimension and seated at a greater depth.

There are many types of filters used for high-pass filtering. Usually the filter is applied to accentuate or sharpen edges and linear features of the image. Two examples are

$$\begin{matrix} -1 & -1 & -1 \\ -1 & 9 & -1 \\ -1 & -1 & -1 \end{matrix} \quad \text{or} \quad \begin{matrix} 0 & -1 & 0 \\ -1 & 4 & -1 \\ 0 & -1 & 0 \end{matrix}$$

The operators of this type of filter are different from low-pass filter operators. Usually, the summation of the operators of the high-pass filter is equal to zero. Furthermore, high-pass filters can be divided into several types by their functions.

The linear edge enhancement filter is simple but useful. The direction filter calculates the gradient to enhance the variation of the image in a certain direction. Following are several directional filters that will enhance the information that is parallel to the pass direction, respectively:

$$\text{East–west} \quad \begin{matrix} 1 & 1 & 1 \\ 0 & 0 & 0 \\ -1 & -1 & -1 \end{matrix} \qquad \text{Northwest} \quad \begin{matrix} 0 & 1 & 2 \\ -1 & 0 & 1 \\ -2 & -1 & 0 \end{matrix}$$

$$\text{North–south} \quad \begin{matrix} -1 & 0 & 1 \\ -1 & 0 & 1 \\ -1 & 0 & 1 \end{matrix} \qquad \text{Southwest} \quad \begin{matrix} -2 & -1 & 0 \\ -1 & 0 & 1 \\ 0 & 1 & 2 \end{matrix}$$

$$\text{East–west} \quad \begin{matrix} -1 & -1 & -1 \\ 0 & 0 & 0 \\ 1 & 1 & 1 \end{matrix} \qquad \text{Southeast} \quad \begin{matrix} 0 & -1 & -2 \\ 1 & 0 & -1 \\ 2 & 1 & 0 \end{matrix}$$

$$\text{North–south} \quad \begin{matrix} 1 & 0 & -1 \\ 1 & 0 & -1 \\ 1 & 0 & -1 \end{matrix} \qquad \text{Northeast} \quad \begin{matrix} 2 & 1 & 0 \\ 1 & 0 & -1 \\ 0 & -1 & -2 \end{matrix}$$

These filters have the same form of the first derivative in a certain horizontal direction during the data processing in space domain.

For arbitrary direction filtering, a suitable result can be obtained by overlaying two images that were filtered in two perpendicular directions, north and east, respectively, in a sine and cosine relationship. The result can be expressed as

$$IA = IN \cos(A) + IE \sin(A) \qquad (3)$$

where IA is the output of an arbitrary directional filtering,

IN is the result of north–south pass filtering,

IE is the result of east–west pass filtering, and

A equals the clockwise angle between north and the direction.

The feature that is along the A direction will be enhanced. This enhancement will also be discussed with overlaying.

The Laplacian convolution may be applied for enhancing the edge or boundary, which is in an arbitrary direction (Jensen, 1986). In fact, the Laplacian convolution is a second derivative calculation. Therefore, before the Laplacian convolution is applied, the situation of the noise in the image should be considered carefully or a low-pass filter should be applied first. These filter operators are symmetrical to the center of the filter. The examples are

$$\begin{matrix} -1 & -1 & -1 \\ -1 & 8 & -1 \\ -1 & -1 & -1 \end{matrix} \qquad \begin{matrix} 1 & -2 & 1 \\ -2 & 4 & -2 \\ 1 & -2 & 1 \end{matrix}$$

Sometimes the linear feature of data is required for enhancement to detect faults, axis of folds, or other linear structures. In this situation the line enhancement filter can be applied. Some line enhancement filters are

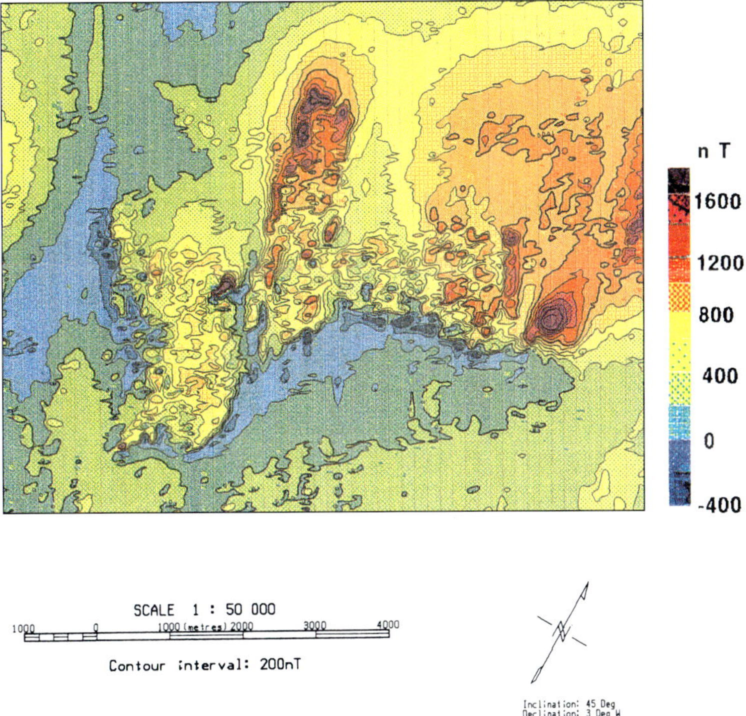

FIGURE 14.7. Reduction to the pole magnetic anomaly in the Yiesan area.

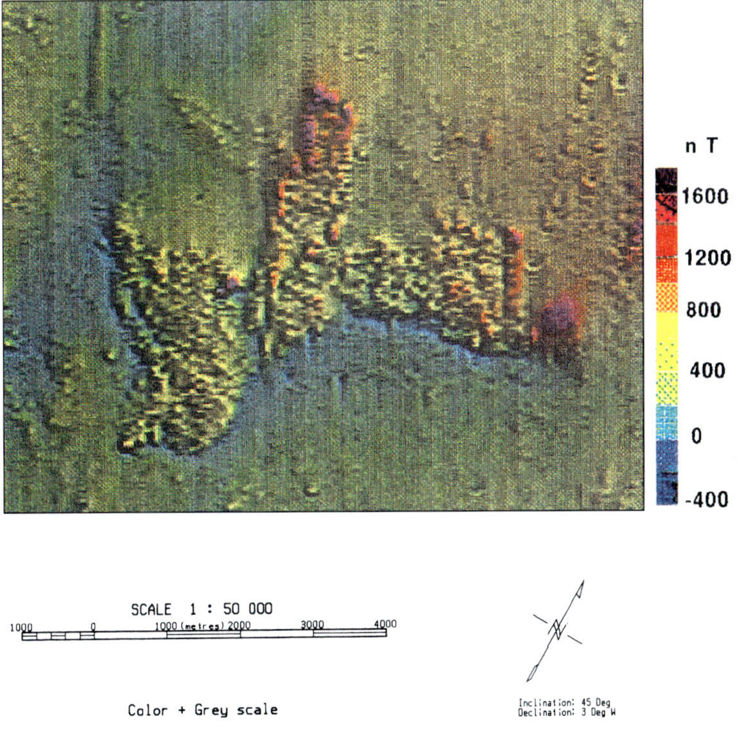

FIGURE 14.9. Color shadow map of magnetic data in the Yiesan area.

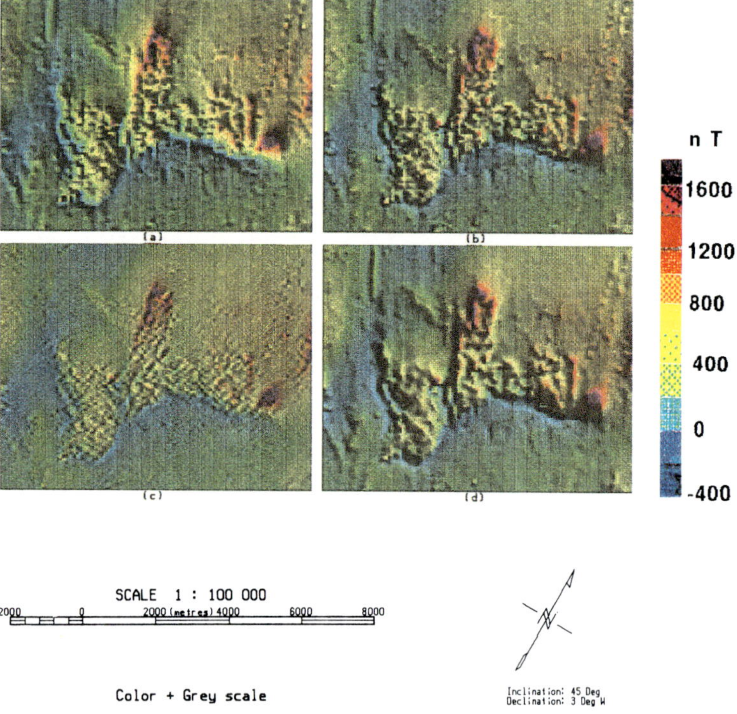

FIGURE 14.10. Comparison of different filtered images of magnetic data in the Yiesan area (a) NW pass, (b) edge enhancement, (c) line enhancement, (d) rank and NW pass.

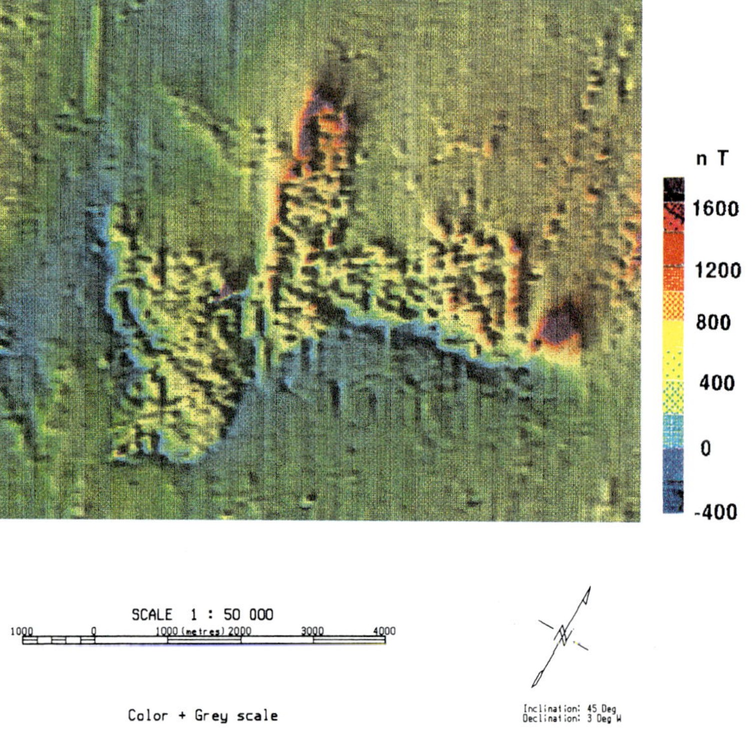

FIGURE 14.12. Color image of magnetic data after NW filtering in the Yiesan area.

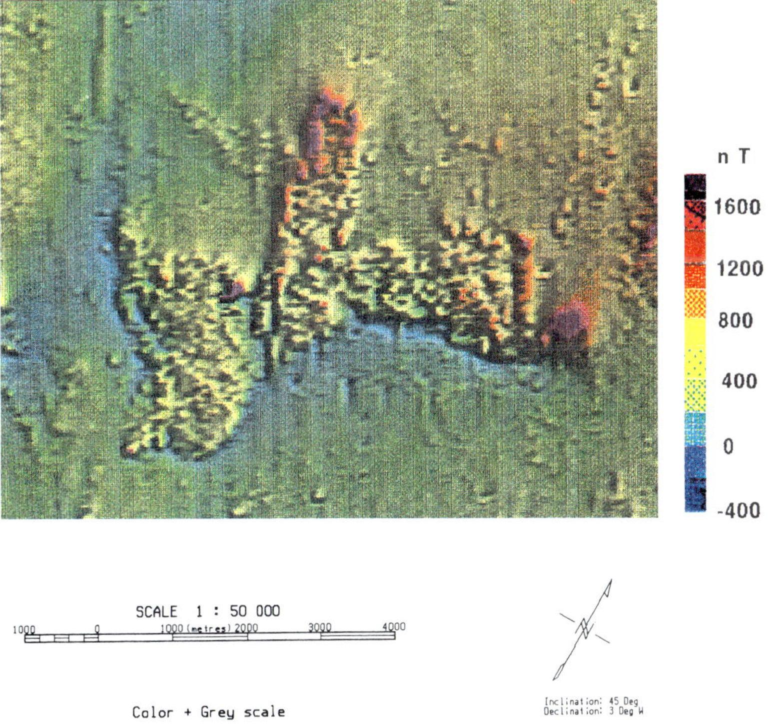

FIGURE 14.13. Color image of magnetic data after edge enhancement in the Yiesan area.

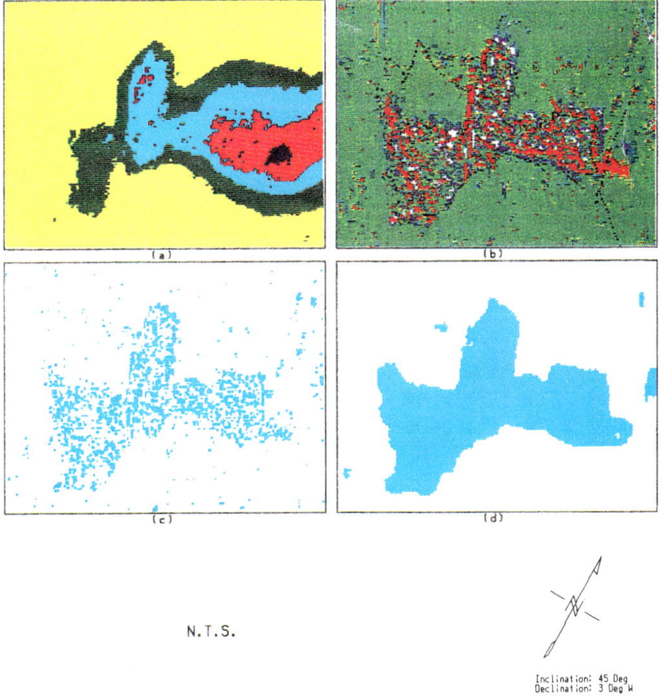

FIGURE 14.16. Hard-copy images in the Yiesan area (printed by color jetprinter 3852-2): (a) pseudo-isodepth of intrusion, (b) overlay of filtered image and geological map, (c) density slicing after line enhancement, (d) density slicing after relief filtering.

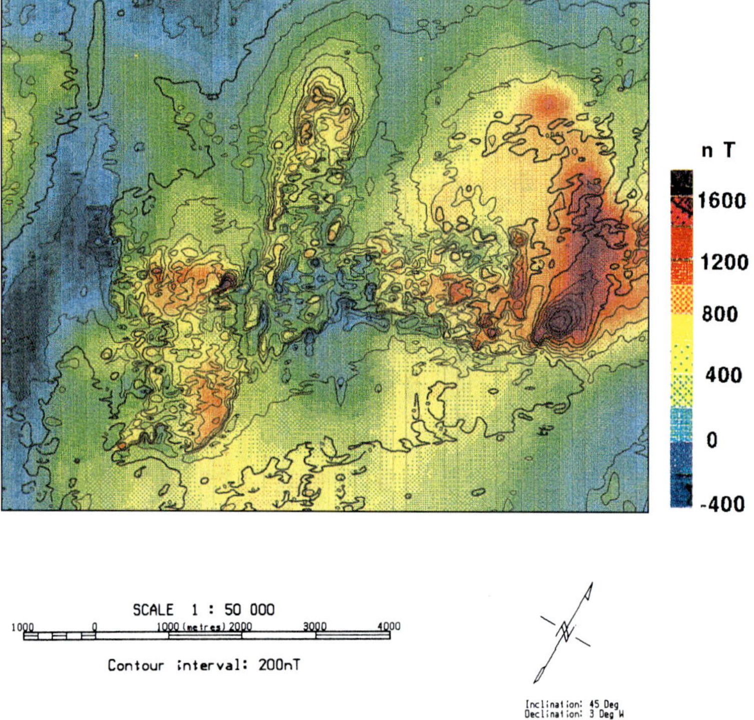

FIGURE 14.17. Color map of magnetic data and related geochemical anomaly in the Yiesan area.

$$
\begin{array}{rrr}
-1 & -1 & -1 \\
2 & 2 & 2 \\
-1 & -1 & -1
\end{array}
\qquad
\begin{array}{rrr}
-1 & -1 & 2 \\
-1 & 2 & -1 \\
2 & -1 & -1
\end{array}
\qquad
\begin{array}{rrr}
-1 & 2 & -1 \\
-1 & 2 & -1 \\
-1 & 2 & -1
\end{array}
$$

If necessary, the nonlinear edge enhancement filters can be employed. These filters are based on using a nonlinear combination of image pixels and are calculated according to a certain relationship, such as Sobel edge enhancement and Robert edge detector (Jensen, 1986).

Local Contrast Stretch

When the causative body is relatively deeper or smaller, the width of the magnetic anomaly will be broader or the amplitude of the magnetic anomaly will be lower, respectively. Sometimes we are interested in these types of targets. In this case, local contrast stretch is a useful tool to enhance the original low contrast.

The important point of local contrast stretch is to stretch the image in a variable weight and the weight varies according to the local pixel value in the boxcar window. Of course, the result is related to the size of the boxcar window and some given parameters.

The Wallis algorithm (Fahnestock and Schowengerdt, 1983) performs a local contrast stretch by determining the mean value and standard deviation in a specified window. The expected mean and standard deviation should be supplied by the user before processing.

Some simple calculations also can stretch local contrast. For example, a local contrast stretch can be performed according to the local mean value in a specified window and the difference between the mean value and the value of the calculating point. The contrast stretch can be expressed by the following formula:

$$
LS_{ij} = \sum_{k=-K}^{K} \sum_{l=-L}^{L} (LC_{i+k,j+l} - MEAN_{ij})
$$

$$
MEAN_{ij} = \sum_{k=-K}^{K} \sum_{l=-L}^{L} (LC_{i+k,j+l})/(2K+1)(2L+1)
$$

$$
i = K+1, K+2, \ldots, M\text{-}K
$$
$$
j = L+1, L+2, \ldots, N\text{-}L
$$

where LS_{ij} is the stretched value of the image,
LC_{ij} is the original value of the image,

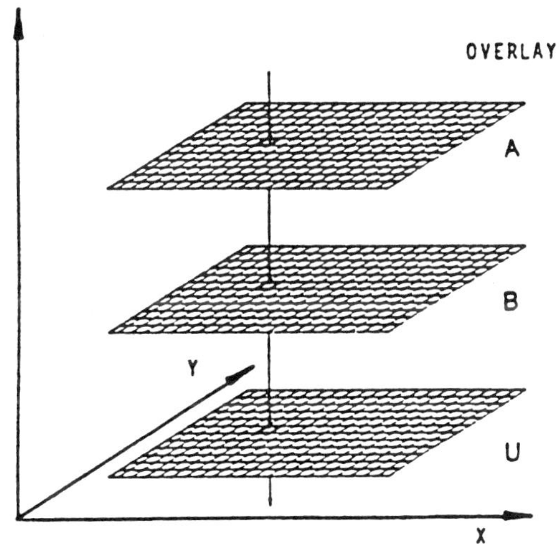

FIGURE 14.2. The concept of overlay (after P.A. Burrough, Principles of Geographical Information Systems for Land Resources Assessment, 1986, by permission of Oxford University Press).

i, j are the row number and column number of the calculating pixel, respectively,
M, N are the numbers of lines and points of the original image, respectively,
K, L are the half size of the boxcar window. For example, if window is a 3×3 matrix, then $K=1$ and $L=1$.

Overlaying

The enhanced image of geophysical data or other geodata is like different maps that concentrate on different information. For integrated interpretation, the correlation among these different kinds of data and the pattern of different data combinations are very important.

Instead of complicated computations, the overlaying technique is the most convenient and direct way. This technique overlies several images in some relationship and same coordinate (see Figure 14.2). It can be expressed in the following equation:

$$
U = f(A, B, \ldots) \tag{5}
$$

where $A, B, \ldots$ are overlaid images,
U is the result of overlaying,
f is the relationship of combination.

Naturally, the relationship can be a simple arithmetical, mathematical, logical, or conditional

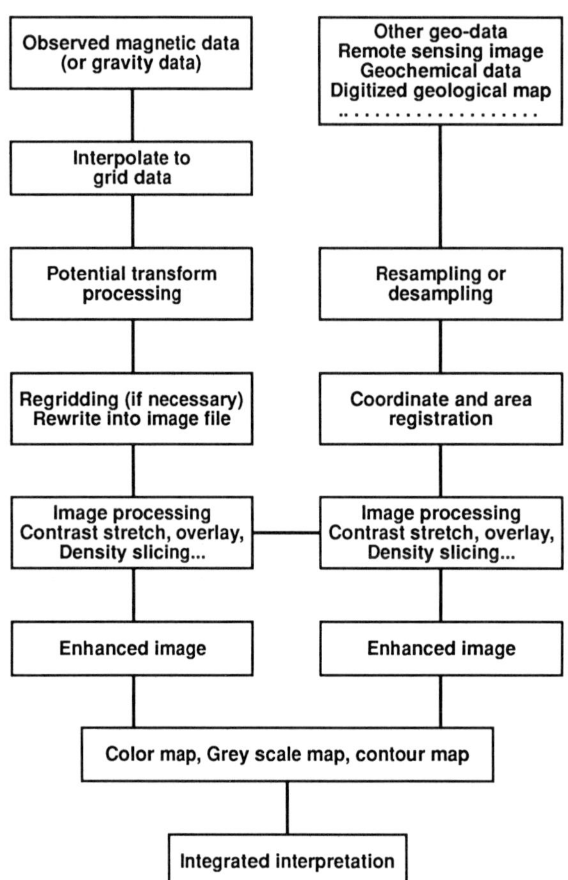

FIGURE 14.3. The flow chart of image processing of magnetic data.

operation. The following map calculation relationships are in the ILWIS system, +, −, *, /, & (logical and), | (logical or), ~ (logical not), mod, div, =, < >, <, <=, >, >=, if, clfy, and, or, xor, not, sqrt, exp, log, sin, cos, atan. Obviously overlaying has a big advantage for integrated interpretation. However, overlaying constrains the parameters of an inverse problem from different data and makes the results of qualitative interpretation more reasonable.

Because of the function relationship among overlaid images, overlaying also performs high-pass, low-pass, and other enhancement. For example, to subtract the image of upward continuation from the observed image a high-pass filter is used to enhance the anomalies caused by shallow sources.

Since there are sine and cosine function relationships, a directional filtering is easy to realize by Eq. (3).

The concept of overlaying can be extended into map presentation. Overlaying the gray scale shadow map on the color map of the same area gives a more effective visual sensing.

Sometimes there is no exact physical meaning for the values of the result image pixels after overlaying, but the values present the correlation among the different data files. Therefore, a reasonable and interpreted map can be obtained from the overlaying image.

Density Slicing

In qualitative interpretation, the range of anomaly values and the pattern of anomaly combination are often used for recognizing different rock types and different structures. But it is not easy to determine the threshold values and to see the result immediately.

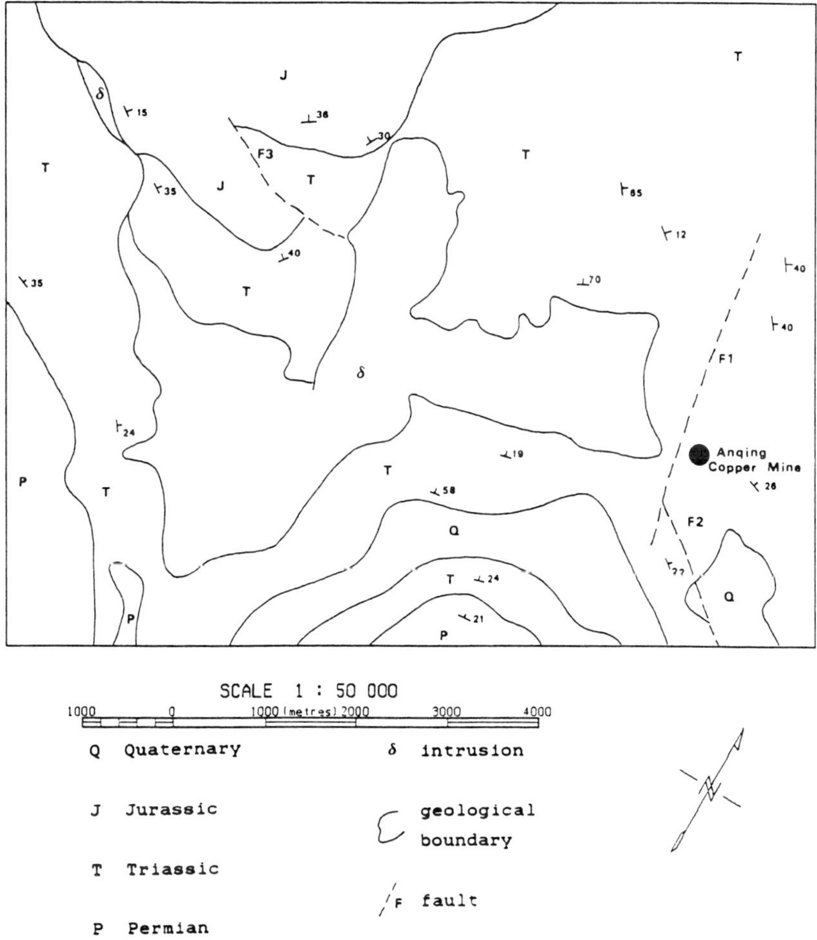

FIGURE 14.4. Sketch map of geology in the Yiesan area.

Density slicing performs this work effectively by determining the divisions of pixel values and colors, respectively, according to the histogram of image, by monitoring the result on the screen, and by adjusting the threshold values immediately. After several repetitions a neat sketch map can be obtained quickly.

Density slicing also provides a bright future for automatic geology mapping from magnetic data. If a proper method is employed and the geophysical condition is sufficient, it is possible to obtain a geological sketch map automatically.

In general, the image processing steps of magnetic data can be simply presented by the flow chart that is shown in Figure 14.3.

The following example illustrates the result and the procedure of a practical attempt.

Application in Yiesan Area

Following the discussion of theoretical principles of image processing techniques for magnetic data, a practical example is given showing the application of these procedures and the results of image processing.

This practical example is the Yiesan area in east China where after many years of exploration, all the shallow ore bodies had been explored. A series of research projects were designed in

TABLE 14.1. Physical property.

Rock or mineral type	Susceptibility SI	Density t/m³	Resistivity Ω·m
Magnetite, pyrrhotite	1430	4.3	1
Diorite	246	2.7	1000–10,000
Limestone	0	2.7	>10,000

recent years to detect new but deeper ore bodies. One research project uses image processing to integrate interpretation of known magnetic data and the other geodata.

Part of the data processing and image processing used the MV/10000 computer and M75 image processing machine in the China University of Geosciences (Wuhan) in 1989. Most data were processed and printed again using GEOSOFT, a geophysical data processing software system designed by GEOSOFT Inc. Canada, and the ILWIS system.

Geologic and Geophysical Condition of the Yiesan Area

In the Yiesan area, located in the northern bank of the Yangtze river, there are 24 iron and copper deposits and prospects. The minerals of the main copper mine, Anqing Copper Mine, are chalcopyrite, pyrrhotite, and magnetite. The deposits, 400 m below ground, were discovered in 1960 as a result of the aeromagnetic anomaly. From that time on 200,000 m of drilling has been carried out (Zhou, 1984).

The rock types of the sedimentary formation in this area are Permian limestone and sandstone,

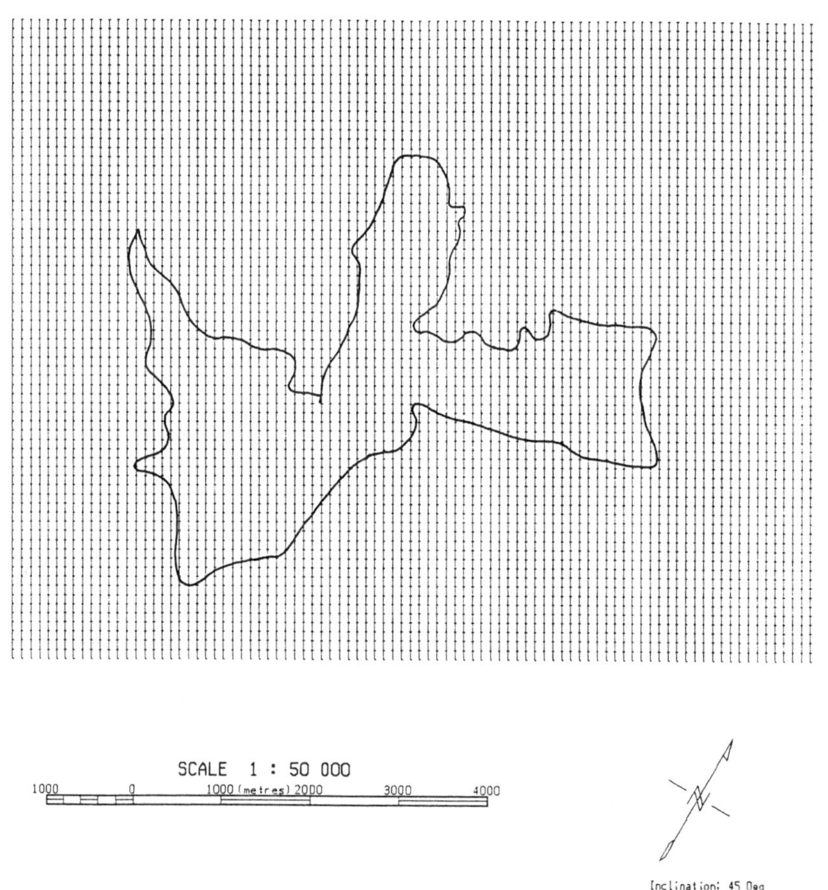

FIGURE 14.5. Magnetic survey stations in the Yiesan area (a tick per 5 stations along lines).

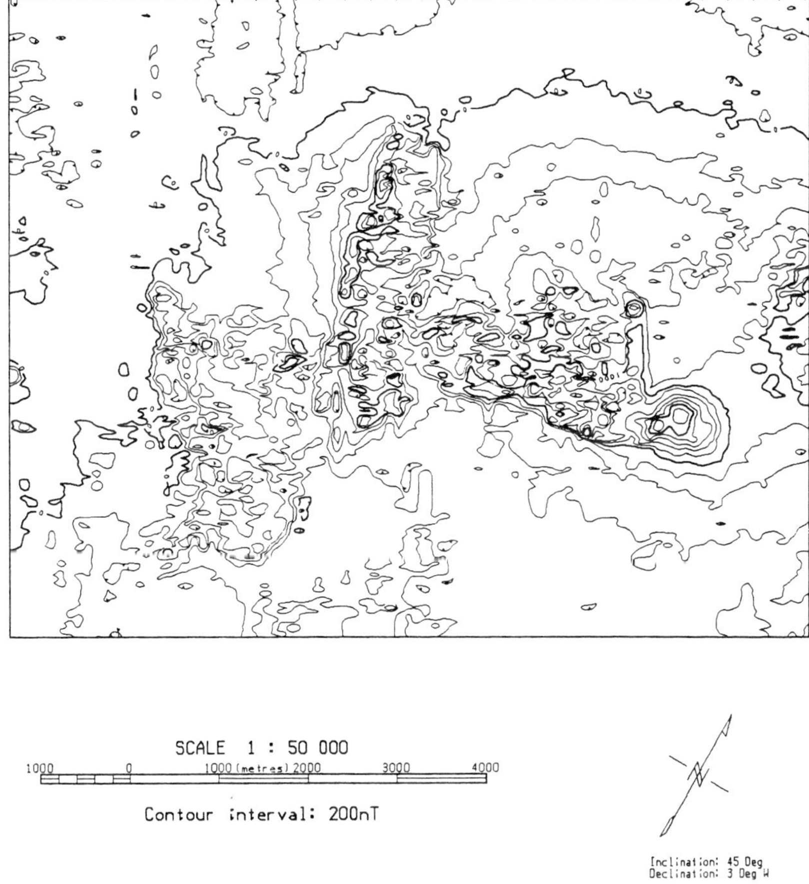

FIGURE 14.6. Contour map of magnetic data in the Yiesan area.

Triassic limestone and calcareous shale, Jurassic calcareous sandstone and siltstone, and Quaternary sedimentary.

The strike of beds is northeast to southwest, and the dip of beds extends to the northeast. The intermediate-acid rock, diorite, cuts through these formations to form an intrusion between two strata. Part of the rock outcropped to the surface, and part is covered by Quaternary sedimentary. Ore bodies were found in the contact zone between the intrusion and the host rock.

There are two sets of faults that were proved by ground geological survey and drilling. The strikes of these faults are from south to north and from northwest to southeast, respectively.

Figure 14.4 is the sketch geological map of this area, where the map frame was chosen perpendicular to the strike of the intrusion.

Prior research determined that the ore body was controlled by intrusion and faults. Therefore, the interpretation of intrusion and faults is very important for ore body detection.

The physical properties of the main rocks and minerals in the Yiesan area are shown in Table 14.1.

No large-scale gravity survey data for ore body detection existed until 1988.

The differences of susceptibility among these rocks and minerals are obvious. Chalcopyrite was formed accompanied with pyrrhotite and magnetite. Therefore, magnetic anomaly also can be used to detect copper ore body.

The inclination of the geomagnetic field is about 43,000 nT.

A ground magnetic survey of an area 9 by 7 km was carried out in 1981 on a 100 by 20 m grid of stations. The magnetic survey stations and the

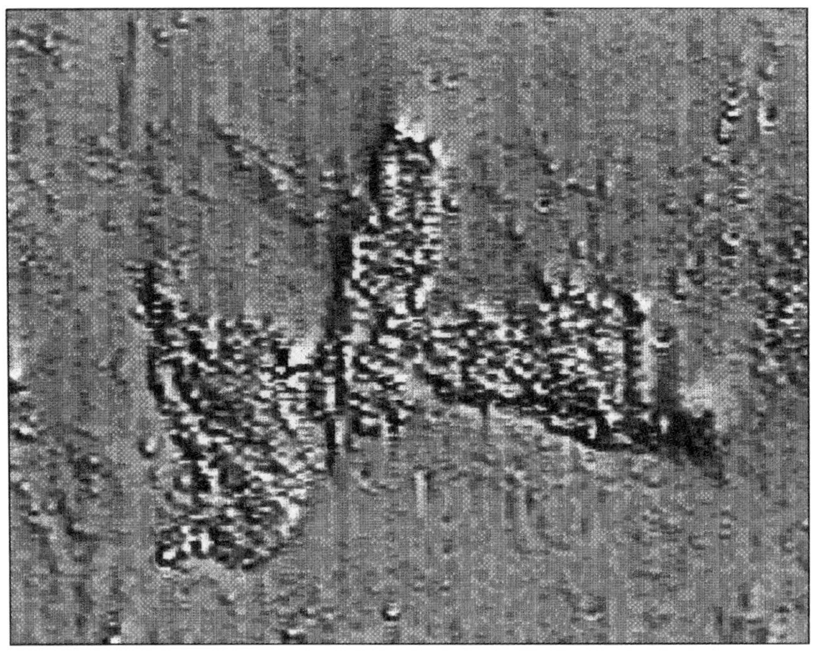

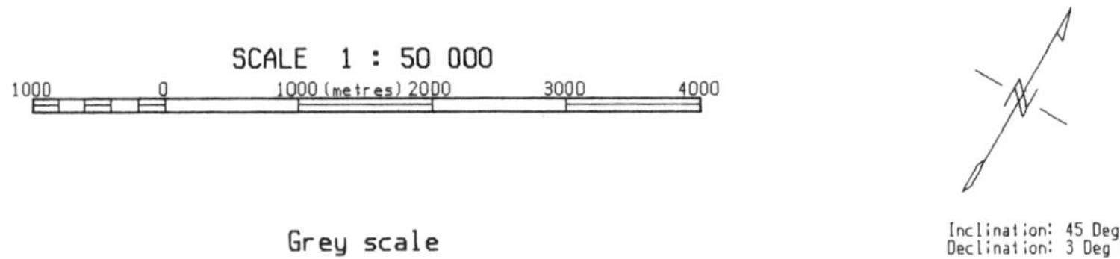

FIGURE 14.8. Shadow map of magnetic data in the Yiesan area.

magnetic anomaly contour map are shown in Figures 14.5 and 14.6, respectively (Team No. 326, 1981).

The magnetic anomalies are caused by diorite, magnetite, and pyrrhotite minerals. These anomalies are compounded together and seem a little bit complicated. But the magnetite and pyrrhotite produced a much higher anomaly than the diorite. Anqing Copper Mine is a part of the large local anomaly (see Figure 14.6).

Potential Processing of Magnetic Data

The observed magnetic data were interpolated into a 40 by 40 m grid data to satisfy potential transformation.

Because of the 45° inclination of the geomagnetic field in this area, reduction to the pole processing was applied first. The result is shown in Figure 14.7 (see Color Plate VI). This result was treated as basic data for all further data processing and image processing. Clearly the anomaly of Anqing Copper Mine shifts to the northeast. The fast gradient of the anomaly and its negative values south of the intrusion show that the dip of the intrusion extends to the north.

Shaded relief of the anomaly from artificial illumination in the 45° inclination and 45° declination was calculated. The gray scale shaded relief map is shown in Figure 14.8. This map shows the boundary between the intrusion and the host rock quite well and is nearly the same as is shown on the geological map (see Figure 14.4). The known faults, F1 and F2, are presented clearly by the linear feature of the map. There are some other linear features that can be recognized on the gray scale shadow map.

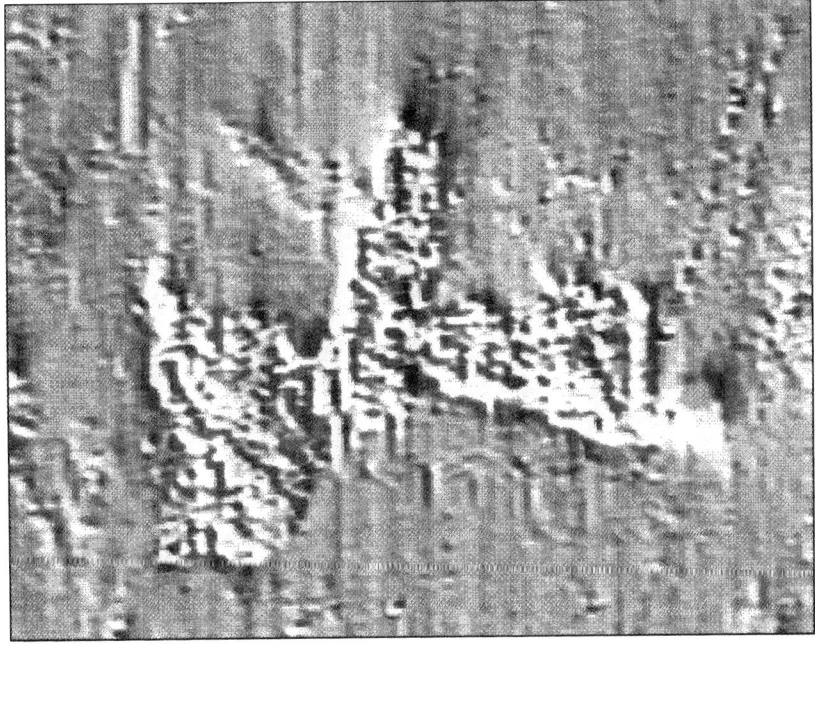

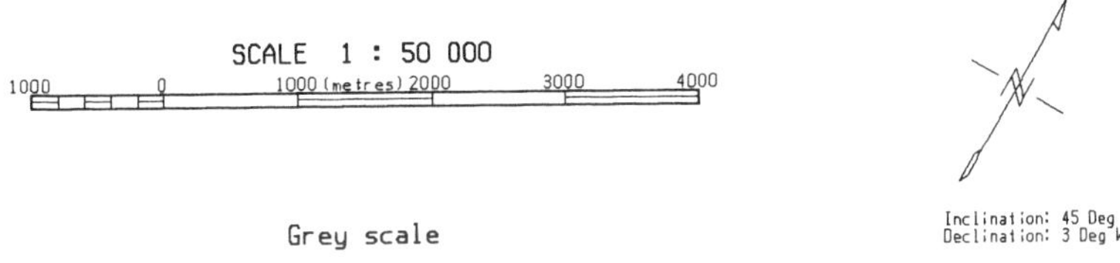

FIGURE 14.11. Image of magnetic data after NW filtering in the Yiesan area.

Figure 14.9 (see Color Plate VI) combines the gray scale shaded relief map and the color map and gives a qualitative concept of the anomaly and stereo visual sensing of the anomaly relief. The linear features are clearer than in Figure 14.8. These features will be interpreted in combination with image processing results.

Several other potential transformations were applied: upward continuation in different heights, apparent susceptibility map, etc. These results were used for further image processing.

Image Processing

Preprocessing was carried out by use of GEOSOFT and other programs applied on the MV/10000 computer. The structure of the resulting data file is different from the structure of the standard image file of the ILWIS system. Therefore, I designed several programs to restructure the data file to satisfy the different systems. All the preprocessed results were transferred as raster image files to ILWIS and then various image processing techniques were applied.

Because of the low resolution of the color jet-printer 3852-2, the hard copy of the resulting images loses a lot of detail and information when compared with the image on the high resolution screen (see Figure 14.16b). The hard copy of the screen, because it is not to scale, partly loses its value for practical interpretation.

To improve this situation and to use the more powerful mapping system of GEOSOFT, the author designed a program to transfer the image of ILWIS into the standard grid file of GEOSOFT. Most images of the results were printed by the mapping

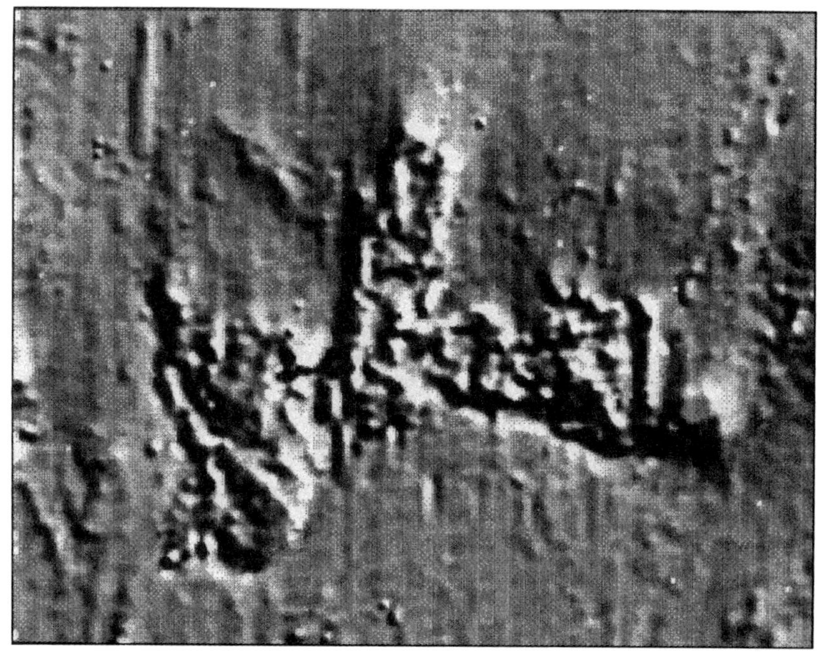

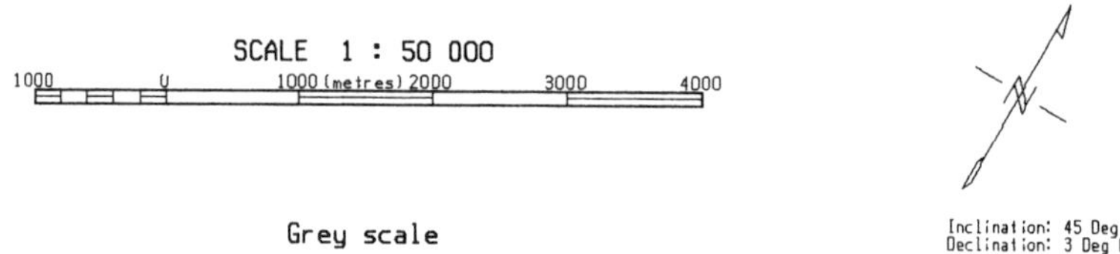

FIGURE 14.14. Filtered image of magnetic data in the Yiesan area.

system of GEOSOFT and FUJITSU DL2600 printer. Even so, there are still disadvantages.

Filtering

Some different filters were applied to enhance the boundary of intrusion and the linear features of faults. Four are shown in Figure 14.10 (see Color Plate VII).

Figure 14.10a is the result of the northwest pass filter and the filter operators are

$$\begin{matrix} 0 & 1 & 2 \\ -1 & 0 & 1 \\ -2 & -1 & 0 \end{matrix}$$

After filtering, the shape of intrusion is clearly seen; and northwest linear features are clearer than in Figures 14.8 and 14.9.

The filter used in Figure 14.10b is an edge enhancement filter,

$$\begin{matrix} -1 & -1 & -1 \\ 1 & 2 & -1 \\ 1 & 1 & -1 \end{matrix}$$

A line filter was applied for edge enhancement and the result is shown in Figure 14.10c. The operators of this Laplacian filter are

$$\begin{matrix} 2 & -1 & -1 \\ -1 & 2 & -1 \\ -1 & -1 & 2 \end{matrix}$$

Figure 14.10d is the result of the combination of rank filtering and northwest pass filtering.

Comparing the four images, we see that the result of line filter in Figure 14.10c is not so clear.

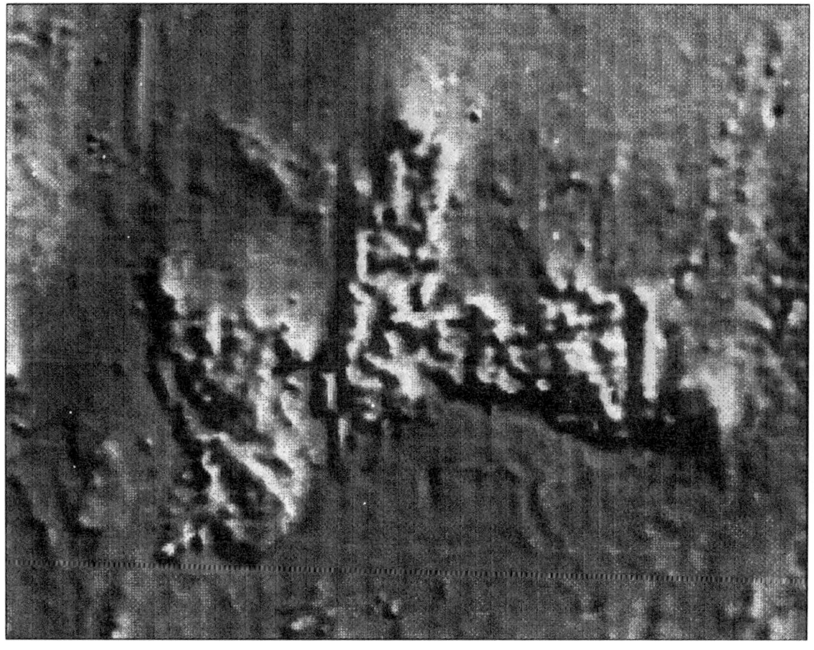

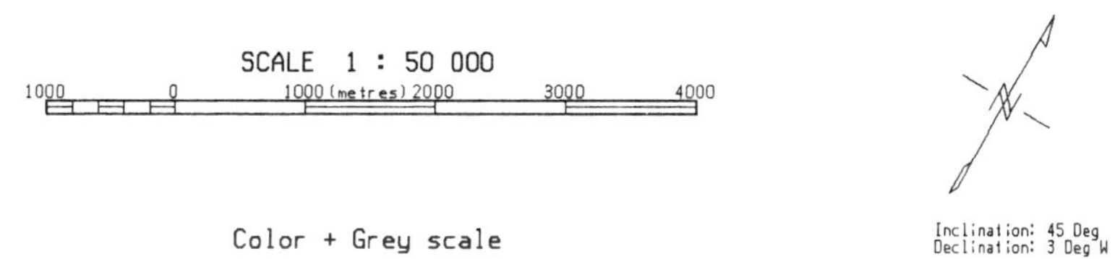

FIGURE 14.15. Map and gray scale filtered image of magnetic data in the Yiesan area.

This means the line filter is either not suitable in this case, or further processing is needed. But good results were obtained by use of the other three filters.

To interpret the filtered data with the results of prior data processing, the image maps were printed at the same scale. Figure 14.11 is a gray scale image of northwest pass filtering, and Figure 14.12 (see Color Plate VII) is a color image of the magnetic data combined with the gray scale image of northwest filtering. Many linear features can be recognized from these two maps, and some of the features cannot be found in other results. The boundary of intrusion is clear also.

Comparing the results of shaded relief and northwest pass filtering, the linear feature is clearer on the NW pass filtered image than it is on the shadow map. The reason is that the direction filtering also enhances the anomalies that have a large gradient but a small relief, while the shaded relief mainly emphasizes the relief. Therefore, in case of enhancing a image that has small relief but large gradient, the direction filter is better than shadow processing.

Edge enhancement filtering makes the boundary of intrusion and ore bodies clearer (see Figure 14.13, see Color Plate VIII). The edge of the known Anqing Copper Mine is clearer than the other images.

Figures 14.14 and 14.15 are the gray scale and the color plus gray scale maps of the results of rank order filtering, which eliminates the small reliefs and enhances the main feature of intrusion and the main faults. The big advantage is strong stereo sensing of these features.

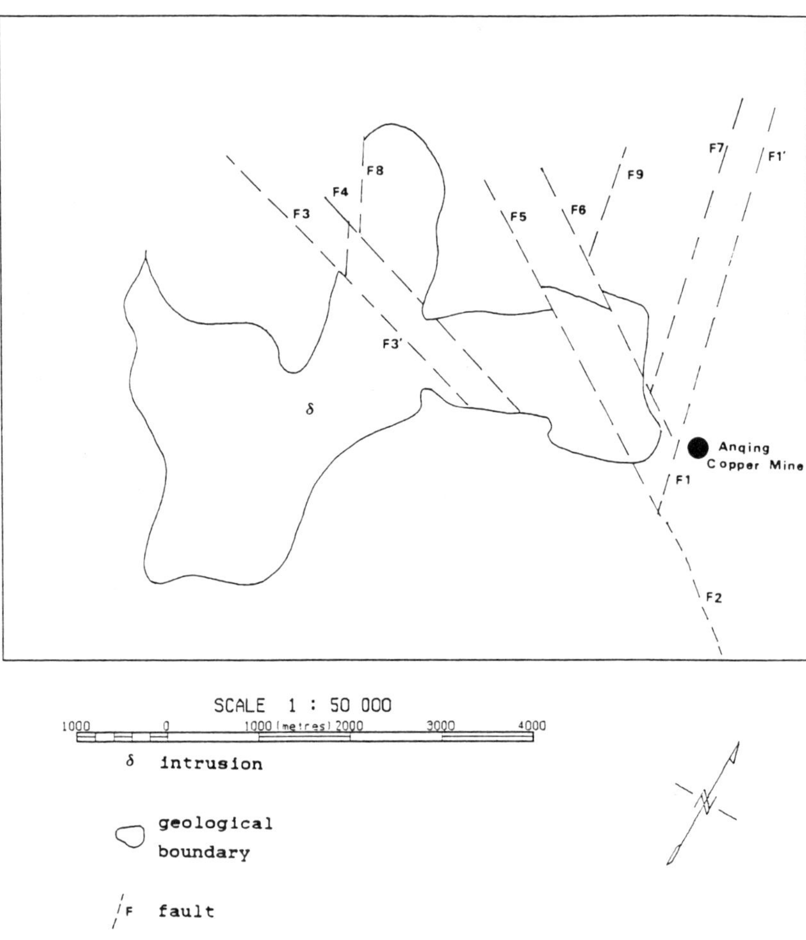

FIGURE 14.18. Interpreted geological map in the Yiesan area.

Overlaying with Different Results of Preprocessing

Figure 14.16a (see Color Plate VIII) is a pseudo isodepth map of intrusion obtained by overlaying and density slicing processing.

Because of the deep position of the intrusive source, three images—observed level, upward continuation to 200 m and upward continuation to 400 m—were overlaid in a simple linear combination and density slicing was applied later. This image gives a rough idea about the thickness of the intrusion and the deep source of intrusive rock.

According to the nature of potential field, the amplitude of an anomaly caused by a small and shallow magnetic body will decrease quickly when the height of the observation level is increasing, while the amplitude of the anomaly of deeper and larger magnetic source will decrease slowly. Therefore, in this specific case the upward continuation emphasizes the information of the deeper magnetic sources. In other words, magnetic data from different heights of observation level mainly present the magnetic source from different depths. Overlaying of these data will show the change of magnetic source at different depths.

From Figure 14.16a we can see that the near surface or outcropped intrusion (green color) is similar to the known geological map. When the depth is increased the intrusion shifts to northeast (blue, red, and black color). The deep source of intrusive rock is located at the northeastern part of the intrusion.

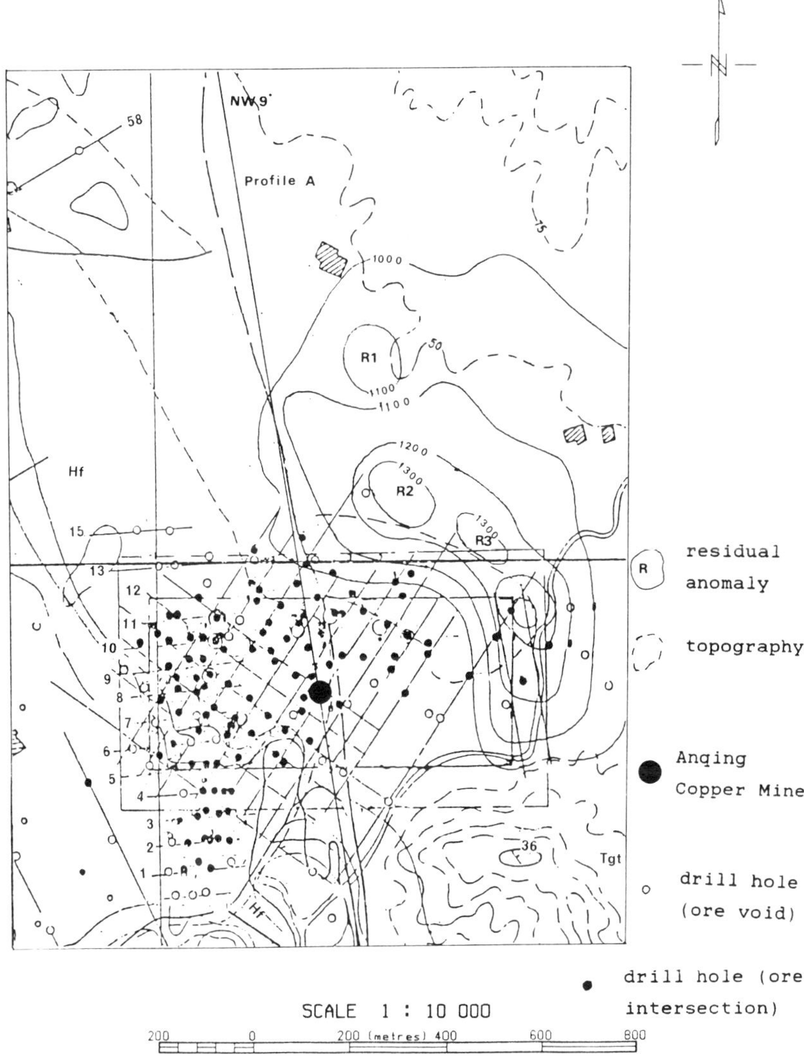

FIGURE 14.19. Residual anomalies in Anqing Copper Mine.

The thickness of the intrusion in the southwestern part is thin and in the northeastern part is thick. This fact may provide some initial geometric parameters of the intrusion for modeling of intrusive rock. The interpretation, combined with some known geological information, is that the intrusion came from the northeast and from a great depth, then extended to the southwest. This interpretation differs from the prior interpretation.

This revised interpretation also provides some information for studying the forming of an intru-sion and is useful information for mineral resources detection.

Overlaying of Geochemical Data

From geochemical surveys carried out several years ago, more than 200 rock samples were collected randomly in this area from drilling and outcrops and were analyzed. The analytical results for potassium and sodium show hydrothermal alternation zones around the known ore bodies.

These haloes are characterized by high potassium and low sodium values. This special relationship of the halos was noticed previously, but was not connected with magnetic anomaly because of the techniques for combining these data together.

The images of these two geochemical anomalies were obtained by digitizing from the known contour maps. Overlaying the maps shows the relationship between the geochemical data and enhances the areas that satisfy the known situation.

This result was printed together with the magnetic anomaly map (see Figure 14.17, see Color Plate IX). The magnetic anomalies are caused by intrusion and ore bodies, and they are complicated. The area that either has magnetic anomaly or geochemical anomaly may be most interesting for ore body detection. It is clear from Figure 14.17 and 14.4 that the north of the Anqing Copper Mine is valuable for further exploration.

Overlaying with Geological Map

The boundary of the intrusion and several known faults in the Yiesan area were digitized from the geological map and registered to the same image size and coordinates of magnetic data.

The result of overlaying this geological map image and the northwest pass filtered image is shown in Figure 14.16b. The black dots on the image are the digitized geological map. This image shows good northwest pass filtering results and gives a new interpretation of the area by comparing the filtering result and the known information. For example, fault F2 can be extended to the northwest and cut through the intrusion and fault F3 can be extended to the southeast and cut through the intrusion. Comparison with fault F1 shows that there is a linear feature parallel to fault F1.

Density Slicing

The enhancement of density slicing can be used to obtain the interpreted sketch map automatically. Slicing the filtered magnetic data image, the result of a line filter, is shown in Figure 14.16c. The pattern of intrusion gives the feeling of an interpreted sketch of the geology map, but is still different from the real sketch.

According to the concept of anomaly relief (Paterson and Reeves, 1985) and the anomaly pattern of the intrusion in the Yiesan area, a 3×3 relief filter was designed to enhance the mapping of intrusion. The main principle is to calculate the sum of the absolute difference between every pixel value and the mean value in a 3×3 window. The calculation of this filter can be expressed by the following equation:

$$RF_{ij} = \sum_{k=-1}^{1} \sum_{l=-1}^{1} ABS(RO_{i+k,j+l} - MEAN_{ij})$$

$$MEAN_{ij} = \left(\sum_{k=-1}^{1} \sum_{l=-1}^{1} RO_{i+k,j+l} \right) /9$$

$$i = 2,3,\ldots,M-1$$
$$j = 2,3,\ldots,N-1 \qquad (6)$$

where RF_{ij} is the relief value of the image,
   RO_{ij} is the original value of the image,
   i,j are the row number and the column number of calculating pixel, respectively, and
   M,N are the numbers of lines and points of original image, respectively.

The search radial distance of the window is an important factor for relief filter. Therefore, a program that permits the user to choose the interval between operators was designed to realize this filtering. The northwest pass filtered image was processed again by the use of a relief filter. The density slicing image of this result is shown in Figure 14.16d, which is really like a sketch map of geology.

Density slicing is useful and effective for the recognition of different rock types according to the different anomaly amplitudes or different values of apparent susceptibility. It has a good future in automatic geology mapping processing.

Integrated Interpretation and Quantitative Interpretation

From the results of data processing and image processing, an integrated qualitative interpretation, shown in Figure 14.18, was obtained.

The boundary of intrusion was drawn on the map. Beside the known faults F1, F2, and F3, there are several new interpreted faults, F4, F5, F6, F7, and F8. Fault F3 extends forward to the southeast as F3' to cut through the intrusion instead

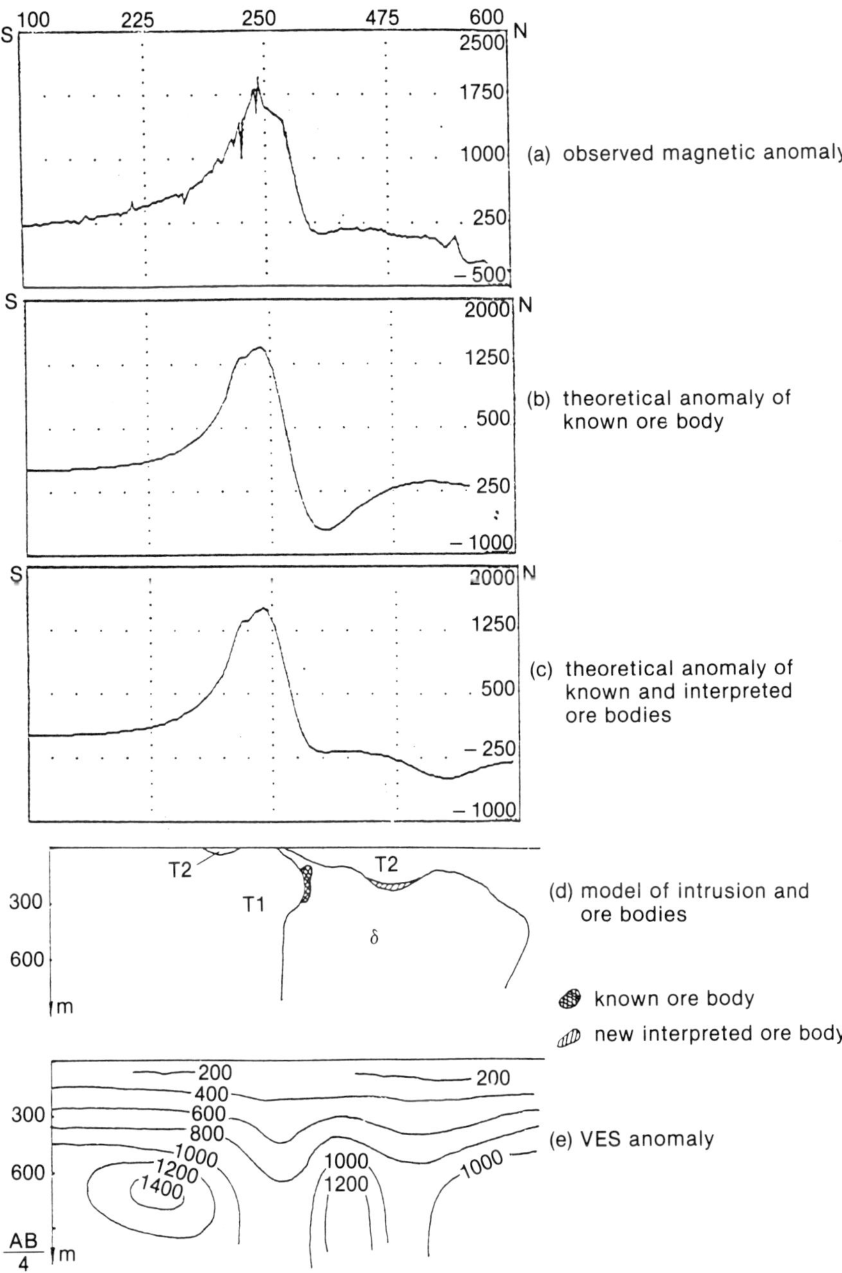

FIGURE 14.20. Quantitative interpretation of Profile A.

of ending at the boundary of the intrusion as shown in a prior map.

As a result of additional linear features, F4, F5, and F6 were reinterpreted. These faults are parallel to F3 and have nearly the same northwest strike direction as the known fault F2. Fault F5 may be defined as the extension of fault F2 progressing forward northwest.

Another faults set is F1–F1′ and F7. F1′ is the extension of F1, which is in a nearly north–south direction and F7 is a new interpreted fault, which is parallel to fault F1–F1′.

Fault F8 can be recognized by the linear feature, which is perpendicular to the strike of intrusion, and which was cut by F3, and later by F4.

These two sets of faults in different strike directions are the main control factor of the ore bodies. For example, the Anqing Copper Mine is located at the intersection point of F1 and F6. Therefore, the other intersection points of F1 and F5, and of F7 and F6 are interesting areas. These areas are also satisfied with the result shown in Figure 14.17. They were chosen for further investigation.

The subarea, the Anqing Copper Mine, was chosen for further quantitative calculation and interpretation to investigate if any residual ore body exists that was not discovered before. This area is shown in Figure 14.19.

A theoretical magnetic anomaly was calculated according to the known ore bodies by use of the line element method, and the residual anomaly was obtained from the difference between observed data and the theoretical magnetic anomaly. The residual anomalies, R1, R2, and R3 are shown in Figure 14.19.

Figure 14.19 shows that less attention was given to the northeast part of the Anqing Copper Mine during the last 20 years. The residual anomalies R1, R2, and R3 are valuable anomalies for further drilling.

High-resolution magnetic survey and vertical electrical sounding were carried out on Profile A. Unfortunately, it did not pass trough R1, but does still have some interpreted value for R1. The magnetic modeling result and the VES data are shown in Figure 14.20.

The magnetic anomaly of the known ore body and of the intrusive rock was calculated and is shown in Figure 14.20b. Comparing the anomaly with the observed data (Figure 14.20a), there was a residual anomaly in the northern part of the anomaly. According to the magnetic anomaly and the VES anomaly a new model of ore body was designed and added at the contact zone between the intrusion and the host rock in the northern part of this profile (Figure 14.20d). Finally, the theoretical anomaly of intrusive rock, and of the known and new interpreted ore body can fit the observed data very well (see Figure 14.20c).

This new ore body also agrees with the low-resistivity anomaly of VES data (see Figure 14.20e).

Integrated interpretation from magnetic data and other geodata using image processing provides useful and new interpreted information concerning the Yiesan area. This information has been used for a new drilling plan.

Conclusion

Much remains to be done for image processing of magnetic data. However, some conclusions from the theoretical study of image processing and the practical applications in the Yiesan area are summarized.

A proper image or shadow map gives a direct even stereo visual sensing of magnetic anomalies. Therefore, the image processing technique is a convenient tool to use in making qualitative interpretations. The advantage of image processing is that real-time processing displays results on the screen immediately and makes it possible for the interpreter to monitor the results and to change the method in time.

Image processing is different from classical data processing of geophysical data. Sometimes image processing just enhances the information in a certain way. Additional experience or known information is needed from the interpreter, but temporal results are obtained quickly and directly.

Overlaying technique provides an effective way to make integrated interpretation with other geodata. The nonunique solution problem of merely geophysical data is improved and a more reasonable interpretation obtained.

Potential transform data processing is the basis of image processing for geophysical data. It cannot be replaced by image processing. Before using the image processing technique, the observed data must be processed properly and some significant potential data processing should be applied to obtain reasonable results.

Some enhancement of image processing only improves the information in which the user is interested. Therefore, the result depends on the experiences of the interpreter. For example, the user can choose the proper filter to obtain a good result according to his understanding of known geological information, such as strikes of geological bodies, and the target of interpretation.

An interpreted sketch map can be obtained automatically by use of overlaying and density slicing. This is a good way to realize automatic geology mapping from magnetic data, but sometimes the physical meaning of this processing is not so clear.

All geophysical data or other data should be transferred and stretched into the standard image file. The real anomaly value will be stretched to integer values from 0 to 255. The threshold value for stretching is chosen according to the histogram of the data.

Some types of filters used in image processing are similar to those used in potential data processing in space domain. The latest version of ILWIS system (August, 1989) allows a choice of 3 by 3, or 5 by 5 filter operators for filtering enhancement. This larger filtering window combined with more flexible parameters of the filter operators can perform more potential data processing under the ILWIS system. It is necessary to have a larger filter window for the image processing of magnetic data.

The printed presentation of the image in ILWIS should be improved to satisfy the practical interpretation both in high resolution and in real scale.

Combining quantitative interpretation with image processing avoids receiving only qualitative results. From the results obtained in the Yiesan area a valuable interpretation is possible.

In this chapter I deal with magnetic data, but the principles and the methods can be applied as well to gravity data.

Pattern recognition by use of image processing techniques should be introduced for gravity data and magnetic data. It will be the practical way to realize autogeological mapping using the results of integrated interpretation.

Acknowledgments. This work is sponsored by the China University of Geosciences (Wuhan) and the Geological Survey and Exploration Team No. 326. The original calculations were completed in the Computer Center and Remote Sensing Laboratory of China University of Geosciences. Most of the map presentations and image processing were completed or calculated again by use of the facility—hardware and software—from ITC, Delft, The Netherlands. The author wishes to thank them and his colleagues.

The author expresses special thanks to Professor C.V. Reeves of ITC, and Dr. Ibrahim Palaz for reading and editing this manuscript, and for their valuable suggestions and discussions.

References

Burrough, P.A., 1986, Principles of Geographical Information System for Land Resources Assessment: Oxford University Press, New York.

Fahnestock, J.D., and R.A. Schowengerdt, 1983, Spatially variant contrast enhancement using local range modification: Opt. Eng. No. 22, 378–381.

Geological Survey and Exploration Team No. 326, 1981, Report of ground magnetic survey in Yiesan area (unpublished paper).

Jensen, J.R., 1986, Introductory Digital Image Processing: Prentice-Hall, Englewood Cliffs, NJ.

Kowalik, W.S., and Glenn, W.E., 1987, Image processing of aeromagnetic data and integration with landsat images for improved structure interpretation: Geophysics 52, 875–884.

Meijerink, A., Valenzuela, M., and Stewart, J., 1988, ILWIS: The integrated land and watershed management information system: ITC Publication number 7, ITC, The Nederlands.

Paterson, N.R., and Reeves, C.V., 1985, Application of gravity and magnetic survey: The state of the art in 1985: Geophysics 50, 2558–2594.

Zhou Yunsheng, 1984, The report of geology and mineral resources in Yiesan area East China (unpublished paper).

15
Interactive Three-Dimensional Seismic Display by Volumetric Rendering

Robert H. Wolfe, Jr. and C.N. Liu

Generally, seismic data have been visualized through display of two-dimensional (2-D) sections, and three-dimensional (3-D) display has come into play only for graphically rendering structural features, once they have been interpreted. We consider the display of the 3-D data directly, before any interpretation is done. We call the direct display procedure volumetric rendering, and we describe the procedure and illustrate its utility by application to a migrated 3-D seismic data set. Our intent is to peer into the volume of seismic samples. We form an image of such a view by representing each seismic sample by a colored cell in the viewed volume. We find that we can identify 3-D reflection features present in the data, much as we have identified reflection horizons in 2-D sections. As in the case of 2-D sections, we discover that sidelobes can obscure somewhat the definition of the reflective surfaces, and we devise a nonlinear filter to suppress the sidelobes. The resulting clarified display provides 3-D views of synclines, anticlines, faults, and pinchouts. We consider stereoscopic viewing and data volume partitioning to further facilitate interpretation. Based on these results, we conclude that volumetric rendering can prove useful for identifying and mapping potential reservoirs.

Introduction

Recent advances in computer graphics technology have certainly made their impact on the way we handle seismic data. Yet, in spite of all the improvements introduced, the visualization of seismic structure data has remained primarily a display of 2-D sections. Ultimately, we want to understand the 3-D structure of the reflective surfaces, and increasingly we are confronted with 3-D seismic surveys of ever more complex geological structures. We describe an alternative method of seismic visualization, wherein we can discern the 3-D structure from the seismic data directly, before doing interpretation on the 2-D sections. The approach, called "volumetric rendering," has been investigated extensively for visualization of 3-D medical data, as described in Farrell and Zappulla (1989). The technique has been tried in several seismic examples (Wolfe and Liu, 1988; and Sabella, 1988), and the success of the technique earns it consideration as a complement to the more traditional methods of interpretation.

While we propose a departure from the traditional analysis of 2-D sections, we should keep in mind what has made the latter so instrumental to geophysical interpretation. The traditional trace plot, illustrated in Figure 15.1 by a section of the data we analyze later, exhibits a duality in visualization, characterized by the positions of waveforns in a section and by the form of the waves themselves. The wave positions imply the structure of reflective surfaces, and the waveforms imply such information as bed thickness, composition, and induced errors. Geophysicists have worked with the classical 2-D section of Figure 15.1, with its visualization duality, for many years. Now, the newer computer graphics displays afford a tradeoff in the duality. The waveform attribute is de-accentuated in deference to the spatial attribute, as shown in Figure 15.2 (see Color Plate X). The waveform/spatial duality is preserved to some extent by the color, but the

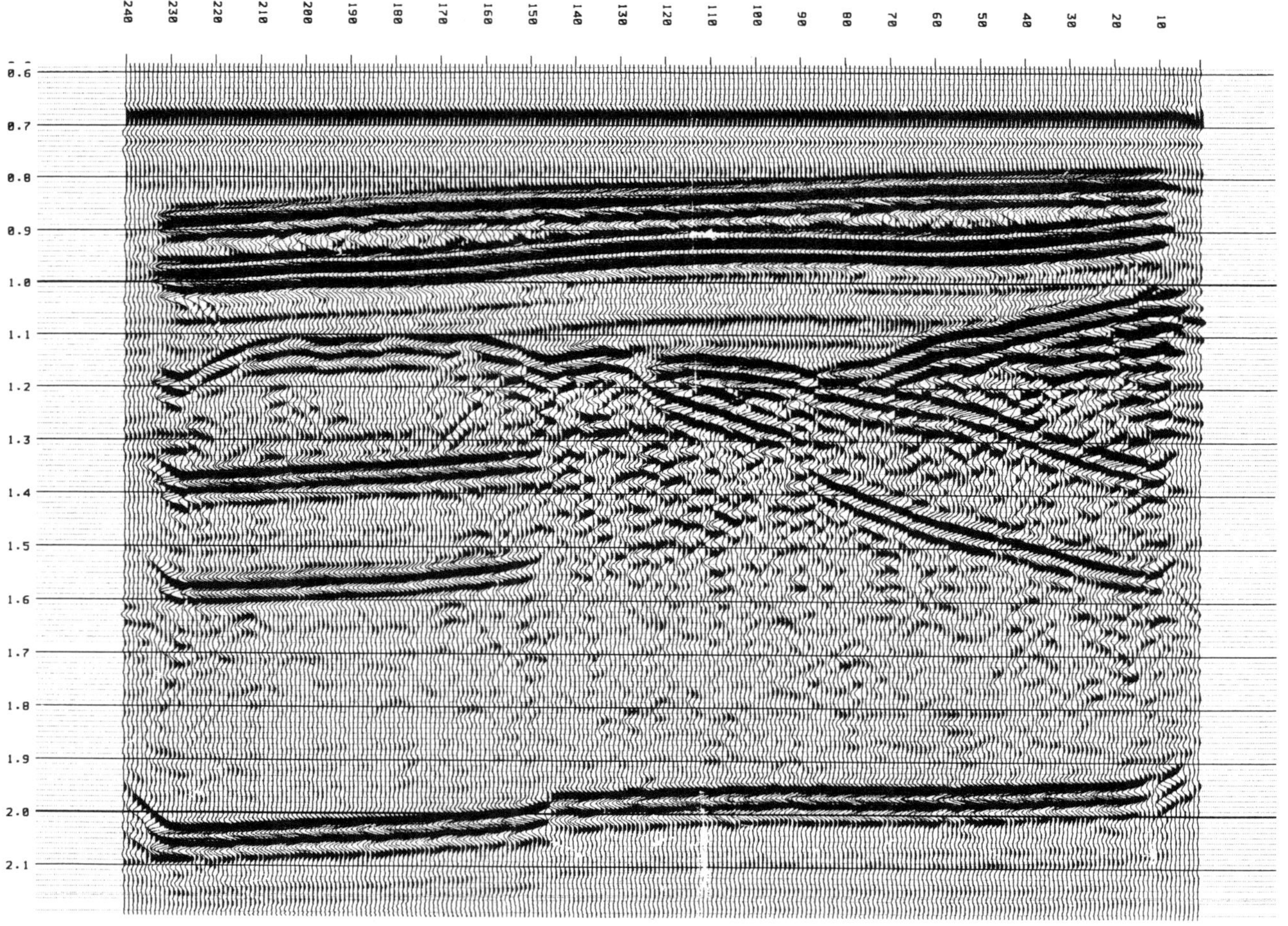

FIGURE 15.1. Traditional representation of seismic section (from University of Houston Seismic Acoustics Laboratory): There are 240 traces, each spaced 100 ft from the adjacent one, in the north–south direction. Notice that the third reflection event is an unconformity, and pinchouts are apparent. This section is one of a total of 240 available in this 3-D data set.

direct view of the waveform, preserved in the trace plots, is gone.

The consequence for 3-D data is that when we attempt to display the whole 3-D volume, we arrive at the display shown in Figure 15.3 (see Color Plate X). We have not given up entirely our hold on the waveform display, but we also cannot see the data beyond the surface of our volume. In fact, our 3-D display is really just three 2-D section displays (top, front, and side) merged in one figure and appropriately skewed so that their adjacent edges match. To view data in the interior, we must slice away intervening data; that is, we choose an interior 2-D section to be displayed and skew it to match the others. Modern computer technology allows this to be done fast enough to enable the interpreter to change the view interactively. Examples of this technique abound, and some representative applications to interpretation are provided in Gerhardstein and Brown (1984), Curtis et al. (1986), and Chakravarty et al. (1986). The latter have expanded the display beyond planar sections to display on interior surfaces of fairly arbitrary shape, defined interactively. The interpreter can move through a sequence of such sections rapidly enough that one can imagine building up a mental picture of the 3-D features, without sacrificing visualization of the waveforms.

Thus, the standard way to visualize 3-D data at present is to see the waveform and two dimensions of wave position and imagine the third from animation. Although one can gain a mental picture of a complicated formation by stepping through a series of 2-D images, it is difficult to convey that picture to others. In some cases it has been necessary to resort to hand-drawn pictures (Curtis et al., 1986, illustrate this point) for that purpose.

Computer graphics techniques, as used in CAD/CAM, are coming into increased use to display 3-D reflection surfaces that are sought in 3-D structure data. In the language of computer graphics, we want to represent each reflection surface as a 3-D graphic object, whether manifested as a mental abstraction, a drawing, a set of contours, or a mesh of surface points. The latter is the characterization of computer graphics, viz. a computer data base of surface coordinates interrelated in the fashion of a mesh or as an ensemble of contiguous polygons. The interpreter interactively constructs a 2-D graphic object of each reflection feature on each section, after which 3-D graphic objects can be constructed by combining together the appropriate 2-D objects. This CAD/CAM approach to analyzing seismic data is still expensive in terms of time and tedium.

As we see in the following sections, volumetric rendering can produce 3-D displays comparable to graphic techniques in less time with less tedium. In terms of the visualization duality we have spoken of, the volumetric approach puts aside the waveform attribute (except for sign), and all three dimensions of wave location are considered at one time. The idea is to display only certain data values present in the data volume so that we can peer into the volume without complete obstruction by the intervening data. The features present in the 2-D sections suggest that the high-amplitude seismic values should be displayed to pinpoint the reflective surfaces and that the low-amplitude values should be excluded to remove obscuration. We describe the process as follows. First, we describe our seismic data and preparation for display and then the volumetric presentation technique and its application to our data. Next, we demonstrate the improvement obtained by suppressing undesirable sidelobes in the data and finally consider display alternatives to minimize obscuration and further facilitate interpretation.

Data Preparation

Our visualization experiment was performed with seismic data physically simulated at the University of Houston Seismic Acoustics Laboratory. The data were generated by recording ultrasound reflections from a scale model composed of various silicone rubber compounds and plastics. The materials and ultrasound frequency have been chosen so that the model and transmitted waves are at a scale of 12,000:1. The laboratory and modeling techniques have been described in Nelson (1983, Chapter 6), with special attention to the particular model we used. That model, called SAL-NOR, represents a 24,000-ft (7.2-km) square, delta formation in the North Sea, referred to as the "Golden Block" and studied extensively in Saeland and Simpson (1982). The area contains a relatively complex structure with considerable faulting and is regarded as an excellent example for testing

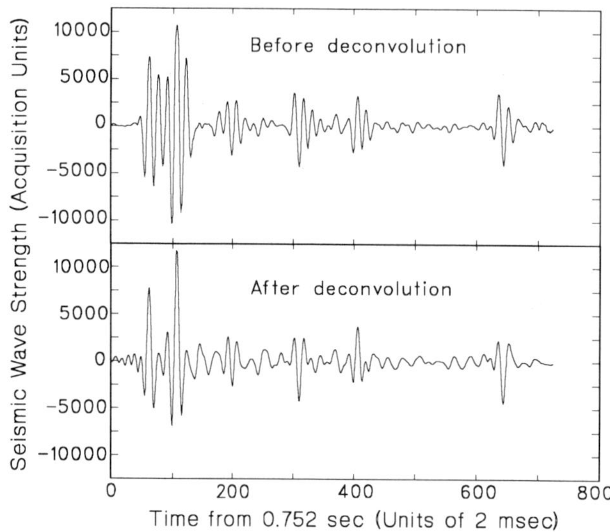

FIGURE 15.4. SALNOR7 Section 16 Trace 180 before and after deconvolution: The filtering has sharpened the waveform, as evident in the strong reflections around 850 ms. The negative reflection waveform at about 1600 ms illustrates well the characteristics of the pulse reshaping. After deconvolution, it has a main lobe surrounded by single sidelobes, characteristic of the Ricker wavelet. The amplified noise at about 800 ms shows that our filter is not causal (© 1988 IEEE).

seismic interpretation techniques. Several surveys, representing a variety of acquisition strategies, were made on the model; the one we selected, called SALNOR7, consists of 240 lines of 240 traces each of 600-ft-offset data. The SALNOR7 data were subjected to Kirchhoff migration at the Seismic Acoustics Laboratory prior to our receipt. The data subvolume of interest consists of 700 four-byte time samples by 200 traces by 200 lines, or 112 MB, a data volume not unusual for modern seismic acquisitions.

The data contain a considerable amount of ringing caused by problems in the ultrasound source. By analogy to real seismic data, we constructed a predictive deconvolution filter to transform the observed waveform into a Ricker wavelet. Since we needed to perform the filtering anyway, why not transform the waveform into something closer to an ideal spike to yield a sharper display? Our rationale for choosing a Ricker wavelet is 2-fold. First, we wanted to investigate visualization of typically processed seismic data, and the Ricker wavelet is more representative of the typical filtering objective. Second, we wanted to avoid the risk of amplified noise encountered in attempting to reconstruct too sharp a pulse from typical data, and the Ricker wavelet, being more band-limited than a sharper waveform, tends to suffer less in that respect. We constructed our filter as follows. Figure 15.1 shows two strong reflections at the top of the section. We assumed their waveforms should

be representative of the incident waveform, and our filtering strategy was to transform those waveforms into Ricker wavelets. The filter can be thought of as comprising two parts: a convolution that reconstructs a unit pulse waveform, followed by a convolution to reshape the unit pulse into a Ricker wavelet. Following the procedure of Liu et al. (1983), we chose a Wiener filter to reconstruct the unit pulse. The second filter is simply a convolution with the Ricker waveform. In the frequency domain, our overall filter transfer function is the product of the Fourier transforms of the convolution functions and is

$$H = H_{\text{Ricker}} \frac{W^*}{p + |W|^2}$$

where H_{Ricker} is the Fourier transform of the Ricker wavelet, W is the Fourier transform of the observed waveform, and p is a parameter, the selection of whose value effects a tradeoff between noise amplification and filter strength. In his survey of seismic deconcolution and inversion techniques, Berkhout (1986) shows that p is related to the expected variance of any added noise in the data. We have no a priori knowledge of the noise statistics, but we found that setting p to 10% of the frequency average of $|W|^2$ gave a good response with sufficiently small residual noise.

Figure 15.4 shows an example of the resulting deconvolution. It is apparent that the reflections (the high-amplitude events) are better resolved

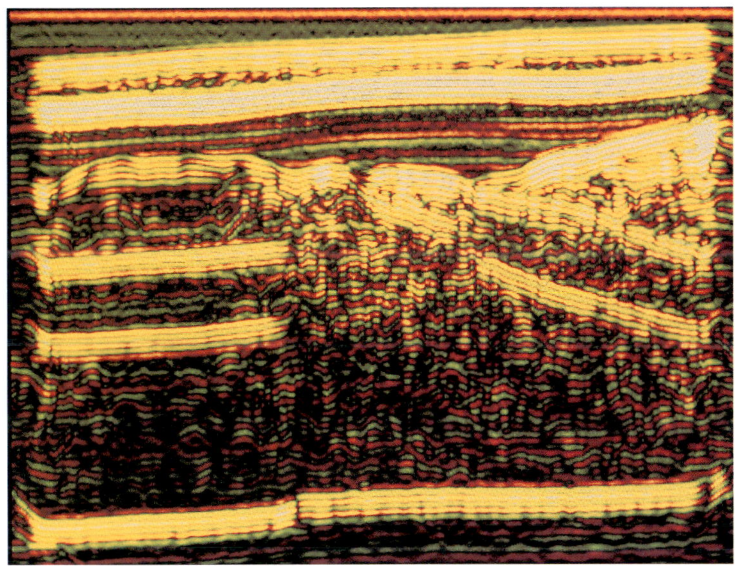

FIGURE 15.2. Color intensity representation of the section in the previous figure: The positive amplitude waves appear in green, and the negative waves appear in orange. The lighter shades of each color correspond to higher intensity; black is zero amplitude.

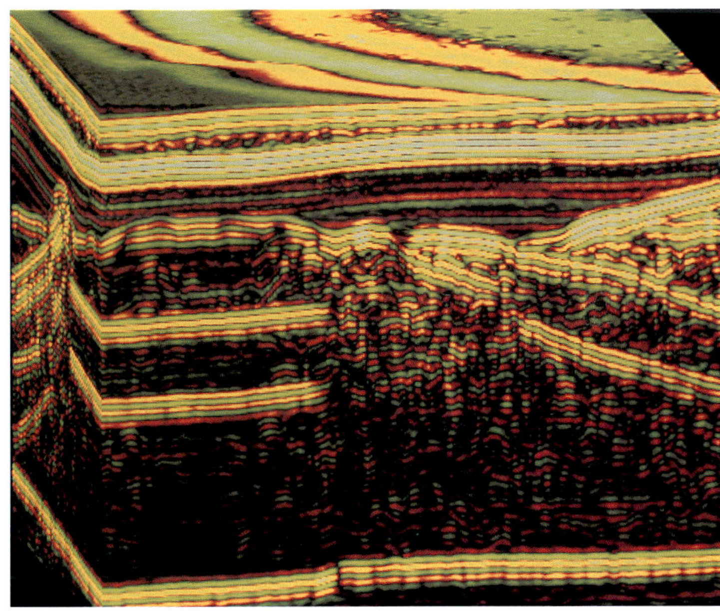

FIGURE 15.3. A view of the total 3-D volume of data: As in Figure 15.2, the positive amplitude waves appear in green, and the negative waves appear in orange. The lighter shades of each color correspond to higher intensity. The problem with this display is that the data inside the "block" cannot be seen. We are really looking at three 2-D sections: two sides and the top.

FIGURE 15.5. The effect of removing low-amplitude data from the 3-D display: Green and orange correspond to positive and negative reflection, respectively. In (a) all wave values are shown with darker shades of color for lower amplitude, as in Figure 15.2. In (b) some of the weaker data values are removed, and the darker shades are gone. The shapes of the reflection surfaces are beginning to become visible. In (c) all data values between -3000 and $+3700$ (in the units in Figure 15.4) have been removed. (d) is (c) shaded, done by darkening the colors of the data in the sections further from the observer. The vertical range shown is 824–2100 ms, and the horizontal span is 20,000 ft (6 km) north–south by 18,000 ft (5.4 km) east–west. North (along-section) is to the left, and east (section-to-section) is into the paper (© 1988 IEEE).

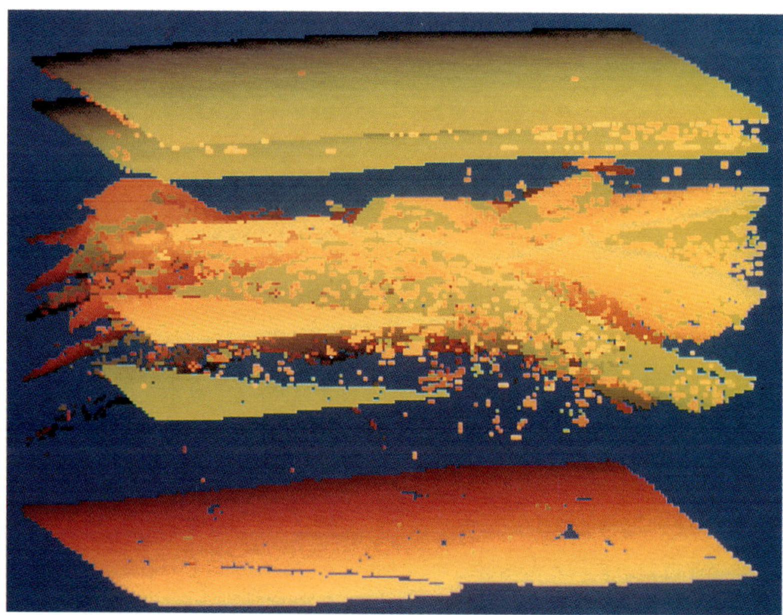

FIGURE 15.7. Seismic data after sidelobe data have been effectively removed. The top surfaces are now green, as they should be, with the orange stripped away, and the intermediate surfaces have become somewhat clearer. It is these latter surfaces that hold the most geophysical interest, and the need for further clarification is evident (©1988 IEEE).

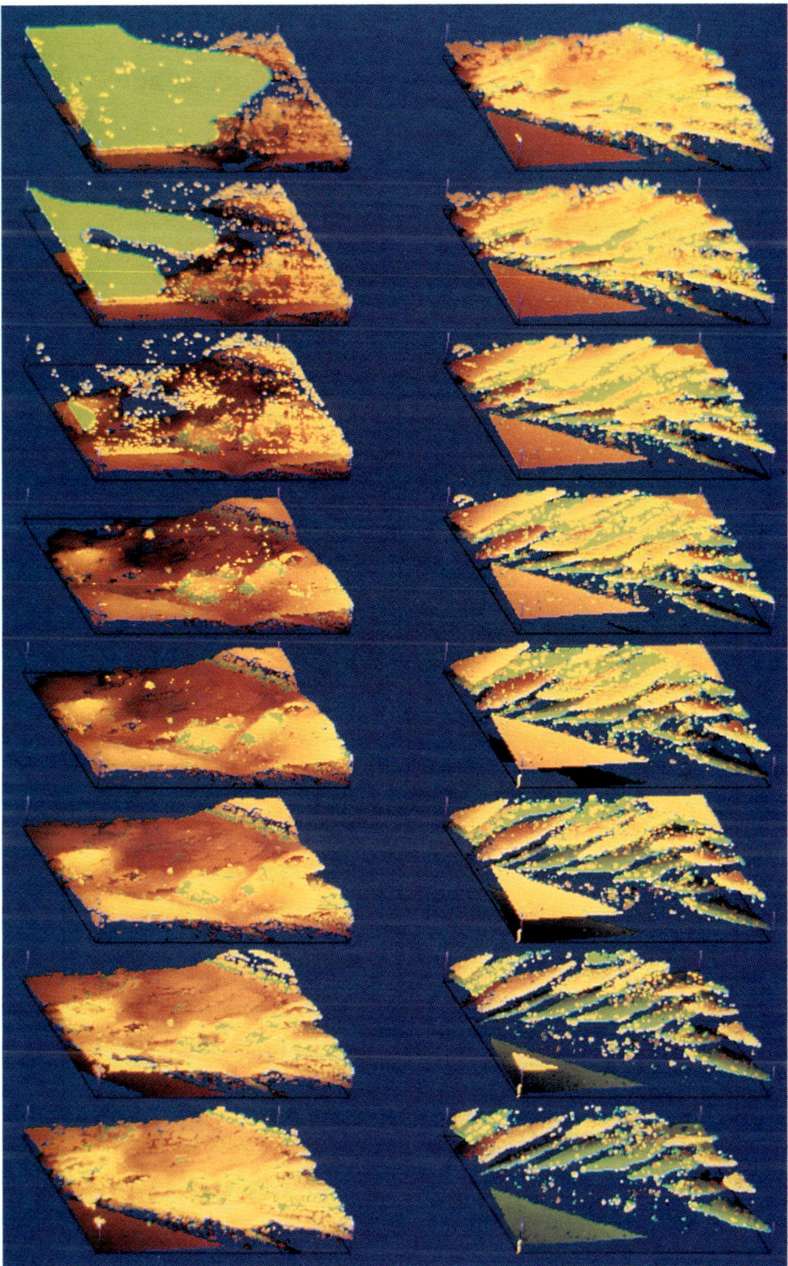

FIGURE 15.8. Movie frames showing a progressive descent of a 3-D window in the data: A 3-D window of 356 ms height descends in the total volume of data, dropping 32 ms from each of the displayed frames to the next. The green portion existing at the top of the first frame belongs to the second horizon in Figure 15.7. As it leaves the top, we see the Jurassic unconformity come into view. The surface moves upward, and as portions strip away, the faulted, dipping sublayers appear. Compare this sequence to the synoptic view in Figure 15.7 and to Figure 15.9 to see how the sequence relates to the total volume.

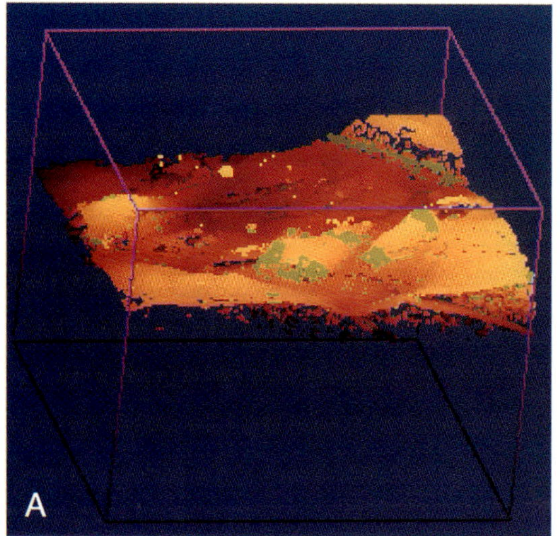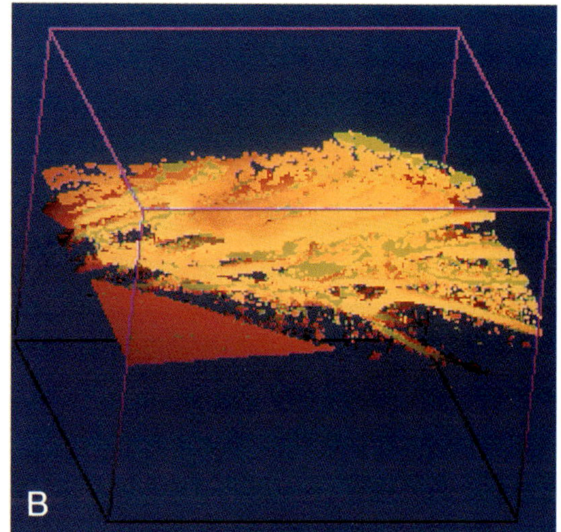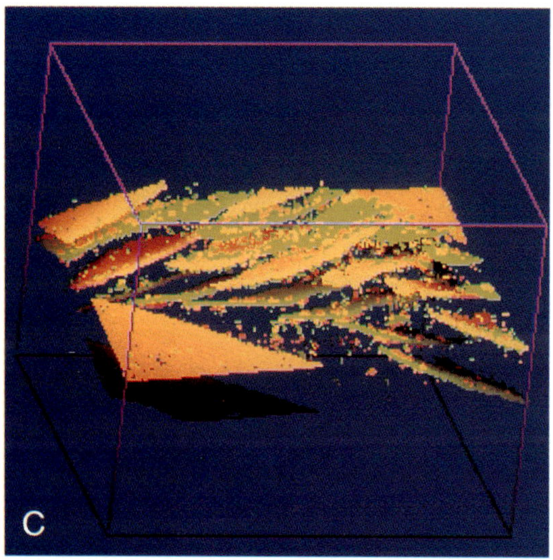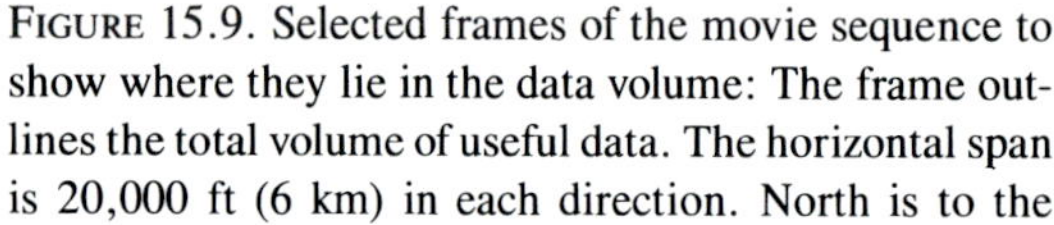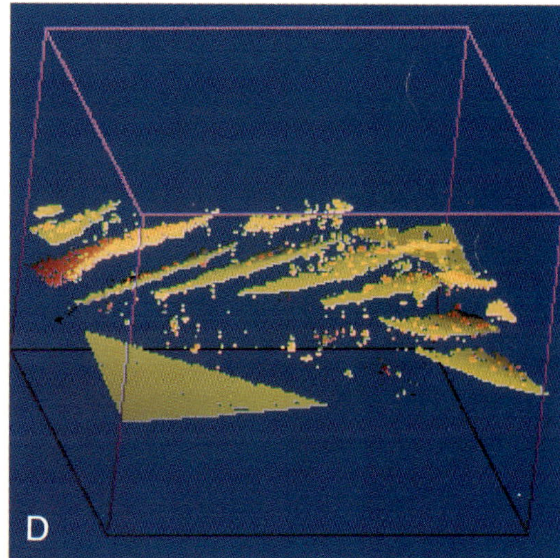

FIGURE 15.9. Selected frames of the movie sequence to show where they lie in the data volume: The frame outlines the total volume of useful data. The horizontal span is 20,000 ft (6 km) in each direction. North is to the left, and east is into the paper. (a) 1080–1336 ms. (b) 1208–1464 ms. (c) 1336–1592 ms. (d) 1464–1720 ms (© 1988 IEEE).

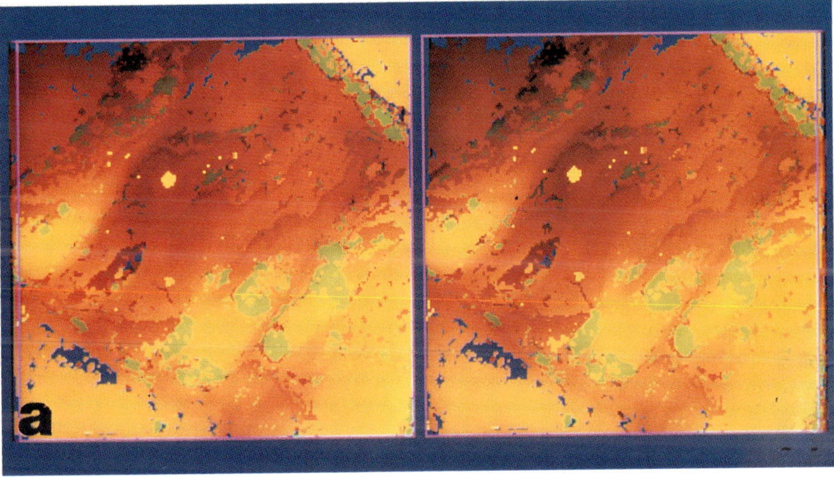

FIGURE 15.10. Stereo horizontal views of the sidelobe-suppressed data for two contiguous depths: The area shown is 20,000 ft (6 km) square. The left images are for the left eye. North is to the left, and east is up. (a) 1080–1336 ms. (b) 1336–1592 ms (© 1988 IEEE).

after deconvolution but that the noise has been amplified somewhat. We can tolerate some noise amplification, so long as it is not large enough to spill into the data ranges we wish to display. In the course of our spectral analysis for filtering, we found that the data bandwidth was less than half our sample frequency. Therefore, we could drop alternate time samples to yield deconvolved data with a 4-ms spacing, thereby effecting a saving in data volume.

Data Presentation

Our seismic data are four-byte floating point numbers with a dynamic range of about $-15,000$ to $+15,000$, and to display the data we must make a decision that confronts all analysts who use a computer display for visualization. Our display system can handle eight bits; so we must somehow channel the interesting information in those numbers through those eight bits. Display hardware generally limits the number of bits to lie between 8 and 24, the latter being eight bits in each of the three primary colors. We will avoid the argument of just how many bits one can actually see, rather letting our displays demonstrate that eight bits are enough for our purpose. Several strategies exist for compressing the data into the display range, depending on the expected dynamic range of the interesting part of the data and the number of display bits of significance. We could truncate our data so that $-15,000$ lies at 0, $+15,000$ lies at 255, and intermediate values are linearly spaced in between. Or, we could linearly stretch out selected subranges, or we could effect a nonlinear transformation, such as logarithmic conversion. We chose a linear transformation by fitting our dynamic range between 0 and 254 with the seismic signal value of 0 at 128.

The volumetric approach to data presentation treats each seismic sample as a colored cell in the 3-D volume enclosing the data set. A 2-D picture of the volume is made for display on the screen. As Figure 15.3 illustrates, if all data samples are given a color, we can see only the exterior of the volume. To see into the volume, we wish to "weed out" uninteresting data, as follows. If the data value falls in a range of our choosing, we give the cell an opaque color; otherwise the cell is transparent. If there are enough transparent cells to let us peer deep into the volume, we hope that the clouds of opaque cells that we see will exhibit some meaningful form. Specifically, if we select large-amplitude ranges, we might isolate the surfaces giving rise to the reflections. Figure 15.5 (see Color Plate XI) illustrates this point. The first image is the block we showed earlier, with weaker waves shown in darker colors. In Figure 15.5(b), we choose data ranges to exclude lower amplitude wave values. Already, we see somewhat into the volume, and we can discern reflective surfaces at the top and bottom. In Figure 15.5(c), we raise further our display thresholds and eliminate more of the weaker values. More of the reflective surfaces become apparent. Finally, in Figure 15.5(d), we apply a very simple shading by making the cell colors darker for data values further away from us. The shading has a dramatic effect, both conveying the impression of depth of view and clarifying somewhat the intermediate surfaces.

The method of building these images is quite simple. The volume of data is treated one section at a time. A 2-D color map of a section is constructed with green or orange for high-amplitude wave values (positive and negative, respectively) and transparency for all others. The color map is overlaid on the display field, the color map for the next section is overlaid slightly displaced relative to the first, and so on. Colors in the underlying maps show through the transparent regions. The amount of displacement can be adjusted to render different view angles. To obtain the shading, dark colors are used on the first color map, and progressively lighter colors are used for successive sections.

The top two reflective surfaces in Figure 15.5(d) should exhibit positive reflection, yet they appear in orange. Closer inspection indicates a layer of green sandwiched between the layers of orange in those surfaces. This artifact is created by the sidelobes inherent in the Ricker wavelet; that is, the sidelobes of some reflection events are large enough to fall in the range of peaks of opposite sign. It is apparent that the effect causes blending of closely-spaced reflections, and we must find a way to reduce the sidelobes if we are to make our simple thresholding scheme work satisfactorily.

Sidelobe Suppression

The presence of sidelobes or reverberations in general has long been a problem in interpreting displayed seismic data, and numerous techniques

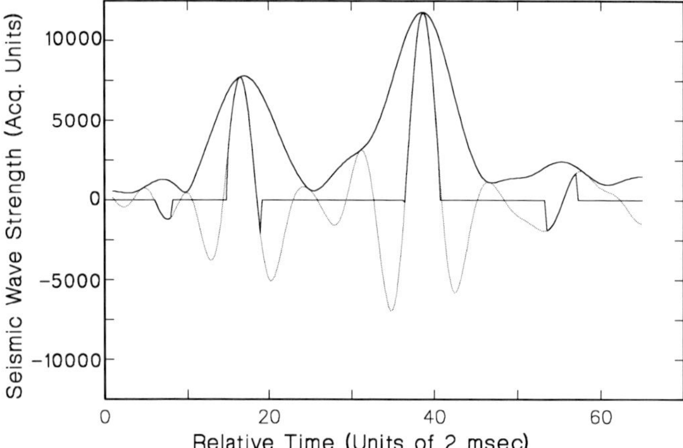

FIGURE 15.6. Use of the amplitude envelope for waveform masking to suppress sidelobes: The amplitude (smooth, solid curve) envelops the seismic waveform (dotted/solid line). The solid segments overlaying the waveform show intervals in which the enveloping amplitude is at least 80% of its local peak. The solid zero segments are where the waveform has been masked out.

have been invoked to sharpen the waveform for a cleaner display. Berkhout (1986) has provided a comprehensive survey of digital deconvolution and other inversion techniques that are applicable in different circumstances. In choosing a pulse-shaping technique we were guided by our desire to test the efficacy of the display technique, given that a spike waveform had been obtained by whatever method, rather than concentrating on the pulse-shaping itself. Guided by that desire and by the desire for expediency and simplicity, we developed a nonlinear filter, as follows. First, we compute the amplitude envelope of the seismic oscillations, the root-sum-square of the seismic signal and its Hilbert transform. Figure 15.6 shows the relationship between the amplitude (upper solid curve) and the seismic signal (dotted curve) for a small portion of a trace. The amplitude is never negative and is generally devoid of the sidelobes present in the trace, and these properties suggest using it for volumetric rendering, rather than using the seismic trace. However, we found that the broad mainlobes of the amplitude tend to produce volumetric images with unacceptably thick surface features. Nevertheless, we found the amplitude to serve as a good criterion for suppressing the sidelobes by using the crests of its mainlobes for filtering the seismic signal, as shown by the narrow solid curves in Figure 15.6. We effect this filtering by determining all local maxima, the peaks, in the amplitude envelope function and the corresponding regions bounded by local minima. Within each such interval, we find the subinterval for which

the envelope remains above a certain fraction of its local peak. We then retain the seismic signal in that subinterval and set seismic values outside to zero. The solid segments along the waveform in Figure 15.6 show where the enveloping amplitude is close enough to its local peak to admit the waveform itself; otherwise the waveform is nullified. Notice that some low-amplitude waves are preserved in the left and right portions of the figure because they fall at local, albeit small, peaks in the envelope amplitude. They are generally too weak to pass the display thresholds and will not appear in the 3-D display. After trial and error, we arrived at 85% of the envelope local peak as our masking criterion.

The result is shown in Figure 15.7 (see Color Plate XI), wherein the same data ranges were chosen as for Figure 15.5(d). Comparison of the two images reveals that the sidelobes have been effectively suppressed without loss of surface definition. A number of the surfaces are relatively clear. The scattered points and holes in the surfaces are caused by noise in the data. The simplicity of the volumetric rendering makes it relatively fast. We found we could build these images in less than 15 s in an IBM 7350 Image Processing System (the use of this system is described in Farrell et al., 1985). We observed comparable performance in an IBM RS/6000 Workstation. Implementation on an IBM PC/AT with an advanced display indicated that images of comparable quality could be generated on a personal computer. Based on the observed computation time, we estimate that in general a personal computer will build these images in a time ranging from a few minutes on a relative slow model to less than a minute on a high performance model of the IBM PS/2. Thus, it is apparent that in a range of systems, the volumetric method can facilitate interactivity, an attribute that is important in geophysical interpretation.

Now, we must consider whether the image in Figure 15.7 facilitates interpretation. The top two horizons are clearly discernible, and the upper one is fairly featureless. By choosing another view angle for rendering, we could see that the second horizon is similar to the first. We would presume that the third surface is an unconformity, evident by the merging surfaces on the right side of the figure. Otherwise, it is difficult to analyze the unconformity because there is too much obstruction in the display, even if other view angles are chosen for rendering. Of course, we know the design characteristics of the model from which the data were collected. The model was designed according to the field interpretation in Saeland and Simpson (1982); hence we can label the third surface as an unconformity of the Jurassic period (J-unconformity). The upper two surfaces separate post-Jurassic sediments.

Further Display and Interpretation

It is apparent that we need a rendering strategy that will separate the reflective surfaces. We are further compelled in that pursuit by the expectation that real seismic data will often display more densely packed reflections than our example data show. We have developed a strategy for viewing a sequence of horizontally sectioned subvolumes as follows. We extract and display a subvolume containing 64 contiguous time samples from the entire 200×200 set of traces. Then, we advance downward several time samples and repeat the process again and again to produce a sequence of images. By rapidly stepping through the sequence, we produce a movie in which the data appear to rise up through the bottom of our 3-D display window and vanish at the top. In Figure 15.8 (see Color Plate XII) we show some frames of the movie sequence for the most interesting part of the data volume. We found that we could further improve the clarity of the surfaces by changing our shading strategy. Rather than darkening the sections further into the screen, we have darkened the

horizontal sections deeper in time in Figure 15.8. The displayed surfaces then resemble aerial views of the earth's surface near sunrise or sunset. Four of the frames in Figure 15.8 are also shown in Figure 15.9 (see Color Plate XIII) to relate the sequence to the total volume. Individual surfaces are more clearly seen than in Figure 15.7, and their shapes are quite discernible.

The J-unconformity can be seen quite clearly in the fifth frame of the sequence. The reversal in sign of the wave value, evident as the green areas on the orange surface, occurs generally where underlying reflective surfaces form pinchouts with the Jurassic surface and cause wave interference effects. The subsequent horizontal slabs in Figures 15.8 and 15.9 show the dipping parallel sublayers, and the final frames of Figure 15.8 clearly illustrate the extent of the faulting, evident in the green segments disjoined by a number of roughly parallel faults.

Stereoscopic views can be made by building two images of slightly different look angles for simultaneous viewing, and we found that they enhance the observer's perception of the surface shapes. Stereoscopic examples are shown in Figure 15.10 (See Color Plate XIV) for horizontal views of two depth intervals. The horizontal view is valuable for relating features, such as the faults, to a map. We also found that a reasonable substitution for stereoscopy can be realized by building a sequence of views of slightly different view aspect and rapidly cycling through the sequence to simulate rotation. The volume appears to rock back and forth slightly, and closer features move opposite to more distant ones. Hidden regions can be brought into view by using a wider range of viewing angles or by cutting out portions of the data.

Summary and Conclusions

We have described a volumetric approach to seismic imaging that enables us to gain 3-D views of underground structures from the data directly, while sidestepping the conventional task of poring over a multitude of 2-D sections. We have found that the method facilitates interactivity on a moderate system and that it can be implemented even on personal computer systems. By masking out the seismic wave function for envelope amplitude values below 85% of their local maxima, we improve the visualization by suppressing undesirable sidelobes. However, the visualization approach is not limited to that transformation. Depending on the characteristics of the particular data set investigated, the exploration geophysicist has access to an arsenal of data enhancement techniques, ranging from sophisticated deconvolution methods, to inversion for the reflection function directly, and even to the use of derived attributes, when preparing the data for volumetric rendering.

We have demonstrated the efficacy of volumetric rendering with opaque colors in 3-D seismic visualization. Sabella (1988) has experimented with the use of translucency, whereby deeper reflections can be viewed through shallower ones, much as one might look through an arrangement of sheer curtains. Based on the results to date, we feel that we have only begun to recognize the potential of volumetric visualization in seismic interpretation. In contrast, medical interpretation has gone far in exploring the use of both opaque and translucent rendering, and those methods, as reviewed in Farrell and Zapulla (1989), should be examined for applicability to seismic data. Our analysis employed a very simple shading for accentuating 3-D. More sophisticated shading methods should be tried. The use of color and intensity variation across a reflective surface should be explored for presenting reflection strength or other attributes. The volumetric display of multivariate data, such as complex trace attribute sets or shear wave components, should be explored. The rendering that we described should be applied to a variety of real seismic acquisitions. Alternative deconvolution and inversion techniques should be tried for applicability. As volumetric rendering is demonstrated on a widening range of seismic data and their derivatives, and improvements and enhancements are discovered, this visualization technique should experience an exciting future in exploration geophysics.

Acknowledgments. We wish to thank G. H. F. Gardner of the University of Houston Allied Geophysical Laboratories for helpful suggestions during our analysis of the SALNOR7 data.

References

Berkhout, A.J., 1986, The seismic method in the search for oil and gas: current techniques and future developments: Proc. IEEE, 74, 1133–1159.

Chakravarty, I., Nichol, B.G., and Ono, T., 1986, The integration of computer graphics and image processing techniques for the display and manipulation of geophysical data, in Advanced Computer Graphics, T.L. Kunii, (Ed.): Springer-Verlag, Tokyo, pp. 318–333.

Curtis, M.P., Gerhardstein, A.C., and Howard, R.E., 1986, Interpretation of large 3-D data volumes: 56th Annu. Int. Mtg., Soc. Expl. Geophys., Expanded Abstracts, 497–499.

Farrell, E.J., and Zappulla, R.A., 1989, Three-dimensional data visualization and biomedical applications: CRC Crit. Rev. Biomed. Eng. 16, 323–363.

Farrell, E.J., Yang, W.C., and Zappulla, R.A., 1985, Animated 3D CT imaging: Comput. Graphics Appl. 5, 26–32.

Gerhardstein, A.C., and Brown, A.R., 1984, Interactive interpretation of seismic data: Geophysics 49, 353–363.

Liu, C.N., Fatemi, M., and Waag, R.C., 1983, Digital processing for improvement of ultrasonic abdominal images: IEEE Trans. Med. Imaging 66–75.

Nelson, H.R., Jr., 1983, New technologies in exploration geophysics: Gulf Publ. Co.

Sabella, P., 1988, A rendering algorithm for visualizing 3D scalar fields: Comput. Graphics 22, 51–58.

Saeland, G.T., and Simpson, G.S., 1982, Interpretation of 3D data in delineating a subconformity trap in Block 34/10, Norwegian North Sea: in The Deliberate Search for the Subtle Trap: Am. Assn. Petr. Geol., AAPG Memoir 32, 207–216.

Wolfe, R.H., Jr., and Liu, C.N., 1988, Interactive visualization of 3D seismic data: A volumetric method: Comput. Graphics Appl. 8, 24–30.

Index

AAPG, 34
acoustic impedance, 167
amplitude correction, 88, 95
analog televiewer, the, 224
analysis of fine scale features, 235
analytic signal analysis, 122
Archie's equation, 35
array design, 71
array theory, 71
artificial intelligence, 1, 2, 33, 61
attributes for an interpretation
    model, 18
augmented transition trees (ATT),
    102, 108
automaton, 121
  error correcting, 121
  finite state, 121
  tree, 121

backtracking, 21
backward chaining, 83, 99
  inference engine, 83
  system, 110
Bayes
  classification, 121
  decision rule, 122
best-fit ellipse, the, 220
BHTV, 44, 51, 223
  data analysis, 225
  file format, 226
  generated logs, 242
blackboard, 56
borehole visualization tool (BVT),
    54
boundary effects, 16
brightness, the, 267

capillary pressure, 256
CAR, 73

CDR, 73
centering and sphering, 191
classification of sedimentary struc-
    tures, 46
CLUSTER, 47
cluster analysis, 261
COBOL, 33
combination
  constraints, 16
  rules, 16
confidence factor, 102
constant pressure fault, 16
constraints
  absolute, 20
  quantitative, 19
  relative, 20
control strategies, 109
COREX, 41
correlation, 47

dashed line, 12
DASHES, 72
data input and trace setting, 87
deconvolution, 88, 95
  minimum entropy, 173
density slicing, 265
differentiation interval, 5
digital evaluational model, 203
digitizing, 213
dip computation, 47
dipmeter analysis, 48
domain knowledge, 35, 82
driller's log, 34

edge calculation, Kirsh, 52
edge detection, 52
editing and stacking, 88
eigenvalues, computation of, 191
ELAS, 35

electric log, 34
EMYCIN, 82
EPT, 44
ESCAT, 48, 50
expert system, an, 37, 87
  for design of array parameters,
    71
  for seismic interpretation, 99
  to assist in processing VSP, 81
EXPLOR, 61
EXPLORER, 81
extrema, 15

feature extraction, 47
filtering, 95
  event enhancement, 164
  f-k, 90
FMS, 44, 51
formation evaluation, 33
  computer based, 34
formation factor, 33, 255
formation permeability, 34
FORTRAN, 33
fractal
  concepts, 254
  dimension, 201
  dimension measuring, 204
  mathematics, 249
  mountains, 201
fracture aperture, 237

Gaussian-Kernel function, 180
GECTA, 147
GEOLOGIX system, 43
geometric correction, 233
geomorphology, 202
geophysical logs, 241
Gestalt movement, 5
Gestalt school of psychology, 5

grammatical inference, 138
graphical method, 1

Havsdorff-Besicovtch dimension, 204
Heave
  compensation, preprocessing for, 158
  motion modeling, 159
  phenomenon, 158
heterogeneous reservoir behavior, 16, 18
heuristics, 35, 36
  measurement calculations, 102
horizon linking processing, 137
Hough transformation, 53, 135, 136
human interface, 227
HYPOTHESIS, 37

IF-THEN, 39, 73, 83, 101, 168
ILRV, 35
ILWIS, 266
image
  acquisition, of an, 250
  analysis of particle shape, 213
  core like, 229
  frame, 3D wire, 229
  manipulation, 215
  processing, 275
  processing of magnetic data, 265
  segmentation, 250
impedance log estimation, 89, 95
inference, 63
  engine, 63, 101
  mechanism, 63, 64
intelligent knowledge base system, 61, 64
INTELLOG, 36
interestingness, 173
interpretation
  computer assisted, 62
  fully automated, 2
  integrated, 280
  models, 1, 15
iterative inversion process, 161

Kalman filtering, 155
Kamb method of contouring, 238
Kernel density estimator, 192
Knowledge
  acquisition, 82
  based approach, thc, 63
  based systems, 54
  base maintenance, 97

component models, of the, 16
  engineer, 81
  human, 34
  organization, 16, 84
  representation, 101
  representation, rule-based, 82
  what constitutes, 63

Laplacian convolution, the, 268
Levenshtein distances, 145
linear classification, 121, 124
linearized delay model detection, 163
LISP, 3, 33, 39, 62, 73
logging
  high resolution, 44
  quality control, 37
logical structure, 73
LOGIX, 36, 37

MacApp, 223
Marquardt inversion, 234
matching, 5, 18, 28
  an automated type curve, 2, 3, 26
  dipmeter curves, 45
  procedure, 21
  qualitative, 19, 29
  quantitative, 19, 29
  string-to-string, 35
  symbolic pattern, 47, 99
microdesimeter, 214
migration, 89, 95
minimum principle, 6
MISE, 187
model identification, 2, 3
  components of, 15
multichannel velocity filtering, 95
MYCIN, 61

natural language interface, 102, 108
NMO, 89
nonaccidentalness argument, 6

object oriented framework, 245
object Pascal, 246
observation, 5, 27
optimization, 182
outlier and robustification, 182
overlaying, 265, 269

parameters, 83
parametric method, 126
parsing, finite-state error correcting, 139

partitioning method, 121
pattern recognition
  decision theoretic, 121
  marine seismic, for, 155
  syntactic, 121
pattern representation process, 145
perception of linear structures, 6
petrographic image analysis (PIA), 249
petrophysical parameters, 251
physical symbol system assumption, 3
physicochemical properties, 34
picture description language, 121
plug permeability, 263
potential transform processing, 266
primitive recognition, 138
prism surface area method, 206
projection pursuit, 173
  basics of, 176
  development of, the, 176
PROLOG, 3, 62, 74
PROSEIS, 71, 74
PROSPECTOR, 61
pseudooutliers, 183

recognition, dipmeter curve, 47
regime description, 18
reservoir characterization, 173
robustification, 176
roughness ranking, 204
rules, 83
  base, 34
  confirmation, 39
  correlation, 39
  nondestructive combination, 17
  petrophysical heuristic, 36

SALNOR7, 288
sampling chamber, 257
SCAT, 50
sedimentary structures, 45
segmentation and primitive extraction, 144
seismic display, 3D, 285
seismic sequence analysis, 66
sequential classification, 121, 132
  using divergence, 133
  using Karhunen-Loene expansion, 133
shape description, 14, 15
shell, 33
sidelope suppression, 289
SIES, 99, 108, 123

similarity coefficient, 47
sizing concepts, 251
sketch, 9
SPE, 34
SPWLA, 34
Startzman and Kuo approach,
    39
stratification, 45
string representation, 144
subtractive smoothing, 241
supervised linear classification
    technique, 122
symbols, 2
syntax rules, 4, 14
systolic array architectures, 162

Tauberian approximation, 162
training patterns, 138
transfer function computation, 89
transient testing, 1
tree automaton for recognition of
    seismic patterns, 146
tree grammars, construction of
    expanded, 150
true reservoir response, 4, 5, 13
type curves, 3

VELXPERT, 99, 114–119
visual diagnosis, 3
visual porosity, 253
volumetric rendering, 285

VSEIS, 8
VSPPA, 82

water saturation, 34, 259
    critical, 260
    irreducible, 260
wavelet shaping, 88, 94
well bore imaging, 51
well logging, 34
    correlation, 38
    interpretation, computerized, 35
well test interpretation, 1
WLAI system, 35

Yiesan area, East China, 265